SEVENTH EDITION

Intentional Interviewing and Counseling

Facilitating Client Development in a Multicultural Society

Allen E. Ivey
University of Massachusetts, Amherst

Mary Bradford Ivey
Microtraining Associates

Carlos P. Zalaquett
University of South Florida, Tampa

BROOKS/COLE
CENGAGE Learning

Australia • Brazil • Japan • Korea • Mexico • Singapore • Spain • United Kingdom • United States

BROOKS/COLE
CENGAGE Learning™

**Intentional Interviewing and Counseling:
Facilitating Client Development in a
Multicultural Society, Seventh Edition**
Allen E. Ivey, Mary Bradford Ivey,
Carlos P. Zalaquett

Acquisitions Editor: Seth Dobrin

Development Editor: Julie Martinez

Assistant Editor: Allison Bowie

Editorial Assistant: Rachel McDonald

Technology Project Manager: Andrew Keay

Marketing Manager: Trent Whatcott

Marketing Assistant: Ting Jian Yap

Marketing Communications Manager:
Tami Strang

Content Project Manager: Rita Jaramillo

Creative Director: Rob Hugel

Art Director: Caryl Gorska, Vernon Boes

Print Buyer: Paula Vang

Text Permissions Editor: Mardell
Glinski Shultz

Production Service: Anne Draus,
Scratchgravel

Copy Editor: Patterson Lamb

Cover Designer: Laurie Anderson

Cover Image: Photonica, Getty Images;
Photographer: Ralph Mercer, VEER

Compositor: Macmillan Publishing Solutions

For product information and technology assistance, contact us at
Cengage Learning Customer & Sales Support, 1-800-354-9706.

For permission to use material from this text or product,
submit all requests online at **www.cengage.com/permissions.**
Further permissions questions can be e-mailed to
permissionrequest@cengage.com.

Library of Congress Control Number: 2008940673

ISBN-13: 978-0-495-60123-4

ISBN-10: 0-49560123-3

Brooks/Cole
10 Davis Drive
Belmont, CA 94002-3098
USA

Cengage Learning is a leading provider of customized learning solutions
with office locations around the globe, including Singapore, the United
Kingdom, Australia, Mexico, Brazil, and Japan. Locate your local office at
www.cengage.com/global.

Cengage Learning products are represented in Canada by
Nelson Education, Ltd.

To learn more about Brooks/Cole, visit **www.cengage.com/brookscole.**

Purchase any of our products at your local college store or at our
preferred online store **www.ichapters.com.**

Printed in the United States of America
3 4 5 6 7 12 11

Love is listening.

—Paul Tillich

◆

ABOUT THE AUTHORS

Allen E. Ivey is Distinguished University Professor (Emeritus), University of Massachusetts, Amherst, and Professor of Counseling at the University of South Florida, Tampa (courtesy appointment). He is president of Microtraining Associates, an educational publishing firm. Allen is a Diplomate in Counseling Psychology and was honored as a "Multicultural Elder" at the National Multicultural Conference and Summit. Allen is author or coauthor of more than 40 books and 200 articles and chapters, translated into 20 languages. He is the originator of the microskills approach, which is fundamental to this book.

Mary Bradford Ivey is Vice President of Microtraining Associates and Courtesy Professor of Counseling, University of South Florida, Tampa. A former school counselor, she has served as visiting professor at the University of Massachusetts, Amherst; University of Hawai'i; and Flinders University, South Australia. Mary is the author or coauthor of 12 books, translated into multiple languages. She is a Nationally Certified Counselor (NCC) and a licensed mental health counselor (LMHC), and she has held a certificate in school counseling. She is also known for her work in promoting and explaining development counseling in the United States and internationally. Her elementary counseling program was named one of the ten best in the nation. She is one of the first 15 honored Fellows of the American Counseling Association.

Carlos P. Zalaquett is Associate Professor and Coordinator of Mental Health Counseling in the Department of Psychological and Social Foundations at the University of South Florida. He is also the Director of the USF Successful Latina/o Student Recognition Awards Program and Executive Secretary for the United States and Canada of the Society of Interamerican Psychology. Carlos is the author or author of more than 50 scholarly publications and 4 books, including the Spanish version of *Basic Attending Skills*. He has received many awards, including the USF's Latinos Association's Faculty of the Year and Tampa's Hispanic Heritage's Man of Education Award. He is an internationally recognized expert on mental health, diversity, and education and has conducted workshops and lectures in seven countries.

CONTENTS

LIST OF BOXES

PREFACE

Welcome to the seventh edition of *Intentional Interviewing and Counseling: Facilitating Client Development in a Multicultural Society*. Our goal is to present the critical basics of interviewing, counseling, and therapy with sensitivity to diversity, ethics, and a positive approach to the interview. This book will ground students in skills and provide a solid introduction to how skills are used in several theories of interviewing, counseling, and therapy.

The information here is based on more than 40 years of extensive teaching, research studies, and counseling and clinical practice. Each of the authors comes from a different background in practice and research. Designed for the skills course in counseling, human relations, psychology, and social work, this edition continues our focus on clear presentation of the specifics of effective and accountable interviewing that enables students to:

▲ Learn and master the key interviewing skills and strategies one by one in a step-by-step framework.

▲ Work from a solid basis of understanding ethics, multicultural competence, and wellness/positive psychology. Multicultural and diversity issues, as always, are thoroughly integrated throughout the text.

▲ Draw out client stories, issues, and problems through the basic listening sequence, and facilitate developing new stories, leading to client change and action.

▲ Complete a full interview using only listening skills by the time students are halfway through the book.

▲ Learn and master the influencing skills of confrontation, interpretation/reframing, psychoeducation, and more. Understand how these skills can be used in varying theoretical approaches.

▲ Integrate key skills and strategies of ethical coaching in their interviewing, counseling, and therapy practice.

▲ Develop basic competence in five approaches to the interview: decisional, person-centered, cognitive-behavioral, brief counseling, and motivational interviewing.

▲ Examine and self-evaluate their own performance through working with the optional and popular interactive CD-ROM. On the CD-ROM students will find case studies where they can make decisions on what they would do with clients, video examples of many skills, flashcards, practice examinations, downloadable Portfolios of Competence, plus many other features. Students can email completed exercises from the CD-ROM to their professors. Complementing the CD-ROM is CengageNOW, the interactive Web site. There students will find pre- and post-tests that connect to the textbook plus extensive practice exercises. Students who use these materials attest that they perform better on examinations.

Suggestions and specifics for a Portfolio of Competence are presented in each chapter. Our experience has been that a well-done portfolio can help students obtain practicum positions and, at times, professional jobs as well. If students work on their portfolios and meet basic objectives regularly, they may complain about the workload, but at the end of the course, the portfolio makes it clear how much they have learned. It even increases our course ratings!

We have heard from several instructors that they would prefer a shorter version of this text, to better align with the needs of their courses. Many instructors teach the skills course at an undergraduate or early graduate level. Others are looking for an abbreviated text for use in practicum or field experience courses, where students may have limited knowledge of what actually makes the interview work. For these instructors, we have created a 300-page essentials version of this text, entitled *Essentials of Intentional Interviewing*.

Together, the comprehensive and essentials versions of this text—*Intentional Interviewing and Counseling* and *Essentials of Intentional Interviewing*—provide the flexibility to meet your teaching needs in skills training. Because they are backed by over 450 databased, research-based studies, both books have also been used with excellent results in multicultural courses and in field experience, as either the main text or a supplemental text.

FEATURES NEW TO THE SEVENTH EDITION

The 21st century brings with it many new challenges—the changing ethnic and racial demographics of society, an up-and-down economy in which many individuals suffer, technological innovations, and continuing wars, terrorism, and traumas. These and other issues make counseling ever more important and prominent in our society. This seventh edition of *Intentional Interviewing and Counseling* continues the tradition of the past but also seeks to prepare students for an unpredictable future. Former users of this book will find that the basics are the same, but we have instituted several changes that will enrich the concepts for all levels of students.

Further streamlining of a text that is research and training based. *Intentional Interviewing and Counseling* is the most thoroughly researched and classroom-tested counseling skills text available. In this latest edition, every concept and sentence has been reviewed to ensure clarity and relevance. This streamlining makes the text easier to read while ensuring that specific information is provided. Also, updated research findings are highlighted in the text.

A new "tone" for our scientific base. Throughout this edition, you will note a stronger emphasis on relationship and the working alliance. This emphasis has always been part of the listening skills, but in this edition we have given this central area much more attention. You will also find increased emphasis on the words *here and now* and *immediacy.* Counseling and neuroscience research reveals the importance of the here and now for successful and healthy living. We have added quotations to each chapter that emphasize the uniqueness and importance of thinking about each skill more broadly. "Love is listening," the quotation by Paul Tillich on our dedication page, captures the essence of the hope and goal of this revision.

Relationship—story and strengths—goals—restory—action.[1] This is a new formulation of the popular five-stage interview structure. This language change integrates these concepts and helps students understand and utilize microskills more effectively. The model also makes it easier for students to generalize the five stages to multiple theories and practices in human relations, social work, counseling, and psychotherapy. Nonetheless, the concepts of the five-stage model are still retained within this new, more understandable, language.

---

[1] The terminology "*relationship—story and strengths—goals—restory—action*" is copyrighted © 2009 by Allen E. Ivey and is released to Cengage Learning for this seventh edition of *Intentional Interviewing and Counseling* for use throughout.

More information on certain skills. Reflection of meaning and interpretation/reframing are now presented in a new Chapter 11. The added depth gives students a better opportunity to understand and practice these two central influencing skills. More attention has been given to the work of Viktor Frankl and the positive reframe.

The logical consequences strategy has been given more emphasis with a focus on its relationship to decisional counseling and the emotional side of decision making.

What was previously termed the *advice/information* skill is now reframed as *information/ psychoeducation.* Giving details to clients on where to find career information, how to work their way through the bureaucracy, and how to relate to family members is indeed an important undertaking that we previously have not addressed fully. When we add that significant word *psychoeducation,* it reminds us that we have a crucial role in teaching clients how to be healthy and exercise sufficiently, how to cope with a difficult boss, how to communicate more effectively in the family, and how to examine values and goals.

Increased integration of cutting-edge neuroscience with counseling skills. We now know that interviewing and counseling change the brain and build new neural networks in both client and counselor through neural plasticity. The discussion of neuroscience and its specific impact on interviewing practice has been enhanced, including an appendix with additional practical implications. Students will find that virtually all their learning in the counseling field is supported by biopsychological research. This material will better enable students to plan the type of interventions likely to be most successful and help them to understand and communicate better with other professionals. Appendix II provides diagrams and an overview of neuroscience and counseling.

Cognitive-behavioral therapy and how to's of practice. These have been added to Chapter 14, where you will find a complete transcript on how to use stress management in the session. This interview demonstrates several cognitive-behavioral strategies, such as automatic thoughts, self-management techniques, and how to use information about the brain during the session.

Predicting skill and interview outcome. We can predict how microskills will affect client conversation. Each microskill is clearly defined with its predicted outcome in the session. Needless to say, the root concept of intentionality reminds us that predictions are never perfect and that it is critical to have another response ready for the unexpected we often find in interviewing and counseling.

Multicultural issues and competencies. Diversity is constantly emphasized in this pathbreaking text, the first to recognize cultural differences in the counseling process. We have updated the coverage of multicultural issues and added the RESPECTFUL model of diversity dimensions in Chapter 2 with a new interactive exercise. As always, we continue to integrate diversity issues throughout the text.

An interactive, dialogical view of the interview. The interview affects both client and counselor. New and special attention is given to this interaction in which both the language and brains of both counselor and client are changed throughout the process. Students will understand the concepts of consciousness, short-term memory, and how their skills can help the client move new thoughts, feelings, and behaviors to long-term memory.

The* creative New *and the Client Change Scale. The *creative New* concept, drawn from the work of theologian Paul Tillich, is introduced in Chapter 9. In interviewing and counseling, the concept of the *creative New* means that when we help empower clients to solve problems, resolve issues, and restory their lives, something *New* has been created. This concept provides more depth to the Client Change Scale and also enables us to use creativity research and practice as part of the skills course.

The Client Change Scale (CCS) represents a change of language and expansion of the Confrontation Impact Scale (CIS), so that students and professionals are aware that the measurement of change flows across all interviewing and counseling. The CCS can be used to assess client change both in the here and now of the interview and over several sessions.

TEACHING TOOLS

An expanded array of teaching aids supplement *Intentional Interviewing and Counseling*, which provide you and your students with many alternatives for instruction.

Book Companion Web site. The Companion Web site, accessed from www.cengage.com/counseling/ivey, includes chapter-by-chapter study and review resources for students, such as chapter outlines, flashcards, weblinks, quizzes, and essay questions. In addition, instructors can access and download password-protected resources such as the Instructor Resource Guide and two PowerPoint® presentations.

Optional CD-ROM package. The popular and effective CD-ROM has been updated to include a variety of learning activities and more than 30 interactive exercises. Each CD-ROM chapter includes most or all of the following: flashcards, interactive exercises, case study, video activity, weblink critique, quiz, Portfolio of Competence, client feedback form, and specific skills forms. Each feature is intended to improve students' learning and practice of the skills. Flashcards encourage rehearsal of key chapter concepts, short movie vignettes bring to life specific issues regarding the interview, and quizzes allow students to test their level of achievement. The CD-ROM helps students work through case studies, interactive exercises, and video activities, and puts all feedback forms, key training documents, and handouts right at the student's and Instructor's fingertips. These updated forms are central for self-assessment and for skill practice and feedback. Students can access the Portfolio of Competence and use the reflections on personal style and the self-evaluation of chapter competencies checklist to develop a personalized portfolio that will prove invaluable in their journey to become effective helpers. Furthermore, students can e-mail their assignments directly from the CD-ROM to their instructors if requested. The completion of these assignments can be noted in course management platforms such as Blackboard.

We have endeavored to provide choices for our readers by offering the book alone (ISBN 0-495-59974-3) or the book and CD-ROM prepack (ISBN 0-495-60123-3). We are pleased that our publisher is able to offer the CD-ROM for a nominal additional fee, as we believe that the interactivity and learning potential available through this technology are invaluable.

CengageNOW (Printed Access Card ISBN 0-495-83258-8). This interactive Web site brings students into the virtual world of education. The comprehensive online learning environment offers the following important features:

▲ *eBook.* The complete textbook is available online in CengageNOW.

▲ *A flexible menu.* Instructors can assign online chapter readings and assignments according to their own teaching preferences. Students can move freely between chapters.

▲ *Pretests and posttests of textbook material for student self-evaluation.* Incorrect answers immediately indicate to students the specific pages in the text where they can find why their answer needs further consideration and should be changed.

▲ *Study plan.* CengageNOW offers students a personalized plan of study based on their responses to pretests. Students can use this plan to focus on specific content areas.

▲ *Interactive case studies.* CengageNOW offers more than 30 interactive case studies. When presented with transcripts of interviews, students respond to client statements by selecting specific interview responses. They receive immediate feedback on their choices. Professionals from around the world also present real cases. Students are asked to think through their case management plans and then can compare those plans with what the experts actually did.

▲ *Flashcards* are used to reinforce student learning and understanding.

▲ *Video clips* are used in most electronic chapters to illustrate counseling skills. Follow-up questions allow students to further reflect on their observations.

▲ *Weblinks.* Links to Web sites related to chapter contents are used throughout this virtual learning environment. Follow-up questions allow students to further reflect on the content of these Web sites.

▲ *Important forms and exercises* can be downloaded. By the end of the term, each student will have a complete Portfolio of Competence that can be presented for field site placements and even for professional positions.

Instructor Resource Guide (ISBN 0-495-60332-5). Available online to adopters, the Instructor Resource Guide (IRG) includes chapter goals and objectives, suggested class procedures, additional discussion of end-of-chapter exercises, and microskills practice exercises. The IRG also includes in the appendices a chapter on developmental counseling and therapy (DCT) that many professors find useful in beginning skills courses. Students also profit from examining their theoretical/practical preferences via the inventory titled "What Is Your Preferred Style?" This informal instrument provides a framework for looking at how each student relates to clients. The IRG is available for download at the password-protected Companion Web site (www.cengage.com/counseling/ivey). To obtain the password, contact your Cengage Learning representative or call 1-800-354-9706.

eBank Test Bank and ExamView® (Windows/Macintosh, ISBN 0-495-60216-7). An electronic test bank is available upon request from your Cengage Learning representative. The Test Bank is also available in the flexible and user-friendly ExamView software, which allows instructors to create and edit tests easily and effectively.

Two sets of PowerPoint® **slides.** These are available on the book's companion Web site at www.cengage.com/counseling/ivey. One set is quite detailed, covering all the concepts of each chapter. The second is abbreviated and covers the main concepts. You may download either or both sets and change and sort/reorder the slides according to your teaching preferences. You can then project them as PowerPoint presentations from your computer.

Microtraivning supportive Web site. At www.emicrotraining.com, students will find interviews with leaders of the field such as Patricia Arredondo, Michael D'Andrea, Janet E. Helms,

Jane Myers, Paul Pedersen, and Derald Wing Sue. With more than 100 weblinks, Microtraining Associates, a privately owned company independent of Cengage Learning, is known for its wide array of multicultural training videos and now has the most complete set in the nation of supplementary materials on multicultural concerns as well as many videos on counseling and therapy skills and strategies.

DVDs illustrating the microskills. Several DVDs that can supplement this text are available from Microtraining Associates (phone/fax 888-505-5576, or visit www.emicrotraining.com). A new *Basic Attending Skills* video is now available featuring Deryl Bailey and Azara Santiago-Rivera as well as Mary and Allen Ivey and Norma Gluckstern Packard. These videos and the accompanying text have been translated into Spanish by Carlos Zalaquett. Thus, it is now possible to provide students with supplemental Spanish language interviewing training. The *Basic Influencing Skills* video can be obtained to supplement the last half of this book. A new video, *Microcounseling Supervision: Classifying Interview Behavior,* has recently been released with a supplementary CD-ROM. This should be helpful to students in classifying and working with skills. Those with an orientation to theoretical approaches should find the new skill and strategy videos useful.

ALTERNATIVE INSTRUCTIONAL SEQUENCES

Each instructor has her or his own view on how to present material. Student backgrounds and experiences vary from campus to campus. Thus, we'd like to speak to some issues of reordering ideas in the text to match student needs and interests. The order of the chapters in this book remains basically the same as in the past, but we have separated from the influencing skills a new chapter on the skills of reflection of meaning and interpretation/reframing. However, some instructors will want to reorder chapters to meet their own instructional goals. We have tried to organize the chapters in such a way as to make alternative sequencing easy.

Questioning questions. Some instructors prefer to teach questioning after the listening skills of encouraging, paraphrasing, and reflection of feeling. They point out that some students have difficulty "going beyond" questions and really listening to clients. This more person-centered approach is certainly effective and a good way to emphasize the importance of active listening.

Challenging confrontation. Another major sequencing issue concerns the placement of confrontation. In Allen and Mary's book with Paul Pedersen, *Intentional Group Counseling: A Microskills Approach* (Microtraining Associates, 2007), we place confrontation skills as the last set of microskills to be learned. We do this because confrontation in groups is particularly complex. We are aware that it can be equally complex with individuals. One possible approach is to have the students read just the Client Change Scale information and then apply it to the skills that follow. Then confrontation can be brought in later. We chose to discuss confrontation in Chapter 9, because we find that the emphasis on attending, observing, and basic listening skills in the first half of the book allows effective and early teaching of basic confrontation.

Empathy and reflection of feeling. It may be wise to ask students to read the material on empathy along with the chapter on reflection of feeling. They really do fit together well. This may be a particularly apt approach for instructors who like to spend three to six hours of class time on this area.

Dealing with five theories of counseling. Many instructors choose from decisional, person-centered, CBT, brief, and motivational interviewing, selecting the theories that make the most sense to their program. Others have groups in each class study a single theory and present it to their classmates. Advanced students will be able to engage in all five theories by the end of the course if they are diligent and work hard.

The sections in Chapter 14 can be paired with earlier chapters. Instructors can combine reading on person-centered interviews with the first eight chapters and cognitive-behavioral theory with influencing skills in Chapter 13. Brief solution-oriented approaches could be paired with Chapter 4 on questions, particularly if questions are taught after the other listening skills. Motivational interviewing is a variety of decisional counseling and could be paired with Chapters 8 and 13.

Teaching in a two-semester course. Some community colleges and universities have used *Intentional Interviewing and Counseling* over two semesters, supplemented by other texts. This enables handling the skills and theories in a more unified plan. Another possibility is to use the book in both the skills and multicultural courses. These alternatives could be used in either a single semester or over a two-semester sequence.

Have it your way! Each instructor needs to shape and adapt textbooks to meet the students' needs and her or his own approach to teaching. Other sequences of skills can be arranged, and we welcome your feedback on this important and challenging instructional issue. We'll give you credit for you contributions.

ACKNOWLEDGMENTS

Thomas Daniels, Memorial University, Cornerbrook, has been central to the development of the microskills approach for many years, and we are pleased that his summary of research on over 450 databased studies is available on the CD-ROM that accompanies this book. We are appreciative of one of our students, Penny John, for permission to use her interview as an example in Chapter 13. Amanda Russo, a student at Western Kentucky University, also allowed us to share some of her thoughts about the importance of practicing microskills.

Weijun Zhang's writing and commentaries remain central to this book. We also thank Owen Hargie, James Lanier, Courtland Lee, Robert Manthei, Mark Pope, Kathryn Quirk, Azara Santiago-Rivera, Sandra Rigazio-DiGilio, and Derald Wing Sue for their written contributions. Robert Marx and Joseph Litterer were important in the early development of this book. Discussions with Otto Payton and Viktor Frankl have clarified the presentation of reflection of meaning. William Matthews was especially helpful in formulating the five-stage model of the interview. Lia and Zig Kapelis of Flinders University and Adelaide University are thanked for their support and participation while we served as visiting professors in South Australia.

David Rathman, Chief Executive Officer of Aboriginal Affairs, South Australia, has constantly supported and challenged this book, and his influence shows in many ways. Matthew Rigney, also of Aboriginal Affairs, was instrumental in introducing us to new ways of thinking. These two people first showed us that traditional, individualistic ways of thinking are incomplete, and therefore they were critical in the development of the focusing skill with its emphasis on the cultural/environmental context.

The skills and concepts of this book rely on the work of many different individuals over the past 30 years, notably Eugene Oetting, Dean Miller, Cheryl Normington, Richard Haase, Max Uhlemann, and Weston Morrill at Colorado State University, who were there at the inception of the microtraining framework. The following people have been especially important personally and professionally in the growth of microcounseling and microtraining over the years: Bertil Bratt, Norma Gluckstern, Jeanne Phillips, John Moreland, Jerry Authier, David Evans, Margaret Hearn, Lynn Simek-Morgan, Dwight Allen, Paul and Anne Pedersen, Lanette Shizuru, Steve Rollin, Bruce Oldershaw, Oscar Gonçalves, Koji Tamase, and Elizabeth and Thad Robey.

The board of directors of the National Institute of Multicultural Competence—Michael D'Andrea, Judy Daniels, Don C. Locke, Beverly O'Bryant, Thomas Parham, and Derald Wing Sue—are now part of our family. Their support and guidance have become central to our lives. Many of our students at the University of South Florida, Tampa, University of Massachusetts, the University of Hawai'i, Manoa, and Flinders University, South Australia, also contributed in important ways through their reactions, questions, and suggestions.

Fran and Maurie Howe have reviewed seemingly endless revisions of this book over the years. Their swift and accurate feedback has been really important in our search for authenticity, rigor, and meaning in the theory and practice of interviewing, counseling, and therapy.

Jenifer Zalaquett has been especially important throughout this process. She not only navigates the paperwork but is instrumental in holding the whole project together.

We are grateful to the following reviewers for their valuable suggestions and comments: Victoria Bacon, Bridgewater State College; Stephanie Hall, Eastern Kentucky University; Garrett J. McAuliffe, Old Dominion University; Graham Neuhaus, University of Houston–Downtown; Uchenna Nwachuku, Southern Connecticut State University; John Patrick, California University of Pennsylvania; Sandy Perosa, University of Akron; Tiffany Rush-Wilson, Walden University; Holly Seirup, Hofstra University; and Heather Trepal, University of Texas at San Antonio. They shared ideas and encouraged changes that you see here, and they also pushed for more clarity and a practical action orientation.

Machiko Fukuhara, Professor Emeritus, Tokiwa University, and president of the Japanese Association of Microcounseling, has been our friend, colleague, and coauthor for many years. Her understanding and guidance have contributed in many direct ways to the clarity of our concepts and to our understanding of multicultural issues. We give special thanks and recognition to this wise partner.

Lisa Gebo and Claire Verduin guided the development of this book for many years, and they are present on every page. Julie Martinez and Marquita Flemming added their wisdom to the process and helped us deal with the complexities of the publishing world. These four experts have become valued friends and consultants. Seth Dobrin, new to the support team, is a "quick study," and we have been vastly impressed with his ideas and contributions to this new version. Without these five individuals, this seventh edition would never have seen the light of day.

Finally, it is always a pleasure to work with the rest of the group at Brooks/Cole, notably Trent Whatcott, Andrew Keay, Allison Bowie, Rachel McDonald, Rita Jaramillo, Vernon Boes, and their associates. Our manuscript editor, Patterson Lamb, has become an important adviser to us. Anne and Greg Draus of Scratchgravel Publishing Services always do a terrific job. We thank all of the above.

We would be happy to hear from readers with your suggestions and ideas. Please use the form at the back of this book to send us your comments. Feel free to contact us also via e-mail. We appreciate the time that you as a reader are willing to spend with us.

Allen E. Ivey
Mary Bradford Ivey
Carlos P. Zalaquett
e-mail: info@emicrotraining.com

Interviewing and Counseling as Science and Art

WELCOME!

Allen: My first courses in counseling were fascinating. I liked the theoretical ideas and the information about testing and careers, but what I enjoyed most was the course on theories of counseling. To me, this was the foundation of the whole process.

Then came the second semester and my first real opportunity to practice what I had learned in my field internship. I really cared, and I wanted to help clients grow and resolve their issues. But I found myself overwhelmed by the amount of information shared by clients making complicated decisions or facing difficult issues. The theories in the books I had read did not easily apply to real people. How was I to survive and help? Somehow, I made it through, but I know I could have done a better job with those early sessions if I had been more skilled before I started.

Intentional Interviewing and Counseling: Facilitating Client Development in a Multicultural Society is designed to teach you specific skills that you can use immediately in the session. The book seeks to "demystify" the art of helping. As you move through your practice interviews, you will find that each step of the microskills hierarchy presents the specifics of counseling in clear and usable form. Whereas I learned from a "guess and try" framework, this book will enable you to enter the reality of counseling with understanding and expertise. You'll encounter many practice exercises that allow you to test out your understanding and competence. Later in this book, you will discover that you can apply these skills with multiple theories of helping.

Mary: I arrived at the University of Wisconsin shortly after Carl Rogers had left, but his influence remained. Rogers came back several times and shared his ideas and his being with us. His impact on all of us was profound. His person-centered theory emphasized client/counselor relationship, positive regard, and the ability of clients to solve their own problems. He described our role as counselors as one that focused on listening and reflecting feelings. Asking questions or influencing a client through directives was something we should not do.

My first experience in counseling found me focusing on a reflective approach. My relationships with clients seemed fine, but something was missing. While listening seemed critical, it often wasn't enough, especially with less verbal clients. Over time, I learned that many clients needed a more active stance from the counselor. Gradually, I learned the skills of interviewing and I found a new, more balanced approach in which relationship and listening are combined with what we call the influencing skills of helping.

Moreover, I was lucky to work with an outstanding behavioral psychologist, Ray Hosford, who passed away much too young from ALS, Lou Gehrig's disease. Ray helped me see that although person-centered Rogerian methods are critical foundations, there are many methods and theories that can help clients. He also taught me that an interviewer's personal style is highly influential in client growth.

I still recognize solid relationships and listening as fundamental, particularly when we use the basic listening sequence to hear client stories fully and accurately. The precision of the microskills helps me be a better listener and also to be more flexible in using varying approaches to meet the needs of an extremely diverse set of clients.

Carlos: I have had the privilege of learning from truly outstanding professors and mentors. They did their best to educate my classmates and me in the art and science of helping. From them I learned theories and techniques in great length, was mesmerized by each, and eagerly practiced their different therapeutic approaches. I have also strived to know myself while attempting to absorb the competencies needed to transform our clients.

I have taught interviewing, counseling, and psychotherapy in four different countries. In each, my students eagerly acquired the knowledge of our profession and our personal experience. But the real issue is taking the knowledge you learn and applying it to the unique individuals that you will meet. This requires flexibility and your ability to continuously change and learn *with* the client.

Working with people, especially those different from ourselves, taught me about the limitations of my knowledge and training. This led me to incorporate a respectful and intentional approach to my clinical and educational work. I learn something new in every session and class that I teach. My ultimate concern is with those who receive the services that we all offer.

Mary, Allen, and Carlos: Together, the three of us highlight the importance of using listening and influencing skills, conceptualizing cases using a respectful and ethical approach, and having a structure to guide our interventions. Furthermore, we try to demystify the art and science of helping by providing concrete tools you can use to become successful professionals.

Welcome! We are delighted to have you join us. We believe that there are multiple ways that we can help clients. Some of you who read this book will become committed to the person-centered approach; others will move toward the cognitive-behavioral, brief, and perhaps even psychodynamic/interpersonal orientations. We know that you will incorporate multicultural and diversity issues in your work. All theories and methods have value, particularly if we match them to client needs. Our own orientation is developmental/integrative for we believe that there are several routes toward the "truth" of effective interviewing.

One of your important tasks as a beginning professional is to develop your own system for integration of skills and theories. We suggest that you start immediately to identify your own natural style and positive strengths and then use these as a base as you work through this book. Each of us has a natural gift that enables us to reach others and help them achieve their goals. We hope that you will take what we present here and then shape the material to fit your natural style and the needs of those whom you would serve.

WHAT DOES THIS BOOK OFFER FOR YOUR DEVELOPMENT?

Throughout this book, you will be examining intentional interviewing and counseling—an interviewing approach that is concerned with flexibility and competence. You will learn specific skills that will enable you to help others find new ways to understand their thoughts, feelings, and behavior. In addition, you will learn how to help them understand the meaning of what happens to them and their vision of deeper lifetime goals.

Many concepts, ideas, and skills are presented here, but there are some important general goals that you can expect to achieve. Through step-by-step study and practice you will encounter and master specific interviewing skills that will enable you to achieve the following. You will be able to

▲ Engage in the basic microskills of the interview: listen, influence, and structure an effective session with individual and multicultural sensitivity. In addition, you will accomplish this with a full awareness of ethics and the importance of a positive wellness approach to the interview.

▲ Predict the likely impact of your helping interventions on client conversation and be able to assess the overall impact of your interview and interventions. When clients do not respond as you expected, your intentional use of microskills will enable you to return to listening more carefully or choose an alternative strategy.

▲ Conduct a full interview using only listening skills.

▲ Master a basic structure of the interview, *relationship—story and strengths—goals—restory—action,* that can be applied to many different theories. You will become skilled in decisional interviewing and counseling—a foundation theory that will enable you to better understand and work with all theories of helping.

▲ Engage in four additional interviewing, counseling, and psychotherapy theories and important strategies in each—person-centered, cognitive behavioral, brief counseling, and motivational interviewing. Included in this is active case management and treatment planning.

▲ Learn about and build on your natural helping style. As you complete the practice exercises, you will continue to learn about yourself and how to integrate new skills within your natural style.

▲ Develop the needed foundation for learning skills and theories through ethical practice, multicultural competence, and a wellness and positive psychology approach to the session.

▲ Develop an understanding of how recent work in neuroscience supports your practice and skills in interviewing and counseling. In addition, you will have an introduction to the cutting edge of counseling.

▲ Generate your own story of the practice of interviewing, counseling, and therapy by constructing your personal theory about the helping process. As you encounter client uniqueness and cultural complexity, anticipate that your story and theory will be one of constant change, growth, and development.

THE MICROSKILLS MODEL

The foundation of this book is microskills—communication skill units of the interview that will help you interact more intentionally with clients. You will learn these single skills in a clearly outlined step-by-step model. Prior to the introduction of the microskills to the helping field, students learned interviewing by what could best be called "guess and try." Counseling was a mystical procedure, but gradually new practitioners would develop expertise, although we still worry about what happened (and still happens) to clients during that learning period.

Microskills offer concrete tools for interviewing, counseling, and therapy. As presented here, they are based on the original single-skills microskills model developed at Colorado State University (Ivey, Normington, Miller, Morrill, & Haase, 1968). In 1974, multicultural differences in communication styles were identified (Ivey, Gluckstern, & Ivey, 1974/2006), and issues of diversity have been central in practice ever since. *Intentional Interviewing and Counseling* is based on original research and teaching practice and also was the first to place multicultural issues at the center of interviewing, counseling, and psychotherapy.

The extensive research on the microskills model is summarized in Chapter 1. The CD-ROM accompanying this book provides a comprehensive research report on more than 450 databased studies on the model. Neuroscience concepts that back up the approach of this book are summarized throughout the book, especially in the research sections. Appendix II provides a brief discussion of key neuroscience issues and suggestions for future reading.

THE SCIENCE AND ART OF INTERVIEWING

The scientific basis of counseling forms much of the foundation of what we share in this book. Our ideas and suggestions are drawn from close to 450 scientific studies and 40 years of clinical practice in the microskills. Research shows that competent interviewing, counseling, and therapy make a difference in the lives of clients and that awareness of diversity is critical. With help, clients can change their thoughts, feelings, behaviors, and the meanings and visions that guide their future.

Research and experience in the world help you determine what to do with each unique client. In the now moment of the session, you decide which piece of research, multicultural knowledge, or your own personal skills will empower your clients toward changes. *You* are a person with a special set of knowledge and skills who must make choices as you seek to help clients grow, change, and create what we call the *New.*

Thus, in the interview, counseling is ultimately an art. You are the artist with the brush who must draw from a palette of knowledge and carefully developed techniques to help the client discover or create a new portrait.

Through interviewing techniques provided in this book and considerable practice, you will gain the technical skills, strategies, and theories that can empower your client to grow, create change, and resolve issues and life challenges.

Yet you are not alone in the change process—the client is equally and perhaps even more important in creating the New. You are a facilitator, and while you make critical decisions, it is also essential that you work *with,* not *on,* the client. Mutuality in relationship, goals, and planning for the present and the future is basic to effective, culturally intentional interviewing, counseling, and therapy.

As you become an intentional interviewer, counselor, or therapist, you will have many options for helping others. You will also be responsible for competencies in interviewing skills and strategies so that you can best serve a diverse population of clients. Understanding what comes naturally to you, building on it, and practicing are vital in achieving this goal.

BUILD ON YOUR NATURAL STYLE OF HELPING

People enter the counseling, psychology, social work, and other human service fields intending to help others. This is a caring profession, and your warmth and ability to establish rapport and relationships with many types of people is critical. Perhaps you, like most who enter our field, have been told that you are a good listener. You may have had friends or family come to you for advice or simply to tell you their problems. This is not counseling, but it is helping. You enter this book with some degree of social skills and natural abilities to assist others.

Natural style is defined here as your spontaneous way of working with others to help them achieve their goals. As you work through this book, look at yourself and respect your own natural competence as the foundation for your growth in interviewing skills. However, your natural style may not "work" with everyone—it may be necessary to shift your style to be fully effective. The effective interviewer gradually develops a blend of natural style and learned competencies.

As you begin work with *Intentional Interviewing and Counseling,* focus on your own natural abilities as a foundation for growth and further development. Competency Practice Exercise 1 presents an exercise in self-understanding that you may wish to complete before moving further in your reading. Let us start with your natural expertise.

COMPETENCY PRACTICE EXERCISE

Exercise 1: What Are the Strengths of Your Natural Style?

A good place to start in interviewing and counseling is awareness of yourself as a person of capability. List below specifics in which you are a person of competence. This is your foundation for growth.

When have you helped someone else? Be as specific as possible.

*What are your
thoughts?*

_____

_____

What, specifically, did you do that was helpful?

_____

_____

Be honest—what strengths do you think you bring to a course of study in helping and interviewing skills? (Saying good things about oneself is not always easy, but do it!)

_____

_____

Ask a friend or family member to identify your natural qualities and skills that might make you an effective helper. Record what they say below:

_____

_____

Once our strengths are identified, it is much easier to face up to our personal challenges and limitations. Your base of positive skills and qualities will enable you to develop further as an interviewer or counselor.

SELF-UNDERSTANDING AND EMOTIONAL INTELLIGENCE

Self-understanding is the broad concept of knowledge about oneself. Closely related is emotional intelligence. Self-understanding and emotional intelligence are essential to interviewing competence and to enhancing your natural style. Interviewing and counseling themselves can be described as exercises in emotional intelligence. Emotional intelligence was first defined by Peter Salovey and John D. Mayer in 1990 and was first brought to wide attention by Daniel Goleman (1998, 2005).

Several domains of emotional intelligence have been defined. Below you will see how the skills and strategies of this book relate to self-understanding and self-development (Goleman, Boyatzis, & McKee, 2002). As you read each one of these, please stop for a moment and think through where you are in each of the following:

1. *Self-awareness.* Throughout this book you will have continual opportunities to examine yourself and your work with others in the interview. You will have the opportunity to hear your communication style on audiotape and, most likely, you will see yourself in action on videotape as well. What are your strengths and limitations? Do you have a positive image of yourself? Unless you feel good about yourself, you may have struggles helping clients.

What are your thoughts?

_____

_____

2. *Self-regulation.* When an interview is challenging, can you handle your feelings and avoid allowing your "buttons to be pushed" too easily? Do you have self-control, flexibility, and the ability to generate new ideas on the spot? The skills and strategies of this book are designed to provide you with an array of possibilities for dealing with those challenging situations.

_____

_____

3. *Motivating yourself and using your abilities.* Each chapter in this book contains practice exercises designed to help you master interviewing skills and concepts, listen actively to others, and learn how to deal with challenging situations. Understanding a concept is not mastery. The emotionally competent interviewer or counselor is motivated for peak performance. Are you persistent in achieving your goals or do you give up easily?

_____

_____

4. *Empathy.* Empathic understanding and listening skills are intimately intertwined. Are you interested in others, can you be empathic to their concerns, can you pick up small signs of the many emotions your clients will have? Do have an understanding of people different from yourself and can you see their perspective?

_____

_____

5. *Social skills.* Relationship is central to the helping process. How effective have you been in working with others? How competent will you be in establishing rapport, trust, and confidence in the helping interview? Can you listen? Do you have the ability to empower clients through your knowledge and ideas?

_____

_____

Throughout this book, and in the accompanying CD-ROM, we will provide many exercises and skill practice ideas that can serve as avenues for self-exploration and a clearer definition of your own competencies. Empathy and social skills, of course, are central to this book and the interviewing and counseling professions. Self-awareness, self-regulation, and motivation may be described as qualities that are vital for effective work in the helping field.

PRACTICE LEADS TO MASTERY AND COMPETENCE

All of us can get better, no matter how good we think we are. The many skills and concepts of this book will be mastered to full competence only if you work actively with them. Role-plays in class or workshops followed by audiotape or videotape practice will help you develop expertise and mastery of skills. Practice will also be vital in your development of personal self-understanding and emotional intelligence.

You can "practice" by going through a skill once and saying to yourself, "That was easy." Or you can really practice by aiming to see if you can get specific and predictable results in the session as a result of your skills. Many students find that several practice sessions are useful with as much feedback as possible. Feedback has been called the "breakfast of champions." For example, you can practice the skill in a situation in which a friend tells you a positive story that he or she is eager to share. More challenging practice follow-ups might include role-plays in which you work with a variety of difficult issues that you might face—clients who are less verbal, people who are hostile and aggressive, and those with complex concerns.

If you are motivated, you will find that several alternative practice sessions increase your skills, confidence, and competence. Practice will determine where you stand in terms of your abilities, skills, and expertise. Four levels of competence are identified in each chapter in this book:

▲ *Level 1: Identification and classification.* Elementary competence and mastery occur when you can identify and classify interviewing behavior. You can observe others' behavior on audiotape or videotape and know what they are doing. A quiz or an examination measures your ability to understand.

▲ *Level 2: Basic competence.* This involves being able to perform the skills in an interview, most often a practice role-play. You may, for example, demonstrate in an audiotaped or videotaped session that you can use both open and closed questions—even though you may not use the skills at a high level.

▲ *Level 3: Intentional competence.* You will find that you can use a skill with predictable results. For example, effective use of attending behavior skills increases client talk-time, while the lack of them reduces client conversation. Intentional competence means that you can help clients talk about their issues in specific ways, and you can even predict what clients will say if you use a certain skill. But clients are real people and not always predictable. If you act intentionally and the predicted result does not occur, you can move to an alternative skill or strategy, which may facilitate client growth in a different way.

▲ *Level 4: Psychoeducational teaching competence.* One way for you to acquire greater mastery is to teach the skill to someone else. Others can profit from your knowledge about interviewing—paraprofessional community volunteers, firefighters and police, or teens in a church, mosque, or synagogue. You will also find that your clients can benefit in the interview through direct teaching of skills. The communication skills emphasized in this book are also basic social skills, useful in daily interactions. The microskills are counseling and therapeutic strategies in themselves, useful in enhancing client self-understanding and efficacy.

Vital to effectiveness as an interviewer is a clear sense of how ethical and multicultural understanding relate to daily practice. In addition, ethical and multicultural competence can be considered vital aspects of self-understanding and emotional intelligence.

A FINAL WORD

What can this book do for you? The key idea is to focus on yourself as a developing interviewer, counselor, or therapist. This book helps you do that by increasing your knowledge and skills, strategies, and theories. Ultimately, you will generate your own theory of interviewing, counseling, or psychotherapy.

We know that the concepts, skills, strategies, and theories presented here make a difference in intentional interviewing. There are many ways to be an effective interviewer, counselor, or therapist. We know that what is here "works," but we also believe that it will be valuable only if it fits with your natural style. We suggest that even if a concept doesn't feel right, try it. What seems awkward at first may become a favored method later in your interviewing practice. And a particular style you may not favor may indeed be what the client needs!

The many practice exercises are designed to remind us all that counseling is an art as well as a science. While we can supply ideas and suggestions, it is you who will apply these concepts to real people's lives. As you learn skills, you may also find it helpful to think of yourself as an artist who is going to put things together in new ways. You will paint the pictures *with* your client. Recognize and respect your natural uniqueness and that of those with whom you work.

Over the years, we have learned much from student comments and suggestions. Please feel free to contact us with your issues and questions. We would welcome your ideas.

Allen E. Ivey
Mary Bradford Ivey
Carlos P. Zalaquett
info@emicrotraining.com
www.emicrotraining.com

Introduction

Listening is the foundation of counseling, interviewing, and psychotherapy. Our first goal is to enable clients to tell their stories. Through this narrative exploration, we can help them rewrite and act on their stories and problems in new ways. Our task is to expand client possibilities for intentional response and action. This is true whether we work in private practice, community agencies, behavioral health organizations, schools, or universities; whether we are counselors, human service professionals, psychologists, social workers, teachers, or others in settings ranging from human relations in management to medicine to those who work across the world in war-torn areas.

The first section of this book is oriented to joining clients where they are. The issues here focus on listening, hearing clients tell their stories, and reflecting with them on what they have experienced. Building on this listening foundation, later chapters on influencing skills will provide you with skills and strategies that will further enable clients to restory their lives and move toward change and action.

Chapter 1. Toward Intentional Interviewing and Counseling   We begin with definitions of interviewing, counseling, and psychotherapy. Important for your understanding and mastery of interviewing skills is the microskills hierarchy, which provides you with an outline of the competencies you will achieve in the book. You will also be introduced to a basic structure of the effective interview, which includes five stages: *relationship—story and strengths—goals—restory—action*. You'll find that using this structure will help you become more competent in all theories of counseling and therapy as well as increase your sense of how to use skills and strategies with clients.

Chapter 2. Ethics, Multicultural Competence, and Wellness   Professional ethical standards are central to the helping professions of counseling, family therapy, human services, psychology, and social work. You will also find essential aspects of ethical standards important for interviewing and key Web sites where the complete standards are available.

In our increasingly global work, we interact with persons who are different from us in many ways, such as age, ethnicity/race, gender, geographical location or community, language, sexual orientation, spiritual/religious beliefs, socioeconomic situation, physical ability, and experience with traumatic situations. The terms *diversity*, *cross-cultural*, and *multicultural* refer to all of these dimensions and will be used interchangeably. The chapter provides you with concrete information to enhance your ability to work with people from diverse cultural backgrounds.

Wellness and the positive psychology movement focus on strengths that clients bring to the interview. We solve our problems and life challenges through what we *can* do rather than on what we *can't* do. Yes, we need to listen carefully to clients' stories and understand their difficulties, but an important part of the counseling process is helping clients discover and understand their own strengths and resources. Counseling and therapy, by their very nature, are optimistic professions based on the belief that people can change and be fully involved in their own growth.

Chapter 3. Attending Behavior: Basic to Communication   Attending is like the air we breathe, and we can easily miss its importance. It is also the first and most basic listening skill of interviewing, counseling, and psychotherapy. Many beginning helpers inappropriately strive to solve the client's issues and challenges in the first 5 minutes of the interview by giving premature advice and suggestions. Please set one early goal for yourself: Allow your clients to talk. Your clients may have spent several years developing their problems before they consult you. Listen first, last, and always.

Later portions of this book emphasize action skills of helping such as confrontation, interpretation, and directives. However, virtually all who work in interviewing, counseling, and psychotherapy consider the ability to listen to and enter the world of the client the most important part of effective helping.

Begin this book with a commitment to yourself and your own natural communication expertise. We recommend that you use the skills and concepts outlined here to intentionally enhance your natural style and expand your alternatives for working with clients.

CHAPTER 1

Toward Intentional Interviewing and Counseling

We humans are social beings. We come into the world as the result of others' actions. We survive here in dependence on others. Whether we like it or not, there is hardly a moment of our lives when we do not benefit from others' activities. For this reason it is hardly surprising that most of our happiness arises in the context of our relationships with others.

—The Dalai Lama

How can intentional interviewing and counseling help you and your clients?

Chapter Goals

This chapter is designed to identify key ideas of the microskills approach and show how the step-by-step model relates to broad concepts of interviewing, counseling, and psychotherapy. These skills are used by all professionals. Intentional interviewing is designed to facilitate the drawing out of client stories, enabling clients to find new ways of thinking about these stories and new ways of acting. It is important that the interviewer have multiple techniques for responding to clients in a culturally sensitive fashion.

Competency Objectives

Awareness, knowledge, and skills in the foundational concepts of *Intentional Interviewing and Counseling* presented in this chapter will enable you to

▲ Identify the similarities and differences among interviewing, counseling, and psychotherapy.

▲ Understand the step-by-step microskills framework for mastering the interview.

▲ Recognize the varying patterns of microskills used by different theories of counseling and psychotherapy.
▲ Define *intentionality*, *cultural intentionality*, and *intentional competence*.
▲ Anticipate the impact of your comments on client conversation by learning and using the basics of intentional prediction.
▲ Outline and define the elements of the counseling and therapy model: *relationship—story and strengths—goals—restory—action.*
▲ Develop awareness of the impact on the brain of interviewing, counseling, and psychotherapy.
▲ Examine your own natural helping style and use personal expertise as a base for further development as you work through this text.

INTRODUCTION: WHAT IS THE "CORRECT" RESPONSE TO OFFER A CLIENT?

Imagine that you are the interviewer, counselor, or psychotherapist and a new client comes in. Immediately after you have discussed the boundaries of the situation, she starts talking rapidly with a list of multiple issues. What might you say or do next that could be helpful?

Client: I'm overwhelmed. My husband was let go in the latest downsizing and is impossible to live with. My job is going okay, but I worry about making the next car payment. Our ancient washer broke, flooded our basement, and ruined a box of family photographs. Our daughter came home crying because the kids are teasing her, and my mother-in-law is coming to visit next week. What should I do?

How would you respond? What would you say? Take a moment to think before reading on, and even better, write down your response so that you can compare it with what others might say. You'll probably find their responses are different from yours. A key question is who made the "correct" response?

What are your thoughts?

How do you manage your feelings of overwelmness?

_____

_____

The answer, of course, is that there are many potentially useful responses in any interviewing situation. Reflecting the client's emotions can be helpful ("You feel you're snowed under with all that's happening"). Selecting one aspect to focus on can be useful, and then later you can examine other dimensions by asking an open question ("What, specifically, is happening between you and your husband right now?" or "Could you share a bit more about your financial situation?"). You might even say, "I hear the stress and tension in your voice. . . . Let's slow down, take a deep breath, and start from the beginning." Among many other possibilities, you could direct attention to the job loss, the daughter's school problem, or the mother-in-law's forthcoming visit. You might even choose to sit silently and see what happens next. Basic to any of these responses is an understanding that the client is *stressed.* Stress is now seen as an underlying dimension of virtually all client issues. Our task is to respect the client, use appropriate skills and strategies, and seek to alleviate stress. Stress and its consequences will be discussed frequently in later chapters.

The aim of this book is to expand your possibilities for responding to people in need. Rather than "What is the correct response?" seek to develop multiple possibilities for helping the client deal with the world. At the same time, there are many commonalities among interviewing, counseling, and therapy and these will be examined next.

INTERVIEWING, COUNSELING, AND PSYCHOTHERAPY

The terms *counseling, interviewing,* and *psychotherapy* are often used interchangeably in this book. Though the overlap is considerable, interviewing may be considered the most basic process used for information gathering, problem solving, and psychosocial information giving. Interviewing is usually short term with only one or two sessions. A human services staff member may interview a client about financial needs and planning. Managers interview potential employees, and college admissions staff interview students applying for admission. After a major disaster (terrorist bombing, hurricane, or flood) a crisis worker may interview a family about their needs and plans for recovery, and then give them advice about what they can do to meet tomorrow's needs.

Ethical coaching is a relatively new term and conception of helping. It focuses on living life more fully and effectively. It falls most closely in the interviewing area. Ethical coaches work from a strength-based foundation and empower individuals, families, and organizations to help them make more effective plans. Life coaching, college coaching, and executive coaching are three examples.

Counseling is a more intensive and personal process. It is generally concerned with helping people cope with normal problems and opportunities, although these "normal problems" often become quite complex. Though many people who interview may also counsel, counseling is most often associated with the professional fields of social work, school counseling, psychology, mental health and clinical counseling, pastoral counseling, and, to a limited extent, psychiatry. Clients with relationship difficulties may need several sessions of counseling to straighten out their situation. The employee or college student facing challenges often needs help in understanding issues and making decisions. In a crisis, the family experiencing a major disaster often needs both short- and long-term counseling.

We can clarify the overlapping differences and similarities of interviewing and counseling with some examples. A personnel manager may interview a candidate for a job but in the next hour counsel an employee who is deciding whether to take a new post in a distant town. A school counselor may interview each class member for 10 minutes during a term to check on course selection but will also counsel many students later about personal concerns and college choice. A psychologist may interview a person to obtain research data, but in the next hour be found counseling a client concerned about an impending divorce. Even in the course of a single contact, a social worker may interview a client to obtain financial data and then move on to counseling about personal relationships.

Both interviewing and counseling may be distinguished from psychotherapy, which is a more intense process, focusing on deep-seated personality or behavioral difficulties. Psychotherapists must interview clients to obtain basic facts and information as they start work with an individual. The skills and concepts of intentional interviewing are equally important for the successful conduct of longer term psychotherapy, which was once almost the exclusive province of psychiatry. As many psychiatrists have turned to prescribing medication, practicing mental health and clinical counselors, clinical social workers, and clinical and counseling psychologists have taken on psychiatry's former role.

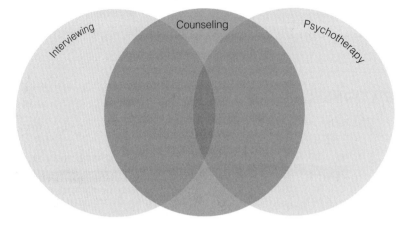

FIGURE 1-1   The interrelationship of interviewing, counseling, and psychotherapy.

Despite relatively clear differences among interviewing, counseling, and psychotherapy, overlap remains (see Figure 1-1). Effective interviewing can help clients make decisions, and that in itself is therapeutic.

THE CORE SKILLS OF THE HELPING PROCESS: THE MICROSKILLS HIERARCHY

Interviewing, counseling, and psychotherapy require a relationship with the client; they all seek to help clients work through issues by drawing out and listening to the client's story. *Intentional Interviewing and Counseling* presents the key skills and strategies used by all three approaches.

Microskills are the foundation of intentional interviewing. They are communication skill units of the interview that provide specific alternatives for you to use with many types of clients and all theories of counseling and therapy. You master these skills one by one and then learn to integrate them into a well-formed interview.

When you are fully competent in the microskills, you are able to listen effectively and help clients change and grow. Effective use of microskills enables you to anticipate or predict how clients will respond to your interventions. And if clients do not do what you expect, you will be able to shift to skills and strategies that match their needs.

The microskills hierarchy (see Figure 1-2) summarizes the successive steps of intentional interviewing as you will encounter them in this book. The skills of the interview rest on a base of ethics, multicultural competence, and wellness. On this foundation lies the first microskill discussed in this text: attending behavior. This culturally and individually appropriate skill includes patterns of eye contact, body language, vocal qualities, and verbal tracking. Through this book, you will have the opportunity to define this skill further, see attending demonstrated in an interview, read about further implications, and finally, master the skill in practice and real interviews. The book is supplemented by rich case studies in both the accompanying interactive Web site CengageNOW and the interactive CD-ROM.

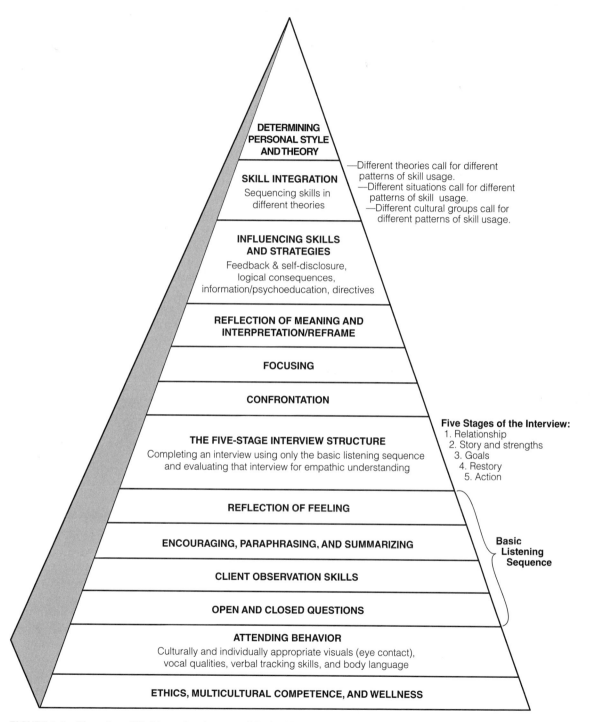

FIGURE 1-2 The microskills hierarchy: A pyramid for building cultural intentionality. (Copyright © 1982, 1987, 2003, 2007, 2010 Allen E. Ivey. Reprinted by permission.)

Once you have mastered attending behavior, you will move up the microskills pyramid to the basic listening skills of questioning, client observation, encouraging, paraphrasing, summarizing, and reflecting feelings. Higher is not necessarily better in this hierarchy. Unless you have developed skills of listening and respect, the upper reaches of the pyramid are meaningless. The foundational skills are critical parts of the practice of even the most experienced professional. Develop your own style of being with clients, but always with respect for this grounding. With a solid background in these central skills, you will be able to conduct a complete interview using only listening skills.

You will then encounter the influencing skills and help clients explore personal and interpersonal conflicts. Confrontation is considered critical for client growth and change. The microskills of focusing, reflection of meaning and interpretation/reframing come next in the hierarchy, followed by other key skills of interpersonal influence—self-disclosure, feedback, logical consequences, information/psychoeducation, and directives.

With a mastery of listening, the ability to conduct an interview using only listening skills, and a command of the influencing skills, you will be prepared to master alternative theories of helping. You will find that these microskills can be organized into different patterns utilized by different theories. For example, if you have mastered the listening skills and the structure of the interview, you have an important beginning in becoming fully competent in Rogerian, person-centered theory. Later, as you move on to other systems of counseling and therapy, you will find that the basic skills are a foundation for mastering those theories as well.

At the apex of the microskills pyramid is determining personal style and theory. It isn't enough just to master skills and theories. You will eventually have to determine your own theory and practice of counseling, interviewing, and psychotherapy. Interviewers, counselors, and psychotherapists are an independent lot; the vast majority of helpers prefer to develop their own styles, and through eclecticism move toward their own blend of skills and theories.

As you gain a sense of your own expertise and power, you will learn that each client has a totally unique response to you and your natural style. While many may work well with you, other clients will need you to adapt to their style of being. You will want to be flexible and have many alternatives ready to help your varying clientele.

The model for learning microskills is practice oriented and follows a step-by-step progression, which will appear throughout the chapters of this text as a basic learning framework.

1. *Warm up.* Focus on a single skill and identify it as a vital part of the holistic interview.
2. *View.* View a DVD or observe a live demonstration.
3. *Read.* Read about the skill or hear a lecture on the main points of effective usage. Cognitive understanding is vital for skill maintenance.
4. *Practice.* Ideally, use video or audio recording for skill practice; however, role-play practice with observers and feedback sheets is also effective.
5. *Generalize.* Complete a self-assessment. Integrate the skills and contract for action into the "real world" of interviewing, counseling, and therapy.

You can "go through" the skills quickly and understand them, but practicing them to full mastery makes for real expertise. We have seen many students "buzz" through the skills, but end with little in the way of mastery and expertise. Also, as the microskills are dimensions of emotional intelligence and social competence, teaching these skills to clients has proven to be an effective counseling and therapeutic technique (Daniels, 2009).

DRAWING OUT CLIENT STORIES

Interviewing, counseling, and psychotherapy are concerned with client stories. You will hear many different story lines—for example, stories of procrastination and the inability to take action, tales of depression and abuse, and most important, narratives of strength and courage. Your first task is to listen carefully to these stories and learn how clients come to think, feel, and act as they do. Sometimes, simply listening carefully with empathy and care is enough to produce meaningful change.

You will also want to help clients think through new ways of approaching their stories. Through the conversation that is interviewing, counseling, and psychotherapy, it is possible to rewrite and rethink/restory old narratives into new, more positive and productive stories. The result can be deeper awareness of emotional experience, more useful ways of thinking, and new behavioral actions.

Development and growth are the aim of all that we do. Expect your clients to have enormous capacity for change. In the midst of negative and deeply troubling stories, one of your tasks is to search for strengths and resources that will empower the client. If you can develop exemplary models from the client's past and present, you are well on the way to combating even the most difficult client situation or story.

One brief example: Imagine that an 8-year-old child comes to you in tears, having been teased by friends. You listen and draw out the story. Through your warmth and caring interest you provide a relationship and the child calms down. The child has strengths, and you point out some of them. You talk about the wisdom of coming to talk about problems with a person who is there to help. You comment on a concrete example of the child's strengths such as verbal or physical ability, a time you noticed the child helping someone else, or perhaps family members who support the child. You may read a short book or tell a story that metaphorically illustrates that problems can be overcome through internal strength. Next time, you notice that the child responds differently to friends' teasing.

How might your brief interaction with the child have effected the change described above? As a result of listening, developing positive assets and strengths, and gaining new perspectives through storytelling, you and the child rewrote the event and planned new narratives and action for the future. The basic treatment structure used with the child can be expanded for counseling with adolescents, adults, and families: Listening to the story, finding positive strengths in that story or another life dimension, and rewriting a new narrative for action are what interviewing, counseling, and psychotherapy are about. In short: *relationship—story and strengths—goals—restory—action.*

Please take a moment to now review Box 1-1, which explores how traditional counseling too often focuses only on client problems. James Lanier suggests positive ways to draw out clients' stories and focus more on strengths.

Of course, many times you may encounter very complex issues. In that case, as you listen, be careful not to minimize the story. Behind the tears of the child may be a history of abuse or other serious concerns. So as you listen to stories, simultaneously search for more complex, unsaid stories that may lie behind the initial narrative.

RELATIONSHIP—STORY AND STRENGTHS—GOALS—RESTORY—ACTION

Narrative theory is a relatively new model for understanding counseling, interviewing, and psychotherapy sessions (Holland, Neimeyer, & Currier, 2007; Monk, Winslade, Crocket, & Epston, 1997; White & Epston, 1990; Whiting, 2007). Narrative theory emphasizes storytelling

BOX 1-1 NATIONAL AND INTERNATIONAL PERSPECTIVES ON COUNSELING SKILLS

Problems, Concerns, Issues, and Challenges—How Shall We Talk About the Story?

James E. Lanier, University of Illinois, Springfield

Counseling and therapy historically have tended to focus on client problems. The word "problem" implies difficulty and the necessity of eliminating or solving the problem. Problem may imply deficit. Not all problems can be solved. Traditional diagnosis such as that found in the *Diagnostic and Statistical Manual of Mental Disorders*-TR (American Psychiatric Association, 2000) carries the idea of problem a bit further, using the word disorder with such terms as panic disorder, conduct disorder, obsessive-compulsive disorder, and many other highly specific disorders. The way we use these words often defines how clients see themselves.

I'm not fond of problem-oriented language, particularly that word "disorder." I often work with African American youth. If I asked them, "What's your problem?" they likely would reply, "I don't have a problem, but I do have a concern." I've found that considering client issues as "problems" can get in the way of a positive relationship. The words "concern" or "issue" suggest something we all experience constantly. These words also suggest that we can deal with it—often from a more positive standpoint. Defining concerns as problems or disorders leads to

placing the blame and responsibility for resolution almost solely on the individual.

Unfortunately, you'll often find that a problem/disorder language predominates in your textbooks and in many interviewing and counseling settings. Finding a more positive way to reframe and discuss stories is relevant to all your clients, regardless of their background. Carrying this idea further, terming a "problem" *a challenge is* a call to our strengths and an opening for change. We all have to struggle through pain and complex issues at times and, as we overcome these challenges to our being, we become stronger and more in touch with what we *can do,* rather than what we *can't do.* And, whatever language you use, problem, concern, or challenge, they all represent *opportunities for change!*

As you work with clients, please consider that change, restorying, and action is more possible if we help clients maintain awareness of already existing personal strengths and external resources. Help clients define their goals clearly and the positive assets they already have to resolve their issues. Then you can help them restory with a *can do* self-image. Out of this will come action and generalization of new ideas and new behaviors to the real world.

and the generation of new meanings. The concepts of narration, storytelling, and conversation are useful frameworks as we examine skills, strategy, and theory in interviewing, coaching, counseling, and psychotherapy.

Theories of counseling as varying as person-centered, cognitive behavioral, brief counseling, and psychoanalytic/interpersonal approaches can be considered narratives or stories of the helping process. Each of these theoretical stories can be helpful to your client at varying times. The narrative and microskills tradition will allow you to develop expertise in multiple theories and strategies. This approach will enable you to understand and become more competent in the multiple theories and strategies that you will encounter.

The narrative model of *Intentional Interviewing and Counseling* may be described as follows: First we need to hear client stories. We also need to listen for strengths and assets. Empowerment and wellness, as well as positive psychology, are an increasingly vital part of interviewing and counseling (cf. Myers & Sweeney, 2005; Peterson & Seligman, 2004). With an understanding of client issues and personal power, we have a positive strength-based foundation for change. Restorying is about developing client stories in new directions. The new story often makes action and change possible.

Relationship

No one wants to tell a story to someone who is not interested or who is not warm and welcoming. Unless you can develop rapport and trust with your client, expect little to happen. The relationship in every interview will be different and will test your social skills and understanding. Basic to this is being your own natural self and your openness to others and to differences of all types. Your attending and empathic listening skills are key to understanding and will play a part throughout all sessions.

Another term for relationship is *working alliance,* which in turn is based on what is now called the *common factors approach.* It is consistently estimated that 30% of successful counseling and therapy outcome is due to relationship or common factors consisting of caring, empathy, acceptance, affirmation, and encouragement (Hubble, Duncan, & Miller, 1999, p. 9). Your ability to listen and be with the client is the starting point for the interview.

Story and Strengths

The listening skills described in Section I are basic to learning how clients make sense of their world—the stories clients tell us about their lives, their problems, challenges, and issues. Let us help them tell their stories in their own way. Attending and observation skills are critical, while encouraging, paraphrasing, reflection of feeling, and summarization will help fill out the story. Regardless of the theory used, these listening skills are central, but different counseling systems and theories may draw out different aspects of stories that lead in varying directions.

The listening and observational skills of Chapter 3 through integrative Chapter 8 will be key in drawing out clients' difficulties, concerns, and issues and clients' strengths to solve these problems. At times, counseling and interviewing can spiral down into a depressing repetition of negative stories—and even whining and complaining. Seek out and listen for times when clients have succeeded in overcoming obstacles. Listen for and be "curious about their competencies—the heroic stories that reflect their part in surmounting obstacles, initiating action, and maintaining positive change" (Duncan, Miller, & Sparks, 2004, p. 53).

Goals

If you don't know where you are going, you may end up somewhere else. Too many interviews wander and never have a focus. Once you have heard the story, and you and the client see the need for a new and more effective story, how would you and the client like the story to develop? What is an appropriate ending? If the client does not have a goal in mind, the new story may be irrelevant. This area is considered so important in brief counseling that counselors often start the interview right here—"What do you want to happen today as a result of our conversation?"

Restory

If you understand client stories, strengths, and goals, you are prepared to help clients restory— generate new ways to talk about themselves. One important strategy for restorying is provided in Chapter 8, in which you will demonstrate your ability to conduct a full interview using only listening skills. Many times effective listening is sufficient to provide clients with the strength and power to develop their own new narratives.

Focusing, confrontation, and the influencing skills presented in Section II are important parts of helping clients generate new stories. The five-stage interview with its many adaptations can be used to enable clients to find new ways of making meaning. The stages and skills are also important in multiple theories of helping. But each counseling theory provides us with alternative ways to think and talk about client stories.

For example, you will find more emphasis on self-discovery, emotion, and meaning in person-centered counseling, whereas cognitive-behavioral methods will actively seek to change clients' ways of thinking and behaving. Also, the new story from a psychoanalytic perspective will vary greatly from the narrative of brief counseling. And in this process of restorying, each theory will have a different language system. Yet all systems gain clarity when you view them as stories that offer clients new ways of being.

Awareness of theoretical diversity is also central. All counseling theories seek to help clients find new ways to think about their concerns and problems. But each counseling theory provides us with alternative ways to think and talk about client stories. For example, you will find more emphasis on self-discovery, emotion, and meaning in person-centered counseling, whereas cognitive-behavioral methods will actively seek to change clients' ways of thinking and behaving. And in this process of restorying, each theory will have a different language system. As you define your own natural style, remain open to the multitude of possibilities offered by the professional helping field.

Action

Pay special attention to the final interviewing stage: *action*. All of the above efforts will be useless if the client does not take action on the new ideas that you develop together in the interview. Contract with the client to *act and think* in new ways during the coming week. Generalization to the "real world" can lead to success, meaning you have made a difference.

If your work with this book is successful, in the end you will have developed a solid understanding of foundation skills and strategies. Competence in the skills and strategies of this book will enable you to conduct interviews using several theories of counseling . You will be able to write your own narrative, your own personal story/theory of interviewing, counseling and psychotherapy. You too will have new ways to *act and think*.

INCREASING SKILL AND FLEXIBILITY: INTENTIONALITY, CULTURAL INTENTIONALITY, AND INTENTIONAL PREDICTION

There are many ways to facilitate client development. As you become increasingly competent, you will learn to blend what is natural for you with new interviewing skills and theory. Intentionality asks you to be yourself but also to realize that if you are to reach a wide variety of clients, you will need to be flexible and constantly learn new ways of being in the interview.

Intentionality

Clients come to us with multiple issues and concerns. How you listen and how you respond may say as much about you and your style as it says about the person you are trying to help. One of the goals of this book is to encourage you to look at yourself and your style of listening.

Beginning interviewers are often eager to find the "right" answer for the client. In fact, they are so eager that they often give quick patch-up advice that is inappropriate. For example, your own personal issues or cultural factors such as ethnicity, race, gender, lifestyle, or religious orientation may have biased your response and interview plan for the client. How ideal it would be to find the perfect empathic response that would unlock the door to the client's world and free the individual for more creative living! However, the tendency to search for a single "right" response and to move too quickly can be damaging.

Intentional interviewing is concerned not with which single response is correct, but with how many potential responses may be helpful. Intentionality is a core goal of effective interviewing. We can define it as follows:

> Intentionality is acting with a sense of capability and deciding from among a range of alternative actions. The intentional individual has more than one action, thought, or behavior to choose from in responding to changing life situations. The intentional individual can generate alternatives in a given situation and approach a problem from multiple vantage points, using a variety of skills and personal qualities, adapting styles to suit different individuals and cultures.

The culturally intentional interviewer remembers a basic rule of helping: If something you try doesn't work, don't try more of the same. Try something different!

Cultural Intentionality

> *All interviewing, counseling, and psychotherapy are multicultural.*
>
> —Paul Pedersen

One of the critical issues in interviewing is that the same skills may have different effects on people from varying cultural backgrounds. Intentional interviewing requires awareness that racial and ethnic groups may have different patterns of communication. Eye contact patterns differ. For example, in U.S. culture, middle-class patterns call for rather direct eye contact, but in some cultural groups direct eye contact is considered rude and intrusive. Some find the rapid-fire questioning techniques of some North Americans off-putting and would prefer more time to develop a relationship before being questioned. Spanish-speaking groups have more varied vocal tones and sometimes a more rapid speech rate than do English-speaking people. Some people consider English as monotonous and slow, showing little expression and emotion.

Also remember that the word *culture* can be defined in many ways. Religion, class, racial/ethnic background (for example, Irish American and African American), gender, and lifestyle differences as well as the degree of a client's developmental or physical disability also represent cultural differences. There is a youth culture, a culture of those facing imminent death through AIDS or cancer, and a culture of the aging. Any group that differs from the "mainstream" of society can be considered a subculture. All of us are part of many cultures. In fact, some propose that diversity is what constitutes the mainstream. This suggests that respecting and honoring our differences is what will bring us together as one people.

Avoid stereotyping. Individuals differ as much as or more than cultures. Attune your responses to the unique human being before you. Lack of intentionality shows in the interview when the helper persists in using only one skill, one definition of the problem, or one theory of interviewing, even when that approach isn't working.

Intentional Prediction

This text is action and results oriented; it is founded on research revealing that you may expect results when you use a specific microskill or strategy in the interview. If you work intentionally in the interview, you can anticipate predictable client responses. And when the expected does not happen, you can intentionally flex and come up with a helpful alternative skill or strategy.

Let us briefly define two important skills discussed in later chapters. You are likely very familiar with questions, but intentionally using the specific microskill of questioning will enable you to be more effective in obtaining information economically and respectfully. Another critical skill is reflection of feeling, which helps us clarify client emotions. Reflection of feeling may by itself facilitate meaningful resolution of client's issues, concerns, and problems.

If you use questioning skills, you can *predict* how clients respond. If you reflect feelings, you can *predict* clients will focus on their emotions. Note below the brief definitions of these two skill areas and the predictions that you can make when you use these skills intentionally.

Questions: Begin open questions with often useful *who, what, when, where,* and *why.* Closed questions may start with *do, is,* or *are. Could, can,* or *would* questions are considered open, but have the additional advantage of being somewhat closed, thus giving more power to the client, who can more easily say that he or she doesn't want to respond.	*Predicted Result:* Clients will give more detail and talk more in response to open questions. Closed questions may provide specific information.
Reflection of Feeling: Identify the key emotions of a client and feed them back to clarify affective experience.	*Predicted Result:* Clients will experience and understand their emotional states more clearly and talk in depth about feelings.

Each microskill is coupled with a general set of predicted results (Appendix I), but it is important to stress that predictability and ability to anticipate results of your interventions will never be perfect. If the first skill does not produce the expected result, be ready with another skill or concept to enable clients to grow in their own direction.

THEORY AND MICROSKILLS

The single skills microskills model was developed in 1966–1968 by a group at Colorado State University, and in 1974 multicultural differences in communication styles were identified (Ivey, Gluckstern, & Ivey, 1974/2006; Ivey, Normington, Miller, Morrill, & Haase, 1968). The extensive research on the microskills model is summarized in Box 1-2, and the CD-ROM accompanying this book provides a summary of key research studies.

Many students ask about the theory underlying the microskills approach. To this question there are two responses. The first is that interviewing and counseling are informed by more than 250 theories and that certainly we don't need another one. As you begin working with this book, we'd prefer that you focus on skills and not emphasize theoretical implications until later.

BOX 1-2   RESEARCH EVIDENCE THAT YOU CAN USE

Validating the Microskills Approach

More than 450 databased microskills studies have been completed to date. The model has been tested nationally and internationally in over 1,000 clinical and teaching programs in the past 35 years (Daniels & Ivey, 2007; Daniels, Rigazio-DiGilio, & Ivey, 1997; van der Molen, Hommes, Smit, & Lang, 1995). Microcounseling was the first systematic video-based counseling model to identify specific observable skills (Ivey et al., 1968) and the first emphasizing multicultural issues (Ivey, Gluckstern, & Ivey, 1974/2006).

Specific research implications and applications of the microskill model will be identified throughout the book. Included will be suggestions of ways you can apply research in your own work. Some of the most important findings relevant to your work with clients include the following:

1. **Expect results.** Several critical reviews have been conducted and they have all found microtraining to be an effective framework for teaching skills to a wide variety of groups. There are also consistent data attesting to the effectiveness of teaching microskills to clients and patients. The skills of the microskills hierarchy have been shown again and again to be clear, useful in the interview, and teachable. You will find that the step-by-step model makes it possible for you to identify more clearly what you are already doing and how you are affecting clients.
2. **Practice is essential.** Practicing the skills to mastery (intentional competence) is important if the skills are to be maintained and used after training. If you do not use the skills and strategies of this book, you will lose them over time. Completing the practice exercises included and working actively to generalize this knowledge to daily practice are essential. Use it or lose it!
3. **Multicultural differences are real.** People from different multicultural groups (e.g., ethnicity/race, gender) prefer different patterns of skill usage. There is a lifelong need for all of us to learn about groups other than our own and adapt skill usage in a culturally appropriate manner without stereotyping any client. Each of us is unique.
4. **Different counseling theories have varying patterns of skill usage.** Expect person-centered counselors to focus almost exclusively on listening skills whereas cognitive-behaviorists use more questions and influencing skills.
5. **Mastery of counseling communication can be assessed in a reliable and valid manner.** Instruments for the assessment of counseling skills, such as the Communication Skills Progress Test (CSPT; Kuntze, van der Molen, & Born, 2007) use video tests to examine trainees' progress.
6. **If you use a specific microskill, then you can expect a client to respond in predictable ways.** A special value of this approach is that you know what may happen in the interview. But each client is different and predictability is not perfect. Cultural intentionality enables you to react to changes in the interview and prepares you for the unexpected.

Second, your expertise with the basic skills can be very helpful with your understanding and practicing multiple theoretical approaches. For example, if you become proficient in attending skills, basic listening skills, and influencing strategies, you lay the groundwork for developing competence in many different theories ranging from person-centered to cognitive-behavioral to multicultural counseling and therapy. All theories require practitioners to listen to client stories and thus all use the listening skills, even though they may listen to different things. However, influencing skills are used quite differently from theory to theory, but the idea and structure of confrontation, focusing, and interpretation/reframing will be similar. From this perspective, the microskills presented here become a theory—an integrative theory that will enable you to practice effective counseling *and* to move more quickly and effectively into the many theories of helping that you will encounter.

Table 1-1 presents the microskill patterns of some popular theoretical approaches and strategies. The first six listed (decisional, person-centered, logotherapy, cognitive-behavioral, brief

TABLE 1-1 Microskills patterns of differing approaches to the interview

MICROSKILL LEAD	Decisional counseling	Person-centered	Logotherapy and meaning	Cognitive-behavioral assertiveness training	Brief solution-oriented	Motivational interviewing	Coaching (GROW Model)	Psychodynamic	Gestalt	Rational-emotive behavioral therapy	Feminist therapy	Business problem solving	Medical diagnostic interview	Eclectic/metatheoretical
BASIC LISTENING SKILLS														
Open question	●	○	●	◒	●	●	●●	◒	●	◒	◒	◒	●	◒
Closed question	◒	○	●	●	◒	◒	◒	○	◒	◒	◒	◒	●	◒
Encourager	●	●	●	●	●	●	●	●	●	◒	◒	◒	◒	◒
Paraphrase	●	●	●	◒	●	●	●	◒	○	◒	◒	◒	◒	◒
Reflection of feeling	●	●	●	◒	◒	●	●	◒	○	◒	◒	◒	◒	◒
Summarization	◒	◒	●	◒	●	◒	●	◒	○	◒	◒	◒	◒	◒
INFLUENCING SKILLS														
Reflection of meaning	◒	●	●	○	○	◒	◒	●	○	◒	●	○	○	◒
Interpretation/reframe	◒	○	◒	○	◒	●	○	●	◒	●	●	◒	◒	◒
Logical consequences	◒	○	○	◒	○	◒	◒	○	○	●	◒	●	◒	◒
Self-disclosure	◒	◒	◒	○	○	◒	◒	○	○	◒	●	◒	○	◒
Feedback	◒	◒	◒	◒	◒	◒	●	○	●	◒	◒	●	○	◒
Information/ psychoeducation	◒	○	○	●	○	◒	◒	○	○	●	●	●	●	◒
Directive	◒	○	◒	●	○	◒	◒	○	●	●	◒	●	●	◒
CONFRONTATION (Combined Skill)	◒	◒	◒	◒	◒	●	◒	◒	●	●	●	◒	◒	◒
FOCUS														
Client	●	●	●	●	●	●	●	●	●	●	◒	◒	◒	◒
Main theme/problem	●	○	◒	◒	●	●	●	○	◒	◒	◒	●	●	◒
Others	◒	○	◒	◒	◒	◒	◒	◒	◒	○	◒	○	○	◒
Family	◒	○	◒	◒	◒	◒	○	◒	○	○	◒	◒	○	◒
Mutuality	○	◒	◒	○	◒	○	○	○	○	○	◒	○	○	◒
Counselor/interviewer	○	◒	◒	○	○	○	○	○	○	◒	◒	○	○	◒
Cultural/ environmental context	◒	○	◒	◒	◒	○	◒	○	○	●	◒	○	○	◒
ISSUE OF MEANING (Topics, key words likely to be attended to and reinforced)	Problem solving	Relationship	Discernment	Behavior problem solving	Problem solving	Change	Strengths and goals	Unconscious motivation	Here-and-now behavior	Irrational ideas/logic	Problem as a "women's issue"	Problem solving	Diagnosis of illness	Varies
AMOUNT OF INTERVIEWER TALK-TIME	Medium	Low	Medium	High	Medium	Medium	Medium	Low	High	High	Medium	High	High	Varies

LEGEND

● Frequent use of skill ◒ Common use of skill ○ Occasional use of skill

© 2000 Allen E. Ivey. Adapted from *Developmental Counseling and Therapy*, pp. 54–55. Framingham, MA: Microtraining Associates. By permission of the author for this 7th edition of *Intentional Interviewing and Counseling*.

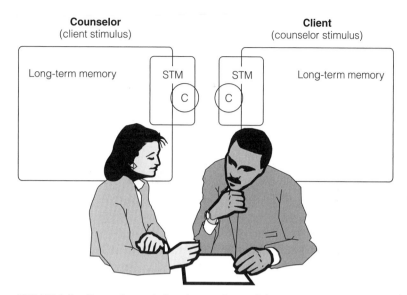

FIGURE 1-3   Counselor and client interaction and their impact on memory and change. (Adapted from Ivey, 2000, pp. 54–55.)

counseling, and motivational interviewing) are presented in this book. These theories have been selected because of their wide use. Observe that virtually all theories give considerable attention to the listening skills. However, the influencing skills vary widely in their frequency.

Once you become skilled with the microskills hierarchy you will be able to use skills and strategies in a variety of settings, and you can master the complex theories of counseling more easily, particularly when it comes to applying these theories in the actual interview.

Ultimately, counseling is an interactive process of influencing the client in positive ways, but counselors are also influenced by the client. Your task is to develop a working relationship (therapeutic alliance) with the client. Effective counseling asks us to enter client consciousness, and then work to restory memories (feelings, thoughts, behaviors, and meanings), which will lead to change (Figure 1-3).

Consciousness (C) represents the psychological present, which ranges in length from 100 to 750 milliseconds and has access to short- and long-term memory. Short-term memory (STM) holds impressions immediately accessible to consciousness for approximately 10 seconds and can hold 100 items. *Here and now* immediacy, being in the moment, occurs here. Reflection and talking and influencing are critical, but we want our clients more able to live and breath immediate experience.

Long-term memory (LTM) is associated with the holistic brain, particularly the hippocampus and prefrontal cortex. Consciousness and short-term memory have easy access to what is often called the *preconscious* and is illustrated by the overlap in the figure. Deeper long-term memories (the unconscious) are less accessible and may require external stimuli to draw them out.

What Does This Mean for You?

A key task of counseling is to help the client restory past experience and develop new memories and connections (behaviors, thoughts, feelings, meanings). Thus we see the importance of immediate *here-and-now* interaction in the conscious present—gaining the client's attention.

If we are successful, new ways of being will enter short-term and then long-term memory. Successful interviewing and counseling change the client and LTM in significant ways and even build new neural networks in the brain (brain plasticity).

The microskill of attending (Chapter 3) is particularly important in *here-and-now* consciousness. Without attention, little changes. Listening skills, especially reflection of feeling, provide the cognitive/emotional "charge" to promote understanding and change itself. The influencing skills of the latter half of this book are oriented to starting and solidifying the change process, which ultimately requires restorying in long-term memory.

In addition, you as interviewer go through a parallel process in which attending to the client in the conscious present ultimately leads you to experiencing new thoughts, feelings, and meanings. Experience and practice with more and more clients provides you with an increasingly sophisticated way of viewing the counseling process. At the same time, we all must be aware of the danger of imposing preexisting theories and mind-sets on our clients.

BRAIN RESEARCH AND NEUROSCIENCE: IMPLICATIONS FOR THE INTERVIEW

One pivotal concept underlies our understanding of the new brain: plasticity . . . the brain's capacity for change. . . . Thoughts, feelings, and actions . . . determine the health of our brain . . . the brain never loses the power to change itself on the basic of experience, and this transformation can occur over very short periods.

—R. Restak, 2003, pp. 7–8

Why do we discuss the brain in an interviewing and counseling text? Daily we see new and exciting developments in neuroscience and neuropsychology on television and in our newspapers and magazines. Research on the brain over the past decade has reached a state of precision that now has important implications for interviewers, counselors, and psychotherapists. Perhaps the most important discovery is neurogenesis and neuroplasticity—the brain develops new neurons and neural connections throughout the lifespan, and it changes in response to new situations or experiences in the environment. "Neuroplasticity can result in the wholesale remodeling of neural networks . . . a brain can rewire itself" (Schwartz & Begley, 2003, p. 16).

Information on neuroscience, counseling, and microskills will be found in Appendix II. There you can see some key regions of the brain in three figures. You will find a chart illustrating how the microskills and counseling strategies relate to neuroscience and change. There is also a chart illustrating the key neurotransmitters that are likely to be influenced by interviewing, counseling, and psychotherapy. We hope that you will visit this Appendix from time to time as it provides further data on several aspects of helping discussed in this book. Please take a moment and notice what is there, but now is not the time to read it in detail.

What Does This Mean for You and the Helping and Interviewing Process?

The most dramatic example of neuroplasticity is evidence that effective counseling can produce new neurons in the brain. When you interact with clients, both your and your client's brain functioning can be measured through a variety of brain-imaging techniques, especially functional magnetic resonance imaging (fMRI) (Arehart-Treichel, 2001). Both you and your client may learn, change, and develop new neural connections as a result of your interaction. Successful interviewing and counseling help clients develop new and useful connections. Neuropsychology's research on emotion validates past counseling and therapy theory and research from a new perspective. In chapters 7 and 8, you will read about mirror neurons, central in empathic understanding of the client.

We have already learned from neuroscience that here and now meditation and exercise help rewire long-term memory; they have become critical aspects of mental practice. We have evidence that a positive approach is more helpful than a "problem-centered" system as the new connections developed with positive thinking enable the brain to deal with stress more effectively. Children who experience abuse, neglect, and deprivation typically have long-term damage to their brains. This finding suggests that social justice and community action for prevention are a critical part of a complete counseling practice.

Particularly important for you as you build counseling microskills is the following from brain imaging research by Restak (2003, p. 9):

> [This research] carried out weekly fMRIs on volunteers while they learned a sequence of finger movements. Within three to four weeks of training, [the researcher] could discern changes in activity patterns in three progressive parts of the brain: the prefrontal cortex, which is responsible for the intention to carry out movements; the supplementary motor cortex, responsible for organizing the sequencing and coordination of the muscles involved in carrying out the action; and the primary motor cortex that gives the "orders" for the movements to take place.

The basis of microskills training is systematic step-by-step learning. Step-by-step skills training is used in ballet, music, golf (Tiger Woods uses it!), and many other settings. If there is sufficient practice, new connections in the brain may be expected, and increased ability in demonstrating these skills will appear in areas ranging from finger movements to dance—and from the golf swing to interviewing skills. "Predictions of whether a particular therapy will work for a given patient may depend more on the individual's brain than how an individual may be diagnosed" (Etkin et al., 2005, pp. 146–147).

What are your thoughts?

_____

_____

As you go forward to further study and professional work, you can expect brain research to be an increasingly important part of our field. We have placed some key information about neuroscience within several of the research boxes in each chapter. Consider what you read here a beginning and an opening to a future in which the relationship between microskills, helping theory, and brain functioning will become increasingly clear.

SUMMARY: MASTERING THE SKILLS AND STRATEGIES OF INTENTIONAL INTERVIEWING AND COUNSELING

Welcome again to the fascinating field of interviewing, counseling, and psychotherapy! You are being introduced to the basics of the individual counseling session, but the same skills are essential in group and family work. And now we know that skills training and effective interviewing, counseling, and therapy affect brain development, thus resulting in longer term change in our clients.

Moreover, you will find that physicians and nurses, managers in business settings, peer counselors, and many others have adopted this skills training format as part of their profession and/or training. The original microskills format presented here has been translated into over 20 languages and used in many varied settings, such as by AIDS and refugee counselors in Africa and Sri Lanka; top-line managers in Sweden, Germany, and Japan; helpers working with trauma survivors from floods and hurricanes; and Aboriginal social workers in Australia and Inuits in the Canadian Arctic. The system works and is constantly changing and growing.

This first chapter frames the entire book. It will help you to study the following key points as these are the things we particularly want you to remember.

The first competency practice exercise in this chapter asks you to examine yourself and identify your strengths as a helper. But, in the middle of all this, *you* are the person who counts, and we hope that you will develop your counseling skills based on your natural expertise and social skills. Good luck!

Key points of Chapter 1 are presented below.

▪ KEY POINTS

Interviewing, counseling, and psychotherapy	These are interrelated processes that sometimes overlap. Interviewing may be considered the more basic and is often associated with information gathering and providing necessary data to help clients resolve issues. Coaching operates from a strength framework and helps plan for immediate and long-term change. Counseling focuses on normal developmental concerns, whereas psychotherapy emphasizes treatment of more deep-seated issues.
Microskills	Microskills are the single communication skill units of the interview (for example, questions, reflection of feelings). They are taught one at a time to ensure mastery of basic interviewing competencies.
Microskills hierarchy	The hierarchy organizes microskills into a systematic framework for the eventual integration of skills into the interview in a natural fashion. The microskills rest on a foundation of ethics, multicultural competence, and wellness. The attending and listening skills are followed by confrontation, focusing influencing skills, and eventual skill integration.
Microskills teaching model	Five steps are used to teach the single skills of interviewing: (1) *warm up* to the skill; (2) *view* the skill in action, (3) *read and learn* about broader uses of the skill; (4) *practice;* and (5) *generalize* learning from the interview to daily life. The model is useful to teach social skills to clients in the interview.
Relationship—story and strengths—goals—restory—action	Our first task is to help clients tell their stories. To facilitate development, we need to draw out narratives of their personal assets. With a positive foundation, clients may learn to write new stories with the possibility of new actions. James Lanier reminds us that language stressing a problem or disorder may get in the way of effective interviewing and counseling.
Intentionality	Achieving intentionality is the major goal of this book and a central goal of the cultural intentionality interviewing process itself. Intentionality is acting with a sense of capability and deciding from among a range of alternative actions. The intentional individual has more than one action, thought, or behavior to choose from in responding to life situations.

(continued)

KEY POINTS (continued)

Cultural intentionality	The culturally intentional individual can generate alternatives from different vantage points, using a variety of skills and personal qualities within a culturally appropriate framework.
Intentional prediction	When you use specific skills in the interview, you can predict what the client is likely to say next. However, each person is different and often will not behave exactly as predicted. You will shift style and change skills to continue the interview smoothly.
Theory and microskills	All counseling theories use the microskills but in varying patterns with differing goals. Mastery of the skills will facilitate your becoming able to work with many theoretical alternatives. The microskills framework can also be considered a theory in itself in which interviewer and client work together to enable the construction of new stories, accompanied by changes in thought and action.
Research validation	The microskills model has been validated by more than 450 databased studies and over 40 years of clinical practice. The skills can be learned, and they do have an impact on clients, but they must be practiced constantly or the user will lose them.
Brain research and neuroscience	Interviewing and counseling will be increasingly informed by research in this area in the coming years and you will want to keep abreast of new developments. Research relevant to interviewing and counseling will be presented in the research boxes throughout this book. Of particular importance is neuroplasticity. "Neuroplasticity can result in the wholesale remodeling of neural networks . . . a brain can rewrite itself" (Schwartz & Begley, 2003, p. 16). Successful interviewing may be expected to help clients develop new and useful connections.
You, microskills, and the interview	Microskills are useful only if they harmonize with your own natural style in the interview. Before you proceed further with this book, audiorecord or videorecord an interview with a friend or a classmate, and make a transcript of this interview. Later, as you learn more about interview analysis, examine or study your behavior in that interview. You'll want to compare it with your performance in an interview some months from now.

YOUR NATURAL STYLE: AN IMPORTANT AUDIO OR VIDEO EXERCISE

At the beginning of this chapter, you were asked to give your own response to an interviewee experiencing multiple issues. These responses must be genuinely your own. If you adopt a response simply because it is recommended, it is likely to be ineffective for both you and your client. Not all parts of the microtraining framework are appropriate for everyone. You have a natural style of communicating, and it is that natural style these concepts should supplement. In effect, learn new skills and be yourself.

Coupled with your natural style will be awareness of, knowledge of, and facility with using skills with the clients with whom you work. How do they respond to your natural style? You will find that you work more effectively with some clients than others. Due to life experience in a "gendered" society, a woman may feel uncomfortable with a male counselor or a man may feel more comfortable with another man. But, also many clients will prefer a person of the opposite sex. Some clients lack trust with interviewers who come from a different race or ethnicity or a different religious/spiritual orientation from their own. In all these cases, you will need to expand your competence and add new methods and information to your natural style.

You are about to engage in a systematic study of the interviewing process. We recommend that you complete Exercise 1: Your Natural Helping Style before proceeding too far into this book. The exercise is explained in the next section. When you complete the exercise, don't forget to request feedback from your "client."

The first audio- or videorecording of yourself using your natural communication style during an interview will help you obtain an accurate picture of where you are as you start this course. You will want to compare your interview with later work as you progress through this text. Your present natural style is a baseline you will want to keep in touch with and honor.

COMPETENCY PRACTICE EXERCISES AND SELF-ASSESSMENT

Individual Practice

Exercise 1: Your Natural Helping Style: An Important Audio or Video Exercise
We believe this exercise is one of the most important exercises in the book. You have a natural style of communicating, and it is that natural style and social skills that you need to build on. In effect, learn these helping skills while still using your natural style. We want authenticity, not actors playing a role.

Also, develop awareness of the natural style of the clients with whom you work, particularly if they are culturally different from you. What do you notice about their style? How do they respond to individual differences? Use your observations to expand your competence and add new methods and information to your natural style.

Find someone who is willing to role-play a client with a concern, problem, opportunity, or issue. Interview that "client" for at least 15 minutes using your own natural style.

Read pages 36–38 and follow the ethical guidelines as you work with a volunteer client. Ask the client, "May I record this interview?" Also inform the client that the tape recorder may be turned off at any time. Common sense demands ethical practice and respect for the client.

You can select almost any topic for the interview. A friend or classmate discussing a school or job problem may be appropriate. A useful topic is interpersonal conflict—for example, concerns over family tensions or decisions about a new job opportunity.

When you complete the interview, ask your client to fill out the Client Feedback Form (Box 1-3). In practice sessions, it is very helpful to get immediate feedback. As you practice the microskills, we encourage you to use the Client Feedback Form. You may even find it helpful to continue the use of this form or some adaptation of it in your work in the helping profession.

Please transcribe this session for later study and analysis. You'll want to compare your first performance with that at the end of this course of study.

Self-Assessment

Review your audio- or videotape and ask yourself and the volunteer client the following questions:

1. What did you do that you think was effective and helpful?
2. What stands out for you from the Client Feedback Form and any other comments the client may have said to you about the session?

BOX 1-3   CLIENT FEEDBACK FORM

_____ (Date)

_____   _____

(Name of Interviewer)   (Name of Person Completing Form)

Instructions: Rate each statement on a 7-point scale where 1 = strongly agree, 7 = strongly disagree, and 4 = neutral. You and your instructor may wish to change and adapt this form to meet the needs of varying clients, agencies, and situations.

	Strongly Agree		Neutral		Strongly Disagree	
1. (Awareness) The session helped you understand the issue, opportunity, or problem more fully.	1	2	3	4	5	6	7
2. (Awareness) The interviewer listened to you. You felt heard.	1	2	3	4	5	6	7
3. (Knowledge) You gained a better understanding of yourself today.	1	2	3	4	5	6	7
4. (Knowledge) You learned about different ways to address your issue, opportunity, or problem.	1	2	3	4	5	6	7
5. (Skills) This interview helped you identify specific strengths and resources you have to help you work through your concerns and issues.	1	2	3	4	5	6	7
6. (Skills) You will take action and do something in terms of changing your thinking, feeling, or behavior after this session.	1	2	3	4	5	6	7

What did you find helpful? What did the interviewer do that was right? Be specific. For example, not "You did great," but rather, "You listened to me carefully when I talked about _____."

What, if anything, did the interviewer miss that you would have liked to explore today or in another session? What might you have liked to have happen that didn't?

Use this space or the other side for additional comments or suggestions.

3. Can you identify one thing you would like to improve?
4. What strengths do you bring to the study of interviewing? Include the natural skills observed in the session plus personal strengths and qualities that you believe will be helpful in your future growth. In what areas would you like to grow and learn more about yourself?

Exercise 2: Diversity, Multiculturalism, and You—Culture Counts!

Cultural and social influences are not the only influences on mental health service and delivery, but they have been historically underestimated—and they do count. Cultural differences must be accounted for to ensure that minorities, like all Americans, receive mental health care tailored to their needs. (Office of Surgeon General, 1999)

This quotation is from the U.S. Surgeon General's Report entitled *Mental Health, Culture, Race and Ethnicity*. We encourage you to visit the Web site for the full report: www.mentalhealth.org/cre/toc.asp.

Part of cultural intentionality and multicultural competence is your awareness of yourself as a cultural being and your ability to work empathetically with people different from you.

Consider the list below and the intersecting multiple cultural identities we all have as part of our being:

Language	Physical ability/disability
Race/ethnicity	Socioeconomic status
Gender	Age (young, old)
Sexual orientation	Significant life experience (e.g., rape, abuse, cancer, war)
Spirituality	Area of the country, nationality

Review the list of dimensions of diversity above and identify yourself from this list as a multicultural being.

Then examine yourself for personal preferences and biases. How much experience do you have with people who are different from you? How able are you to work with those who may be different from you? For example, if you are heterosexual, how able are you to work with clients from the gay or lesbian culture? If you are gay or lesbian, how able are you to work with clients from the heterosexual culture? What developmental steps do you need to take to increase your understanding and awareness?

DETERMINING YOUR OWN STYLE AND THEORY: CRITICAL SELF-REFLECTION ON YOUR FIRST INTERVIEW

We recommend that you keep a journal of your path through this course and your reflections on its meaning to you. Your first session is a critical foundation on which to build. Here are just a few of many possible questions that you can consider.

1. We build on strengths. What did you do right in this session? What did the client notice as helpful?
2. What was the essence of the client's story? How did you help the client bring out his or her narrative/issues/concerns/problems?
3. How did you demonstrate intentionality? When something you said did not go as anticipated, what did you do next?
4. Name just one thing on which you would like to improve in the next session you have.

CHAPTER 2

Ethics, Multicultural Competence, and Wellness

Ethics, Multicultural Competence, and Wellness

I *am (and you also)*
D*erived from family*
E*mbedded in a community*
N*ot isolated from prevailing values*
T*hough having unique experiences*
I*n certain roles and statuses*
T*aught, socialized, gendered, and sanctioned*
Y*et with freedom to change myself and society.*

—Ruth Jacobs*

How can ethics, multicultural competence, and wellness help you and your client?

Chapter Goals   This chapter emphasizes that effective interviews build on professional ethics, multicultural competence, and a positive wellness approach. It is designed to provide specifics for action in the interview.

---

*R. Jacobs, *Be an Outrageous Older Woman*, 1991, p. 37. Reprinted by permission of Knowledge, Trends, and Ideas, Manchester, CT.

Competency Objectives	Awareness, knowledge, and skills in applying ethics, multicultural competence, and wellness will enable you to

▲ Understand ethical principles in interviewing, counseling, and psychotherapy.
▲ Apply these ethical principles in developing your own informed consent form.
▲ Appreciate the importance of multicultural competence and develop awareness of multiple cultural identities in your clients.
▲ Examine your own multiple cultural identities.
▲ Define wellness and positive psychology, and apply these concepts in an assessment interview.
▲ Distinguish between self and self-in-relation and the importance of placing the client in cultural/environmental context.

Kendra, age 25, enters your office, and after some preliminaries in which you establish rapport, she comments on her concerns:

> I'm really upset. I've got a child at home with my mother, and I'm trying to work my way through community college, and my boss at the nursing home has been hitting on me. I want to leave, but I can't afford to stay in school without this job.

In working with all our clients we need a sense of ethical practice, an awareness of their multicultural backgrounds, and an emphasis on their positive strengths. Each client we encounter is one of a kind. Kendra's uniqueness stems from her biological background and the way she has lived her life in connection to others. Family, community, and culture have deeply influenced Kendra's values and socialization. Our task, as interviewers, counselors, and therapists, is to facilitate her growth within this broad context and perhaps even to encourage her to work on social justice issues in her community, if she wishes, but her time is limited, and this might be discussed for the future, after she resolves her issues.

Even from her brief personal statement, you already have some ideas about Kendra. What are some possible personal wellness strengths you imagine that Kendra might already possess that would help lead to problem resolution?

What are some family and community resources that Kendra might draw on for help? First assume that she is Irish American. Then assume that she is African American. What differences do you note in terms of support likely to be available to her?

What are your thoughts?
_____
_____

You may wish to compare your thoughts with ours presented on page 61.

ETHICS IN THE HELPING PROCESS

It is essential that you read and understand the ethical code of your profession. If you practice ethically, you can *predict* how clients may respond. Note the following brief description of ethical behavior and the predictions that you can make.

Ethics	Predicted Result
Observe and follow professional standards, and practice ethically. Particularly important issues for beginning interviewers are *competence, informed consent, confidentiality, power,* and *social justice.*	Client trust and understanding of the interviewing process will increase. Clients will feel empowered in a more egalitarian session. When you work toward social justice, you contribute to problem prevention in addition to healing work in the interview.

All major helping professions throughout the world have codes outlining guidelines for ethical practice. "The codes promote professional empowerment by assisting professionals and professionals-in-training to: (a) keep good practice, (b) protect their clients, (c) safeguard their autonomy, and (d) enhance the profession" (Pack-Brown & Williams, 2003, p. 4). Ethical codes would also be summarized by the words "Keep the best interest of your clients in mind; do no harm to your clients; treat them responsibly with full awareness of the social context of helping." As interviewers, counselors, and therapists, we are responsible for the client before us and for society as well. At times these responsibilities conflict, and you will especially want to seek more detailed guidance and consultation from your supervisor, professional colleagues, and written ethical codes.

Box 2-1 presents Internet sites of some key ethical codes in English-speaking areas of the globe. All codes contain information on competence, informed consent, confidentiality, and diversity. Issues of advocacy, power, and social justice are implicit in all codes, but most explicitly in counseling, social work, and human services. We strongly recommend you review the complete text of the ethical code for your intended profession.

Competence

Regardless of the human services profession with which you identify, competence is central. Competence has been defined as "the habitual and judicious use of communication, knowledge, technical skills, clinical reasoning, emotions, values, and reflections in daily practice for the benefit of the individual and community being served" (Epstein & Hundert, 2002, p. 227).

The American Counseling Association's ethics statement (2005) provides a good example of a statement on competence and its relationship to issues of diversity. Note the emphasis on continuing to learn and expand one's qualifications over time.

> C.2.a. Boundaries of Competence. Counselors practice only within the boundaries of their competence, based on their education, training, supervised experience, state and national professional credentials, and appropriate professional experience. Counselors will demonstrate a commitment to gain knowledge, personal awareness, sensitivity, and skills pertinent to working with a diverse client population.

Part of interviewing competence is recognizing that you can't do it all. Very few of us can work with all clients and all issues. When you sense a challenging problem, seek appropriate supervision from qualified professionals. Don't use referral as a way to avoid certain clients. When you do refer a client, make sure the client does not feel rejected.

BOX 2-1   PROFESSIONAL ETHICS CODES WITH WEB SITES

Listed below are some important ethics codes. Web site addresses are correct at the time of printing but can change. For a keyword Web search, use the name of the professional association and the words "ethics" or "ethical code."

American Association for Marriage and Family Therapy (AAMFT) Code of Ethics — *http://www.aamft.org*

American Counseling Association (ACA) Code of Ethics — *http://www.counseling.org*

American Psychological Association (APA) Ethical Principles of Psychologists and Code of Conduct — *http://www.apa.org*

American School Counselor Association (ASCA) — *http://www.schoolcounselor.org*

Australian Psychological Society (APS) Code of Ethics — *http://www.psychology.org.au*

British Association for Counselling and Psychotherapy (BACP) Ethical Framework — *http://www.bacp.co.uk*

Canadian Counselling Association (CCA) Codes of Ethics — *http://www.ccacc.ca*

National Association of Social Workers (NASW) Code of Ethics — *http://www.naswdc.org*

National Career Development Association (NCDA) — *http://www.ncda.org*

New Zealand Association of Counsellors (NZAC) Code of Ethics — *http://www.nzac.org.nz*

Ethics Updates provides updates on current literature, both popular and professional, that relate to ethics. — *http://ethics.sandiego.edu*

For example, you may be able to help Kendra work out her difficulties with her supervisor who is trying to get too intimate, but in the process you may discover that she would benefit from a women's support group. You could continue to work with her on her individual issues while she attends the support group meetings. However, if Kendra demonstrates severe distress and you feel that her issues are beyond your competence, then referral is essential.

Informed Consent

Informed consent is one of the most important elements in counseling. The counselor tells the client the goals, procedures, benefits, and risks of the counseling process, and the client agrees to what has been outlined. Even as a learner, when you enter into role-plays and practice sessions, it is important to inform your volunteer "clients" of their rights, your own competence, and what is likely to happen. For example, you might say,

Kendra, I'm taking an interviewing course, and I appreciate your being willing to help me out. Obviously, I'm beginning this type of work, so only talk about things that you want to talk about. I'll audiotape the interview, but if you want me to turn the recorder off, I'll do so immediately and erase it as soon as possible. I'll type out a transcript of this session and share it with you before passing it in to the instructor. I'll take out anything that might identify you personally. Remember, we will stop any time you wish. Do you have any questions?

Use the sample contract in Box 2-2 as an ethical starting point and eventually develop your own approach to this critical issue. The American Psychological Association (2002) stresses that psychologists should inform clients if the interview is to be supervised and provides additional specifics:

Standard 10.01 . . . When the therapist is a trainee and the legal responsibility for the treatment provided resides with the supervisor; the client/patient, as part of the informed consent procedure, is informed that the therapist is in training and is being supervised and is given the name of the supervisor.

Standard 4.03 Recording. Before recording the voices or images of individuals to whom they provide services, psychologists obtain permission from all such persons or their legal representatives.

Counseling is an international profession. The Canadian Counselling Association (1999) approach to informed consent is particularly clear.

B4. Client Rights and Informed Consent. When counselling is initiated, and throughout the counselling process as necessary, counsellors inform clients of the purposes, goals, techniques, procedures, limitations, potential risks and benefits of services to be performed, and other such pertinent information. Counsellors make sure that clients understand the implications of diagnosis, fees and fee collection arrangements, record keeping, and limits to confidentiality. Clients have the right to participate in the ongoing counselling plans, to refuse any recommended services, and to be advised of the consequences of such refusal.

Box 2-2 presents a Sample Practice Contract you can use or adapt for use for the purposes of your own practice exercises.

Confidentiality

As a learner or beginning professional, you usually do not have legal confidentiality. Nonetheless, your academic faculty expects you to honor the confidential nature of your client's communication. This means that what you hear in class role-plays or what is said to you in a practice session needs to be kept to yourself. Trust is built on your ability to keep confidences. Be aware that each state has varying laws on confidentiality.

The American Counseling Association's Ethical Code (2005) states:

Section B: Introduction. Counselors recognize that trust is the cornerstone of the counseling relationship. Counselors aspire to earn the trust of clients by creating an ongoing partnership, establishing and upholding appropriate boundaries, and maintaining confidentiality. Counselors communicate the parameters of confidentiality in a culturally competent manner.

BOX 2-2   SAMPLE PRACTICE CONTRACT

The following is a sample contract that you adapt for your own practice sessions with volunteer clients.

Dear Friend,

I am a student in interviewing skills at [insert name of class and college/university]. One of the requirements of this course is that I practice counseling skills with volunteers. I appreciate your willingness to work with me on my class assignments.

You may wish to talk about real concerns that you may have or you may prefer to role-play a problem or issue that does not necessarily relate to you. Please let me know, however, which of these two possibilities you chose.

Here are some important dimensions of our work together:

Confidentiality. As a student, I cannot offer any form of legal confidentiality. You may rest assured, however, that what you tell me in real or role-played situations will remain confidential and remain with me except for the following important exceptions, which must be reported as required by state law: 1. A serious issue of harm to yourself; 2. Indications of abusing or neglecting children; 3. Other special conditions as required by our state [insert as appropriate].

Audio- and/or Videotaping. An important part of interviewing training is making a recording and listening to my own work. This may be shared with the supervisor [insert name and phone number of professor or supervisor] and/or students in my class. You'll find that recording does not affect our practice session so long as you and I are comfortable. If you wish, we can turn off the recorder at any time. Recordings or written transcripts of the recording will be destroyed at the end of the course unless I have additional written permission from you.

Boundaries of Competence. As I am beginning as an interviewer, I obviously cannot do counseling and therapy. This is a practice session so that I can learn more about the interview. In fact, I'd appreciate feedback from you as to my performance and what you find helpful.

_____   _____
Volunteer Client   Interviewer

Date _____

Professionals encounter many challenges to this issue. State law often requires you to inform parents before counseling a child and specifies that information from the interview must be shared with the parents. If issues of abuse should appear, you must report them to the authorities. If the client is dangerous to self or others, then rules of confidentiality change. You will want to study this important professional and legal issue in much more detail.

Technology

The Internet, telephone, and text messaging are becoming more frequently used in counseling, which has led professional organizations to discuss the use of technology in their codes of ethics. For example, the ACA Code of Ethics states that practitioners should inform clients of the benefits, limitations, and potential risks of using these communication devices. As part of informed consent, clients need to be made aware of the difficulties of maintaining confidentiality, as hackers, phone taps, or simple errors in pushing the wrong key can result in information going to others. In addition, the ethical guidelines offer standards for the use of electronic communications over the Internet to provide online counseling services.

Many people now use the Internet to find a partner. But results are not always positive, and serious stalking or sexual acting out does happen. Teens spend an immense amount of time on Facebook and similar Web sites. Internet bullying has become a popular "sport" with damaging results. Your counseling skills will be important in helping clients who face these issues.

Internet addictions, characterized by excessive, compulsive, and out-of-control use of the Web, has become a significant counseling issue. Web-based activities such as excessive gaming, sexual preoccupations, and overuse of e-mail and text messaging have been termed *addictions* because of the similarity to physical and emotional addictions such as smoking, drinking, and gambling (Mallen, Vogel, & Rochlen, 2005).

Different strategies have been suggested for the assessment and treatment of Internet addictions. Preliminary analyses of cognitive behavioral therapy (see Chapter 14) indicate that clients are able to manage addiction by the eighth session and at 6-month follow-up (Young, 2007). Also, reality therapy group counseling seems to be effective to reduce Internet addiction levels in university students (Kim, 2008). You will find that motivational interviewing (Chapter 14), closely allied with microskills, will also be a useful intervention technique.

Power

> *If [race] is not brought up at the beginning, then it becomes a problem for the person without the power, then they don't feel comfortable bringing it up.*
>
> *I went back to the client and asked her what it is like for her to talk with me as I'm White. So we were able to process our relationship and I felt like it was a real meaningful point in that session and in the course of therapy.* (Gillen, Barton, Cane, Tomko, Fetherson, & Anderson, 2008)

The National Organization for Human Service Education (2000) comments on power, an ethical issue that often receives insufficient attention or is ignored in the session:

> Statement 6. Human service professionals are aware that in their relationships with clients power and status are unequal. Therefore, they recognize that dual or multiple relationships may increase the risk of harm to, or exploitation of clients, and may impair professional judgment.

The very act of helping has power implications. The client or helpee starts in a position of lesser power than the counselor. Power differentials occur in a society where privilege goes with skin color, gender, sexual orientation, or other multicultural dimensions. Awareness of and openness about these issues facilitate working toward a balance of power in helping sessions. For example, if you are a male counseling a woman, you might say, "How does it feel, being a woman, to talk about this issue with me?" If your client or you are uncomfortable, it is wise to discuss this issue further. Referral may be necessary at times.

Dual relationships—having more than one relationship with a client—can cause problems. If the client is a classmate or friend, you are engaged in a dual relationship in your practice session. This situation may occur if you work in a small town and counsel a member of your church or school community. Dual relationships can become a complex issue in the helping profession, and you will want to examine this issue in more detail in the ethical codes.

Social Justice and Advocacy

Is the problem, concern, or challenge "in the client," "caused by the environment," or in some balance of the two? Counseling and psychotherapy focus on the individual, but it is also critical to consider the client's social context. Too many interviewers, counselors, and therapists fail to consider external issues that may be the real "cause" of the client's problem. For example, some therapists might see Kendra's issue with the harassing boss differently. Some therapists might say, "That is often part of a job—you'll just have to live with it." Other therapists might ask how she dresses and if she engages in any provocative behavior. These negative approaches are called "blaming the victim." They potentially lead Kendra to feel guilt and think that she has been doing something "wrong." Harassment is harassment! Our task is to support the client.

Is the interviewer's task completed when the session is over? The National Association of Social Workers (1999) ethical code suggests that awareness of the environment and action beyond the interview may be critical if the client is ever going to resolve problems. The social justice approach demands action from you to prevent problems by acting as an advocate for your client. When appropriate and with the client's consent, you work to examine potential barriers and obstacles that prevent the growth and development of your client at an individual, group, or societal level.

> Ethical Principle: Social workers challenge social injustice. Social workers pursue social change, particularly with and on behalf of vulnerable and oppressed individuals and groups of people. Social workers' social change efforts are focused primarily on issues of poverty, unemployment, discrimination, and other forms of social injustice. These activities seek to promote sensitivity to and knowledge about oppression and cultural and ethnic diversity. Social workers strive to ensure access to needed information, services, and resources; equality of opportunity; and meaningful participation in decision making for all people.

As Kendra talks about the unwelcome advances her boss is making toward her at work, the issue of oppression of women has arisen and should be named as such. The social justice perspective requires that you help her understand that the problem is not her fault, and you can support her in efforts to change the working environment. At a broader level, you can work outside the interview to promote higher standards in nursing home care.

This book is about skills and strategies for human change. At the same time, we need to remember that our clients live in relationship to the world. The microskill of focusing stresses that we need to be aware of the cultural/environmental/social context of our clients (Chapter 10).

DIVERSITY AND MULTICULTURAL COMPETENCE

Our field is rapidly becoming competency based. Specifically, it is no longer enough to pass a test or to understand what effective counseling and therapy are. The major question is *can you do counseling for the benefit of your clients?* Multicultural competency raises additional challenging questions—*can you work for the benefit of clients who are culturally different from you? Are you able to provide competent counseling for men? Women? A person who is of a different race or ethnic group from you? How effective are you with heterosexuals, gays, lesbians, bisexuals, and transgendered individuals?* When these issues are combined with religious and spiritual orientation, age, and many other factors, you can see that becoming multiculturally competent is a process that likely will continue throughout your helping career. Also view Box 2-3 for a further discussion of the importance of culture.

If your practice takes multicultural aspects into consideration, you can *predict* how clients may respond. Note below the predictions that you can make.

Multicultural Competence	Predicted Result
Base interviewer behavior on an ethical approach with an awareness of the many issues of diversity. Include the multiple dimensions described in this chapter. All of us have many intersecting multicultural identities.	Anticipate that both you and your clients will appreciate, gain respect, and learn from increasing knowledge in intersecting identities and multicultural competence. You, the interviewer, will have a solid foundation for a lifetime of personal and professional growth.

The American Counseling Association (2005) focuses the Preamble to their Code of Ethics on diversity as a central ethical issue.

The American Counseling Association is an educational, scientific, and professional organization whose members work in a variety of settings and serve in multiple capacities. ACA members are dedicated to the enhancement of human development throughout the lifespan. Association members recognize diversity and embrace a cross-cultural approach in support of the worth, dignity, potential, and uniqueness of each individual within their social and cultural contexts.

The Ethical Standards of Human Service Professionals (National Organization of Human Service Professionals, 2000) include the following three assertions:

Statement 17. Human service professionals provide services without discrimination or preference based on age, ethnicity, culture, race, disability, gender, religion, sexual orientation or socioeconomic status.

Statement 18. Human service professionals are knowledgeable about the cultures and communities within which they practice. They are aware of multiculturalism in society and its impact on the community as well as individuals within the community. They respect individuals and groups, their cultures and beliefs.

Statement 19. Human service professionals are aware of their own cultural backgrounds, beliefs, and values, recognizing the potential for impact and their relationships with others.

BOX 2-3   NATIONAL AND INTERNATIONAL PERSPECTIVES ON COUNSELING SKILLS

Multiculturalism Belongs to All of Us
Mark Pope, Cherokee Nation and Past President of the American Counseling Association

Multiculturalism is a movement that has changed the soul of our profession. It represents a reintegration of our social work roots with our interests and work in individual psychology.

Now, I know that there are some of you out there who are tired of culture and discussions about culture. You are the more conservative elements of us, and you have just had it with multicultural this and multicultural that. And, further, you don't want to hear about the "truth" one more time.

There is another group of you that can't get enough of all this talk about culture, context, and environmental influences. You are part of the more progressive and liberal elements of the profession. You may be a member of a "minority group" or you have become a committed ally. You may see the world in terms of oppressor and oppressed.

Perhaps now you are saying, "good analysis" or alternatively, "he's pathetic" (especially if you disagree with me). I'll admit it is more complex than these brief paragraphs allow, but I think you get my point.

Here are some things that perhaps can join us together for the future:

1. We are all committed to the helping professions and the dignity and value of each individual.
2. The more we understand that we are part of multiple cultures, the more we can understand the multicultural frame of reference and enhance individuality.
3. Multicultural means just that—many cultures. Racial and ethnic issues have tended to predominate,

but diversity also includes gender, sexual orientation, age, geographic location, physical ability, religion/spirituality, socioeconomic status, and other factors.

4. Each of us is a multicultural being and thus all interviewing and counseling involve multicultural issues. It is not a competition as to which multicultural dimension is the most important. It is time to think of a "win/win" approach.
5. We need to address our own issues of prejudice—racism, sexism, ageism, heterosexism, ableism, classism, and others. Without looking at yourself, you cannot see and appreciate the multicultural differences you will encounter.
6. That said, we must always remember that the race issue in Western society is central. Yes, I know that we have made "great progress," but each progressive step we make reminds me how very far we have to go.

All of us have a legacy of prejudice that we need to work against for the liberation of all, including ourselves. This requires constantly examining yourself, honestly and painfully. You are going to make mistakes as you grow multiculturally; but see these errors as an opportunity to grow further.

Avoid saying, "Oh, I'm not prejudiced." We need a little discomfort to move on. If we realize that we have a joint goal in facilitating client development and continue to grow, our lifetime work will make a significant difference in the world.

Source: Mark Pope, Elder of the St. Francis River Band of Cherokees and Past-President of the American Counseling Association.

Diversity and multiculturalism have become central to the helping professions throughout the world. Many of the persons we interact with professionally come from cultural backgrounds that are different from ours. For example, if Kendra has a concern related to diversity issues, you will need to be competent to address them. Otherwise, you may need to refer her to someone else. You have the responsibility to engage in constant learning to minimize the possibility for the need of referral. Referral to others cannot be an ethical excuse over the long term. You have a responsibility to build your multicultural competence through constant study and supervision.

Multicultural Practice

The American Psychological Association's Multicultural Guidelines begin with this statement: "All individuals exist in social, political, historical, and economic contexts and psychologists are increasingly called upon to understand the influences of these contexts on individuals' behavior" (APA, 2003, p. 377). Multicultural counseling competencies have been developed to provide specifics for culturally sensitive helping (Roysircar, Arredondo, Fuertes, Ponterotto, & Toporek, 2003; Sue et al., 1998). This section reviews some of the most important of these practice guidelines for the beginning professional and you are urged to consult the ACA and other resources for more details and information. Box 2-4 presents a summary of the competencies. For a more complete discussion of the multicultural competencies, please visit http://www.counseling.org/Resources.

Expect the issue of multicultural competence to become increasingly important to your professional helping career. For example, cultural competency training is now required for medical licensure in New Jersey, while at least four other states have pending legislation with similar bills (Adams, 2005).

In this section, the words "multiculturalism" and "diversity" are defined broadly to include race/ethnicity, gender, sexual orientation, language, spiritual orientation, age, physical ability/disability, socioeconomic status, geographical location, and other factors. The multicultural competencies talk about awareness, knowledge, and skills. They ask you to become aware that specific issues exist, to learn about them, and to develop skills that can be used with clients. Given that each client, as a cultural being, has a life experience different from everyone else's, you face a lifetime of multicultural learning.

Awareness of Your Own Assumptions, Values, and Biases

Awareness of yourself as a cultural being is vital as a beginning. Unless you have this awareness, you will have difficulty developing awareness of others. You need to understand your own cultural background and the differences that may exist between you and people from different cultures. You will be constantly learning about other cultures and thus developing new skills. An important skill is recognizing your limitations and the need in certain cases for referral.

Privilege is related to power as discussed in the preceding section. Privilege is power given to people through cultural assumptions and stereotypes. The concept of White privilege was originated by McIntosh (1998) when she pointed out that White people enjoy certain benefits simply by the color of their skin. She spoke of the "invisibility of Whiteness," commenting that White European Americans tend to be unaware of their color or the advantages that come to them because of it. Five examples of White privilege are presented below, and you can use your computer search function to access the complete list (keywords: White, privilege, McIntosh):

▲ If I should need to move, I can be pretty sure of renting or purchasing housing in an area that I can afford and in which I would want to live.
▲ I can go shopping alone most of the time, pretty well assured that I will not be followed or harassed.
▲ Whether I use checks, credit cards, or cash, I can count on my skin color not to work against the appearance of financial reliability.
▲ I can do well in a challenging situation without being called a credit to my race.

BOX 2-4   GUIDELINES FOR MULTICULTURAL COMPETENCE

The American Counseling Association and the American Psychological Association have developed guidelines of multicultural competence in practice, research, and training (American Psychological Association, 2003; Arredondo et al., 1996; Sue et al., 1998).

The following guidelines are a summary of some key issues related to interviewing and counseling practice. At a later point, you should examine the full statements provided by professional associations.

Guideline 1—Awareness

The intentional interviewer or counselor will make a lifetime commitment to developing increased cultural expertise. Interviewers strive to demonstrate the following:

1. They are aware of themselves as cultural beings, paying special attention to developing awareness of personal preferences and biases that might enhance as well as impede or work against the effective delivery of services.
2. They are aware of how contextual problems outside a person's control affect the way that an individual discusses his or her concerns. For example, external issues such as oppression or discrimination (sexism, racism, failure to recognize disability) may deeply affect a client without his or her conscious awareness. Is the problem "in the individual," "in the environment," or in some balance of the two?

Guideline 2—Knowledge

The intentional interviewer or counselor strives to make a lifelong commitment to learning the multicultural base of practice.

1. Interviewers strive to learn about multicultural groups, their history, and their present concerns as a constantly ongoing process.
2. Interviewers learn about helping processes in non-Western cultures and seek to include them, as appropriate, in their own practice. For example, how can spiritual or community leaders supplement counseling practice?
3. Interviewers learn their own limitations in cultural expertise and seek supervision as necessary. In addition, they learn when and how to refer a client for more appropriate assistance.

Guideline 3—Skills

The intentional interviewer or counselor strives to develop effective multicultural practice in these ways:

1. Through developing skills that are attuned to the unique worldview and culture of a widely varying base of clients. Each skill, strategy, or helping theory is examined for its cultural appropriateness.
2. By respecting the first language of the client and/or ensuring that careful translation is available. Obtain informed consent about the language in which interviewing is to be conducted.
3. By ensuring that contextual and diversity factors such as level of education, socioeconomic status, physical ability/disability, acculturation stress, and others are considered a part of overall treatment planning and action. In addition, clients are assisted in learning how their "individual" concerns are related to these contextual issues.

▲ I am never asked to speak for all the people of my racial group.
▲ I can choose blemish cover or bandages in flesh color and have them more or less match my skin.

McIntosh's concepts have been extended to privileges enjoyed by men, the benefits of middle-class economic status, and descriptions of many other groups who hold less power and privilege in our society. Here are five examples from a male privilege checklist (Deutsch, 2002):

▲ My odds of being hired for a job, when competing against female applicants, are probably skewed in my favor. The more prestigious the job, the larger the odds are skewed.
▲ I can be confident that my co-workers won't think I got my job because of my sex even though that might be true.
▲ If I am never promoted, it's not because of my sex.
▲ The odds of my encountering sexual harassment on the job are so low as to be negligible.
▲ I am not taught to fear walking alone after dark in average public spaces.

Following are five example items from the middle-class privilege list (Liu, Pickett, & Ivey, 2007):

▲ I can be assured that I have adequate housing for my family and myself. I have the resources to make choices regarding my medical care.
▲ I can buy not only what I need to have, but also what I want.
▲ When politicians speak of the middle class, I know they are referring to me.
▲ If my child runs into a problem in school, I feel that my concerns as a parent will be heard.
▲ I have enough financial reserves that I can handle a major car problem without a large financial crisis.

Whites, males, heterosexuals, upper- and middle-class people, and others who enjoy power and privilege all have the additional privilege of not being aware of their privilege. The physically able see their capacities as "normal" with little awareness that most of them are "temporarily able" until old age or a trauma occurs. When Christians are the dominant religion, as in the United States and much of the Western world, they often are not aware of the privileges that exist for them. Similarly, when other religions are the majority in other parts of the world, they also may be unaware of their privilege and power.

The interviewer is in a power situation and thus enjoys some privileges that the client does not have; for example, counselors and therapists often can decide whether they want to work with a particular client. Clients, in turn, are often expected to work with the person they are referred to. With little knowledge, clients are in a "one down" position and can be subject to abuses of power.

You as the interviewer also face some challenges. If you are of White European descent, male, middle class, and heterosexual, and the client is female, working class, and of a different race from you, she is less likely to trust you and thus establishing rapport may be more difficult. Your task is to improve your awareness, knowledge, and skills about who you are if you are to work with clients different from you.

Understanding the Worldview of the Culturally Different Client

Worldview is formally defined as the way we interpret humanity and the world. For interviewing purposes, think of worldview as the way your clients see themselves and the world around them. "Research shows that even subliminal presentation of Black faces activates the amygdala (seat of emotions) in Whites" and "African Americans know when someone is liberal on the outside but uncomfortable on the inside" (Weston, 2007, p. 65). There is a general tendency to view those who are "different" from you with caution. This same phenomenon shows when African Americans view White faces. Racism and prejudice manifest themselves in brain reactions, and this illustrates the long journey we have if we are to eliminate intolerance of all types. Due to varying multicultural backgrounds and life experiences, each client views the world differently. The next chapters of this book on attending, observation, and listening skills are designed to help you learn the skills of listening to the many worldviews of clients.

The multicultural competencies stress the importance of our being aware of negative emotional reactions we may have to groups different from us. Our own multicultural backgrounds sometimes taught us to view certain groups through stereotyped perceptions that often are inaccurate and demeaning. Thus, it is doubly important to listen and learn the worldview of the client and not impose your own way of thinking on him or her. Diversity also means that each individual is unique.

We work toward multicultural competence when we develop knowledge of the many cultural groups that we will meet. Some traditional approaches to counseling theory and skills may be inappropriate and ineffective. We also need to learn about socioeconomic influences on the client and give special attention to how racism, sexism, heterosexism, and other oppressive forces may act on a client's worldview.

You will develop skills in understanding various worldviews through academic study and reading. But perhaps more important, you will be become actively involved in the community where your clients reside. This means having friends from diverse backgrounds, attending community events, social and political functions, and celebrations and festivals. Most important, it means getting to know those who are culturally different from you on a personal basis.

The RESPECTFUL Model

The RESPECTFUL model will help you further develop your multicultural competence. As you review the list below, first identify your own multicultural identities, for all of us belong to multiple cultural groups. Then examine your beliefs and attitudes toward those who are similar to and multiculturally different from you on each issue below.

R Religion/spirituality. What is your religious and spiritual orientation? How does this affect you as an interviewer or counselor?

E Economic/class background. How will you work with those whose financial and social background differs from yours?

S Sexual identity. How effective will you be with those whose gender or sexual orientation differs from yours?

P Personal style and education. How will your personal style and educational level affect your interviewing practice?

E Ethnic/racial identity. The color of our skin is one of the first things we notice. What is your reaction to different races and ethnicities?

C   Chronological/lifespan challenges. Children, adolescents, young adults, mature adults, and older persons all face different issues and problems. Where are you in the developmental lifespan?

T   Trauma. It is estimated that 90% or more of the population experiences serious trauma(s) in their lives. Trauma underlies the issues faced by many of your clients. War, flood, rape, and assault are powerful examples, but divorce, loss of a parent, or being raised in an alcoholic family are more common sources of trauma. The constant repetition of racist, sexist, and heterosexist acts and comments can also be traumatic. What is your experience with life trauma?

F   Family background. We learn culture in our families. The old model of two parents with two children is challenged by the reality of single parents, gay families, and varying family structures. How has your life experience been influenced by your family history (both your immediate family and your intergenerational history)?

U   Unique physical characteristics. Become aware of disabilities, special challenges, and false cultural standards of beauty. Help clients think about themselves as physical beings and the importance of nutrition and exercise. How well do you understand the importance of the body in the interview, and how will you work with others different from you?

L   Location of residence and language differences. Whether in the United States, Great Britain, Canada, or Australia, there are marked differences between the south and north, the east and west, urban and rural. Remember that a person who is bilingual is advantaged and more skilled, not disadvantaged. What languages do you know, and what is your attitude toward those who use a different language from you?

As you can see, all of us are multicultural beings (D'Andrea & Daniels, 2001), and all interviewing, counseling, and psychotherapy is multicultural. Broaden your definition of diversity beyond race and ethnicity to include the other factors in the RESPECTFUL model. All your sessions will involve intersecting and interacting identities. As you can see, becoming multiculturally competent is a lifetime endeavor that will enhance your interviewing skills.

DEVELOPING APPROPRIATE INTERVENTION STRATEGIES AND TECHNIQUES

The multicultural movement in counseling grew out of the dissatisfaction of minorities, women, gays/lesbians, people with disabilities, and other groups who felt that traditional counseling and psychotherapy were not working effectively for them. A classic study found that 50% of minority clients did not return to counseling after the first session (cited in Sue & Sue, 2008). A general theory of multicultural counseling and therapy (MCT) has been developed to address this problem (Sue, Ivey, & Pedersen, 1996). Feminist therapy is an example of a culturally specific approach to new ways of working with women.

Over time, you will want to expand your knowledge and skills in learning traditional theory and strategies as well as embracing newer methods. Counseling and therapy theory is now being adapted to show how to use existing traditional theory in a culturally respectful manner (Ivey, D'Andrea, Ivey, & Simek-Morgan, 2007). It is particularly important that you become aware of the history of cultural bias in assessment and testing instruments and the impact of discrimination on clients.

Boys and men, in particular, now need consideration as an important part of multicultural counseling. Recent data show that boys are more likely to drop out of school and substantially more women are attending college than men. Depression among men is becoming a more significant issue (Levant, 2008). While men may participate in sexist acts, counselors need to consider the implications of these society changes. Many men are not in positions of power, particularly when we consider social class issues.

Cultural intentionality—the ability to engage in many and varied verbal and nonverbal helping responses—is a basic skill for multicultural work. *Intentional Interviewing and Counseling* seeks to address intentionality by providing you with ideas for multiple responses to your clients. If your first response doesn't work, you need to be ready with another. As an example, for clients who may be blaming themselves for a problem with friends or family, the skill of focusing (Chapter 10) can help determine whether their issues are actually related to what the clients have done, or if the real source of the problem is with the friends or family.

Adapting present methods and theories to be more culturally sensitive, of course, is a major objective for us all. Psychoeducation, the direct teaching of skills and knowledge to clients, can be a useful intervention, and this book gives considerable attention to that area. In addition, you may wish to add social action to your skills as an interviewer. Through work in the community and schools, you may be able to prevent some of the problems your clients face.

WELLNESS AND POSITIVE PSYCHOLOGY

Wellness counseling is a way of life oriented toward optimal health and well-being, in which body, mind, and spirit are integrated with the goal of living life more fully.
—Jane Myers, Thomas Sweeney, & Joe Witmer

This section presents the background of a strength-based approach to the interview. Wellness models and positive psychology seek to present an alternative approach to clients' problems through a wellness orientation. If you help clients recognize their strengths, you can *predict* how they may respond. Note the predictions that you can make.

Wellness	Predicted Result
Help clients discover and rediscover their strengths through wellness assessment. Find strengths and positive assets in the clients and in their support system. Identify multiple dimensions of wellness.	Clients who are aware of their strengths and resources can face their difficulties and discuss problem resolution from a positive foundation.

Wellness models and positive psychology do not deny human problems and difficulties. Rather, they suggest that if issues are discussed in an atmosphere of strengths, resources, and possibilities for growth, we will enhance our chances for enabling our clients to work through both common daily problems and more severe complex issues. Compare this with the older approach, which focuses almost solely on client deficits and difficulties.

Wellness and Positive Psychology: The Search for Strengths

Clients come to us to discuss their problems, their issues, and their concerns. They are talking with us about what is wrong with their lives and may even want us to fix things for them. There is no question that our role is to enable clients to live their lives more effectively and meaningfully. But an important part of this process of problem solving is helping clients discover their strengths and resources.

Leona Tyler, one of the first women presidents of the American Psychological Association, developed a system of starting counseling based on strengths:

> Initial stages of . . . therapy include a process that might be called exploration of resources. The counselor pays little attention to personality weaknesses . . . [and] is most persistent in trying to locate . . . ways of coping with anxiety and stress, already existing resources that may be enlarged and strengthened once their existence is recognized. (Tyler, 1961, p. 213)

Then, once the client's strengths and resources are known, the counselor and the client can better approach complex issues and problems. Tyler's positive ideas have been central to the microskills framework since its inception. The strength and resource-oriented *relationship—story and strengths—goals—restory—action* model is an elaboration of her ideas. Chapters 13 and 14 of this book discuss five concrete theoretical approaches that all focus on human strengths—Carl Rogers's person-centered theory, decisional counseling, cognitive-behavioral therapy, brief counseling, and motivational interviewing.

Recently, the term *positive psychology* has become important, and the field has developed an extensive body of knowledge and research supporting the importance of a strengths-based approach, also oriented to moral character (Petersen & Seligman, 2004; Snyder & Lopez, 2002). Positive psychology is defined as the scientific study of the strengths and virtues that enable individuals and communities to thrive. Psychology has overemphasized the disease model—"We've become too preoccupied with repairing damage when our focus should be on building strength and resilience" (Seligman, 2002, p. 1). Positive psychology brings together a long tradition of emphasis on positives within counseling, human services, psychology, and social work. Note the similarity of this approach to that originated by Tyler 50 years ago. Sometimes the field is slow in coming to terms with vital concepts.

In the case of Kendra, presented at the beginning of this chapter, we see a client distressed by several possible issues—living at home, caring for her child while she works, the stresses of community college, and facing sexual harassment in a needed job (which likely does not pay a living wage). The positive approach reminds us to look for Kendra's personal strengths and environmental resources, but the question remains, "How?" A wellness assessment is one route toward finding strengths, positive assets, and resources for later problem solving. And we need to constantly look for positive assets and strengths in the client.

How to Apply Wellness in Assessment, Interviewing, and Counseling

The counseling profession's wellness approach is a way of life oriented toward optimal health and well-being in which body, mind, and spirit are integrated by individuals, so they can live life more fully and in harmony with others. The wellness model is holistic and refers to a self-in-relation, the person-in-community, and individual-in-social

CONTEXTS:

Local (safety)
Family
Neighborhood
Community

Institutional (policies & laws)
Education
Religion
Government
Business/Industry

Global (world events)
Politics
Culture
Global Events
Environment
Media

Chronometrical (lifespan)
Perpetual
Positive
Purposeful

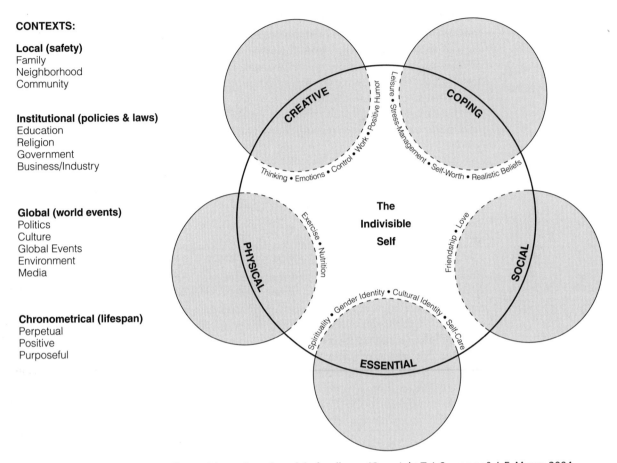

FIGURE 2-1 The indivisible self: An evidence-based model of wellness. (Copyright T. J. Sweeney & J. E. Myers, 2004. Reprinted with permission.)

context (Myers & Sweeney, 2004, 2005). This means that the individual is fully connected to family and community, and even the world as a whole. In effect, we are one. Similarly, while we break individuals into smaller pieces to discuss specific issues, the person is a whole, in which all parts affect all other parts—*the indivisible self* (Myers and Sweeney, 2005). (See Figure 2-1.)

A Contextual/Holistic View of Wellness

The Sweeney and Myers Wellness Model speaks of the "indivisible self" in which 17 dimensions of wellness have been found to group in five factor-analyzed categories, discussed in detail below. Each category has practical implications for assessing clients and facilitating their growth and development.

The indivisible self holistic model stresses the importance of context. As appropriate to the individual client before you, it may be helpful to explore the multiple contexts of human development. For example, what is going on locally (family, neighborhood, community)?

Problems here obviously affect the individual, and even more important, strengths and wellness assets may be found here as well.

Change in any part of the wellness system can be beneficial through the whole person—or it may damage many of the 17 dimensions of wellness. A problem or a positive change in one part of the total system affects all others. For example, a person may have all dimensions of wellness operating effectively but then encounter a difficult contextual issue such as parents divorcing, a major flood or hurricane, or a major personal trauma. On the other hand, the individual may use wellness assets to surmount these challenges and come out of them stronger.

Other contextual issues that may be important to clients' wellness include the institutions that define so much of their experience, such as education, government, and business/industry. At an even broader level, political events, culture, environmental events, global events, and the media can deeply affect clients. An economic recession, change in governmental social services, global warming, and a call-up for military service are three examples of contextual issues that affect the individual.

A final contextual issue is lifespan development. Issues for a child entering the teenage years are very different from those for a teen entering the military, work, or college. Marrying or selecting a life partner, raising a family, and approaching older maturity all present different contextual issues that need to be considered.

In the case of Kendra, some contextual institutions may affect her situation. The nursing home setting is one obvious example, as is her community college. What other supports does the community college offer? If she is involved in a church, what resources does it have to help her? Government and business programs sometimes are available as helpers (and sometimes as hindrances). World events can affect Kendra, particularly if she has a friend serving overseas in the armed forces. Time and key generational events can be a factor in how a client thinks about the world. Those who experienced the Vietnam period and the 1960s may have a different worldview from those born later. In future years, expect to find a significant difference between those who experienced the Iraq war as an adult and younger people coming up who do not have these vivid memories.

For a more detailed presentation of wellness research, see Myers and Sweeney (2005). Below are brief definitions of the 17 personal dimensions, grouped into five categories, and some beginning wellness questions for exploration with your clients.

Wellness Assessment: Identifying Client Strengths

This section presents 17 dimensions of wellness basic to optimal health. These factors, organized into five categories, have direct implications for helping clients become aware of their capacities and strengths.

Seventeen dimensions are a lot to deal with, but if you first focus on yourself and your own wellness strengths, you will have an initial understanding of the power of the positive in the interview. Later, you can work through this wellness assessment with a volunteer client. You will not ordinarily conduct a full wellness assessment as presented here, but—as appropriate to the situation—you will want to use portions of the assessment with most clients. And in longer-term counseling situations, a full session given to wellness assessment can prove extremely valuable in the long run. Also, please consider adding in brief form some or all of these wellness dimensions to your intake form. It can be most helpful to clients to present strengths before they enter the interviewing office.

Let's focus on you as we first consider wellness assessment. We suggest that you write down your specific strengths relating to each dimension presented below. You may wish to photocopy these pages. Throughout the questioning process, try to draw out concrete examples and specifics of strengths available.

What are your thoughts about each issue below?

_____

_____

Category 1—The Essential Self

Dimension 1—*Spirituality.* What strengths and supports do you gain from your spiritual/religious orientation? Be as specific as possible. How could you draw on this resource when faced with life challenges?

Dimension 2—*Gender identity.* Two elements are important here. First, what strengths can you draw on as a female or male? Do you have positive models who help you live your life? Second, what strengths do you draw from your sexual orientation? What models support you as a heterosexual person, a gay, a lesbian, a bisexual, or a transgendered person?

Dimension 3—*Cultural identity.* What strengths do you derive from your multicultural background, including race, ethnicity, gender, sexual orientation, ability/disability, spirituality, and other types of diversity? Who are your heroes?

Dimension 4—*Self-care.* This refers to how well people take care of themselves—cleanliness, healthy habits, avoidance of drugs, safety habits (wearing seat belts) are all examples that lead to a longer life. Those clients who do not engage in self-care may be depressed or have other serious issues. How are you doing in this area? What are your strengths?

Category 2—The Social Self

Dimension 5—*Friendship.* This focuses on your ability to be a friend and have friends and have healthy long-term relationships. We are people in connection and not meant to be alone. It takes time to nurture relationships. Give examples of yourself as a social being and think of the positive feelings that come from a good relationship with friends.

Dimension 6—*Love.* Caring for special people such as family and a loved one results in intimacy, trust, and mutual sharing. Give examples of family members and role models from whom you can gain some support. Sexual intimacy and sharing with a close partner are key areas of wellness. What strengths from the past and present can you identify?

Category 3—The Coping Self

Dimension 7—*Leisure.* This is a topic that interviewing and counseling often forget, but it is essential to wellness. People who take time off to enjoy things can return to the world full of stress more effectively. What leisure time activities do you enjoy? Equally important—do you take time to do them? When was the last time, and how did it feel?

Dimension 8—*Stress management.* Life in the modern world provides us with endless opportunities to become "stressed out." What do you do when you encounter stress? What works best for you, and do you remember to use these strategies? Give at least one example when you managed stress well.

Dimension 9—*Self-worth.* Self-esteem—feeling good about yourself—is needed for personal comfort and effective living. We need to accept our imperfections as well as acknowledge our strengths. How do you evaluate yourself on this important dimension?

Dimension 10—*Realistic beliefs*. Seeing the situation as *it is* is part of mental health. Facts are friendly, even when they are uncomfortable. On the other hand, some of us have come to believe negatively about our capabilities and ourselves; these unrealistic beliefs need to be challenged. How good are you at seeing what *is* in a realistic fashion? Give specific examples when you have made adequate and realistic self-assessments.

Category 4—The Creative Self

Dimension 11—*Thinking*. This refers to effective problem solving and overall personal adjustment. What are some things that have gone well for you in the past? The present? What might you anticipate in a positive way for the future?

Dimension 12—*Emotions*. Ability to experience emotion appropriately is vital to a healthy lifestyle. When have you expressed emotion appropriately with a good result? Can you allow yourself to feel positive emotions? Be specific.

Dimension 13—*Control*. People who feel in control of their lives see themselves as making a difference—they are in charge of their own "space." When have you been able to control difficult situations in a positive way? How are you in control of your own destiny? Again, provide positive, concrete examples.

Dimension 14—*Work*. We all need to work to sustain ourselves and to give meaning to our lives. In fact, we spend most of our time in this activity. What work habits do you have that are particularly strong? What do you enjoy about work? Give examples of each.

Dimension 15—*Positive humor*. A sense of humor certainly helps an individual cope with the world. Can you laugh easily? What things occur to you around this area? What is fun?

Category 5—The Physical Self

Dimension 16—*Exercise*. These last two dimensions need to become a more important part of our interview practice. Evidence is now clear that exercise not only helps the body, it increases brain blood flow and is even a recommended treatment for potential Alzheimer's disease (Ratey, 2008). One useful treatment for clients who may be depressed is helping them get their bodies moving. What is your exercise routine? What works for you? If your physical activity is limited at present, what are you going to do about it?

Dimension 17—*Nutrition*. This is another dimension that needs more attention from interviewers and counselors. What are your eating habits? Especially, what strengths have you developed in your understanding and ability to apply concepts of effective nutrition?

Intentional Wellness Plan

Assessment is not enough. We need to develop an intentional wellness plan for ourselves and also with our clients. The first step is a concrete assessment of wellness strengths. The second step is an honest appraisal of areas that you and your clients need to examine to determine what could be done to develop a healthier lifestyle. Effective wellness assessment provides a summary of strengths and areas where improvements might be made (Myers & Sweeney, 2005).

Again, we will remind you that you can use the assessment questions above with clients. You will not often have time to go through them all, and you may wish to have clients fill out their responses in writing. Then, you can go over important issues of strengths before turning

BOX 2-5 RESEARCH EVIDENCE THAT YOU CAN USE

Wellness

Wellness has been studied in various populations; this research has involved many cross-cultural and cross-national studies involving participants of all ages, resulting in a database containing information on more than 12,000 individuals (Myers & Sweeney, 2005). Among the major findings have been that wellness measures are associated with general psychological well-being, a positive body consciousness as contrasted with body shame, healthy love styles, job satisfaction, ethnic identity, and acculturation.

The essential conceptualizations gleaned from these studies thus far affirm that wellness is indivisible—that is, positive choices for living well in one area of life have implications for other areas of a person's physical, mental, emotional, and spiritual well-being. For example, if you recommend an exercise program as part of treatment, clients will likely be healthier, feel better emotionally, suffer less stress, and develop feelings of self-worth. Wellness over the lifespan is equally important regardless of gender, culture, race, or geographical location. And for your practice—wellness is measurable in meaningful ways suitable for counseling and other educational interventions.

Wellness and Neuropsychology

Brain research supports the wellness approach. Fitness training and exercise affect blood flow, cognitions, and neurogenesis. These impact the structures that support the brain's executive functions (Hillman, Erickson, & Kramer, 2008; Ratey, 2008).

The left hemisphere is associated more with positive emotions such as happiness and joy, whereas the right hemisphere and amygdala are associated more often with negative feelings (see Appendix II). In depression and deep sadness, brain scans reveal that the positive areas are less active (Davidson, Pizzagalli, Nitschke, & Putnam, 2002). Happiness involves physical pleasure, the absence of negative emotions, and positive meanings (Carter, 1999; Davidson, 2001). In effect, positive thoughts and action can help override

fear, anger, and sadness. This finding is directly parallel to the wellness research cited above.

What, specifically does this mean for your practice? When a client focuses solely on problems and negative emotions, we can help the client through a wellness approach that strengthens and nourishes the individual. In this way we can help clients "build a tolerance for negative emotions and gradually acquire a knack for generating positive ones" (Damasio, 2003, p. 275). So, the wellness approach is not just "window dressing." This strengthening can be both psychological (reminding clients of positive experiences and personal strengths) and physical (exercise and sports, nutrition, and adequate sleep). A base of strengths facilitates problem solving and working through the many complex issues we all face.

Importance of Exercise

The physical and mental health benefits of exercise are universally acknowledged (Aittasalo, 2008; Crone & Guy, 2008; Ratey, 2008). Unfortunately, only half of adults in developed countries fulfill the recommended minimum physical activity. Studies with clients with mental health problems show that sports therapy is a beneficial adjunct to usual treatment (Crone & Guy, 2008). Furthermore, involvement in sports and exercise seems to play a role in helping clients to restory aspects of their lives through creating and sharing personal stories in which they rebuild or maintain a positive sense of self and identity (Carless & Dougless, 2008). Current research shows positive effects of aerobic exercise in reducing posttraumatic stress disorder (PTSD) and trauma symptom severity in adolescents (Diaz & Motta, 2008). Finally, current reviews of meta-analyses from peer-reviewed journals show that exercise is more effective than no treatment and as effective as many traditional interventions for combating depression (Daley, 2008).

Exercise is now considered a fundamental aspect of effective interviewing, counseling, and therapy. Work with virtually all your clients to facilitate a healthy mind and body through exercise.

toward a wellness plan for future growth and development. Interestingly, you may even find that a wellness plan serves as a good outline for treatment and follow-up. Effective counseling and psychotherapy can result from a well-conceptualized wellness plan.

It is particularly important not to overwhelm the client with the many things that could be done to improve overall wellness. Work with the client to select one or two items from the wellness assessment as a start and negotiate a contract for action and follow-up. Check with your client on a regular basis to see how the plan is working.

Briefly, indicate here your early ideas for a personal wellness plan for yourself.

What are your thoughts?

_____

_____

This wellness plan can become part of one session or a major portion of a longer-term interviewing and counseling plan. You can facilitate the entire process of interviewing, counseling, and psychotherapy with portions of the wellness approach. The growing interest in positive psychology further supports these practices in a variety of settings and with clients of all ages. Box 2-5 presents relevant research on wellness.

SUMMARY: INTEGRATING WELLNESS, ETHICS, AND MULTICULTURAL PRACTICE

As we review this chapter, let us return to Kendra, our fictional client presented at the beginning of this chapter. Her IDENTITY has been formed through multiple relations in her family, community, and broader society. These relationships have socialized her by placing her in certain roles and statuses. Kendra is a person-in-community, a self-in-relation to others. Yet she still has freedom to change herself and society, and our task is to facilitate that process.

Key ethical issues in the interview include making sure that you are competent to work with her, obtaining appropriate informed consent, preserving confidentiality, and using counselor power responsibly. The possible sexual harassment needs to be explored, and Kendra will need your support as she moves to a decision here. At this moment in her life, Kendra appears to have real financial needs. Those who have a lower income clearly do not have the same possibilities and privileges as those who are more economically stable.

Every client we meet has a unique multicultural background, and we constantly need to develop and improve our awareness, knowledge, and skills in many areas. The areas discussed in this chapter have been ethnicity/race, gender, sexual orientation, spirituality, ability/disability, socioeconomic status, language, and age. Kendra at 25 has differing needs and life experience than if she were 45. If she is White, she has access to some privileges despite her economic situation. If she is a Person of Color, she may face discrimination for race as well as for gender.

The challenge of multiculturalism is a broad one. The route toward multicultural competence first lies in understanding yourself as a multicultural being. Armed with this beginning, you are better prepared to learn how to work with people who are different from you. The task is humbling, but one we should take on with joy and enthusiasm, for we constantly will learn how difference enriches all our lives. It would not be very interesting if we all had the same experiences, behaved the same, and had similar values.

We have suggested that Kendra's issues (attending college, caring for a child, living at home with her mother, working, experiencing possible sexual harassment, and struggling with financial problems) can best be addressed if we include dimensions of positive psychology and wellness in our work with her. While it might be helpful to conduct a full wellness assessment, this will not always be possible. What you can do is search constantly for strengths that you can then use to help her address the challenges she faces.

Key points of the chapter are presented below.

■ KEY POINTS

Identity	"**I** am (and you are also)—**D**erived from family—**E**mbedded in a community—**N**ot isolated from prevailing values—**T**hough having unique experiences—**I**n certain roles and statuses—**T**aught socialized, gendered, and sanctioned—**Y**et with freedom to change myself and society." (Ruth Jacobs comments on identity at the beginning of this chapter.)
Ethics and competence	Practice only within the bounds of your competence, seek supervision when necessary, and refer appropriately, while supporting the client with a solid relationship as much as you can. You will continue to gain ethical knowledge and competence throughout your career.
Ethics and informed consent	Obtain consent from role-played and real clients in which you tell them the goals, procedures, benefits, and risks of counseling *and* the client agrees to what has been outlined.
Ethics and confidentiality	Keep confidences so far as legally possible and in accord with state law. As beginning counselors in role-plays, you do not have legal confidentiality.
Ethics and power	Maintain awareness of power differentials in the interview and seek to avoid dual or multiple relationships. Power differentials occur in many ways— economic status, gender, other multicultural variables. The interviewer is generally in a more powerful position than the client.
Ethics and social justice	Maintain awareness that client problems and issues may be the result of oppressive environments and, where possible, you will actively seek to enhance and protect the rights of your clients.
Multicultural competence	Recognize that all individuals have dignity, and you will embrace an awareness of cultural differences of many types. You will not discriminate, and you will seek increasing knowledge of multicultural issues. You will become aware of your own multicultural background.
Power and privilege	Become aware that certain groups have more privileges and entitlements than others and consider these issues in your practice. Examples included in the text are White, male, and middle-class power and privilege. These three do not cover all forms of power and privilege, which are present and all countries and cultures.
Wellness and positive psychology	Clients come to us with many strengths and positive assets from their life experience and their own unique personal competencies. They have family and friends, cultural resources, and many others that need to be recognized in the interview. Once positive strengths are identified clearly, problem solving and working through issues can be expected to work more smoothly.
The indivisible self	A wellness model developed by Sweeney and Myers. It includes the creative, coping, social, essential, and physical selves. Within the five categories are 17 specific dimensions of wellness.

(continued)

KEY POINTS (continued)

Wellness assessment	Client strengths can be assessed in 17 dimensions: spirituality, gender identity, cultural identity, self-care, friendship, love, leisure, stress management, self-worth, realistic beliefs, thinking, emotions, control, work, positive humor, exercise, and nutrition. As needed, conduct a full wellness assessment with clients. Realistically, however, a quick survey can help you and the client select one or two key issues.
Wellness plan	From the assessment of strengths and wellness assets, the client can examine areas where more effort and planning might be helpful. A balance of strengths and areas for growth is identified.

COMPETENCY PRACTICE EXERCISES AND PORTFOLIO OF COMPETENCE

Intentional interviewing and counseling is achieved through practice and experience. It will be enhanced by your own self-awareness, emotional competence, and ability to observe yourself, thus learning and growing in skills.

The competency practice exercises on the following pages are designed to provide you with learning opportunities in three areas:

1. *Individual practice.* A short series of exercises gives you an opportunity to practice the concepts.
2. *Group practice.* Practice alone can be helpful, but working with others in role-played interviews or discussions is where the most useful learning occurs. Here you can obtain precise feedback on your interviewing style. And if videotapes or audiotapes are used with these practice sessions, you'll find that seeing yourself as others see you is a powerful experience.
3. *Self-assessment.* You are the person who will use the skills. We'd like you to look at yourself as an interviewer and counselor via some additional exercises.

Individual Practice

Exercise 1: Review an Ethical Code

Select the ethical code from Box 2-1 that is most relevant to your interests and review it in more detail. Then visit the ethical code of another country or another helping profession and note similarities and differences on competence, informed consent, confidentiality, social justice, and diversity. What is your own position on these issues? Write your observations and comments in a journal.

Exercise 2: You as a Multicultural Being, Your Multiple Intersecting Identities

Again, we all have multiple multicultural identities, although many of us are unaware of them. Please take a moment to review the RESPECTFUL model and find your identities. After you have finished this process, indicate those areas where people typically have privileges. Usually along with those privileges they have some implicit power over others. In what areas do you lack privilege and power? How privileged are you?

RESPECTFUL Dimension	*What are your multicultural identities on these dimensions?*	*Do you have privilege and implicit power because of your identity(ies) within this dimension?*
R Religion/spirituality		
E Economic/class background		
S Sexual identity		
P Personal style and education		
E Ethnic/racial identity		
C Chronological/ lifespan challenges		
T Trauma experience		
F Family background		
U Unique physical characteristics		
L Location of residence and language; differences/capabilities		

As you review the above, what are your thoughts about your multiple identities and your implicit power or lack of power?

Exercise 3: Personal Wellness Assessment

Review the Sweeney-Myers Wellness Model contextual issues on pages 50–55. What strengths and resources do you find in your context and in your own personal wellness?

Strengths and resources from your context:

Local	*Institutional*	*Global (World Events)*
Family	Education	Politics
Neighborhood	Religion	Culture
Community	Government	Global Events
	Business/Industry	Environment
	Media	

Review the individual personal strengths of pages 52–53. Then review your own strengths and resources.

Essential	*Social*	*Coping*	*Creative*	*Physical*
Spirituality	Friendship	Leisure	Thinking	Exercise
Gender identity	Love	Stress management	Emotions	Nutrition
Cultural identity		Self-worth	Control	
Self-care		Realistic beliefs	Work	
			Positive humor	

Group Practice

Exercise 4: Conduct a Wellness Assessment and Develop a Wellness Plan

Now that you have engaged in a wellness assessment for yourself, meet with two of your class members and engage in a wellness assessment with one of them. Conclude this practice with discussion of a plan for the future. The third person will be an observer and provide comments and give feedback on the process. We recommend that your volunteer client fill out the Client Feedback Form from Chapter 1. Alternatively, do this as a homework assignment with a volunteer.

Exercise 5: Develop an Informed Consent Form

Box 2-2 presents a sample informed consent form, or practice contract. With your small group, develop your own informed consent form that is appropriate for your particular school situation and for your state or commonwealth.

Exercise 6: Exploring Multicultural Competence

Again in a small group situation, review the major concepts of multicultural competence presented here. Please maintain awareness that this is a very brief summary and does not cover the full richness of this topic.

▲ Being aware of your own assumptions, values, and biases including dimensions of privilege
▲ Understanding the worldview of the culturally different client
▲ Developing appropriate intervention strategies and techniques

As your group reviews these issues, what goals for learning can you establish as you work through this and further study in interviewing, counseling, and psychotherapy?

Portfolio of Competence

How would you assess your understanding and competencies in the concepts of this chapter? We are asking you to assess yourself and your ability to use the ideas presented here.

Self-Assessment

Reflecting on yourself as a future interviewer, counselor, or psychotherapist via a written journal can be a helpful way to review what you have learned, evaluate your understanding, and think ahead to the future. Here are three questions that you may wish to consider.

1. What stood out for you personally in the section on ethics? What one thing did you consider most important? Ideas of social justice and action in the community are considered by some a controversial topic. What are your thoughts?
2. How comfortable are you with ideas of diversity and working with people different from you? Can you recognize yourself as a multicultural person full of many dimensions of diversity?
3. Wellness and positive psychology have been stressed as a useful part of the counseling and psychotherapy interview. At the same time, relatively little attention has been given so far to the very real problems that clients bring to us. While many difficult issues will be covered throughout this text, what are your personal thoughts at this moment on wellness and positive psychology? How comfortable are you with this approach?

Self-Evaluation of Chapter Competencies

Use the following as a checklist to evaluate your present level of mastery. As you review the items below, ask yourself, "Can I do this?" Check those dimensions that you currently feel able to do. Those that remain unchecked can serve as future goals. Do not expect to attain intentional competence on every dimension as you work through this book. You will find, however, that you will improve your competencies with repetition and practice.

Level 1: Identification and classification. You will need this minimal level of mastery for those coming examinations.

❑   Define and discuss the key aspects of ethics as they relate to the interview: competence, informed consent, confidentiality, power, and social justice.
❑   Define and discuss the three dimensions of multicultural competence: awareness of your own assumptions, values, and biases; understanding the worldview of the culturally different client; developing appropriate intervention strategies and techniques.
❑   Define and discuss positive psychology and wellness.
❑   Define and discuss the contextual factors of the wellness model.
❑   Define and discuss the five personal dimensions of the wellness model: essential self, coping self, social self, creative self, and physical self.

Level 2: Basic competence. Here you are asked to perform the basic skills in a more practical context such as self-evaluation or an actual interview. This is an initial level of competence and can be built on and improved throughout your use of this text.

❑   Write an informed consent form.
❑   Define myself as a multicultural being.
❑   Evaluate my own wellness profile, both personal and contextual.
❑   Take another person through a wellness assessment.

Levels 3 and 4, intentional competence and teaching competence, will not be presented in this chapter. You will encounter them in the following chapters.

DETERMINING YOUR OWN STYLE AND THEORY: CRITICAL SELF-REFLECTION ON ETHICS, MULTICULTURAL COMPETENCE, AND WELLNESS

Regardless of what any text on interviewing, counseling, and psychotherapy says, the fact remains that it is you who will decide whether to implement the ideas, suggestions, and concepts. What single idea stood out for you among all those presented in this chapter, in class, or through informal learning? What stands out for you is likely to be important as a guide toward your next steps. What points in this chapter struck you as most important? What might you do differently? How might you use ideas in this chapter to begin the process of establishing your own style and theory? Write your ideas here or in a journal.

What are your thoughts?

OUR THOUGHTS ABOUT KENDRA

Kendra presents her central concern as dealing with sexual harassment on the job, complicated by her need to finance her education. A quick solution is to quit the job, but then the loss of income while searching for another job can be a serious issue. Beyond that, how are things going with her mother and child? Single parents lead complex lives.

Personal resources. While we need to address Kendra's problems, we also need to be fully aware of her strengths and make sure that Kendra is conscious of them as well. As part of getting to know Kendra, we'd discuss several of her positive assets, particularly her ability to balance work, parenting, and school. Any client who does all this should be recognized as a person of strength, full of personal wellness assets. Further discussion will bring out other strengths. The skills she has in all these areas will combine to help her find a suitable solution to the current issue.

Community resources. The most obvious resource that Kendra has may likely be found in her family and her mother's support. Kendra's child is also an important emotional support, even though likely challenging at times. Exploration of resources may reveal friends and the church, mosque, or synagogue as a place of emotional and physical support. There may be scholarships and other support options available at the college. Finally, let us hope that there is someone in her work setting who can serve as a friend and advocate.

Social context and cultural background. We believe that individual counseling is best conducted when we are able to see the client in a social and cultural context. All the major helping professions now stress the importance of culture, even making it an ethical matter. There are two issues that we'd like you to consider at this time, although more could be commented on. First is our own awareness of multicultural issues and context that might affect Kendra. We need to be aware that Kendra is a woman, a single parent, from a specific ethnic/racial background (which is unstated in this case). These three factors (plus others, of course) help us understand Kendra more completely. Our awareness and sensitivity to possible culture issues is vital, even if they are never discussed. Second, as appropriate, we need to help Kendra understand the broader context of her issues. Most clearly, harassment tends to be a women's issue, and issues of sexual discrimination may need to be discussed. Other multicultural factors can be brought into the session as the interview progresses. What is a solution that is personally satisfying to Kendra? This may require a social justice approach and may involve helping her develop action plans to change the work-setting rules. Or it may simply be recognizing the nature of the cultural context as Kendra makes a personal decision.

CHAPTER 3

Attending Behavior: Basic to Communication

Attending Behavior

Ethics, Multicultural Competence, and Wellness

Attention and consciousness are the foundations on which we create an understanding of the world. Together, they form the ground upon which we build a sense of who we are. [Attention and consciousness enable us to] define ourselves in relation to the myriad physical and social worlds we inhabit. They are also the basic functions that give rise to "the mind."

—John Ratey, M.D.

How can attending behavior be used to help your clients?

Chapter Goals

The principal focus in this chapter is the most basic communication strategy—attending behavior. We consider this the most important skill in this book. Attending skills include visuals (eye contact), vocal tone and emphasis, verbal following, and body language. Well-developed skills assist the interviewer in two ways. First, they help interviewers demonstrate that they care and are involved. Second, they enable counselors to observe unspoken messages and respond more appropriately to immediate client needs.

Competency Objectives

Awareness, knowledge, and skills in attending behavior will enable you to

▲ Communicate to the client that you are interested in what he or she is saying.
▲ Increase your awareness of the client's pattern of focusing on certain topics.
▲ Modify your patterns of attending to establish rapport with each individual.

> ▲ Recognize that different people and different cultural groups sometimes have different patterns of attending.
> ▲ Develop recovery skills that you can use when you are lost or confused in the interview. Even the most advanced professional doesn't always know what is happening. When you don't know what to do, attend!
> ▲ Examine the use of attending skills as a psychoeducational treatment.

Attention is a powerful reinforcer. It is the connective force of conversations, so central that we are seldom aware of its presence. But we usually know when someone is not attending to us. The way one attends and listens deeply affects what is talked about in the interview. It often helps to observe *what not to do* and then *what might be better and more effective* becomes more clear. All of the following are clearly ineffective counselor responses. Take time to write a more useful helping statement that actually attends to the client.

Anish: My world was turned upside down by 9/11. I'm a dark-skinned Indian, and everyone looks at me suspiciously now. Going to an airport is sheer hell. They usually stop me, and one time I was even strip-searched.

Counselor: I can imagine. I was really upset when those planes hit the towers.

What would you say? _____

Tarni: I'm having a really hard time studying. My boyfriend and I just broke up.

Counselor: What's your major?

What would you say? _____

Sarah: Why do I have to take biology? The professor keeps trying to sell the students on evolution. My pastor told me that a state university would not be fair.

Counselor: Well, it is a requirement, and it was listed in the catalog before you came here.

What is your response? _____

What not to do. Note the first counselor fails to listen and focuses on her or his own feelings and thoughts. Sometimes clients bring up issues that affect us so deeply we start thinking of ourselves and forget the client. The second counselor jumps topics and fails to see Tarni's issues at all. The third counselor starts problem solving far too quickly and misses the challenges that the client is facing. *All of them fail to attend and listen!*

What might help Anish. The words "Anish" and "you" need to be the foundation rather than the "I" focus of the counselor. For example, "Anish, since 9/11, your life has turned upside down. You look angry and hurt at what your country is doing to you at the airport." Another attending response that stays with the topic might be: "You sound hurt and angry at what's happened. Could you tell me about other situations that have hurt you as well? As I understand more, we can then look to solutions."

What might help Tarni. The counselor needs to help the client tell the story more fully. "I hear the studying problem, and I also hear that you are upset about the breakup. Which would you like to talk about first?" This summary gives the client the choice of where to go next. Another possibility is to focus on the breakup: "So this breakup has you really upset. Could you explore that a bit more?" Here the counselor draws out the story of the breakup. Later they can explore study issues.

What might help Sarah. The counselor is there for the client. Focus on her concern. Avoid taking an authoritarian position, and listen to the client's reasons and issues. Biology is a requirement, but by attending to the client the counselor can be supportive. An alternative response might be: "I think I understand. You said earlier than you came from an evangelistic tradition. Courses like biology can be a challenge and can seem incompatible with your beliefs. I'd like to know a bit more about you and the course, and then we can discuss what to do next. Is that okay?" The counselor seeks to attend to the client and to draw out the story before discussing a solution.

INTRODUCTION: THE BASICS OF LISTENING

Effective attending behavior may be considered the foundation skill of this book. If you are to develop a trusting relationship with a client, the skills of attending are essential. Foremost in any interview or counseling situation is the ability to make contact with another human being. We make this contact through listening and talking as well as by nonverbal means. Listening to other people is most critical, as it enables them to continue to talk and explore. How can we define effective listening more precisely?

One of the best ways to understand good quality listening is to experience the opposite—poor listening. Find a partner to role-play an interview or, if no partner is available, think back on some of your bad listening experiences. Perhaps a family member or friend failed to hear your concerns, a teacher or employer misunderstood your actions and unfairly blamed you, or you had that all-too-familiar experience of calling a help desk and didn't get the answer you needed.

Spend about 3 minutes role-playing a poor and ineffective interview. Feel free to exaggerate ineffective helping skills (or recalling the specifics of the time someone failed to listen to you).

After the role-played session, list what went wrong and worked against effective listening. Or, if you recall the failed listening attempt, describe what went wrong and how you felt when you were not heard.

List behaviors, traits, and qualities that are characteristic of the ineffective interviewer or counselor who fails to listen. What did you see and hear in the role-played session? What about your own past experience with people who fail to listen? Lists of ineffective interviewing behaviors sometimes run to 30 items or more.

What are your thoughts?   _____

_____

You may recall with laughter the bored interviewer, the teacher, or family member who misinterpreted what you said, or the ineffective person on the help line. Yet as you remember your own experience of not having your story heard, your strongest memory may be of a feeling of emptiness, disappointment, and perhaps even anger. Think of your frustration with the insensitivity of the interviewer. If you are to be effective and competent, do the opposite of the ineffective counselor. If you use attending as indicated below, you can make the following predictions.

Attending Behavior	Predicted Result
Support your client with individually and culturally appropriate visuals, vocal quality, verbal tracking, and body language.	Clients will talk more freely and respond openly, particularly around topics to which attention is given. Depending on the individual client and culture, anticipate fewer eye contact breaks, a smoother vocal tone, a more complete story (with fewer topic jumps), and a more comfortable body language.

Obviously, you can't learn all the necessary qualities and skills immediately. Rather, it is best to select the most basic. The first interviewing skill of the microskills hierarchy is attending behavior, which consists of four central dimensions and is critical to all other helping skills. The four aspects of attending behavior listed below are first described from a North American/Northern European perspective. As you review these highly researched dimensions, recall that many cultural groups have varying patterns of attending behavior; we will outline variations in this chapter. Consider this list only a start on your multicultural journey.

1. *Visual/eye contact*. If you are going to talk to people, look at them.
2. *Vocal qualities*. Your vocal tone and speech rate also indicate clearly how you feel about another person. Think of how many ways you can say "I am really interested in what you have to say" just by altering your vocal tone and speech rate.
3. *Verbal tracking*. The client has come to you with a topic of concern. Don't change the subject; stick with the client's story.
4. *Body language: attentive*. Clients know you are interested if you face them squarely and lean slightly forward, have an expressive face, and use facilitative, encouraging gestures. In short, allow yourself to be yourself—authenticity in attending is essential.

Think of these as the three V's + B of listening—**V**isuals, **V**ocals, **V**erbals, and **B**ody language.* If you are going to listen you need to reduce your talk-time and provide clients with an opportunity to tell their stories as concretely and with as much detail as needed. The ability to listen to clients—simply attend to their being—is the foundation of successful interviewing, counseling, and therapy.

The attending behavior concepts above were first introduced to the helping field by Ivey, Normington, Miller, Morrill, and Haase (1968). Cultural variations in microskills usage were first identified as central to the model by Allen Ivey, when he worked with native Inuits in the

---

* We thank Norma Block Gluckstern for the three V's acronym.

BOX 3-1   ATTENDING BEHAVIOR AND PEOPLE WITH DISABILITIES

Attending behaviors (visuals, vocals, verbals, and body language) all may require modification if you are working with people who are disabled. It is your role to learn their unique ways of thinking and being, for these clients will vary extensively in the way they deal with their issues. Focus on the person, not on the disability. For example, think of a person with hearing loss rather than "a hearing impaired client," a person with AIDS rather than "an AIDS victim," a person with a physical disability rather than "a physically handicapped individual." So-called handicaps are often societal and environmental rather than personal.

People Who May Have Limited Vision or Be Blind	Eye contact is so central to the sighted that initially you may find it very demanding to work with clients who are blind or partially sighted. Some may not face you directly when they speak. Expect clients with limited vision to be more aware of your vocal tone. People who are blind from birth may have unique patterns of body language. At times, it may be helpful to teach them the skills of attending nonverbally even if they cannot see their listener. This may help them communicate more easily with the sighted.
People Who May Have Hearing Loss or Deafness	An important beginning is to realize that some people who are deaf do not consider themselves impaired in any way. Many of this group were born deaf and have their own language (signing) and their own culture, a culture that often excludes those who hear. You are unlikely to work with this type of client unless you are skilled in sign language and are trusted among the deaf community. You may counsel a deaf person through an interpreter. This experience will not prove to be positive without some training in the use of an interpreter in addition to a basic understanding of deaf culture. Too many counselors speak to the interpreter instead of to the client, and often use phrases such as "tell him . . ." This certainly will cut off the client. Also, eye contact is vital, whether in direct counseling with a deaf client or while using an interpreter. Those who have moderate to severe hearing loss will benefit if you extensively paraphrase their words to ensure that you have heard them correctly. Speak in a natural way, but not fast. Speaking more loudly is often ineffective, as ear mechanisms often do not equalize for loud sounds as they once did. In turn, teaching those with hearing loss to paraphrase what others say to them can be helpful to them in communicating with others.
People With Physical Disabilities	First, each person is unique. We cannot place people with physical disability in any one group. Consider the differences among the following: a person who uses a wheelchair, an individual with cerebral palsy, one who has Parkinson's disease, one who has lost a limb, or a client who is physically disfigured by a serious burn. They all may have the common problem of lack of societal understanding and support, but you must work with each individual from her or his own perspective. Their body language and speaking style will vary. What is important is to attend to each one as a complete person.
People Who Are Temporarily Able	Most people reading this section will have experience with some but not all of the issues above. We suggest that you consider yourself one of the many who are temporarily able. Age and life experience will bring most of you some variation of the challenges above. For older individuals, the issues discussed here may become the norm rather than the exception. Approach all clients with humility and respect.

central Canadian Arctic. He found that sitting side by side with them was more appropriate than direct eye contact (body language and visuals vary among cultures), and that developing a solid relationship was equally important as staying on the verbal topic. Nonetheless, smiling, listening, and a respectful and understanding vocal tone are all things that help us work through cultural differences. *In short, attending behavior and listening are essential for human communication, but we need to be prepared for and expect individual and multicultural differences.*

However, some authors, researchers, and the popular press still discuss attending and listening skills without attention to cultural differences. These skills are infinitely more effective when they are used within an individually and culturally appropriate framework. Diversity and multiculturalism are defined broadly in this text. For example, Box 3-1 presents the use of attending skills with persons with disabilities, a cultural group that gets insufficient attention.

One more critical foundational point about attending behavior: Observe your clients' style of attending and listening. You can discover much of importance if you note their patterns of eye contact, the emotions in their vocal tone, and the topics that they are willing (or not willing) to discuss. Their body language will tell you a good deal about how well you are doing in the session and when you are "off base." Clients from varying racial and ethnic backgrounds will differ in their patterns of listening to you. Use client observation skills (Chapter 5) to help you adapt your style to meet the needs of the unique person before you.

EXAMPLE INTERVIEWS: I DIDN'T GET A PROMOTION— IS THIS DISCRIMINATION?

Azara, a 45-year-old Puerto Rican manager, was not promoted, although she thinks her work is of high quality. She is weary of being passed over and seeing less competent individuals take the position she feels she deserves.

The first interview segment, presented below is designed as a particularly ineffective interview so that it provides a sharp contrast with the more positive effort that follows. In both cases, the interviewer, Allen, has the task of developing a relationship and drawing out the client's story. Note how disruptive visual contact, vocal qualities, failure to maintain verbal tracking, and poor body language can lead to a poor session.

Negative Example

Interviewer and Client Conversation	Process Comments
1. *Allen:* Hi, Azara, you wanted to talk about something today.	Allen fails to greet Azara warmly. He just starts and does nothing to develop rapport and a relationship, and this is especially important in a cross-cultural session. He remains seated in his chair behind a desk.
2. *Azara:* Yes I do. I've come to you because there's been an incident at my job a couple of days ago. And I'm kind of upset about it.	Azara sits down and immediately moves ahead with her issues regardless of what Allen does. She is clearly ready to start the interview.

(continued)

Interviewer and Client Conversation	Process Comments
3. *Allen:* What is your job?	Allen's voice is aggressive. He ignores Azara's upset feelings and asks a closed question.
4. *Azara:* Well, right now I'm an assistant manager for a company, and I've worked at this company for 15 years.	Azara keeps trying. Allen looks down while she talks.
5. *Allen:* So after 15 years you're still an assistant. When I was in business, I didn't take that long to get a promotion. Let me tell you about what I did to get ahead . . . (he goes on at length about himself).	The focus is taken away while Allen, the counselor, talks about himself. With this long response, he has more talk-time than the client, Azara. The evaluative "put down" is an example of how counselors inappropriately use their power.
6. *Azara:* Yeah, I'm still an assistant after 15 years. But what I want to talk to you about is I was passed over for a promotion.	Is there an issue of discrimination here? By ignoring cultural issues, Allen will eventually lose this relationship.
7. *Allen:* Could you tell me a little bit more about some of the things you might have been doing wrong?	Still looking out the window, he returns to Azara with an open question, but he continues to ignore the main issue and topic jumps with an emphasis on the negative.
8. *Azara:* Well, I don't think I did anything wrong. I've gotten very good feedback from . . .	Azara starts defending herself here, but Allen interrupts.
9. *Allen:* Well, they don't usually pass people up for promotions unless they're not performing up to standards.	The counselor supplies his interpretation, a negative evaluation without any data. He is not drawing out her story or really seeking to define her concerns.
10. *Azara:* But I think I have. I've had great evaluations. I've trained several men, and they all are above me now.	The client is good at defending herself, but she needs support, not criticism.
11. *Allen:* Maybe one thing we can do in here is sort of help you find new ways to work better with your boss.	The counselor heads toward a solution before the client has told her story. Allen is talking firmly and trying to convince her of his opinions.
12. *Azara:* But I really do get along. . . . Now I'm frustrated.	Azara finally realizes that she is in an impossible situation. Her European-American counselor just doesn't "get it."
13. *Allen:* Well I can imagine that you're frustrated. But I think you're going to have to face the fact that there are some things we're going to have to do to work this through.	Allen confronts Azara inappropriately, and it is very unlikely that she will return for another session.

Positive Example

Interviewer and Client Conversation	Process Comments
1. *Allen:* Hi, Azara. Nice to see you.	Allen stands up, smiles, faces the client directly, and shakes hands.
2. *Azara:* Thank you, nice to see you too.	She sits down and smiles in return, but appears tense.
3. *Allen:* Thanks for coming in.	The counselor likes to honor the client's willingness to come to the session. It is a small attempt to equalize the power relationship that exists in counseling.
4. *Azara:* Thanks. I'm hopeful that you can help me.	Azara relaxes a little.
5. *Allen:* Azara, I've been looking at your file, and I see that you'd like to talk a about a problem on the job. Is that right?	Allen has reviewed the forms she completed before entering counseling. He briefly mentions what appears to be her main concern. He maintains direct eye contact and does not refer to the file at that moment.
6. *Azara:* Yes, that's right.	Her mouth is a little tense, and she sits back.
7. *Allen:* Before we start, there are a couple of things I'd like to talk about. One of them is that anything that goes on in here is confidential, and I'm videotaping this particular interview because I do get some supervision. So is it okay to make a tape? If you feel any time you want to turn the tape off, I will turn it off—is that okay? And if it makes you uncomfortable—and some people it can—we'll turn it off. If you wish, later I can give you the tape to take home and watch. Would you sign this so I can tape?	The reason Allen mentioned his awareness of Azara's issue early is that he wants to help her become aware that he is interested and knows a little about what is troubling her. But he also knows that he needs to structure the interview and give some basic information to the client. Each agency or counseling office will have different sets of information that they give the client, but this presentation is fairly typical.
8. *Azara:* Okay.	She signs the permission form and sits back.
9. *Allen:* Another issue as we start is that everything is confidential, and it's not to be shared, except in a situation where you hurt somebody or indicate problems where you might hurt yourself or others. I think we need to share that as well. And then another thing as we start is that, hey, I'm a White male, you're from Puerto Rico, and a woman. What are your thoughts about working with me?	The rules of confidentially need to be shared. But it is important to maintain a warm, supportive voice. The issue of multicultural difference most often needs to be explored early in the session. This can include race/ethnicity, gender, sexual orientation, ability/disability, and other factors of cultural difference. Allen maintains direct eye contact, a warm vocal tone, and attentive body language. Azara is clearly middle-class, but some traditional Latina/o clients, particularly those who are poor, may be uncomfortable with a lot of direct eye contact.

(continued)

Interviewer and Client Conversation	Process Comments
10. *Azara:* Well, I'm glad you brought that up because that's the first thing that came to mind when I asked about who I could talk to about my problem. And when your name was mentioned I thought, oh my goodness, is he going to understand? So I'm glad you brought it up. I think that that helps me a little bit. Well, what do you know about our culture?	The counselor is in a fortunate position. Even though he is a White European American and sensitive to many multicultural issues, Azara does not know this.
11. *Allen:* I've been lucky enough to be down in Puerto Rico—San Juan and Ponce. I've done some work there, and I've counseled several Puerto Ricans here. But perhaps you might want to ask me more. I love fried plantain and chicken *asopoa*.	The counselor follows Azara's lead and shares his experience *briefly*. If you don't have a background in the culture, admit that fact, but without being defensive. Share what you do know and where you are openly.
12. *Azara:* Okay, well that's a good start. Thanks for sharing that. My mother makes wonderful *asopoa* when I visit there. I think that that helps me a little bit, and I think I can kind of begin to talk to you a little bit about what's going on.	The client is ready to start.

13. *Allen:* So if there are issues of my lack of understanding or missing something along the way I want you to feel free to ask me a question or confront me at any time.
 	
    So there is a concern on the job. I'd like to hear about it.
	It is important to let clients know that they can raise issues with you at any point in the session.
	    Allen returns now to the job issue.

14. *Azara:* Okay, well a few days ago I found out that I was passed over for a promotion at my job. And I've been with this company for 15 years. And I was really pretty upset when I first found out, because the person who got the job, first of all, is a male; he's only been with the company for 5 years. And, you know, I think I'm much better qualified than he is for this position. I've gotten really good evaluations from my supervisor; I have a great working relationship with my colleagues. And, you know, I'm like completely shocked to find out that I didn't get this promotion. Cause I was actually encouraged to apply for this job. And, you know, I didn't get it. This is . . . I'm just really, really angry.	Azara says a lot in this comment, and we as counselors sometimes have difficulty in hearing it all. This is where the skills of paraphrasing and summarization of Chapter 6 can be most important. The task of these skills is to repeat what the client has said, but in a more succinct form.
15. *Allen:* 15 years compared to 5 and you are really, really angry. And what I've heard is what makes you angry is that you've had a good record, you were even asked to apply for this job, and finally this White male who hasn't been there that long gets the job. Have I heard you correctly?	The counselor's summary of what has been said indicates that he has been engaging in verbal as well as nonverbal attending. "Have I heard you correctly?" is termed a "check-out" in the microskills framework. If you are accurate, the client will often say "yes" or even "exactly!"

(continued)

Interviewer and Client Conversation	Process Comments
16. *Azara:* Yes, you heard me. . . . Well, I think it's discrimination. Now the problem I'm having, Allen, is that I think this. I think it's discrimination, but now I have to decide what I'm going to do. If I'm going to file a complaint, and deal with the consequences of filing a complaint, you know, will that upset my colleagues, will that get my boss, my supervisor upset with me? I'm really worried about the consequences. I don't want to lose my job, but I think it's discrimination.	Having been heard, the client moves on.
17. *Allen:* Azara, it's a tough decision to make. If you file for discrimination, you set yourself up for a lot of hassles; if you don't file, then you're stuck with your anger and frustration. Could you tell me a bit more about that dilemma you are feeling?	Here Allen paraphrases the main ideas and reflects Azara's feelings as well. This is followed by an open question about the dilemma.
18. *Azara:* Well, it's like I'm stuck. I don't know what to do. On the one hand I think it's important to file the complaint because I think it will show the company that they really need to think about diversity in the workforce, and I'm kind of tired of being the only Latina working in this company for as long as I have, when you know they need to do something different. So I'm torn between that and being afraid of losing my job.	Azara summarizes key aspects of her conflict. The discrepancies or incongruity between herself and the company could be summarized this way: The responsibility to file a discrimination suit because it appears that the company is consistently being unfair versus the fear of losing her job if she takes this on.
19. *Allen:* So you're angry, afraid, frustrated. A lot of stuff comes together for you all at once.	Allen is sitting, leaning forward, using a supportive vocal tone while he reflects her emotions and her dilemma.
20. *Azara:* Yes, that's right. And I don't know what to do about that.	The client provides her own "check-out" and speaks of her puzzlement.
21. *Allen:* One thing I heard you saying that I'd like to understand a little bit more, you had good evaluations, you say you have good relationships, success, a reasonable rate of promotion, at least raises along the line. I'd just like to hear at this point . . . I'd like to just hear about examples of something more specific that's gone right in the past. Something you're proud of. Because when a person talks to me about difficulties, it kind of makes them feel a little embarrassed, and I'd like to understand some of your strengths. I've got a general understanding of your problem, and we will come back to that. Could you tell me a little bit about some of your strengths too?	Now that the issues are clearer, Allen turns to the positive asset search. What are Azara's strengths that we can draw on as we work on these concerns? Note that Allen has avoided the use of the word "problem," as that is a self-defeatist negative view of client issues. You will find that most counseling training books use a problem-centered language.

(continued)

Interviewer and Client Conversation	Process Comments
22. *Azara:* Well, I recently completed my major project, and, you know, my supervisor really liked the quality of the work that I produced. So that's one really positive thing that happened a month ago. And that actually is related to why I decided to apply for the promotion, because I got such a good feedback about this report that I completed. My evaluations have been top notch for years.	Ultimately we resolve our concerns using our strengths. When talking only about difficulties, clients lose power and tend to think in negative terms. We can go on forever talking in detail about negatives. We suggest "accentuating the positive," which will help "eliminate the negative" as in an old popular song.
23. *Allen:* So, we have very specific evidence that you are a successful employee, Azara. I see that you are proud of what you have accomplished. I'd like to hear more about your strengths and things that work well for you. We often solve our issues through our strengths and positiveassets.	Allen provides clear information on "problem solving" while focusing on Azara's strengths. Attending is critical for all sessions. It is a gift to the client to truly listen.

Here we see a much stronger focus on Azara as a person with individual needs and feelings. A relationship has been established, and multicultural issues can now be discussed as appropriate to the moment. Through attending and listening, we see Azara's story and concerns more fully. A positive asset search for strengths has been initiated.

INSTRUCTIONAL READING: GETTING SPECIFIC ABOUT LISTENING AND SOME MULTICULTURAL DIFFERENCES IN STYLE

A frequent tendency of the beginning counselor or interviewer is to try to solve the client's difficulties in the first 5 minutes. Think about it. Clients most likely developed their concerns over a period of time. It is critical that you slow down, relax, and attend to client narratives. Listen before you leap!

Attending and giving clients talk-time demonstrates that you truly want to hear their story and major issues. And it is also important to note and recognize that your clients will have varying styles of attending and interacting with you. The following discussion outlines the specifics of attending behavior and its many individual and cultural variations.

The Three V's + B: Visuals, Vocals, Verbals, and Body Language
Visual: Patterns of Eye Contact
Not only do you want to look at clients but you'll also want to notice breaks in eye contact, both by yourself and by the client. Clients often tend to look away when thinking carefully or discussing topics that particularly distress them. You may find yourself avoiding eye contact while discussing certain topics. There are counselors who say their clients talk about "nothing but sex" and others who say their clients never bring up the topic. Through eye-contact breaks, both types of counselors indicate to their clients whether the topic is comfortable.

When an issue is especially interesting to clients, you may find that their eye contact is more direct. If you watch carefully, their pupils may tend to dilate. On the other hand, when the topic is uncomfortable or boring, their pupils may contract. If you have a chance to observe your own face carefully on videotape, you'll note that you as a counselor or interviewer indicate to your clients your degree of interest in the same subtle fashion.

Cultural differences in eye contact abound. Direct eye contact is considered a sign of interest in European North American middle-class culture. However, even in that culture a person often maintains more eye contact while listening and less while talking. Furthermore, when a client from virtually all cultures is uncomfortable talking about a topic, it may be better to avoid eye contact. Research indicates that some traditional African Americans in the United States may have reverse patterns; that is, they may look more when talking and slightly less when listening. Among some traditional Native American and Latin groups, eye contact by the young is a sign of disrespect. Imagine the problems this may cause the teacher or counselor who says to a youth, "Look at me!" when this directly contradicts the individual's basic cultural values. Some cultural groups (for instance, certain traditional Native American, Inuit, or Aboriginal Australian groups) generally avoid eye contact, especially when talking about serious subjects.

Vocal Qualities: Tone and Speech Rate

Your voice is an instrument that communicates much of the feeling you have about yourself or about the client and what the client is talking about. Changes in its pitch, volume, or speech rate convey the same things that changes in eye contact or body language do. If the client is stressed, you'll note that in vocal tone. And if the topic is uncomfortable for you, your vocal tone or speech rate may change as well. Keep in mind that different people are likely to respond to your voice differently.

Awareness of your voice and of the changes in others' vocal qualities will enhance your skill in attending to their stories. Again, note the timing of vocal changes, as they may indicate comfort or discomfort depending on the culture of a client. Speech hesitations and breaks and sudden rapid talking are another signal, often indicating confusion or stress. Clearing one's throat may indicate that words are not coming easily.

Verbal underlining is another useful concept. As you consider the way you tell a story, you may find yourself giving louder volume and increased vocal emphasis to certain words and short phrases. Clients do the same. The key words a person underlines via volume and emphasis are often concepts of particular importance.

Accent is a particularly good example of how people will react differently to the same voice. What are your reactions to the following accents—Australian, BBC English, Canadian, French, Pakistani, New England, Southern United States? Obviously we need to avoid stereotyping people because their accents are different from ours.

Try the following exercise with a group of three or more people.*

Ask all members of the group to close their eyes while you talk to them. Talk in your normal tone of voice on any subject of interest to you. As you talk to the group, ask them to notice your vocal qualities. How do they react to your tone, your volume, your speech rate, and perhaps even your regional or ethnic accent? Continue talking for 2 or 3 minutes. Then ask the group to give you feedback on your voice. Summarize what you learn here. If you don't have a group easily available, spend some time noting the vocal tone/style of

---

* This exercise was developed by Robert Marx, School of Management, University of Massachusetts.

various people around you. What do you find most engaging? Do some types of speech cause you to move away from the speaker?

What are your thoughts?

_____

_____

This exercise often reveals a point that is central to the entire concept of attending. People differ in their reactions to the same stimulus. Some people find one voice interesting, whereas others find that same voice boring; still others may consider it warm and caring. This exercise and others like it reveal again and again that people differ, and that what is successful with one person or client may not work with another.

Verbal Tracking: Following the Client or Changing the Topic

Staying with your client's topic is critical in verbal tracking. Encourage the full elaboration of the narrative. Just as people make sudden shifts in nonverbal communication, they change topics when they aren't comfortable. And cultural differences may appear as well. In middle-class U.S. communication, direct tracking is most appropriate, but in some Asian cultures such direct verbal follow-up may be considered rude and intrusive.

Selective attention is a type of verbal tracking that counselors and interviewers need to be especially aware of. We tend to listen to some things and ignore others. Over time we have developed patterns of listening that enable us to hear some topics more clearly than others. For example, take the following statement, in which the client presents several issues.

Client: (speaks slowly, seems to be sad and depressed) I'm so fouled up right now. The first term went well, and I passed all my courses. But this term, I am really having trouble with chemistry. It's hard to get around the lab in my wheelchair, and I still don't have a textbook yet. (An angry spark appears in her eyes, and she clenches her fist.) By the time I got to the bookstore, they were all gone. It takes a long time to get to that class because the elevator is on the wrong side of the building for me. (looks down at floor) Almost as bad, my car broke down, and I missed two days of school because I couldn't get there. (The sad look returns to her eyes.) In high school, I had lots of friends, but somehow I just don't fit in here. It seems that I just sit and study, sit and study. Some days it just doesn't seem worth the effort.

There are obviously several different directions in which the interview could go. You can't talk about everything at once. List those several directions here. To which one(s) would you selectively attend? After you have responded please turn to page 89 and read one professional's thoughts about this case.

What are your thoughts?

_____

_____

A useful way to respond would be to reflect the main theme of the client's story; for example, "You must feel like you're being hit from all directions. Which should we work on first?" But there are no correct answers. In such a situation different interviewers would place emphasis on different issues. Some interviewers consistently listen attentively to only a few key topics while ignoring other possibilities. Be alert to your own potential patterning of responses. It is important that no issues get lost, but it is equally important not to attack everything at once, as confusion will result.

The concept of *verbal tracking* may be most helpful to the beginning interviewer or to the experienced interviewer who is lost or puzzled about what to say next in response to a client. Relax, take whatever the client has said in the immediate or near past, and direct attention to that through a question or brief comment. You don't need to introduce a new topic. Build on the client's topics, and you will come to know the client very well over time.

Body Language: Attentive and Authentic

The anthropologist Edward Hall once examined film clips of Native Americans of the Southwest and of European North Americans and found more than 20 different variations in the way they walked. Just as cultural differences in eye contact exist, body language patterns differ.

A comfortable conversational distance for many North Americans is slightly more than arm's length, and the English prefer even greater distances. Many Latin people often prefer half that distance, and those from the Middle East may talk practically eyeball to eyeball. As a result, the slightly forward leaning we recommend for attending behavior is not going to be appropriate all the time. A natural, relaxed body style that is your own is most likely to be effective, but be prepared to adapt and flex according to the individual with whom you are talking.

Just as shifts in eye contact tell us about potentially uncomfortable issues for clients, so do changes in body language. A person may move forward when interested and away when bored or frightened. As you talk, notice people's movements in relation to you. How do you affect them? Note your patterns in the interview. When do you change body posture markedly? Are there patterns of which you need to be aware?

Authenticity is vital to the way you communicate. Whether you use visual, vocal, or verbal tracking or attentive body language, it is vital that you be a real person in a real relationship. Take on the skills, be aware and respectful of cultural differences, but also make sure that you are yourself—for your authentic personhood is a vital presence in the helping relationship.

Box 3-2 presents national and international perspectives on counseling skills.

The Value of Nonattention

There are times when it is inappropriate to attend to client statements. For example, a client may talk insistently about the same topic over and over again. A depressed client may want to give the most complete description of how and why the world is wrong. Many clients want to talk about only negative things. In such cases, intentional nonattending may be useful. Through failure to maintain eye contact, subtle shifts in body posture, vocal tone, and deliberate jumps to more positive topics, you can facilitate the interview process.

The most skilled counselors and interviewers use attending skills to open and close client talk, thus making the most effective use of limited time in the interview.

The Usefulness of Silence

For a beginning interviewer, silence can be frightening. After all, doesn't counseling mean talking about issues and solving problems verbally? Perhaps this is true for many of us, but others like to "sit on" their issues, mull them over, and think carefully before responding. The

BOX 3-2   NATIONAL AND INTERNATIONAL PERSPECTIVES ON COUNSELING SKILLS

Use With Care—Culturally Incorrect Attending Can Be Rude
Weijun Zhang, Management Consultant, Shanghai, China

The visiting counselor from North America got his first exposure to cross-cultural counseling differences at one of the counseling centers in Shanghai. His client was a female college student. I was invited to serve as an interpreter. As the session went on I noticed that the client seemed increasingly uncomfortable. What had happened? Since I was translating, I took the liberty of modifying what was said to fit each other's culture, and I had confidence in my ability to do so. I could not figure out what was wrong until the session was over and I reviewed the videotape with the counselor and some of my colleagues. The counselor had noticed the same problem and wanted to understand what was going on. What we found amazed us all.

First, the counselor's way of looking at the client—his eye contact—was improper. When two Chinese talk to one another, we use much less eye contact, especially when it is with a person of the opposite sex. The counselor's gaze at the Chinese woman could have been considered rude or seductive in Chinese culture.

Although his nods were acceptable, they were too frequent by Chinese standards. The student client, probably believing one good nod deserved another, nodded in harmony with the counselor. That unusual head bobbing must have contributed to the student's discomfort.

The counselor would mutter "uh-huh" when there was a pause in the woman's speech. While "uh-huh" is a good minimal encouragement in North America, it happens to convey a kind of arrogance in China. A self-respecting Chinese would say *er* (oh), or *shi* (yes) to show he or she is listening. How could the woman feel comfortable when she thought she was being slighted?

He shook her hand and touched her shoulder. I told our respected visiting counselor afterward, "If you don't care about the details, simply remember this rule of thumb: in China, a man is not supposed to touch any part of a woman's body unless she seems to be above 65 years old and displays difficulty in moving around."

"Though I have worked in the field for more than 20 years, I am still a lay person here in a different culture," the counselor commented as we finished our discussion.

first thing to do when you feel uncomfortable with silence is to look at your client. If the client appears comfortable, draw from his or her ease and join in the silence break. If the client seems as disquieted by the silence as you are, then rely on your attending skills and ask a question or make a comment about something relevant that has been mentioned earlier in the session.

Counseling and interviewing are talking professions. But sometimes the most useful thing you can do as a helper is to support your client silently. As a counselor, particularly as a beginner, you may find it hard to sit and wait for clients to think through what they want to say. Your client may be in tears, and you may want to give support through your words. However, the best support may be simply being with the person and not saying a word.

Finally, remember the obvious: Clients can't talk while you do. Examine your interviews from time to time for amount of talk-time. Who talks the most, you or your client? With less verbal young children, however, it may be wise for you to talk slightly more, tell stories, and so on, to help them verbalize. Extensive research summarized in Box 3-3 provides ample evidence in support of the attending skills.

BOX 3-3   RESEARCH EVIDENCE THAT YOU CAN USE

Attending Behavior

Empirical research on attending behavior has been extensive over the years (see Daniels & Ivey, 2007, and the CD-ROM accompanying this book for a comprehensive report). Hill and O'Brien (1999, p. 90) present a review of the attending literature and conclude that "smiling, a body orientation directly facing the client, a forward trunk lean, both vertical and horizontal arm movements, and a medium distance of about 55 inches between the helper and client are all generally helpful nonverbal behaviors."

On the other hand, Nwachuku and Ivey (1992) tested a program of culture-specific training and found that variations to meet cultural differences were essential. A particularly important research study found that White counselors' perception of their expressed empathy and listening was not in accord with the perceptions of African American males, who saw them as less effective (Stewart, Jackson, Neil, Jo, Nehring, & Grondin, 1998). The way an interviewer can "be with" a client tends to vary among cultures.

Researchers have documented the importance of communication skills training for physicians. Training improves physicians' communication, self-efficacy, confidence, and satisfaction with the training programs. Furthermore, communication skills training has a positive effect on patient outcomes such as satisfaction and perception that the physician understood their disease (Ammentorp, Sabroe, Kofoed, & Mainz, 2007; Back et al., 2007; Bylund et al., 2008; Libert et al., 2007). Bensing (1999) provides a thought-provoking quote from her extensive review of the literature: "Simply looking at the patient has proven to be very important . . . and even silence can be very therapeutic, at least when it is used effectively" (p. 295). Among other findings, she noted that U.S. physicians are more detached and task-oriented than are Dutch physicians, who are rated as warmer and more involved. We can expect variations among both individuals and cultures in their style of attending.

Implications for your practice: Attending behavior "works," but be sensitive to individual and cultural differences.

The teaching of social skills has been successful with a variety of populations. Early work demon-strated the value of teaching attending behavior to hospitalized patients (Donk, 1972; Ivey, 1973). A study of adult schizophrenics showed that teaching social skills with special attention to attending behavior was successful and that patients maintained the skills over a 2-month period (Hunter, 1984). Shy, withdrawn people (avoidant personality) profited from social skills training in a well-controlled study by van der Molen (2006). Despite these and many other successes, the counseling and therapy fields still give relatively little attention to social skills as a mode of treatment (Sanderson, 2002).

Implications for your practice: Social skills training—teaching your clients communication and other skills—is a promising and effective treatment modality. But the field generally still seems unwilling to use this system fully and effectively.

Attention and Neuropsychology

Attention is not just a psychological concept—it is measurable through brain imaging. When a person attends to a stimulus (e.g., the client's story) many areas of the brain of both interviewer and client become involved (Posner, 2004). Two key factors in attention are arousal and focus. Arousal involves the reticular activating system, at the brain's core, which transmits stimuli to the cortex and activates neurons firing throughout many areas. By giving attention to the client, both interviewer and client thought processes have been activated. Selective attention is a critical aspect of listening—"Focus is brought about by . . . a part of the thalamus, which operates rather like a spotlight, turning to shine on the stimulus" (Carter, 1999, p. 186). If you do not attend to the client's key issues, this focus is lost.

Once we have attention, what happens next? A lot of things, but one example is Siegel (2007, pp. 114–115), who speaks of *executive attention,* which is central to planning and decision making. Within the left brain, the anterior cingulate cortex "allocates attentional resources." Siegel gives special attention to meditation as a way to heighten general focus and health. Meditation is an important type of selective attention.

(continued)

BOX 3-3 (continued)

Implications for your practice: By listening and attending, both your brain and your client's "light up." You won't get any change unless the client attends to you, and you attend to the client. Research described by Siegel indicates that practice in meditation facilitates neural development and focus. Chapter 12 on influencing skills speaks in more detail to this treatment strategy. After physical exercise, meditation and mindfulness practice seem to be the most effective strategies for mental and physical health.

Just as attending behavior underlies all the microskills of this book, attention and selective attention provide the physiological foundation for personal growth. By listening and selective attention, both our brain and the client's are reacting and changing.

PSYCHOEDUCATION, SOCIAL SKILLS, AND ATTENDING BEHAVIOR

Social skills training is training in a specific set of psychoeducational strategies oriented to teaching clients basic communication skills. Rather than talking through issues, social skills training involves educational methods to teach clients an array of interpersonal skills and behaviors. These skills include a wide range of behaviors, such as listening, dating behaviors, drug-refusal skills, assertiveness, mediation, and job-interviewing procedures. Virtually all interpersonal actions can be taught through social skills training.

Training as treatment is a term that summarizes the method and goal of social skills training. The microtraining format of selecting specific skill dimensions for education has become basic to most psychoeducational social skills programs since its inception. As you extend the counseling and interviewing dimensions to skills training itself, think of the following steps: (a) negotiate a skill area for learning with the client; (b) discuss the specific and concrete behaviors involved in the skill, sometimes presenting them in written form as well; (c) practice the skill with the client in a role-play in the interview or group counseling session; and (d) plan for generalization of the skill to daily life.

Shortly after the first work in identifying interviewing microskills, Allen Ivey was working with a first-year college student who suffered a mild depression and complained about the lack of friends. Allen asked the student what he talked about with those in his dormitory. The student responded by continuing his list of complaints and worries. With further probing, the student acknowledged that he spent most of his time with others talking about himself and his difficulties. It was easy to see that potential friends would avoid him. We all tend to move away from those who talk negatively and stay away from those who talk only about themselves and fail to listen to us.

On the spot, Allen talked to the student about attending behavior and its possible rewards. The three V's + B were presented, and the importance of gaining trust and respect from others by listening was emphasized. Allen suggested that the student might profit from actively listening to those around him rather than talking only about himself. The student expressed interest in learning these skills, and a practice session was initiated there in the interview. First, negative attending was practiced, and the student was able to see how his lack of listening might contribute to his isolation in the dormitory. Then positive attending was practiced, and the student discovered that he could listen.

Allen and the client discussed specifics of selecting someone with whom to try these skills. When the student returned the following week, he had a big smile and reported that he had found his first friend at the university. Moreover, he discovered an important side effect: "I feel less sad and depressed. First, I don't feel so alone and helpless. The second thing I noticed was that when I am attending to someone else, I am not thinking about myself, and then I feel better." When you are attending to someone else, it becomes much more difficult to think negatively about yourself. Instruction in attending behavior is one of the foundations of social skills training.

Many types of clients can benefit from learning and practicing these skills. Allen Ivey found that teaching attending and other microskills to veterans at a VA hospital was sufficient to enable them to return to their families and communities (Ivey, 1973). Van der Molen (1984, 2006) used attending behavior and other microskills in a highly successful psychoeducational program in which he taught people who were shy (also known as the "avoidant personality") to become more socially outgoing. As just one other example, children diagnosed with attention deficit disorder (ADD) who receive skills training are less disruptive (Pfiffner & McBurnett, 1997). Effective psychoeducation can help children learn better in the classroom. Perhaps we can use less medication with more competent use of helping skills and strategies?

As you work through this book, think about how educating clients in the various microskills can be beneficial. Although attending behavior is often the foundation, other microskills are useful in treatment. Teaching couples to listen to each other can make a difference rather quickly!

USING ATTENDING IN CHALLENGING SITUATIONS

Some beginning interviewers and counselors may think that attending skills are simple and obvious. They may be anxious to move to the "hard stuff." You also may have wondered how attending behavior can be useful if you plan to work with challenging clients in schools, community mental health centers, or hospitals. The following is designed to illustrate from our personal experience some things that we and others have observed about attending over time.

Mary: Attending is natural to me, and the basic listening sequence has always been central to my work with children, but I do find at times that my natural style is not enough, particularly when I work with distracted or acting-out children. Even with all my experience with children, like other professionals, sometimes I am at a loss as to what to do next. After some analysis, I found that if I moved back to my foundation in attending skills and focused carefully on visuals, vocals, verbal following, and body language, I could regain contact with even the most troubled child. Similarly, in challenging situations with parents, I have at times found myself returning to a focus on attending behavior, later adding the basic listening sequence and other skills. Conscious attending has helped me many times in involved situations. Attending is not a simple set of skills.

I train older students to work as school peer mediators and another group to be peer tutors for younger children. I have found that using the exercise at the beginning of this chapter on poor attending and then contrasting it with good attending works well as an introductory exercise. I then teach attending skills and the basic listening sequence to my student groups.

Allen: One of my most powerful experiences occurred when I first worked at the Veterans Administration with schizophrenic patients who talked in a stream of consciousness "word salad." I found that if I maintained good attending skills and focused on the exact words they were saying, they soon were able to talk in a normal, linear fashion. I also found that teaching communication skills with video and video feedback to even the most troubled patients was effective. Sometimes attending was sufficient treatment by itself to move them out of the hospital. Depressed psychiatric inpatients, in particular, responded well to social skills training. However, I did find that highly distressed patients could learn only one of the four central dimensions of attending. Thus, we would start from the patient's observation of the video and selected choice of behavioral change. Most often, visual/eye contact and body language were chosen first. Later we moved on to other skills selected with the client.

Like Mary, when the going gets rough, I find that it helps me to return to basic attending skills and a very serious effort to follow what the client is saying as precisely as possible. In short, when in doubt, attend. It often works!

Practice, practice, practice—also known as "use it or lose it!"

SUMMARY: BECOMING A SAMURAI*

Japanese masters of the sword learn their skills through a complex set of highly detailed training exercises. The process of masterful sword work is broken down into specific components that are studied carefully, one at a time. In this process of mastery, the naturally skilled person often suffers and finds handling the sword awkward. The skilled individual may even find performance worsening during the practice of single skills. Being aware of what one is doing can interfere with coordination and smoothness in the early stages.

Once the individual skills are practiced and learned to perfection, the samurai retire to a mountaintop to meditate. They deliberately forget what they have learned. When they return they find the distinct skills have been naturally integrated into their style or way of being. The samurai then seldom have to think about skills at all: They have become samurai masters.

Consider driving. When you first sat at the wheel, you had to coordinate many tasks, particularly if you drove a car with a stick shift. The clutch, the gas pedal, the steering wheel, and the gear ratios had to be coordinated smoothly with what you saw through the windshield. When you gave primary attention to the skills of driving, you might have lost sight of where you were going. But practice and experience soon led you to forget the specific skills, and you were able to coordinate them automatically and give full attention to the world beyond the windshield. The mastery of single skills led you to achieve your objectives.

The Samurai effect and the importance of practice to mastery now have a scientific base in brain research. Through single skill practice, new neural networks are developed, which eventually become integrated into even larger networks, and this results in a natural whole for skilled performance. The golf champion Tiger Woods decided that he wanted to improve his swing. He practiced single skills; these skills become integrated with others and eventually formed a natural, automatic whole. Early in this process, his score temporarily dropped, but ultimately his score improved, and he went on to even better play. See Appendix II for a more complete discussion of the brain basis of the Samurai effect.

Learn the skills of this book, but allow yourself time for meditation and/or integrating these ideas into your own natural authentic being. Intentional and relaxed performance of skills and strategies takes time and practice. You may have found discomfort in practicing the single skill of attending. Later you may find the same problem with other skills. This happens

---

* We are indebted to Lanette Shizuru, University of Hawai'i, Manoa, for the example of the samurai.

to both the beginner and the advanced counselor. Improving and studying our natural skills often result in a temporary and sometimes frustrating decrease in competence, just as they do for the samurai and did for Tiger Woods.

Step-by-step training requires conscious attention in the here-and-now present. Repetition of the key skills successfully stores them in long-term memory, and brain plasticity enables new neural connections to develop—the meditative integration of the master samurai. A quick "run through" of microskills without significant practice will be ineffective in the long term. Practice, practice, practice!

There is no need to talk about yourself or give long answers when you attend to someone else. Give clients ownership of "airtime." After all, it is their story we need to hear. Your main responsibility as a helper is to assist others in finding their own answers. You'll be surprised at how able they are to do this if you are willing to attend.

Key points of the chapter are presented below.

■ KEY POINTS

Central goals of listening	When we use attending behavior, we all have one goal in common: to reduce interviewer talk-time while providing clients with an opportunity to examine issues and tell their stories. You can't learn about the other person or the problem while you are doing the talking! Changing the focus of the interview may also be used to stop needless client talk at any time during the interview. But always use attending with individual and cultural sensitivity.
Four aspects of attending	Attending behavior consists of four simple but critical dimensions (three V's + B), but all need to be modified to meet individual and cultural differences. 1. Visual/eye contact. If you are going to talk to people, look at them. 2. Vocal qualities. Your vocal tone and speech rate indicate much of how you feel about another person. Think of how many ways you can say "I am really interested in what you have to say" just by your vocal tone and speech rate. 3. Verbal tracking. The client has come to you with a topic of interest; don't change the subject. Keep to the topic initiated by the client. If you change the topic, be aware that you have and realize the purpose of your change. 4. Body language: attentive and genuine. In general, clients know you are interested in them if you face them squarely and lean slightly forward, have an expressive face, and use facilitative, encouraging gestures. If you are a naturally warm and caring person, this likely comes easily.
Where should you focus your attention?	Attending is easiest if you focus your attention *on the client rather than on yourself.* Note what the client is talking about, ask questions, and make comments that relate to your client's topics. For example:
Client:	I'm so confused. I can't decide between a major in chemistry, psychology, or language.
Interviewer:	(nonattending) Tell me about your hobbies. What do you like to do? or What are your grades?
Interviewer:	(attending) Tell me more, or You feel confused? or Could you tell me a little about how each subject interests you? or Opportunities in chemistry are promising now. Could you explore that field a bit more? or How would you like to go about making your decision?

(continued)

KEY POINTS (continued)

	Note that all attending responses follow the client's verbal statement. Each might lead the client in a very different direction. Interviewers need to be aware of their patterns of selective attention and how they may unconsciously direct the interview. Finally, nonattending responses may sometimes be helpful in discouraging certain client talk and focusing the interview.
With whom?	Attending is vital in all human interactions, be they counseling, medical interviews, or business decision meetings. It is also important to note that different individuals and cultural groups may have different patterns of listening to you. For example, some may find the direct gaze rude and intrusive, particularly if they are dealing with difficult material. Consider the possibility of teaching attending behavior to clients. Social skills training may be useful for those who are shy or depressed and for many others in distress. The business person who is overbearing and talks constantly can also benefit from learning how to attend.
What should you do when you don't know what to do next?	A simple but often helpful rule for interviewing is use attending skills when you become lost or confused about what to do. Simply ask the client to comment further on something just said or mentioned earlier in the interview.

COMPETENCY PRACTICE EXERCISES AND PORTFOLIO OF COMPETENCE

Intentional interviewing and counseling is achieved through practice and experience. Learning and growing in skills will be enhanced by your own self-awareness, emotional competence, and ability to observe yourself.

The competency-practice exercises on the following pages are designed to provide you with opportunities in three areas:

1. *Individual practice.* A short series of exercises is provided to give you an opportunity to practice by yourself aspects of attending behavior.
2. *Group practice.* Practice alone can be helpful, but working with others in role-played interviews is where the most useful learning occurs. You can obtain precise feedback on your interviewing style. And if videotape (or audiotape) is used with these practice sessions, seeing yourself as others see you is a powerful experience.
3. *Self-assessment.* You are the person who will use the skills. We'd like you to look at yourself as an interviewer and counselor via some additional exercises.

Individual Practice

Exercise 1: Generating Alternative Attending Statements

A client comes to you with the following story:

> Our relationship is getting worse. We used to do so well and somehow seemed to strengthen each other. The parents on both sides were always opposed to our getting together, and lately Carlena is listening to her folks. She thinks I'm too rigid and can't let loose. And I just got in trouble at work. Sleeping is becoming difficult.

Some clients begin their interviews with a long list of problems. Write three things below that you might say in response to this client (remember that each might lead the client in a different direction). As a thought question, "How would you plan to work with these multiple issues over time?"

What would you say?

_____

_____

Exercise 2: Identifying Topic Jumps

Review the two Allen and Azara interviews demonstrating negative and positive attending behavior. Identify specific places in both segments when Allen, the interviewer, changed topic. List each topic jump and then compare them to our list on page 89.

Negative session: _____

Positive session: _____

Exercise 3: Behavioral Counts, an Evaluation and Research Strategy

This exercise introduces you to early research in microskills. After training in attending behavior, we have found that beginning interviewers have fewer eye contact breaks, speech hesitations, and topic jumps as well as a smaller number of distracting body gestures or movements.

Observe an interview. This could be a role-played counseling session, a television talk show, or simply an interaction between friends and family. Use the following four-point form to count specific behaviors. Specifically, make a mark for each instance of less effective attending.

As you work through this exercise, recall multicultural and individual differences that can occur in the interview. Your observations and interpretations of what you observe need to be moderated with sensitivity to diversity.

_____ Visuals—number of eye-contact breaks

_____ Vocals—number of speech hesitations or uses of disruptive vocal tone

_____ Verbal tracking—number of topic jumps

_____ Body language—number of distracting movements or gestures

Group Practice

Exercise 4: Group Practice Using Attending Skills

The following instructions are designed for groups of four but may be adapted for use with pairs, trios, and groups up to five or six. Ideally, each group will have access to video- or audiorecorders. However, use of careful observation (see Box 3-4) and the feedback sheets in the book (see Box 3-5) can provide enough structure for a successful practice session without the benefit of recording equipment.

Step 1: Divide into practice groups. Get acquainted with each other informally before you go further.

Step 2: Select a group leader. The leader's task is to ensure that the group follows the specific steps of the practice session. It often proves helpful if the least experienced group member serves as leader first. Group members then tend to be supportive.

Step 3: Assign roles for the first practice session.

▲ Client. The first role-play client will be cooperative and have a story or issue to present, talk freely about the topic, and not give the interviewer a difficult time.
▲ Interviewer. The interviewer will demonstrate a natural style of attending behavior with the client and practice the basic skills.
▲ Observer 1. The first observer will fill out the feedback form (Box 3-5) detailing some aspects of the interviewer's attending behavior. Observation of these practice videos could also be called "microsupervision" in that you are helping the interviewer understand what he or she is doing and the effectiveness of the brief session. Later, when you are working as a professional helper, it is vital that you continue to share your work with colleagues through verbal report or audio/videotape. Supervision is a vital part of effective interviewing and counseling.
▲ Observer 2. The second observer will time the session, start and stop any equipment, and fill out a second observation sheet as time permits.

Step 4: Plan. The interviewer states her or his goals clearly, and the members of the group should take time to plan the role-play. The interviewer helps to open and to facilitate the client's talking about the story or concern. An increased percentage of client talk-time on a single topic will indicate a successful session. The interviewer may also plan to close off client talk and then open it again. It may help to keep the conversation going if he or she elicits both positive and negative comments; this also results in a deeper understanding of the client. The more concrete the plan, the greater is the likelihood of success.

The suggested topic for the attending practice session is "Why I want to be a an interviewer, counselor, or therapist." The client talks about her or his desire to join the helping profession or at least to consider it as a work setting.

The interviewer demonstrates attending skills. Other possible topics for the session include the following:

▲ A job you have liked and a job that you didn't (or don't) enjoy.
▲ A positive experience you have had that led to a new learning about yourself.
▲ As part of the planning process, try at least one of the role-played interviews in which the client portrays a person who is physically challenged. The varying aspects of visuals, vocals, verbals, and body language may help sensitize you to individual and multicultural differences.

The topics and role-plays are most effective if you talk about something meaningful to you. You will also find it helpful if everyone in the group works on the same topic as roles are rotated. In that way you can compare styles and learn from one another more easily.

While the interviewer and the interviewee plan, the two observers preview the feedback sheets and plan their own practice sessions to follow.

Step 5: Conduct a 3-minute practice session using attending skills. The interviewer practices the skills of attending, the client talks about the current work setting or other selected topic, and the two observers fill out the feedback sheets. Do not go beyond 3 minutes. If possible, record the interview.

Step 6: Review the practice session and provide feedback to the interviewer for 12 minutes. As a first step in feedback, the role-play client gives her or his impressions of the session and completes the Client Feedback Form of Chapter 1. This may be followed by interviewer self-assessment and comments by the two observers. As part of the review, ask yourselves the key question, "Did the interviewer achieve her or his own planning objective?" This is critical for assessing the level of mastery obtained.

On the feedback form, note both verbal and nonverbal behaviors and the different effects they had on the client and observers. Giving useful, specific feedback is particularly critical. Note the suggestions for feedback in Box 3-4.

Finally, as you review the audio- or videotapes of the interview, start and stop the tape periodically. Replay key interactions. Only in this way can you fully profit from the recording media. Just sitting and watching television is not enough; use media actively.

BOX 3-4   GUIDELINES FOR EFFECTIVE FEEDBACK

	To see ourselves as others see us. To hear how others hear us. And to be touched as we touch others . . . These are the goals of effective feedback.
Feedback	Feedback is one of the skill units of the basic attending and influencing skills developed in this book; it is discussed in more detail in Chapter 12. However, if you are to help others grow and develop in this program, you must provide feedback to them now on their use of skills in practice sessions. Here are some guidelines for effective feedback:
Guidelines	▲ *The person receiving the feedback is in charge.* Let the interviewer in the practice sessions determine how much or little feedback is wanted. ▲ *Feedback includes strengths,* particularly in the early phases of the program. If negative feedback is requested by the interviewer, add positive dimensions as well. People grow from strength, not from weakness. ▲ *Feedback is most helpful when it is concrete and specific.* Not "Your attending skills were good" but "You maintained eye contact throughout except for breaking it once when the client seemed uncomfortable." Make your feedback factual, specific, and observable. ▲ *Corrective feedback should be relatively nonjudgmental.* Feedback often turns into evaluation. Stick to the facts and specifics, though the word *relatively* recognizes that judgment inevitably will appear in many different types of feedback. Avoid the words *good* and *bad* and their variations. ▲ *Feedback should be lean and precise.* It does little good to suggest that a person change 15 things. Select one to three things the interviewer actually might be able to change in a short time. You'll have opportunities to make other suggestions later. You should check how your feedback was received. The client response indicates whether you were heard and how useful your feedback was. "How do you react to that?" "Does that sound close?" "What does that feedback mean to you?"

BOX 3-5   FEEDBACK FORM: ATTENDING BEHAVIOR

_____ (Date)

_____   _____
(Name of Interviewer)   (Name of Person Completing Form)

Instructions: Provide written feedback that is specific and observable, nonjudgmental, and supportive. As an alternative, use behavioral counts as shown in Exercise 3.

1. *Visual/eye contact.* Facilitative? Staring? Avoiding? Sensitive to the individual client? At what points, if any, did the interviewer break contact? Facilitatively? Disruptively?

2. *Vocal qualities.* Vocal tone? Speech rate? Volume? Accent? Points at which these changed in response to client actions? Number of major changes or speech hesitations?

3. *Verbal tracking and selective attention.* Was the client able to tell the story? Stay on topic? Number of major topic jumps? Did shifts seem to indicate interviewer interest patterns? Did the interviewer demonstrate selective attention in pursuing one issue rather than another? Did the client have the majority of the talk-time?

4. *Attentive body language.* Leaning? Gestures? Facial expression? At what points, if any, did the interviewer shift position or show a marked change in body language? Number of facilitative body language movements? Was the session authentic?

5. *Specific positive aspects of the interview.*

6. *Discussion question:* What areas of diversity do the interviewer and client represent? How does this affect the session?

Step 7: Rotate roles. Everyone should have a chance to serve as interviewer, client, and observer. Divide your time equally!

Some general reminders. It is not necessary to compress a complete interview into 3 minutes. Behave as if you expected the session to last a longer time, and the timer can break in after 3 minutes. The purpose of the role-play sessions is to observe skills in action. Thus, you should attempt to practice skills, not solve problems. Clients have often taken years to develop their interests and concerns, so do not expect to solve one of these problems or obtain the full story in a 3-minute role-play session. Written feedback, if carefully presented, is an invaluable part of a program of interview skill development.

Portfolio of Competence

Skills and strategies are vital in interviewing and counseling, but they are not enough. No two interviewers are the same—you as a person will integrate the skills in your own way. Artistic and integrated interviewing becomes part of the natural style of the person, but all of us need to examine ourselves and what we are doing. We also need feedback on our performance from others, and that is where group practice, supervision, and conversations with our peers are important.

What Is Your Natural Style of Listening and Attending?

An important part of self-awareness is reflecting on yourself as a listener and your natural style of attending. Attending to others is fundamental to empathic understanding, one of the central dimensions of emotional intelligence. How able are you to listen to and be with others?

Following are several alternatives to help you think about your abilities as a listening person. Some are thought questions to which you can respond in journal form. Others are action oriented.

▲ What led you to a course in interviewing skills? Are you a "people-person"? Have you had friends come to you to share their concerns and problems? Do you like to listen to others? What are your motivations?

▲ How comfortable are you with ideas of diversity and working with people different from you? Can you recognize yourself as a multicultural person full of many dimensions of diversity? Include issues of physical challenge as an important part of multiculturalism and diversity.

▲ View yourself on videotape or listen to an audiotape of an interview. One of the best ways to examine your natural style of listening is to observe the video or audiotape you made as you started this book. Again, if you have not made that first video, now is the time! What do you observe in an informal viewing of yourself?

▲ View the video or listen to the audiotape again. This time, use the observation suggestions of the Feedback Form: Attending Behavior, or Exercise 3 of Individual Practice: Behavioral Counts. If you use the behavioral count system, you may use that as a baseline for comparison with later interviews as you work through this book.

Self-Evaluation of Attending Skills Competencies

Use the following as a checklist to evaluate your present level of mastery. As you review the items below, ask yourself, "Can I do this?" Check those dimensions that you currently feel able to do. Those that remain unchecked can serve as future goals. Do not expect to attain intentional competence on every dimension as you work through this book. You will find, however, that you will improve your competencies with repetition and practice.

Level 1: Identification and classification. You will need this minimal level of mastery for those coming examinations!

❑   Discuss issues in diversity that occur in relation to attending behavior.
❑   Write attending statements in response to written client statements.
❑   Make behavior counts of the three V's + B while observing interviews.

Level 2: Basic competence. We suggest that you aim for this level of competence before moving on to the next skill area in this book.

❑   Demonstrate culturally appropriate visuals/eye contact, vocal qualities, verbal following, and body language in a role-played interview.
❑   Increase client talk-time while reducing your own.
❑   Stay on a client's topic without introducing any new topics of your own.

Level 3: Intentional competence. In the early stages of this book, strive for basic competence and work toward intentional competence later. Experience with the microskills model is cumulative, and you will find yourself mastering intentional competencies with greater ease as you gain more practice.

❑   Understand and manage your own pattern of selective attention.
❑   Change your attending style to meet client individual and cultural differences.
❑   Note topics that clients particularly attend to and topics that they may avoid.
❑   Use attending skills with more challenging clients.
❑   Help clients through attention and inattention to move from negative, self-defeating conversation to more positive and useful topics. Conversely, this also includes helping clients who are avoiding issues to talk about them in more depth.

Level 4: Psychoeducational teaching competence. Teaching attending behavior to clients and small groups is something that many counselors and therapists now do. It is a first choice psychoeducational social skills strategy. Clearly, you are not expected to teach attending behavior at this time, but the opportunity may present itself. You may find a client who would benefit from psychoeducational instruction in attending behavior. You may be asked to work with a small group of children or adolescents who need help with social skills. You may present a class in a church group or the dormitory.

❑   Teach clients in a helping session about the social skill of attending behavior. You may either tell clients about the skill or you may practice a role-play with them.
❑   Teach small groups the skills of attending behavior.

DETERMINING YOUR OWN STYLE AND THEORY: CRITICAL SELF-REFLECTION ON ATTENDING BEHAVIOR

This chapter has focused on the importance of listening as the foundation of effective interviewing practice: When in doubt as to what to do—listen, listen, listen! Individual and cultural differences are central—visual, vocal, verbal, and body language styles vary. Avoid stereotyping any group.

What single idea stood out for you among all those presented in this chapter, in class, or through informal learning? What stands out for you is likely to be important as a guide toward your next steps. How might you use ideas in this chapter to begin the process of establishing your own style and theory? Please turn to your journal and write your thoughts. What are your thoughts about using attending behavior in psychoeducational practice?

What are your thoughts?

_____

_____

RESPONSE TO CLIENT ON PAGE 74

Dr. Howard Busby of Gallaudet College has commented on this case (2002, personal e-mail communication to Allen Ivey, January 24). He points out that the problem could be related to the disability, the problems at school, or both. The "wheelchair might be the problem as much as how this client is dealing with it. The client obviously has a disability (terminology), but it does not have to be disabling unless the client makes it so. Although I am categorized as having a disability due to deafness, I have never allowed it to be disabling.

"My interpretation of the client is that this depression, on the surface, could be the result of mobility restriction. However, there are other factors that might have caused the depression, even if there were no need for a wheelchair. It is easy for the client to blame problems on the disability and thus distract from personal issues. The issue could be a poor campus environment, learned helplessness on the part of the client, or a combination of these and other multiple factors."

Allen, Mary, and Carlos comment: All of us would do well to consider Dr. Busby's comments. He is pointing out to us the importance of taking a broad and comprehensive view of all cases. We must avoid stereotyping, but we must also be sensitive to cultural differences.

COMMENTS ON INDIVIDUAL PRACTICE, EXERCISE 2

Allen clearly topic jumped at 5, 7, 9, and 11. Most of the responses in the second segment represent verbal following. At 13, we actually start the interview when Allen asks Azara to share what was on her mind, but even this goes back to earlier in the interview at 5 when he acknowledged that she had a concern about her job. At 15 he summarized what the client has said, clearly demonstrating that he is attending to Azara. At 21, he topic jumps to a discussion of strengths, with the intention of coming back to her major concerns.

Hearing Client Stories: How to Organize an Interview

Attending behavior is basic to all the communication skills of the microskills hierarchy. Without individually and culturally appropriate attending behavior, there can be no interviewing, counseling, or psychotherapy.

This section adds to attending skills by presenting the basic listening sequence that will enable you to elicit the major facts and feelings pertinent to a client's concern. Through the skills of questioning, encouraging, paraphrasing, reflecting feelings, and summarizing, you will learn how to draw out your clients and understand the way they think about their stories.

Look for the following in this second section.

Chapter 4. Questions: Opening Communication   We encounter questions every day. Most theories of counseling now use questions rather extensively. The new and highly influential coaching movement considers them the foundation of assessment and reaching client goals. Brief counseling and motivational interviewing use them continuously. This chapter explains open and closed questions and shows you their place in your communication. But we also point out that the use of questions is sometimes a controversial issue. Several experts would argue that this chapter belongs *after* the critical and central skills of accurate and reflective listening.

Chapter 5. Observation Skills   This chapter gives you the opportunity to practice noting your clients' verbal and nonverbal behavior. You are also asked to observe your own nonverbal reactions in the session. Clients often come in with a dejected and "down" body posture. With your observation and listening skills, you can anticipate that they will show a more positive body language as the session progresses.

Chapter 6. Encouraging, Paraphrasing, and Summarizing: Key Skills of Active Listening
Here we examine the clarifying skills of paraphrasing, encouraging, and summarizing, which are foundational to developing a relationship and working alliance with your client. They are central also for drawing out the story.

Chapter 7. Observing and Reflecting Feelings: A Foundation of Client Experience   Reflecting client feelings is the focus of this chapter. You will learn how to bring out the rich emotional world of your clients. Perhaps even more important than Chapter 6, this skill gets at the heart of the issue and truly personalizes the interview.

Chapter 8. Integrating Listening Skills: How to Conduct a Well-Formed Interview   Once you have mastered observation skills and the basic listening sequence, you are prepared to engage in a full, well-formed interview, comprising five stages. You will be able to use in this interview listening and observation skills. Furthermore, it is important that you be able to evaluate your interviews and those of others for level of empathic understanding. It is vital not only to listen but also to listen empathetically. Some instructors will want to include the empathy readings with Chapters 6 and 7.

This section, then, has ambitious goals. By the time you have completed Chapter 8, you'll have attained several major objectives, enabling you to move on to the influencing skills of interpersonal change, growth, and development. At an intentional level of competence, you may aim to accomplish the following in this section:

1. Master the basic listening sequence, enabling the client to tell the story. In addition, draw out key facts and feelings related to client issues.
2. Observe clients' reactions to your skill usage and modify your skills and attending behaviors to complement clients' uniqueness.
3. Conduct a complete interview using only listening and observing skills.
4. Evaluate that interview for its level of empathy; in effect, examine yourself and your ability to communicate warmth, positive regard, and other, more subjective dimensions of interviewing and counseling.

When you've accomplished these tasks, you may find that your clients have a surprising ability to solve their own problems, issues, concerns, or challenges. Furthermore, you may also gain a sense of confidence in your own ability as an interviewer or counselor. The underlying theme of this book's first eight chapters, then, is: "When in doubt, listen!"

CHAPTER 4

Questions: Opening Communication

Open and Closed Questions

Attending Behavior

Ethics, Multicultural Competence, and Wellness

How you ask questions is very important in establishing a basis for effective communi-
cation. Effective questions open the door to knowledge and understanding. The art of
questioning lies in knowing which questions to ask when. Address your first question to
yourself: if you could press a magic button and get every piece of information you
want, what would you want to know? The answer will immediately help you compose
the right questions.

—**Robert Heller and Tim Hindle**

How can questions help you and your clients?

Chapter Goals	This chapter describes how questions can be used to enhance the interview and draw out the client's story. Two types of questions, open and closed, will be discussed. Open questions elicit more information and give the client more room to respond. Closed questions will elicit shorter responses and provide information and specifics. The chapter also focuses on developing awareness that, like attending behavior, questions can encourage or discourage client talk. With questions, however, the interviewer generally takes the lead. The client is often talking within the interviewer's frame of reference. Questions potentially can take the client away from self-direction.
Competency Objectives	Awareness, knowledge, and skill in questioning will enable you to ▲ Elicit additional specifics relating to the client's world and enrich his or her story.

▲ Make an effective assessment of a client's concern or issue.
▲ Guide the manner in which a client talks about an issue. For example, *what* questions often lead to talk about facts, *how* questions to feelings or process, and *why* questions to reasons and also to meanings and life visions.
▲ Help to open or close client talk according to the individual needs of the interview.
▲ Engage in some of the basics of the influential coaching movement and include key questioning strategies to coach as well as counsel your clients.

Benjamin is in his junior year of high school, in the middle third of his class. In this school, each student must be interviewed about plans after graduation—work, the armed forces, or college. You are the high school counselor and have called him in to check on his plans after graduation. His grades are average, and he is not particularly verbal and talkative but is known as a "nice boy."

Reread the quotation on questions that introduced this chapter. What are some questions you could use to draw him out and help him think ahead to the future? And what might happen if you ask too many questions?

What do you anticipate?

_____

_____

You may want to compare your questions with our thoughts on page 120 at the end of this chapter.

INTRODUCTION: QUESTIONING QUESTIONS

Although attending behavior is the skill and action foundation of the microskills hierarchy, it is questioning that provides a systematic framework for directing the interview. Questions help an interview begin and move along smoothly. They open new areas for discussion, assist in pinpointing and clarifying issues, and aid in clients' self-exploration. If you use questions as described below, you can make the follow predictions.

Open and Closed Questions	Predicted Result
Begin open questions with often useful *who, what, when, where,* and *why.* Closed questions may start with *do, is,* or *are. Could, can,* or *would* questions are considered open but have the additional advantage of being somewhat closed, thus giving more power to the client, who can more easily say that he or she doesn't want to respond.	Clients will give more detail and talk more in response to open questions. Closed questions may provide specific information but may close off client talk. Effective questions encourage more focused client conversations with more pertinent detail and less wandering. *Could, would,* and *can* questions are often the most open of all.

Questions are an essential component in many theories and styles of helping, particularly cognitive-behavioral counseling, brief counseling, and much of career decision-making work. The employment counselor, the social worker conducting an assessment interview, and the high school guidance counselor helping a student work on college admissions all need to use questions. Many in the increasingly influential coaching movement believe that questions are the most important helping skill.

This chapter focuses on two key styles of questioning—open and closed questions:

Open questions are those that can't be answered in a few words. They encourage others to talk and provide you with maximum information. Typically, open questions begin with *what, how, why, or could*: For example, "Could you tell me what brings you here today?" You will find these helpful as they can facilitate deeper exploration of client issues.

Closed questions can be answered in a few words or sentences. They have the advantage of focusing the interview and obtaining information, but the burden of guiding the talk remains on the interviewer. Closed questions often begin with *is, are, or do*: For example, "Are you living with your family?" Used judiciously, they enable you to obtain important specifics.

Some theorists and many practitioners raise important issues around the use of questioning, however. Some experts maintain that questions are best learned *after* expertise is developed in the reflective listening skills of Chapters 6 and 7. There is also the danger that some will like questions so much that they will not give enough attention to the critical listening skills. Certainly, excessive use takes the focus from the client and gives too much power to the interviewer. Your central task in this chapter is to find your own balance in using questions in the interview. We suggest you seek to be egalitarian and share power.

Key Issues Around Questions

Why do some people object to questions? Take a minute to recall and explore some of your own experiences with questions in the past. Perhaps you had a teacher or a parent who used questions in a manner that resulted in your feeling uncomfortable or even attacked. Write here one of your negative experiences with questions and the feelings and thoughts the questioning process produced in you.

My difficult personal experience with questions was as follows:

What is your experience?

_____

_____

The thoughts and feelings this experience produced in me were these:

What are your feelings?

_____

_____

People often respond to this exercise by describing situations in which they were put on the spot or grilled by someone. They may associate questions with anger and guilt. Many of us have had negative experiences with questions. Furthermore, questions may be used to direct and control client talk. School discipline and legal disputes typically use questions to control the person being interviewed. If your objective is to enable clients to find their own way, questions may inhibit your reaching that goal, particularly if they are used ineffectively.

BOX 4-1   RESEARCH EVIDENCE THAT YOU CAN USE

Questions

Research results can be a guide or a confirmation for what types of questions work best in certain situations or with particular theoretical orientations. Not too surprising is confirmation that open questions produce longer client responses than closed ones (Daniels & Ivey, 2007; Tamase, Torisu, & Ikawa, 1991). Sternberg and others (1996) found this to be true also with children. Research shows that different theoretical orientations to helping have widely varying usage of questions. Effective use of open questions can help interviewers avoid leading the child to false statements, as has sometimes happened in forensic work. Clients may most easily talk about feelings when they are directly asked about them (Hill, 2004; Tamase, 1991).

Research on group leadership found that person-centered leaders tend to use very few questions whereas 40% of the leads of a problem-solving group leader were questions (Ivey, Pedersen, & Ivey, 2001; Sherrard, 1973). You will find that using motivational interviewing or brief counseling will require extensive use of questions whereas the person-centered model uses very few (Chapter 14). Your decision about which theory of

helping you use most frequently will be important in your use of this skill.

Questions and Neuroscience

Questions are often a good route to help a client discuss issues from the past residing in long-term memory, lodged primarily in the temporal lobe cortex and hippocampus (Kolb & Wishaw, 2003). The goal of questions is to obtain information that will enhance client growth and development and ultimately lead to the development of new neural networks.

However, questions that lead clients too much can result in their constructing stories of things that never happened. In a classic study, Loftus (1997) found that false memories could be brought out by simply reminding people of things that never happened. In other studies, brain scans revealed that false memories activate different neurons from those that are true (Schachter, 1997), but obviously you don't have that information available. Needless to say, a counselor or therapist may not know whether the memories are true or false. Be careful of putting your ideas into the client's head via probing questions.

It is for these reasons that some helping authorities, particularly those who are humanistically oriented, object to questions in the interview.

Additionally, in many non-Western cultures, questions are inappropriate and may be considered offensive or overly intrusive. Nevertheless, questions remain a fact of life in our culture. We encounter them everywhere. The physician or nurse, the salesperson, the government official, and many others find questioning clients basic to their profession. Many counseling theories espouse using questions extensively. Coaching, cognitive-behavioral, and brief counseling, for example, use many questions. The issue, then, is how to question wisely and intentionally.

The goal of this section is to explore some aspects of questions and, eventually, to determine their place in your communication skills repertoire. Used carefully, questioning is a valuable skill. Box 4-1 provides interesting research regarding the use of questons.

Sometimes Questions Are Essential—"What Else?"

Clients do not always provide you with important information, and sometimes the only way to get at missing data is by asking questions. For example, the client may talk about being depressed and unable to act. As a helper, you could listen to the story carefully but still miss important underlying issues relating to the client's depression. The open question "What important things are happening in your life right now or with your family?" might bring out the fact that a separation or divorce is near at hand, that a job has been lost, or that there is some other important dimension underlying the concern. What you first interpreted as a classical clinical depression becomes modified by what is occurring in the client's life, and treatment takes a different direction.

An incident in Allen's life illustrates the importance of questions. His father became blind after open heart surgery. Was that a result of the surgery? No, rather, it was because the physicians failed to ask the basic open question "Is anything else happening physically or emotionally in your life at this time?" If that question had been asked, the physicians would have discovered that Allen's father had developed severe and unusual headaches the week before surgery was scheduled, and they could have diagnosed an eye infection that is easily treatable with medication.

In counseling, a client may speak of tension, anxiety, and sleeplessness. You listen carefully and believe the problem can be resolved by helping the client relax and plan changes in her work schedule. However, you ask the client, "What else is going on in your life?" Having developed trust in you because of your careful listening and interest, the client finally opens up and shares a story of sexual harassment. At this point, the goals of the session change.

Useful questions from the helper than can provide more complete data include the following:

What else is going on in your life?

Looking back at what we've been talking about, what else might be added? As you think about this session, what might we have missed? You may even have thought about something and not said it.

Could you tell me a bit about whatever occurs to you at this moment? (This question often provides surprising and helpful new information.)

What else might a friend or family member add to what you've said? From a _____ perspective (insert ethnicity, race, sexual preference, religious, or other dimension), how could your situation be viewed? (These questions change the focus and help clients see their issues in a broader, network-based context of friends, family, and culture.)

Have we missed anything?

EXAMPLE INTERVIEW: CONFLICT AT WORK

Virtually all of us have experienced conflict on the job. Angry, difficult customers, insensitive supervisors, lazy colleagues, or challenges from those whom we may supervise give us concern. In the following set of transcripts, we see an employee assistance counselor, Jamila, meeting with Kelly, a junior manager who has a conflict with Peter.

Closed-Question Example

The first session illustrates how closed questions can bring out specific facts but can sometimes end in leading the client, even to the point of putting the counselor's ideas into the client's mind.

Interviewer and Client Conversation	Process Comments
1. *Jamila:* Hi, Kelly. What's happening with you today?	Jamila has talked with Kelly once in the past about difficulties she has had in her early experiences supervising others for the first time. She begins the session with an open question that could also be seen as a standard social greeting.
2. *Kelly:* Well, I'm having problems with Peter again.	Jamila and Kelly have a good relationship. Not all clients are so ready to discuss their issues. More time for developing rapport and trust will be necessary for many clients, even on return visits.
3. *Jamila:* Is he arguing with you?	Jamila appears interested, is listening and demonstrating good attending skills. However, she asks a closed question; she is already defining the issue without discovering Kelly's thoughts and feelings.
4. *Kelly:* (hesitates) Not really; he's so difficult to work with.	Kelly sits back in her chair and waits for the interviewer to take the lead.
5. *Jamila:* Is he getting his work in on time?	See Jamila try to diagnose the problem with Peter by asking a series of closed questions. It is much too early in the interview for a diagnosis.
6. *Kelly:* No, that's not the issue. He's even early.	
7. *Jamila:* Is his work decent? Does he do a good job?	Jamila is starting to grill Kelly.
8. *Kelly:* That's one of the problems; his work is excellent and always there on time. I can't criticize what he does.	
9. *Jamila:* (hesitates) Is he getting along with others on your team?	Jamila frowns and her body tenses as she thinks of what to ask next. Interviewers who rely on closed questions suddenly find themselves having run out of questions to ask. They continue searching for another closed question usually further off the mark.
10. *Kelly:* Well, he likes to go off with Daniel, and they laugh in the corner. It makes me nervous. He ignores the rest of the staff—it isn't just me.	
11. *Jamila:* So, it's you we need to work on. Is that right?	Jamila has been searching for an individual to blame. Jamila relaxes a little as she thinks she is on to something. Kelly sits back in discouragement.
12. *Kelly:* (hesitates and stammers) . . . Well, I suppose so . . . I . . . I . . . really hope you can help me work it out.	Kelly looks to Jamila as the expert. While she dislikes taking blame for the situation, she is also anxious to please and too readily accepts the interviewer's diagnosis.

Closed questions can overwhelm clients and can be used to force them to agree with the interviewer's ideas. While the session above seems extreme, encounters like this are common in daily life and even occur in interviewing and counseling sessions. There is a power differential between clients and counselors. It is possible that an interviewer who fails to listen can impose inappropriate decisions on a client.

Open-Question Example

The interview is for the client, not the interviewer. Using open questions, Jamila learns Kelly's story rather than the one she imposed with closed questions in the first example. Again, this interview is in the employee assistance office.

Interviewer and Client Conversation	Process Comments
1. *Jamila:* Hi, Kelly. What's happening with you today?	Jamila uses the same easy beginning as in the closed question example. She has excellent attending skills and is good at relationship building.
2. *Kelly:* Well, I'm having problems with Peter again.	Kelly responds in the same way as in the first demonstration.
3. *Jamila:* More problems? Could you share more with me about what's been happening lately?	Open questions beginning with "could" provide some control to the client. Potentially a "could" question may be responded to as a closed question and answered with "yes" or "no." But in the United States, Canada, and other English-speaking countries, it usually functions as an open question.
4. *Kelly:* This last week Peter has been going off in the corner with Daniel, and the two of them start laughing. He's ignoring most of our staff, and he's been getting under my skin even more lately. In the middle of all this, his work is fine, on time and near perfect. But he is so impossible to deal with.	We are hearing Kelly's story. The predicted result from open questions is that Kelly will respond with information. She provides an overview of the situation and shares how it is affecting her.
5. *Jamila:* I hear you. Peter is getting even more difficult and seems to be affecting your team as well. It's really stressing you out, and you look upset. Is that pretty much how you are feeling about things?	When clients provide lots of information, we need to ensure that we hear them accurately. Jamila summarizes what has been said and acknowledges Kelly's emotions. The closed question at the end is termed a perception check or check-out. Periodically checking with your client can help you in two important ways: (a) It communicates to clients that you are listening and encourages them to continue; (b) it allows the client to correct any wrong assumptions you may have.
6. *Kelly:* That's right. I really need to calm down.	

(continued)

Interviewer and Client Conversation	Process Comments
7. *Jamila:* Let's change the pace a bit. Could you give me a specific example of an exchange you had with Peter last week that didn't work well?	Jamila asks for a concrete example. Specific illustrations of client issues are often helpful in understanding what is really occurring.
8. *Kelly:* Last week, I asked him to review a book-keeping report prepared by Anne. It's pretty important that our team understand what's going on. He looked at me like, "Who are you to tell *me* what to do?" But he sat down and did it that day. Friday, at the staff meeting, I asked him to summarize the report for everyone. In front of the whole group, he said he had to review this report for me and joked about me not understanding numbers. Daniel laughed, but the rest of the staff just sat there. He even put Anne down and presented her report as not very interesting and poorly written. He was obviously trying to get me. I just ignored it. But that's typical of what he does.	Specific and concrete examples can be representative of recurring problems. The concrete specifics from one or two detailed stories can lead to a better understanding of what is really happening. Now that Jamila has heard the specifics, she is better prepared to be helpful.
9. *Jamila:* Underneath it all, you're furious. Kelly, why do you imagine he is doing that to you?	Will the "why" question lead to the discovery of reasons?
10. *Kelly:* (hesitates) Really, I don't know why. I've tried to be helpful to him.	The intentional prediction did not result in the expected response. This is, of course, not unusual. It is likely too soon for Kelly to know why. This illustrates a common problem with "why" questions.
11. *Jamila:* Gender can be an issue; men do put women down at times. Would you be willing to consider that possibility?	Jamila carefully presents her own hunch. But instead of expressing her own ideas as truth, she offers them tentatively with a "would" question and reframes the situation as "possibility."
12. *Kelly:* Jamila, it makes sense. I've halfway thought of it, but I didn't really want to acknowledge the possibility. But it is clear that Peter has taken Daniel away from the team. Until Peter came aboard, we worked together beautifully. (pause) Yes, it makes sense for me. I think he's out to take care of himself. I see Peter going up to my supervisor all the time. He talks to the female staff members in a demeaning way. Somehow, I'd like to keep his great talent on the team, but how when he is so difficult?	With Jamila's help, Kelly is beginning to obtain a broader perspective. She thinks of several situations indicating that Peter's ambition and sexist behavior are issues that need to be addressed.

(continued)

Interviewer and Client Conversation	Process Comments
13. *Jamila:* So, the problem is becoming clearer. You want a working team, and you want Peter to be part of it. We can explore the possibility of assertiveness training as a way to deal with Peter. But, before that, what do you bring to this situation that will help you deal with him?	Jamila provides support for Kelly's new frame of reference and ideas for where the interview can go next. She suggests that time needs to be spent on finding positive assets and wellness strengths. Kelly can best resolve these issues if she works from a base of resources and capabilities.
14. *Kelly:* First, I need to remind myself that I really do know more about our work than Peter. He is new to it. I worked through a similar issue with Jonathan two years ago. He kept hassling me until I had it out with him. He was fine after that. I know my team respects me; they come to me for advice all the time.	Kelly smiles for the first time. She has sufficient support from Jamila to readily come up with her strengths. However, don't expect it always to be that easy. Clients may return to their weaknesses and ignore their assets.
15. *Jamila:* Could you tell me specifically what happened when you sat down and faced Jon's challenge directly?	This "could" question searches for concrete specifics when Kelly handled a difficult situation effectively. Jamila can identify specific skills that Kelly can later apply to Peter. At this point the interview can move from problem definition to problem solution.

In this excerpt, we see that Kelly has been given more talk-time and room to explore what is happening. The questions focused on specific examples clarify what is happening. We also see that question stems such as *why, how,* and *could* have some predictability in expected client responses. The positive asset search is a particularly important part of successful questioning. Issues are best resolved by emphasizing strengths.

You are very likely to work with clients who have similar interpersonal issues wherever you may practice. The previous case examples focus on the single skill of questioning as a way to bring out client stories. Questioning is an extremely helpful skill, but do not forget the dangers of using too many questions,

INSTRUCTIONAL READING: MAKING QUESTIONS WORK FOR YOU

> *Questions make the interview work for me. I searched through many questions and found the ones that I thought most helpful in my own practice. I then memorized them and now I always draw on them as needed. Being prepared makes a difference.*
>
> —Norma Gluckstern-Packard

Following are several issues around the use of questioning techniques and strategies. Questions can be facilitative or they can be so intrusive that clients want to close up and say nothing. Use the ideas presented here to help you define your own position around questioning, facilitate your memory of key questions, and learn how questions fit with your natural style.

Eight Major Issues Around Questions

1. Questions Help Begin the Interview

With verbal clients and a comfortable relationship, the open question facilitates free discussion and leaves plenty of room to talk. Here are some examples:

What would you like to talk about today?

Could you tell me what you'd like to see me about?

How have things been since we last talked together?

The last time we met we talked about your plan to face up and talk with your partner about not having enjoyed sex for nearly two months. How did it go this week?

The first three open questions provide considerable room, in that the client can talk about virtually anything. The last question is open but provides some focus for the session, building on material from the preceding week.

However, such open questions may be more than a nontalkative client can handle. It may be best in such situations to start the session with more informal conversation—for example, focusing on the weather, a positive part of last week's session, or a current event that you know is of interest to the client. As the client becomes comfortable, you can then turn to the issues for this session.

2. Open Questions Help Elaborate and Enrich the Client's Story

A beginning interviewer often asks one or two questions and then wonders what to do next. Even more experienced interviewers at times find themselves hard-pressed to know what to do. An open question on some topic the client presented earlier in the interview helps the session start again and keep it moving:

Could you tell me more about that?

How did you feel when that happened?

Given what you've said, what would be your ideal solution?

What might we have missed so far?

What else comes to your mind?

3. Questions Help Bring Out Concrete Specifics of the Client's World

If there is one single open question that appears to be useful in practice sessions and in most theoretical persuasions, it is the one that aims for concreteness and specifics in the client's situation. The model question "Could you give me a specific example?" is the most useful open question available to any interviewer. Many clients tend to talk in vague generalities, and specific, concrete examples enrich the interview and provide data for understanding action. Some additional open questions that aim for concreteness and specifics follow:

Client: Ricardo makes me so mad!

Counselor: Could you give me a specific example of what Ricardo does?

What does Ricardo do specifically that brings out your anger?

What do you mean by "makes me mad"?

Could you specify what you do before and after Ricardo makes you mad?

Closed questions, of course, can bring out specifics as well, but they place more responsibility on the interviewer. However, if the interviewer knows specifically the desired direction of the interview, closed questions such as "Did Ricardo show his anger by striking you?" "Does Ricardo tease you often?" and "Is Ricardo on drugs?" may prove invaluable, because they may encourage clients to say out loud what before they were only hinting at. But even well-directed closed questions may take the initiative away from the client.

4. Questions Are Critical in Assessment

Physicians must diagnose their clients' physical symptoms. Managers may have to assess a problem on the production line. Vocational counselors and coaches need to assess a client's career history. Questions are the meat of effective diagnosis and assessment. George Kelly, the personality theorist, has suggested for general problem diagnosis the following set of questions, which roughly follow the who, what, when, where, how, why of newspaper reporters:

Who is the client? What is the client's personal background? Who else may be involved?
What is the client's concern? What is happening? What are the specific details of the situation?
When does the issue occur? When did it begin? What immediately preceded the occurrence of the issue?
Where does the problem occur? In what environments and situations? With what people?
How does the client react to the challenge? How does the client feel about it?
Why does the problem occur? Why is the client concerned about it?

Needless to say, the who, what, when, where, how, why series of questions also provides the interviewer with a ready system for helping the client elaborate or be more specific about an issue at any time during a session. To these, again, we suggest adding the *what else* question to encourage openness.

5. The First Word of Certain Open Questions Partially Determines What the Client Will Say Next

Often, but not always, using key-question stems results in predictable outcomes.

What questions most often lead to facts. "What happened?" "What are you going to do?"
How questions often lead to a discussion about processes or sequences or to feelings. "How could that be explained?" "How do you feel about that?"
Why questions most often lead to a discussion of reasons. "Why did you allow that to happen?" "Why do you think that is so?" Finding reasons can be helpful but also can lead into sidetracks. Remember: Many clients associate *why* with a past experience of being grilled. But *why* is also often esssential in meaning, vision, and determining one's life direction. A softer approach to *why* is "I hear your dream and vision. Could you tell me *why* that dream is so important to you?"
Could, would, and *can* questions are considered maximally open and contain some of the advantages of closed questions in that the client is free to say "No, I don't want to talk about that." *Could* questions reflect less control and command than others. "Could you tell me more about your situation?" "Would you give me a specific example?" "Can you tell me what you'd like to talk about today?" As in the example above, the use of *could* softens the *why* and gives the client more room to respond.

6. Questions Have Potential Problems

Questions can have immense value in the interview, but we must not forget their potential problems. Among them are the following.

Bombardment/grilling. Too many questions will tend to put many clients on the defensive. They may also give too much control to the interviewer.

Multiple questions. Interviewers may confuse their clients by throwing out several questions at once. This is another form of bombardment, although at times it may be helpful to some clients as the client can select which question to answer.

Questions as statements. Some interviewers may use questions as a way to sell their own points of view. "Don't you think it would be helpful if you studied more?" "What do you think of trying relaxation exercises instead of what you are doing now?" This form of question, just like multiple questions, can be helpful at times. Awareness of the nature of such questions, however, may allow you to consider alternative and more direct routes of reaching the client. A useful rule of thumb is that if you are going to make a statement, it is best not to frame it as a question.

Why questions. Unless asked in a safe and comfortable relationship, *why* questions can put interviewees on the defensive and cause discomfort. As children, most of us experienced some form of the "Why did you do that?" question. This same discomfort can be produced by any question that evokes a sense of being attacked. Poorly phrased *why* questions also are subtractive. For example, how do you personally respond to questions such as (a) "Why did you do that?" (b) "Why do you talk so fast?" (c) "Why were you rude to my friend?" (d) "Why do you get so anxious?" These and questions like them make us defensive and put us on the spot. These questions lack empathy—they show little respect for the client or anyone else. They are clearly not warm; they show no sign of positive regard.

But knowing reasons behind client actions remains important. Here are some possibilities beyond why. (a) "Could we explore the background of what happened just before you did that?" (b) "People say you talk too fast. Let's explore what's going on around those situations. Is that okay?" (c) "We say things that surprise ourselves and others. What was going on that pushed your buttons?" (d) "How does your body feel when you get anxious?" All these explore reasons behind behavior without producing client defensiveness.

7. In Cross-Cultural Situations, Questions Can Promote Distrust

If your life background and experience are in relative synchrony with those of the client, you may find that you can use questions immediately and freely. On the other hand, if you come from a different cultural background, your questions may be only grudgingly answered. Questions place the power in the interviewer. The rapid-fire North American questioning style is often received less favorably in other societies. In addition, research has revealed that men tend to ask more questions than women. Some interpret this as an indication that men are using questions to control the conversation. A barrage of questions to a poor client if you, the interviewer, are clearly middle class may cause the client not to come back for any more interviews. If you are African American or White and working with an Asian American or a Latino/a, an extreme questioning style can produce mistrust. If the ethnicities are reversed, the same problem could occur. See additional examples in Box 4-2.

BOX 4-2 NATIONAL AND INTERNATIONAL PERSPECTIVES ON COUNSELING SKILLS

Using Questions With Youth at Risk
Courtland Lee, Past President, American Counseling Association, University of Maryland

Malik is a 13-year-old African American male who is in the seventh grade at an urban junior high school. He lives in an apartment complex in a lower middle-class (working-class) neighborhood with his mother and 7-year-old sister. Malik's parents have been divorced since he was 6 and he sees his father very infrequently. His mother works two jobs to hold the family together, and she is not able to be there when he and his sister come home from school.

Throughout his elementary school years, Malik was an honor roll student. However, since starting junior high school, his grades have dropped dramatically, and he expresses no interest in doing well academically. He spends his days at school in the company of a group of seventh- and eighth-grade boys who are frequently in trouble with school officials.

This case is one that is repeated among many African American early teens. But this problem also occurs among other racial/ethnic groups as well, particularly those who are struggling economically. And the same pattern occurs frequently even in well-off homes. There are many teens at risk for getting in trouble or using drugs.

While still a boy, Malik has been asked to shoulder a man's responsibilities as he must pick up things his mother can't do. Simultaneously, his peer group discounts the importance of academic success and wants to challenge traditional authority. And Malik is making the difficult transition from childhood to manhood without a positive male model.

I've developed a counseling program designed to empower adolescent Black males that focuses on personal and cultural pride. The full program is outlined in my book *Empowering Black Males* (1992) and focuses on the central question, "What is a strong Black man?" While this question is designed for group

discussion, it is an important one for adolescent males in general, who might be engaging in individual counseling. The idea is to use this question to help the youth redefine in a more positive sense what it means to be strong and powerful. Some of the related questions that I find helpful include these:

- ▲ What makes a man strong?
- ▲ Who are some strong Black men that you know personally? What makes these men strong?
- ▲ Do you think that you are strong? Why?
- ▲ What makes a strong body?
- ▲ Is abuse of your body a sign of strength?
- ▲ Who are some African heroes or elders that are important to you? What did they do that made them strong?
- ▲ How is education strength?
- ▲ What is a strong Black man?
- ▲ What does a strong Black man do that makes a difference for his people?
- ▲ What can you do to make a difference?

Needless to say, you can't ask an African American adolescent or a youth of any color these questions unless you and he are in a positive and open relationship. Developing sufficient trust so that you can ask these challenging questions may take time. You may have to get out of your office and into the school and community to become a person of trust.

My hope for you as a professional counselor is that you will have a positive attitude when you encounter challenging adolescents. They are seeking models for a successful life, and you may become one of those models yourself. I hope you think about establishing group programs to facilitate development and that you'll use some of these ideas here with adolescents to help move them toward a more positive track.

Allen was conducting research and teaching in South Australia with Aboriginal social workers. He was seeking to understand their culture and their special needs for training. Allen is naturally inquisitive and sometimes asks many questions. Nonetheless, the relationship between him and the group seemed to be going well. But one day, Matt Rigney, whom Allen felt particularly close to, took him aside and gave some very useful corrective feedback:

You White fellas! . . . Always asking questions! Let me tell you what goes on in my mind when a White person asks me a question. First, my culture considers many questions rude. But I know you, and that's what you do. But this is what goes on in my mind when you ask me a question. First, I wonder if I can trust you enough to give you an honest answer. Then, I realize that the question you asked is too complex to be answered in a few words. But I know you want an answer. So I chew on the question in my mind. Then, you know what? Before I can answer the first question, you've moved on to the next question!

Allen was lucky that he had developed enough trust and rapport that Matt was willing to share his perceptions. Many people of color have said that the Australian Aboriginal feedback represents how they felt about many interactions with White people. Moreover, disabled individuals, gays/lesbians/bisexuals, spiritually conservative persons, and many others may be distrustful of the interviewer who uses questions too much if they are from a different group. Finally, anyone who is quizzed too much may feel the same way!

8. Questions Can Be Used to Help Clients Search for Positive Assets, Strengths, and Patterns of Wellness

Stories presented in the helping interview are often negative and full of problems and difficulties. People grow from strength, not from weakness. Carl Rogers, the founder of client-centered therapy, was always able to find something positive in the interview. He considered positive regard and respect for the client essential for future growth. Strength-based questions are a foundation of effective questions.

The positive asset search or a wellness review are concrete ways to approach positive regard and respect for the client. As you listen to the client, constantly search for strengths and positives. Then, share your observations. Of course, you do not want to become overly optimistic and minimize the seriousness of the client's situation. However, it is increasingly clear that if you listen only to the sad and negative parts of the client's story, progress and change will be slow and painful (Peterson & Seligman, 2004; White & Epston, 1990).

As a general rule, we suggest that you use the basic listening sequence to draw out the story in relatively brief form. Then repeat and summarize the story to ensure that you have heard it accurately. We also recommend that you obtain at least one concrete and specific example of the story to ensure mutual understanding.

As appropriate to the client and situation, begin your search for positive assets and strengths. If you develop with your client a list of strengths and assets, you will find you can draw on them later for resolution of concerns and problems. Naturally, do not push strengths against client wishes, as this may appear to minimize her or his concerns. However, seek to make a positive approach part of the interview and later treatment plan.

Some specific, concrete examples of how to engage in a positive asset search include the following.

Personal strength inventory. Clients tend to talk about their problems and what they can't do. This puts them "off-balance." We can help them center and feel better about themselves through a strength inventory. What is the client doing right?

> As part of any interview, I like to do a strength inventory. Let's spend some time right now identifying some of the positive experiences and strengths that you either have now or have had in the past.

▲ Could you tell me a story about a success you have had sometime in the past? I'd like to hear the concrete details.
▲ Tell me about a time in the past when someone supported you and what he or she did. What are your currently available support systems?
▲ What are some things you have been proud of in the past? Now?
▲ What do you do well or others say you do well?

Cultural/gender/family strength inventory. Here we move outside the individual and look at context for positive strengths.

▲ Taking your ethnic/racial/spiritual history, can you identify some positive strengths, visual images, and experiences that you have now or have had in the past?
▲ Can you recall a friend or family member of your own gender who represents some type of hero in the way he or she dealt with adversity? What did that person do? Can you develop an image of her or him?
▲ We all have family strengths despite frequent family concerns. Family can include our extended family, our stepfamilies, and even those who have been special to us over time. For example, some people talk about a special teacher, a school custodian, or an older person who was helpful. Could you tell me concretely about them and what they mean to you?

Positive exceptions to the concern. Searching for times when the problem doesn't occur is often useful. This approach is common in brief counseling. With this information, you can determine what is being done right and encourage more of the same.

> Let's focus on the exceptions. When is the problem or concern absent or a little less difficult? Please give me an example of one of those times.

▲ Few problems happen all the time. Could you tell me about a time when it didn't happen? That may give us an idea for a solution.
▲ What is different about this example from the usual?
▲ How did the more positive result occur?
▲ How is that different from the way you usually handle the concern?

Many clients will be hesitant to say good things about themselves. Your observations and feedback can be helpful to them in developing a new view of themselves. The what else question provides an opportunity for the client to add more strengths and resources to your feedback.

You obviously will not have time for all these possible strengths and positive asset searches. But when you focus only on the negative story, you place your clients in a very vulnerable position. Do not use the positive asset search to cover up or hide basic issues. Rather, wellness strengths are resources for resolving our concerns.

Using Open and Closed Questions With Less Verbal Clients

Generally, open questions are much preferred to closed questions in the interview. Yet it must be recognized that open questions require a verbal client, one who is willing to share information, thoughts, and feelings with you. Here are some suggestions that may encourage clients to talk more freely with you.

Build trust at the client's pace. A central issue with hesitant clients is trust. If the client is required to meet with you or is culturally different from you, he or she may be less willing to talk. At this time, your own natural openness and social skills are particularly important. Trust building and rapport need to come first. With some clients, trust building may take a full session or more. Extensive questioning too early can make trust building a slow process with some clients. Often it is helpful to discuss multicultural differences openly. "I'm wondering how you feel about my being (White/male/heterosexual or vice versa) as we discuss these issues."

Accept some randomness. Your less verbal clients may not give you a clear, linear story of the problem. If they lack trust or are highly emotional, it may take some time for you to get an accurate understanding. A careful balance of closed and open questions to draw out the story and get "bits and pieces" will help you put together a coherent narrative. Keep your words as simple, straightforward, and as concrete as possible. With some clients you will find that briefly disclosing your own stories is helpful.

Search for concrete specifics. Counselors and therapists talk about the abstraction ladder. If you or the client moves too high on the abstraction ladder, things won't make sense to anyone. This is especially so with less verbal or emotionally distraught clients. After some trust is generated, you might begin by making an observation such as "The teacher said you and she had an argument." Then try asking a concrete open question such as "What did your teacher say (or do)?" "What did you say (or do)?" If you focus on concrete events and avoid evaluation and opinion in a nonjudgmental fashion, your chances for helping the client talk will be greatly expanded. Examples of concrete questions focusing on narrow specifics include the following:

(A) What happened first? (B) What happened next? (C) What happened after it was over? (This brings out the linear sequence of the story.)

What did the other person say? What did he or she do? What did you say or do? (This focuses on observable concrete actions.)

What did you feel or think just before it happened? During? After? What do you think the other person felt? (This helps focus emotions.)

Note that each of the preceding questions requires a relatively short answer. These are open questions that are more focused in orientation and can be balanced with some closed questions. Do not expect your less verbal client to give you full answers to these questions. You may need to ask closed questions to fill in the details and obtain specific information. ("Did he say anything?" "Where was she?" "Is your family angry?" "Did they say 'yes' or 'no'?")

A leading closed question is dangerous, particularly with children. In the previous examples, you can see that a long series of closed questions can bring out the story, but it may provide only the client's limited responses to your questions rather than what the client really thought or felt. Worse, the client may end up adopting your way of thinking or may simply stop coming to see you.

Questioning and other listening skills. One effective counseling method is to repeat the client's main words by paraphrasing or reflecting feeling (see Chapters 6 and 7) and then to raise the intonation of your voice at the end in a questioning tone. For example:

Client: I was really upset by my parents. They entered my room when I was gone and searched the whole place. They suspect me of taking drugs.

Some open questions might include "Could you tell me what led your parents to the search?" or "What feelings does this bring out in you?" Another way to help the client keep talking is to repeat what you have heard. A paraphrase might be "Your parents entered your room and suspect you of taking drugs" while a reflection of feeling could be "You sound really angry." These are not questions, but the tone of voice offers the same openness and may enable your client to talk more deeply. This is one of the reasons some humanistic counselors object to questions. Effective rephrasing can often accomplish the same objective.

Working with children. Children, in particular, may require considerable help from the interviewer before they are willing to share at all. With children, a naturally warm, talkative person who likes and accepts children will be able to elicit more information more easily. Thus, it helps to begin sessions with children by sharing something fun and interesting. Games, clay, and toys in the counseling room are useful when dealing with children. You will find that children generally like to do something with their hands while they talk; having a child draw a picture during the conversation can often be useful to the child and to the interviewer. And the drawings often reveal what is going on in the child's life. Many open questions are too broad for young children to understand, but be careful with closed questions and avoid leading too much.

COACHING AND POWERFUL COACHING QUESTIONS*

Professional coaching is an ongoing professional relationship that helps people produce extraordinary results in their lives, careers, businesses, or organizations. Through the process of coaching, clients deepen their learning, improve their performance, and enhance the quality of their lives.

—International Coach Federation, The ICF Code of Ethics

Many of you will want to add coaching methods to your interviewing, counseling, or psychotherapy practice. With further study, some of you will want to apply the skills of this book to your own coaching practice. Coaching is a form of interviewing; it focuses almost solely on a client's strengths and how to use these positives to reach highly specific and doable goals. The word "problem" is seldom used and coaching authorities recommend substituting "challenges" and "goal attainment."

Ethics first. Coaching is not "counseling light." It is not exploitation of clients. It does not include self-declared competence as a coach. Coaches do not operate below the "professional radar." Coaching has certification of competence and standards, but their use is not universal. This raises the possibility that charlatans and dishonest practitioners will call themselves coaches. Also, it means that you should not declare yourself a coach without considerable further study and supervision.

---

* © 2009 by Allen E. Ivey and released to Cengage Learning for this publication. Information and permissions may be obtained from the author. The 13 "Powerful Coaching Questions" are not copyrighted and are credited to Margaret Moore.

The code of professional ethics for coaches may be found at www.coachfederation.org. The five areas of ethics statements are presented in unusually clear and accessible form. As one example of clarity, this code uses "conflict of interest," which may be a more useful term than "dual relationship," the words most often found in ethics statements of counseling associations. Following are examples from each of the areas.

Professional Conduct at Large

2. I will not knowingly make any public statements that are untrue or misleading, or make false claims in any written documents relating to the coaching profession.

5. I will at all times strive to recognize personal issues that may impair, conflict or interfere with my coaching performance or my professional relationships. Whenever the facts and circumstances necessitate, I will promptly seek professional assistance and determine the action to be taken, including whether it is appropriate to suspend or terminate my coaching relationship(s).

Professional Conduct With Clients

10. I will be responsible for setting clear, appropriate, and culturally sensitive boundaries that govern any physical contact that I may have with my clients.

12. I will construct clear agreements with my clients, and will honor all agreements made in the context of professional coaching relationships.

Confidentiality/Privacy

22. I will respect the confidentiality of my client's information, except as otherwise authorized by my client, or as required by law.

Conflicts of Interest

26. Whenever any actual conflict of interest or the potential for a conflict of interest arises, I will openly disclose it and fully discuss with my client how to deal with it in whatever way best serves my client.

GROW   The Basic Coaching Model

Four elements in skilled coaching have been identified as basic: Goal, Reality, Options, Way forward (GROW) (Kauffman, 2008). Coaching uses a different language, but you will find many similarities between the GROW model and the *relationship–story and strengths–goals–restory action* model. The initial emphasis on goals is characteristic of brief solution-oriented counseling and motivational interviewing (both presented in Chapter 14).

GROW	(Relationship is central throughout and assumed.)	G = goal. What positive goal does the client seek?	R = reality. What is the client's present situation and strengths on which goal attainment can be reached?	O = options. How can the client and coach generate an array of possible ways to reach the goal?	W = way forward and will. How do we act on goals, measure steps toward success, and maintain change over time?
MICROSKILLS	Relationship	Goal	Story and Strengths	Restory	Action

Coaching has almost as many theories as counseling and psychotherapy. You'll find person-centered coaching, narrative and storytelling coaching, cognitive-behavioral coaching, and many others. But the positive philosophy and strength-based format will be found in all coaching theories. The focus on goals and strengths with an accompanying desire to avoid looking to the past or present for problems explains why coaching is really an interviewing form of help rather than counseling or therapy.

Powerful Coaching Questions

Nine central and powerful questions used in coaching have been identified by Margaret Moore, director and cofounder of the Coaching and Positive Psychology Insitute at McLean Hospital. She is also a faculty member in the Department of Psychiatry at Harvard Medical School. She kindly gave us permission to list these uncopyrighted questions in this book. Effective coaching assumes a positive working relationship between client and coach. Without that, coaching will almost certainly fail.

To be successful, a coach must be an effective listener, one who uses attending behavior and the important skills of clarifying and summarizing what the client has said. This tells clients that you have heard them. These skills include observation, encouraging, paraphrasing, acknowledging feelings, and the summary, and they are detailed in Chapters 5 through 7. While we emphasize questions in this chapter, we hope that you will use them minimally when you practice interviewing using only the listening skills presented in Chapter 8. Below are the central questions in the GROW model:

G = Goals

1. **What is the ideal person you want to be? Your best self?** This is coupled with such questions as "What's going on your life right now?" "How do you imagine your ideal life?" "What matters most to you?" and "What do you really want?" You will note the person-centered influence of the real and ideal self here, plus some influence from decisional counseling and meaning issues (Chapters 8, 12, 13).

2. **What is the gap between the now and your vision?** "How does your vision of the future differ from the now?" "How does your real self differ from your ideal self?" In the second question, we see the early confrontation of the discrepancy between the expressed or implied goals in question 1 and the client's present position. This is particularly characteristic of brief solution-focused counseling and motivational interviewing (Chapter 13).

3. **Why does this vision really matter to you?** "How does this goal make a difference in your life?" "Could we get more specific as to how the vision or goal is defined?" This is clearly a meaning issue. As we say in Chapter 12, achieving meaning and one's vision in life is often the most important issue we face in counseling. Note how central it is to coaching.

R = Reality (Story and Strengths)

4. **What strengths can you use to help you get there?** This should sound very familiar to you. Use the questions we provide in this chapter and the many examples throughout the book. Coaching has a very strength-based and nonpathological orientation.

5. **What is the key challenge?** "What's getting in your way?" The language of challenge and possibility is used rather than the problem-focused approach so often used in our field.

O = Options (Restory)

6. **What workable strategies can you apply?** "Which strengths, resources, and positive assets can we employ to reach your goal?" Rather than draw on external theories as we often do in counseling, the coach seeks to draw out from clients things that might work and show them how their own existing capacities can be used to reach their goals. The Latin term *educare* describes this process. *Educare* is the root word of education, but the real translation is to draw out answers that already exist in the person or client.

7. **How confident are you that you can reach this vision?** "How do you feel and think about yourself now as we look at this goal?" "What's going on in your life right now?" Here, we move to the area of emotions. And we want to use the executive right brain to focus on positive emotions. Coaching is not a negatively focused approach. We can best eliminate the negative from a strength-based positive psychology and wellness approach.

W = Way Forward and Will (Action)

8. **Are you ready and committed?** "Will you do it?" "How committed are you to change and action?" "On a scale of 1 to 10, how committed are you to actually doing this?" This examines clients in the here and now and their level of motivation for actually reaching their goals.

9. **Will you do it tomorrow?** "Can we write a contract for action?" "Let's select something small enough that you actually want to and feel confident that you will do it." Coaching believes that major change will not happen immediately, but we want to start the process as soon as possible. If the client has a challenging goal, break it down into small, manageable steps.

There is no apology for the many questions in coaching, and perhaps it is the emphasis on client goals and strengths that makes this possible. Those with mental health issues or more complex individual issues are referred as soon as possible.

Research

As a new field, coaching is seeking to expand a relatively small research base. As it has many commonalities with other helping fields, one might suggest that coaching researchers draw on existing research in related fields. For example, the research-related theories such as motivational interviewing and brief solution-oriented counseling support much of what we read here. Many databased studies from microskills will support the skills basis of coaching, including attending and listening.

Each helping system needs its own research for support, justification, and improvement of service delivery. Evidence-based coaching research has found that a solution-focused coaching program has lowered depression and anxiety, reduced stress, and improved the general quality of life for clients (Grant, 2003). A second study produced similar outcomes (Spence & Grant, 2005), and a third shows that the effects of coaching are maintained over 30 weeks (Green, Oates, & Grant, 2005). The Australian group (Grant, 2008) examined mindfulness (see Chapter 12) and coaching. They concluded that short "mindfulness training *before* coaching seems to build 'psychological muscle'" that led to better results in effecting change and reaching goals.

Coaching Summary: Some Challenges

Despite a promising beginning, coaching has given little consideration to multicultural issues, working with social justice, or those who are poor. Many people who would refuse counseling are likely to accept the safer word "coaching" with its positive approach to change and growth. Coaching as an adjunct to interviewing, counseling, and psychotherapy needs serious attention. As clients improve through their counseling and therapy sessions, working with them on a coaching approach likely will be useful in helping them take action and maintain change over time.

Clearly, more research is needed on what makes an effective and ethical coach. At present, anyone can use the title of coach and market himself or herself at a high figure; this is disturbing. Even so, there are many ethical coaches, and groups such as the International Coach Federation (ICF) and the Harvard University and McLean Hospital Coaching and Positive Psychology Initiative aim for certification and competence in the field. The ICF code of ethics is stated in exceptionally clear language (the major helping professions would benefit if they reworked their own ethical standards to make them comparably short and clear). At present, only a few degree programs in coaching are available, although many universities and professional schools are beginning to offer instruction in coaching and coaching skills. Most notable is Coaching Psychology at the University of Sydney, Australia.

The future of coaching seems solid. The popular media give considerable attention to college coaches, executive coaches, personal coaches, retirement coaches, wellness coaches, and many others. It has been said that one cannot make it through today's complex and confusing world without a good coach. The coaching model would likely make for improved services in the elementary and secondary schools due to its positive emphasis and solution-focused orientation. We hope that you will examine this new field in more detail, but with full awareness that you are not a trained coach, and ethically you cannot call yourself a coach. Use some coaching concepts and questions in your interviewing, counseling, and therapy practice, but continue the focus on strengths and positives.

SUMMARY: MAKING YOUR DECISION ABOUT QUESTIONS

We began this chapter by asking you to think carefully about your personal experience with questions. It is clear that their overuse can damage an interview. On the other hand, questions do facilitate conversation and help ensure that important points are brought in. Questions can help the client bring in missing information. Among such questions are "What else?" "What have we missed so far?" and "Can you think of something important that is occurring in your life right now that you haven't shared with me yet?"

Person-centered theorists and many professionals sincerely argue against the use of any questions at all. They strongly object to the control implications of questions. They point out that careful attending and use of the listening skills can usually bring out major client issues. If you work with someone culturally different from you, a questioning style may develop distrust. In such cases, questions need to be balanced with self-disclosure and listening. Coaching challenges those who avoid questions, but there is a person-centered style of coaching that seeks to reduce emphasis on questioning strategies.

Our position on questions is clear—we believe in questions, but we also fear overuse and the fact that they can reduce equality in the interview. We are impressed by the brief solution-focused counselors who seem to use questions more than any other skill but are still able to respect their clients and help them change. On the other hand, we have seen students who have demonstrated excellent attending skills regress to using only questions. Questions can be an easy "fix" but they require listening to the client if they are to be meaningful.

The positive asset search has been a foundation of the microskills program since its beginnings. We believe that Carl Rogers was correct when he focused on positive regard and unconditional acceptance. Ethical coaching echoes Rogers, but uses questions. We have noted again and again that therapy all too often ends in a self-defeating repetition of problems. Questions that bring out strengths and resources often lead clients to specific assets that they can use to help resolve issues and problems.

The most useful chapter summary will be your impressions and decisions. Where do you personally stand on the use of questions?

Key points of Chapter 4 are summarized below.

■ KEY POINTS

Value of questions	Questions help begin the interview, open new areas for discussion, assist in pinpointing and clarifying issues, and assist the client in self-exploration.
Open and closed questions	These questions can be described as open or closed. *Open questions* are those that can't be answered in a few short words. They encourage others to talk and provide you with maximum information. Typically, open questions begin with *what, how, why,* or *could*. One of the most helpful of all open questions is "Could you give a specific example of . . . ?" *Closed questions* are those that can be answered in a few words or sentences. They have the advantage of focusing the interview and bringing out specifics, but they place the prime responsibility for talk on the interviewer. Closed questions often begin with *is, are,* or *do*. An example is "Where do you live?" It is important to note that a question, open or closed, on a topic of deep interest to the client will often result in extensive talk-time if it is important enough. If an interview is flowing well, the distinction between open and closed questions is less important.
Newspaper questions for context	A general framework for diagnosis and question asking is provided by the newspaper reporter framework of *who, what, when, where, how, why.* ▲ *Who* is the client? What are key personal background factors? Who else is involved? ▲ *What* is the issue? What are the specific details of the situation? ▲ *When* does the problem occur? What immediately preceded and followed the situation? ▲ *Where* does the issue occur? In what environments and situations? ▲ *How* does the client react? How does he or she feel about it? ▲ *Why* does the problem or concern happen? ▲ *What else* is there to add to the story? Have we missed anything? Interviewing is about more than problems. The above set of questions could be asked to discover what events and issues surround a positive situation or accomplishment. Interview training often overemphasizes concerns and difficulties. A positive approach is needed for balance.

(continued)

KEY POINTS (continued)

Multicultural issues and questions	These questions may turn off some clients. Some cultural groups find North American rapid-fire questions rude and intrusive, particularly if asked before trust is developed. Yet questions are very much a part of Western culture and provide a way to obtain information that many clients find helpful. If questions are properly structured and your clients know their real purpose is to help them reach their own goals—as in coaching—questions may be used more productively.
Be positive	Emphasizing only negative issues results in a downward cycle of depression and discouragement. The positive asset search, strength emphasis, positive psychology, and wellness need to balance discussion of client issues and concerns. What is the client doing right? What are the exceptions to the problem? What are client personal, family, and cultural/contextual resources?

COMPETENCY PRACTICE EXERCISES AND PORTFOLIO OF COMPETENCE

Individual Practice

Exercise 1: Writing Closed and Open Questions

Select one or more of the following client stories and then write open and closed questions to elicit further information. Can you ask closed questions designed to bring out specifics of the situation? Can you use open questions to facilitate further elaboration of the topic including the facts, feelings, and possible reasons? What special considerations might be important with each person as you consider age-related multicultural issues?

Jordan (age 15, African American): I was walking down the hall and three guys came up to me and called me "queer" and pushed me against the wall. They started hitting me, but then a teacher came up.

Alicja (age 35, Polish American): I've been passed over for a promotion three times now. Each time, it's been a man who has been picked for the next level. I'm getting very angry and suspicious.

Dominique (age 78, French Canadian): I feel so bad. No one pays any attention to me in this "home." The food is terrible. Everyone is so rude. Sometimes I feel frightened.

Write open questions for one or more of the above. The questions should be designed to bring out broad information, facts, feelings and emotions, and reasons.

Could _____ ?

What _____ ?

How _____ ?

Why _____ ?

Now generate three closed questions that might bring out useful specifics of the situation.

Do _____ ?

Are _____ ?

Where _____ ?

Finally, write a question designed to obtain concrete examples and details that might make the problem more specific and understandable.

Exercise 2: Observation of Questions in Your Daily Interactions

This chapter has talked about the basic question stems *what, how, why,* and *could,* and how clients respond differently to each. During a conversation with a friend or acquaintance, try these five basic question stems sequentially:

Could you tell me generally what happened?
What are the critical facts?
How do you feel about the situation?
Why do you think it happened?
What else is important? What have we missed?

Record your observations here. Were the predictions of the book fulfilled? Did the person provide you, in order, with (a) a general picture of the situation, (b) the relevant facts, (c) personal feelings about the situation, and (d) background reasons that might be causing the situation?

Group Practice

Two systematic exercises are suggested for practice with questions. The first focuses on the use of open and closed questions, the second on the nine basic questions of the coaching GROW model. The instructional steps for practice are abbreviated from those described in Chapter 3, on attending behavior. As necessary, refer to those instructions for more detail on the steps for systematic practice.

Exercise 3: Systematic Group Practice on Open and Closed Questions

Step 1: Divide into practice groups.

Step 2: Select a group leader.

Step 3: Assign roles for the first practice session.

▲ Client
▲ Interviewer
▲ Observer 1, who uses the Feedback Form (Box 4-3) and leads the microsupervision process. Remember to focus on interviewer strengths as well as areas for improvement.
▲ Observer 2, who runs equipment, keeps time, and also completes a copy of the form.

Step 4: Plan. The interviewer should plan to use both open and closed questions. It is important in the practice session that the key *what, how, why,* and *could* questions be used. Add *what else* for enrichment.

Discuss a work challenge. The client may share a present or past interpersonal job conflict. The interviewer first draws out the conflict, then searches for positive assets and strengths.

Suggested alternative topics might include the following:

▲ A friend or family member in conflict
▲ A positive addiction (such as jogging, health food, biking, team sports)
▲ Strengths from spirituality or ethnic/racial background

Step 5: Conduct a 3- to 6-minute practice session using only questions. The interviewer practices open and closed questions and may wish to have handy a list of suggested question stems. The client seeks to be relatively cooperative and talkative but should not respond at such length that the interviewer has only a limited opportunity to ask questions. More time will be needed if you decide on a more challenging topic.

Step 6: Review the practice session and provide feedback to the interviewer for 12 minutes. Remember to stop the audio- or videotape periodically and listen to or view key happenings several times for increased clarity. Generally speaking, it is wise to provide some feedback before reviewing the tape, but this sometimes results in a failure to view or listen to the tape at all.

Step 7: Rotate roles.

Exercise 4: The Powerful Questions of the GROW Model

Note: This exercise may be completed more successfully if it comes after Chapters 5, 6, and 7 on observation and the basic listening sequence. On the other hand, this is also an excellent chance to test out positive questions in a more complete session.

Step 1: Divide into practice groups.

Step 2: Select a group leader.

Step 3: Assign roles for the first practice session.

▲ Client, who may think through a possible goal before the session.
▲ Interviewer, who will go through the GROW model and use the nine basic questions. The interviewer should also demonstrate listening skills in addition to the questions. Make the session real by asking suitable questions and showing that you listen and care.
▲ Observer 1, who uses the Feedback Form (Box 4-3) and leads the microsupervision process.
▲ Observer 2, who runs equipment, keeps time, and also completes the form.

Step 4: Plan. The interviewer and client should both have copies of the model and questions from pages 111 and 112 and refer to them as needed. The topic for the session ideally will develop from the questioning process itself. And the goal needs to be established early in the practice session. As necessary, build rapport and relationship as you start.

Step 5: Conduct a 6- to 10-minute practice session using only the nine questions or adaptations of them. The client seeks to be relatively cooperative and talkative but should not respond at such length that the interviewer has only limited opportunity to work through the GROW model and questions. More time will be needed if you decide on a more challenging topic.

Step 6: Review the practice session and provide feedback to the interviewer for 12 minutes. Remember to stop the audio- or videotape periodically and listen to or view key happenings several times for increased clarity. Generally speaking, it is wise to provide some feedback before reviewing the tape, but this sometimes results in a failure to view or listen to the tape at all.

Step 7: Rotate roles.

BOX 4-3   FEEDBACK FORM: QUESTIONS

_____ (Date)

_____   _____
(Name of Interviewer)   (Name of Person Completing Form)

Instructions: On the lines below, list as completely as possible the questions asked by the interviewer. At a minimum, indicate the first key words of the question (*what, why, how, do, are,* and so on). Indicate whether each question was open (O) or closed (C). Use additional paper as needed. In the coaching practice session, is the practicing coach able to use the the nine questions and the GROW model? Does the session focus on strengths and goal attainment? How well did the interviewer listen as well as ask questions?

1. _____
2. _____
3. _____
4. _____
5. _____
6. _____
7. _____
8. _____
9. _____
10. _____

1. Which questions seemed to provide the most useful client information?

2. Provide specific feedback on the attending skills of the interviewer.

3. Discuss the use of the positive asset search, wellness, and the use of questions.

Portfolio of Competence

"Determining Your Own Style and Theory," the apex of the microskills hierarchy, can be best accomplished on a base of competence. Each chapter closes with a reflective exercise asking your thoughts and feelings about what has been discussed. By the time you finish this book, you will have a substantial record of your competencies and a good written record as you move toward determining your own style and theory.

Use the following as a checklist to evaluate your present level of mastery. Check those dimensions that you currently feel able to do. Those that remain unchecked can serve as future goals. Do not expect to attain intentional competence on every dimension as you work through this book. You will find, however, that you will improve your competencies with repetition and practice.

Level 1: Identification and classification.

❑ Ability to identify and classify open and closed questions.
❑ Ability to discuss, in a preliminary fashion, issues in diversity that occur in relation to questioning.
❑ Ability to write open and closed questions that might predict what a client will say next.

Level 2: Basic competence. Aim for this level of competence before moving on to the next skill area.

❑ Ability to ask both open and closed questions in a role-played interview.
❑ Ability to obtain longer responses to open questions and shorter responses to closed questions.

Level 3: Intentional competence. Work toward intentional competence throughout this book. All of us can improve our skills, regardless of where we start.

❑ Ability to use closed questions to obtain necessary facts without disturbing the client's natural conversation.
❑ Ability to use open questions to help clients elaborate their stories.
❑ Ability to use *could* questions and, as predicted, obtain a general client story. ("Could you tell me generally what happened?" "Could you tell me more?")
❑ Ability to use *what* questions to facilitate discussion of facts.
❑ Ability to use *how* questions to bring out feelings ("How do you feel about that?") and information about process or sequence. ("How did that happen?")
❑ Ability to use *why* questions to bring out client reasons. ("Why do think your spouse/lover responds coldly?")
❑ Ability to bring out client concrete information and specifics. ("Could you give me a specific example?")
❑ Ability to use the newspaper sequence for assessment. (*who, what, where, when, why, how*)

Level 4: Psychoeducational teaching competence. As stated earlier, do not expect to become skilled in teaching groups or peer counselors skills at this point. You may, however, find some clients who benefit from direct instruction in open questions focusing on other's thoughts and opinions, rather than their own. Those who talk too much about themselves find this skill important in breaking through their self-absorption. At the same time, please point out the dangers of too many questions, especially that *why* question, which can put others on the spot and make them defensive.

❑ Ability to teach clients in a helping session the social skill of questioning. You may either tell clients about the skill or you may practice a role-play with them.

❑ Ability to teach small groups the skills of questioning.

DETERMINING YOUR OWN STYLE AND THEORY: CRITICAL SELF-REFLECTION ON QUESTIONING

This chapter has focused on the pluses and minuses of using questions in the session. While we, as authors, obviously feel that questions are an important part of the interviewing process, we have tried to point out that there are those who differ from us. Questions clearly can get in the way of effective relationships in interviewing, counseling, and therapy.

Regardless of what any text on interviewing, counseling, and psychotherapy says, the fact remains that it is YOU who will decide whether to implement the ideas, suggestions, and concepts. What single idea stood out for you among all those presented in this chapter, in class, or through informal learning? What stands out for you is likely to be important as a guide toward your next steps. What are your thoughts on multiculturalism and how it relates to your use of questions? How might you use ideas in this chapter to begin the process of establishing your own style and theory?

What are your thoughts?

_____

_____

OUR THOUGHTS ABOUT BENJAMIN

We would probably start the interview by explaining to Benjamin that we'd like to know what he is thinking about his future after he completes school. We would begin the session with some informal conversation about current school events or something personal we know about Ben. The first question might be stated something like this: "You'll soon be starting your senior year; what have you been thinking about doing after you graduate?" If this question opens up some tentative ideas, we'd listen to these and ask him for elaboration. If he focuses on indecision between volunteering for the army or entering a local community college or the state university, we'd likely ask him some of the following questions:

"What about each of these appeals to you?"
"Could you tell me about some of your strengths that would help you in the army or college?"
"If you went to college, what might you like to study?"
"How do finances play a role in these decisions?"
"Are there any negatives about any of these possibilities?"
"How do you imagine your ideal life 10 years from now?"

On the other hand, Benjamin just might say to any of these, "I don't know, but I guess I better start thinking about it" and look to you for guidance. You sense a need to review his past likes and dislikes as possible clues to the future.

"What courses have you liked best in high school?"
"What have been some of your activities?"
"Could you tell me about the jobs you've had in the past?"
"Could you tell me about your hobbies and what you do in your spare time?"
"What gets you most excited and involved?"
"What did you do that made you feel most happy in the past year?"

From questions such as these, we may see patterns of ability and interest that suggest actions for the future.

If Benjamin is uncomfortable in the counseling office, all of these questions might put him off. He might feel that we are grilling him and perhaps even see us as intruding in his world. Generally speaking, getting this type of important information and organizing it requires the use of questioning. But questions are effective only if you and the client are working together and have a good relationship.

The pyramid labels, from top band down:

Client Observation Skills

Open and Closed Questions

Attending Behavior

Ethics, Multicultural Competence, and Wellness

In working with patients [clients], if you miss those nuances—if you misread what they may be trying to communicate, if you misjudge their character, if you don't notice when their emotions, gestures, or tone of voice don't fit what they are saying, if you don't catch the fleeting sadness or anger that lingers on their face for only a few milliseconds as they mention someone or something you might otherwise not know was important— you will lose your patients [clients]. Or worse still, you don't. [And your client continues with you despite your insensitivity.]

—Drew Weston

How can observation skills help you and your clients?

Chapter Goals   The aim of this chapter is to increase your ability to observe what occurs between you and your clients verbally and nonverbally in the interview. Observation skills guide you to key issues in the here and now of the interview. In addition, they will help you respond appropriately to both individual and multicultural differences.

Competency Objectives   Awareness, knowledge, and skills in observation will enable you to observe and understand

▲ Nonverbal behavior. How do we make meaning of nonverbal behavior?
▲ Verbal behavior. How do you and your clients use language?

▲ Discrepancies and conflict. Much of interviewing is about working through conflict and coping with the inevitable stressful incongruities we all face.
▲ Styles associated with varying individual and cultural ways of expression. How can you flex intentionally and avoid stereotyping in your observation?

Allen sometimes testifies in cases involving Social Security disability claims. In one case, the "claimant," Horace, was an African American, about 60 years old, who had southern roots. He had a severe back problem from years of arduous lifting in a local factory. This is a common price that is paid in the bodies of men and women who do work that repeats the same motion again and again.

When called to the witness stand, Horace held his hat in his hand, spoke quietly to the judge, and took his oath. But as the judge questioned him, his eyes started to wander—all about the room. He did not look at the judge at all. The judge became irritated by what he saw as a lack of attention and spoke sharply to Horace about his lack of respect toward the bench. Horace was obviously embarrassed, but didn't know what to do. In fact, as he became increasingly nervous, he only looked away more.

What sense do you make of what is happening here? Is Horace being disrespectful? What would you do in a situation like this if you were sitting in court?

What are your thoughts?

_____

_____

Please turn to the end of the chapter to compare your thoughts with what Allen did.

INTRODUCTION: KEEPING WATCH ON THE INTERVIEW

If you use observation skills as defined here, you will notice more effectively what is going on in the immediate here and now of the session.

Observation Skills	Predicted Result
Observe your own and the client's verbal and nonverbal behavior. Anticipate individual and multicultural differences in nonverbal and verbal behavior. Carefully and selectively feed back your observations to the client as topics for discussion.	Observations provide specific data validating or invalidating what is happening in the session. Also, they provide guidance for the use of various microskills and strategies. The smoothly flowing interview will often demonstrate movement symmetry or complementarity. Movement dissynchrony provides a clear clue that you are not "in tune" with the client.

What should you observe about client behavior in the interview? From your own life experience you are already aware of many things that are important for a counselor or interviewer to notice. Brainstorm from what you already know and make a list.

But there are two people in the relationship. What about you? How are you affecting the client verbally and nonverbally? Looking at your way of being can be equally as important as, or more important than, observing the client. Start by taking a brief inventory of your own nonverbal style. You might begin by thinking back to your natural style of attending, but expand those self-observations. What is your interpersonal style and how might it affect your relationship with others? Here it might be helpful to write and then compare what you find with later examination of your own videos and feedback that you obtain from others.

What is your style?   _____

_____

Observe Attending Patterns of Clients

An ideal place to begin improving your observation skills is by noting your own and your client's style of attending behavior. Clients may break eye contact, shift bodily, and change vocal qualities as their comfort level changes when they talk about varying topics. You may observe clients crossing their arms or legs when they want to close off a topic, using rapid alterations of eye contact during periods of confusion, or exhibiting increased stammering or speech hesitations when topics are difficult. And if you watch yourself carefully on tape, you, as interviewer, will exhibit many of these same behaviors.

Observe Nonverbal Behavior

Jiggling legs, making complete body shifts, or suddenly closing one's arms most often indicate discomfort. Hand and arm gestures may give you an indication of how you and the client are organizing things. Random, discrepant gestures may indicate confusion whereas a person seeking to control or organize things may move hands and arms in straight lines and point fingers authoritatively. Smooth, flowing gestures, particularly those in harmony with the gestures of others, such as family members, friends, or the interviewer, may suggest openness.

Observe Verbal Behavior

Language is basic to interviewing and counseling, and there are many ways to consider verbal behavior, ranging from detailed linguistic examination through the differing language systems of varying counseling and therapy theories. This chapter will consider four dimensions useful for direct verbal observation in the session: patterns of selective attention, client key words, "I" statements and "other" statements, and abstract and concrete conversation.

Observe Conflict, Incongruencies, and Discrepancies

Whether you are helping clients work through problems, deal with issues, encounter challenges, or manage concerns, you will be facilitating the resolution of discrepancies, incongruity, and conflict in their lives. Out of your awareness of verbal and nonverbal behavior will come an increased ability to notice conflicts of many types. Stress comes from internal and external conflict. Examples of internal conflict and discrepancies include indecision, guilt, depression, and anxiety. Problems with interpersonal relations, cultural oppression, and work are three examples of external conflict. Of course, many of your clients will be dealing with both types of conflict.

Careful observation of multiple types of discrepancies give you a deeper understanding of where clients "really are" in terms of their issues. Conflict is literally the "stuff" of counseling and is often where you can help clients the most.

Observing Individual and Multicultural Issues in Verbal and Nonverbal Behavior

As you engage in observation, recall that each culture has a different style of nonverbal communication. A study was made of the average number of times friends of different cultural groups touched each other in an hour while talking in a coffee shop. The results showed that English friends did not touch each other at all, French friends touched 110 times, and Puerto Rican friends touched 180 times (cited in Asbell & Wynn, 1991). Croce (2003) cites a parallel study showing that White students at the University of Florida, Gainesville, touched twice whereas there were 120 instances of touching between friends in San Juan, Puerto Rico.

Smiling is a sign of warmth in most cultures, but in some situations in Japan, smiling may indicate discomfort, but the same in lesser form happens in U.S. culture as well. Eye contact may be inappropriate for the traditional Navajo but highly appropriate and expected for a Navajo official who interacts commonly with European American Arizonans.

Be careful not to assign your own ideas about what is "standard" and appropriate nonverbal communication. It is important for the helping professional to begin a lifetime of study of nonverbal communication patterns and their variations. In terms of counseling sessions, you will find that changes in style may be as important as, or more important than, finding specific meanings in communication style. (See Box 5-1.) Edward Hall's *The Silent Language* (1959) remains a classic. Paul Ekman's work (2007) is the current major reference for nonverbal communication. You can also visit several useful Internet sites devoted to nonverbal communication, one of which is http://nonverbal.ucsc.edu (or use your search engine with the key words *nonverbal communication*). Needless to say, the visuals available on the Internet will provide clearer examples of nonverbal communication than we can provide through the written word.

EXAMPLE INTERVIEW: IS THE ISSUE DIFFICULTY IN STUDYING OR RACIAL HARASSMENT?

Kyle Yellowhorse is a second semester junior business major in a large university in the Rocky Mountains. He was raised in a relatively traditional Lakota family on the Rosebud Reservation in South Dakota. Native American Indians are unlikely to come to counseling unless they are referred by others or the interviewer has established herself or himself previously as a person who can be trusted.*

---

* Many schools, elementary through university, have a small population of Native American Indians. For example, the Chicago public schools have slightly over 1,000 Native Americans mixed in schools throughout a large system. Most often, the Native American Indian population is invisible, and you may never know this cultural group exists unless you indicate through your behavior and actions in the community that you are a person who can be trusted and who wishes to know the community. They are invisible to you, and you are invisible to them.

BOX 5-1   NATIONAL AND INTERNATIONAL PERSPECTIVES ON COUNSELING SKILLS

Can I Trust What I See?

Weijun Zhang

James Harris, an African American professor of education, and I were invited by a national Native American youth leadership organization to give talks in Oklahoma. I attended Dr. Harris's first lecture with about 60 Native American children. Dynamic and humorous, Dr. Harris touched the heartstrings of everyone. But much to my surprise, when the lecture was over, he was very upset. "A complete failure," he said to me with a long face; "they are not interested." "No," I replied, "It was a great success. Don't you see how people loved your lecture?" "No, not at all." After some deliberation, I came to see why he could have such an erroneous impression.

"There were not many facial expressions," Professor Harris said. I said, "You may be right, using African American standards. However, that is not a sign that your audience was not interested. Native American people, in a way, are programmed to restrain their feelings, whether positive or negative, in public; as a result, their facial expressions would be hard to detect. Native American people have always valued restraint of emotion, considering this a sign of maturity and wisdom, as I know. Actually, in terms of emotional expressiveness, African American culture and Native American culture may represent two extremes on a continuum."

"They did not ask a single question, though I repeatedly asked them to," he said. I replied, "Well, they didn't because they respect you." "Come on, you are kidding me." I told him, "Many Native Americans are not accustomed to asking questions in public, probably for the following reasons. (a) If you ask an intelligent question, you will draw attention away from the teacher and onto yourself. That is not an act of modesty and may be seen as showing off. (b) If your question is silly, you will be seen as a laughingstock and lose face. (c) Whether your question is good or bad, one thing is certain: You will disturb the instructor's teaching plan, or you may suggest the teacher is unclear. That goes against the Native American tradition of being respectful to the senior. So you can never expect a Native American audience to be as active as African Americans or Whites in asking questions. In today's situation, some kids probably did want to ask you questions, for you repeatedly asked them to raise questions and issues. But, unfortunately, they still couldn't do it." He asked why. I said, "You waited only a few seconds for questions before you went on lecturing, which is far from enough. With Native Americans, you have to adopt a longer time frame. European Americans and African Americans may ask questions as soon as you invite them to; American Indians may wait for about 20 to 30 seconds to start to do so. That period of silence is a necessity for them. You might say that Native Americans are true believers in the saying 'Speech is silver, silence is golden.'" He asked, "Well, why didn't you tell me that on the spot, then?" "If it is respectful for those Native American kids not to ask you questions, Dr. Harris," I said, "how could you expect this humble Chinese to be so disrespectful as to come to the stage to correct you?"

The director of the Native American organization, who overheard my conversation with James Harris, approached me with the question "How do you possibly know all this about us Native Americans?" "Well, I don't think it is news for you that Native Americans migrated from Asia some thousands of years ago. You don't mind that your Asian cousins still share your ethnic traits, do you?" The three of us all laughed.

Kyle did well in his first two years at the university, but during the fall term his "B" average dropped to barely passing. His professor of marketing has referred him to the campus counseling center. The interviewer in this case is European American.

In this case, you will see a slow start, due at least partially to multicultural differences. See how the interviewer uses observation skills to open Kyle to discussing his issues.

Interviewer and Client Conversation	Process Comments
1. *Derek:* Kyle, come on in; I'm glad to see you.	Derek walks to the door, he smiles, shakes hands, and has direct eye contact.
2. *Kyle:* Thanks. (pause)	Kyle gives the interviewer eye contact for only a brief moment. He sits down quietly.
3. *Derek:* You're from the Rosebud Reservation, I see.	Derek knows that contextual and family issues are often important to Native American Indian clients. Rather than focusing on the individual and seeking "I" statements, he realizes that more time may be needed to develop a relationship. Derek's office decorations include artwork from the Native American Indian, African American, and Mexican traditions as well as symbols of his own Irish American heritage.
4. *Kyle:* Yeah. (pause)	While his response is minimal, Kyle notes that Derek relaxes slightly in the chair.
5. *Derek:* There've been some hard times here on campus lately. (pause, but there is still no active response from Kyle) I'm wondering what I might do to help. But first, I know that coming to this office is not always easy. I know Professor Harrison asked you to come in because of your grades dropping this last term.	During the fall term, the Native American Indian association on campus had organized several protests against the school mascot—"the fighting Sioux." As a result, the campus has been in turmoil with recurring events of racial insults and several fights. Noting Kyle's lack of eye contact, Derek has reduced the amount of direct gaze, and he also looks down. Among traditional people, the lack of eye contact generally indicates respect. At the same time, many clients who are depressed use little eye contact.
6. *Kyle:* Yeah, my grades aren't so good. It's hard to study.	Kyle continues to look down and talks slowly and carefully.
7. *Derek:* I feel honored that you are willing to come in and talk, given all that has happened here. Kyle, I've been upset with all the incidents on campus. I can imagine that they have affected you. But first, how do you feel about being here talking with me, a White counselor?	Derek self-discloses his feelings about campus events. He makes an educated guess as to why Kyle's grades have dropped. As he talks about the campus problems, Kyle looks up directly at him for the first time, and Derek notes some fire of anger in his eyes. Kyle nods slightly when Derek says "I feel honored."
8. *Kyle:* It's been hard. I simply can't study. (pause) Professor Harrison asked me to come and see you. I wouldn't have come, but I heard from some friends that you were okay. I guess I'm willing to talk a bit and see what happens.	People who are culturally different from you may not come to your office setting easily. This is where your ability to get into the community is important. In this case, college counseling center staff have been active in the campus community leading discussions and

(continued)

Interviewer and Client Conversation	Process Comments
	workshops seeking to promote racial understanding. As Kyle talks, Derek notices increased relaxation and senses that the beginning of trust and rapport has occurred. With some clients, reaching this point may take a full interview. Kyle is bicultural in that he has had wide experience in White American culture as well as in his more traditional Lakota family.
9. *Derek:* Thanks, maybe we could get started. What's happening?	We see that Kyle's words and nonverbal actions have changed in the short time that he has been in the session. Derek, for the first time, asks an open question. Questions, if used too early in the session, might have led Kyle to be guarded and say very little.
10. *Kyle:* I'm vice president of the Native American Indian Association—see—and that's been taking a lot of time. Sometimes there are more important things than studying.	Kyle starts slowly, and as he gets to the words "more important things," the fire starts to rise in his eyes.
11. *Derek:* More important things?	The restatement encourages the client to elaborate on the importance of the critical issue that Derek has observed through Kyle's eyes.
12. *Kyle:* Yeah, like last night, we had a march and demonstration against the Indian mascot. It's so disrespectful and demeaning to have this little Indian cartoon with the big teeth. What does that have to do with education? They talk about "liberal education." I think it's far from liberal; it's constricting. But worse, when we got back to the dormitories, the car that belongs to one of our students had all the windows broken out. And inside was a brick with the words "You're next" painted on it.	Kyle is now sitting up and talking more rapidly. Anger and frustration show in his body—his fists are clenched, and his face shows strain and tension. Women, gays, or other minorities who experience disrespect or harassment may feel the same way and demonstrate similar verbal and nonverbal behaviors. The car bashing incident clearly illustrates that the harassing students had a lack of "other" esteem and respect (see page 137).
13. *Derek:* That's news to me. The situation on campus is getting worse. Your leadership of the association is really important, and now you face even more challenges.	Derek shares his knowledge of the situation and paraphrases what Kyle has just said. He sits forward in his chair and leans toward Kyle. However, if Kyle were less fully acculturated in White European American culture, the whole tone of the conversation above would be quite different. It probably would have taken longer to establish a relationship, Kyle would have spoken more carefully, and his anger and frustration would likely not have been visible.

(continued)

Interviewer and Client Conversation	Process Comments
	The interviewer, in turn, would be likely to spend more time on relationship development, use more personal self-disclosure, use less direct eye contact, and—especially—be comfortable with longer periods of silence.
14. *Kyle:* Yeah. (pause) But, we're going to manage it. We won't give up. (pause) But—I've been so involved in this campus work that my grades are suffering. I can't help the association if I flunk out.	Kyle feels heard and supported by the interviewer. Being heard allows him to turn to the reasons he came to the counselor's office. He relaxes a bit more and looks to Derek, as if asking for help.
15. *Derek:* Kyle, what I've heard so far is that you've gotten caught up in the many difficulties on campus. As association vice president, it's taken a lot of your time—and you are very angry about what's happened. I also understand that you intend to "hang in" and that you believe you can manage it. But now you'd like to talk about managing your academics as well. Have I heard you correctly?	Derek has used his observation skills so that he knows now that Kyle's major objective is to work on improving his grades. He wisely kept questioning to a minimum and used some personal sharing and listening skills to start the interview. He also uses a check-out to ensure he's understanding Kyle.
16. *Kyle:* Right! I don't like what's going on, but I also know I have a responsibility to my people back home on the Rosebud Reservation, the Lakota people, and to myself to succeed here. I could talk forever on what's going on here on campus, but first, I've got to get my grades straightened out.	Here we see the importance of self-esteem and self-focus if Kyle is to succeed. But his respect for others is an important part of who he is. He does not see himself as just an individual. He also sees himself as an extension of his group. Kyle has considerable energy and perhaps a need to discuss campus issues, but his vocal tone and body language make it clear that the first topic of importance for him today is staying at the university.
17. *Derek:* Okay . . . If you'd like, later we might come back to what's going on. You could tell me a bit about what's happening here with the mascot and all the campus troubles. But now the issue is what's going on with your studies. Could you share what's happening?	Derek observes that Kyle's mission in the interview right now is to work on staying in school. He asks an open question to change the focus of the session to academic issues, but he keeps open the possibility of discussing campus issues later in the interview.

This interview has now started. Kyle has obviously been observing and deciding if he can trust Derek. Fortunately, Derek has a good reputation on campus. He regularly attends Native American pow-wows and other multicultural events. You will find if you work in a school or university setting that clients will be very aware of the effectiveness and trustworthiness of counseling staff.

Although focused on Native American Indian issues, this interview has many parallels with other native people who have been dispossessed of their land and whose culture has been belittled—Hawai'ians, Aboriginals in Australia, Dene and Inuit in Canada, Maori in New Zealand, and Celtic people in Great Britain. Moreover, the interview in many ways illustrates what might happen in any type of cross-cultural counseling. A European American student meeting an interviewer who is Latina/o, African American, or other Person of Color (or vice versa if the roles of counselor and client are switched) might also have early difficulty in talking and establishing trust.

You will find that many interviews start slowly, regardless of the cultural background of the client. Your skill at observing nonverbal and verbal behavior in the here and now of the session will enable you to chose appropriate things to say. Some alternatives that may help you as you begin sessions include patience; a good sense of humor; and a willingness to disclose, share stories, and talk about neutral subjects such as sports or the weather. You will also find an early exploration of positive assets useful at times—"Before we start, I'd like to get to know a little bit more about you. Could you tell me specifically about something from your past that you feel particularly good about?" "What are some of the things you do well?" "What types of things do you like?" As time permits, consider conducting a full wellness review.

INSTRUCTIONAL READING

Three organizing principles for understanding interview interaction are stressed in this chapter: nonverbal behavior, verbal behavior, and discrepancies and conflict. Over time, you will gain considerable skill in drawing out and working with the client's view of the world and how you as a person relate to the client. The best way to learn observation skills may be to observe yourself and your client in a videotaped session.

Nonverbal Behavior

As you have seen, observing clients' attending behavior patterns is often central. Clients may be expected to break eye contact, exhibit bodily movement, and change vocal qualities when they are talking about topics of varying levels of comfort to them. You may observe clients crossing their arms or legs when they want to close off a topic, using rapid alternations of eye contact during periods of confusion, or exhibiting increased stammering or speech hesitations while pursuing difficult topics. "The voice of the therapist, regardless of what he or she says, should be warm, professional—competent, and free from fear" (Grawe, 2007, p. 411). You can learn about nonverbal behavior research in Box 5-2. Periodically, observe yourself on video and carefully note your own nonverbal style. And continue this, even long after you become a practicing professional. We all fall into bad habits, and observing yourself from time to time will help you correct or avoid them.

Facial Expressions

For you, as the interviewer, smiling is a good indicator of your warmth and caring. Your ability to develop a relationship will often carry you through difficult problems and situations. When it comes to observing the client, here are some things to notice: The brow may furrow, lips may tighten or loosen, flushing may occur, or a client may smile at an inappropriate time. Even more careful observation will reveal subtle color changes in the face as blood flow reflects emotional reactions. Breathing may speed up or stop temporarily. The lips may swell, and pupils may dilate or contract.

BOX 5-2   RESEARCH EVIDENCE THAT YOU CAN USE

Observation

Research on nonverbal behavior has a long and distinguished history. Edward Hall's *The Silent Language* of 1959 is a classic of anthropological and multicultural research and still remains the place to start. Early work in nonverbal communication was completed by Paul Ekman, and his 1999 and 2007 summaries of his work are basic to the field. Eye contact and forward trunk lean were found to be highly correlated with ratings of empathy (Sharpley & Guidara, 1993; Sharpley & Sagris, 1995). Hill and O'Brien's review (2004) noted that clients used fewer head nods when they were reacting negatively. Using an empathic accuracy paradigm, Hall and Schmid Mast (2007) found that participants shift attention toward visual nonverbal cues and away from verbal cues when asked to infer feelings; when asked to infer thoughts, they did the reverse.

A classic study by Mayo and LaFrance remains important today as it highlights issues of cultural change and acculturation. They commented in 1973 (p. 389),

> In interaction among Whites, a clear indication of attention by the listener is that he looks at the speaker. . . . Blacks do not look while listening, presumably using other cues to communicate attention. When Blacks and Whites interact, therefore, these differences may give rise to communication breakdowns. The White may feel he is not being listened to, while the Black may feel unduly scrutinized. Further, exchanges of listener-speaker roles may become disjunctive, leading to generalized discomfort with the encounter.

Many African Americans have acculturated to White standards and may not show the same differences as in the past. At the same time, it is well known that minorities in the United States and Canada often have two communication styles: one for the "outgroup" (White culture) and one within their own family and cultural group. These observations on eye contact remain relevant today to Native American and Latin cultures. Cross-cultural communication occurs most often in a situation in which the majority has power and many minorities have learned to mask their emotions and nonverbal expressions.

LaFrance turned to research on gender differences and power and noted major differences in communication style. For example, LaFrance and Woodzicka (1998) found women's nonverbal reactions to sexist humor are quite marked and different from those of men. Beek and Dubas (2008) found age and gender differences in decoding nonverbal cues. Pre- and adolescent school students showed differences in the perceived emotional intensity of complex and more ambiguous facial expressions. No difference was observed in the perceptions of basic emotional expressions. Older adolescents attributed more negative meaning to them; girls attributed more anger to the facial expressions than boys. These findings are important to understanding the development of relationships.

Miller (2007) studied communication in the workplace and found that "connecting," an important aspect of communication of compassion, included empathy and perspective taking. Two studies by Haskard et al. (2008) report that physicians' nonverbal communication, particularly their tone of voice, plays an important role in the relationship with their patient. It affects patient satisfaction and adherence to treatment. Health care providers need to be aware of the power of their tone of voice that may inadvertently communicate their emotions and affect clients' satisfaction.

Reading about nonverbal communication is helpful, but we suggest that you use an Internet search engine (e.g., www.google.com) and enter "nonverbal communication" as the key search term. You will often find helpful visuals to support what you have read here. Add the word "research" to your search, and you will find several summaries of databased work. The Web site Exploring Nonverbal Communication at http://nonverbal.ucsc.edu is especially useful and well worth your time.

Nonverbal Communication and Neuropsychology
Observation of nonverbal communication is obviously based in physiology and the ability of the brain to recognize what is occurring. Facial expressions of emotions are important in nonverbal communication. Grawe (2007, p. 78) reviews key literature and points out that

(continued)

BOX 5-2 (continued)

the amygdala, critical center of emotional experience, appears to be highly sensitive to "fearful, irritated, and angry faces . . . even when the faces have not been perceived consciously. . . . We can be certain that in psychotherapy the patient will respond to even the tiniest sign of anger in the facial expressions of therapists." Self-awareness of your own being is obviously as important as awareness of client behavior.

Nonverbal Communication and Culture

There are fascinating multicultural findings that have immediate relevance to counseling and therapy process. Among them are the following reported by Eberhardt (2005):

▲ Japanese have been found to be more holistic thinkers than Westerners (Masuda & Nisbett, 2001). Expect the possibility of different cognitive/emotional styles when you work with people who are culturally different from you—but never stereotype!

▲ Brain systems can be modified by life experience (Draganski et al., 2004). As you learn to observe your client more effectively in the interview, your brain is likely developing new connections. Expect your multicultural learning to become one of those new connections.

▲ Blacks and Whites both exhibit greater brain activation when they view same-race faces and less when race is different (Golby et al., 2001). Note here that this could affect your work with a client whose race is different from yours, thus suggesting that discussing racial and other cultural differences early in the session can be a helpful way to build trust.

Vital to our work as counselors and therapists is awareness that brain functioning is not fixed but also relates to cultural factors and environmental issues (e.g., recurring racism and sexism), significant interpersonal events (e.g., experience of trauma, divorce, serious illness), and our own learning of new skills such as those described in this book.

These seemingly small responses are important clues to what a client is experiencing; to notice them takes work and practice. You may want to select one or two kinds of facial expressions and study them for a few days in your regular daily interactions and then move on to others as part of a systematic program to heighten your powers of observation. Ultimately, these careful observation skills become part of your being, and you will not have to think about them again—the samurai effect.

Body Language

People who are communicating well often "mirror" each other's body language. They may unconsciously sit in identical positions and make complex hand movements together as if in a ballet. This is termed *movement synchrony. Movement complementarity* refers to paired movements that may not be identical but are still harmonious. For instance, one person talks and the other nods in agreement. You may observe a hand movement at the end of one person's statement that is answered by a related hand movement as the other takes the conversational "ball" and starts talking.

Some expert counselors and therapists deliberately "mirror" their clients. Experience shows that matching body language, breathing rates, and key words of the client can heighten interviewer understanding of how the client perceives and experiences the world. If you are sufficiently empathic, you will start feeling the client's feelings in your own body.

Particularly important are discrepancies in nonverbal behavior. Be alert for *movement dissynchrony*. Watch for times when clients suddenly change body posture. For example, you may challenge clients, and they will cross their legs or quickly look away. Movement dissynchrony also occurs when a client is talking casually about a friend, for example; one hand may be tightly clenched in a fist and the other relaxed and open, possibly indicating mixed feelings toward the friend. Lack of harmony in movement is common between people who disagree markedly or even between those who may not be aware they have subtle conflicts. You likely have seen this type of behavior in couples that you know have problems in communicating.

But be careful with deliberate mirroring. A practicum student reported difficulty with a client, noting that the client's nonverbal behavior seemed especially unusual. Near the end of the session, the client reported, "I know you guys; you try to mirror my nonverbal behavior. So I keep moving to make it difficult for you." You can expect that some clients will know as much about observation skills and nonverbal behavior as you do. What should you do in such situations? Use the skills and concepts in this book with *honesty* and *authenticity*. And talk with your client about his or her observations of you without being defensive. Openness works!

Acculturation Issues in Nonverbal Behavior: Avoid Stereotyping

We have stressed that there are many differences in individual and cultural styles. Acculturation is a fundamental concept of anthropology with significant relevance for the interview.

Acculturation is the degree to which an individual has adopted the norms or standard way of behaving in a given culture. Due to the unique family, community, economic status, and part of the country in which a person is raised (and many other factors), no two people will be acculturated to general standards in the same way. In effect "normative behavior" does not exist in any single individual. Thus, stereotyping individuals or groups needs to be avoided at all costs.

An African American client raised in a small town in upstate New York in a two-parent family has different acculturation experiences from those of an African American person from a two-parent family raised in Los Angeles or East St. Louis. If one were from a single-parent family, the acculturation experiences would change further. If we alter only the ethnic/racial background of this example client to Italian American, Jewish American, or Arab American, the acculturation experience changes again. Many other factors, of course, influence acculturation—religion, economic bracket, and even being the first or second child in a family. Awareness of diversity in life experience is critical if we are to recognize uniqueness and "specialness" in each individual. If you define yourself as White American, Canadian, or Australian and you think of others as the only people who are multicultural, you need to rethink your awareness. All of us are multicultural beings with varying and singular acculturation experiences.

Finally, consider biculturality and multiculturality. Many of your clients have more than one significant community cultural experience. A Puerto Rican, Mexican, or Cuban American client is likely to be acculturated in both Latina/o and U.S. culture. A Polish Canadian client in Quebec, a Ukrainian Canadian client in Alberta, and an Aboriginal client in Sydney, Australia, may also be expected to represent biculturality. And all Native Americans and Hawai'ians in the United States, Dene and Inuit in Canada, Maori in New Zealand, and Aboriginals in Australia exist in at least two cultures. There is a culture among people who have experienced cancer, AIDS, war, abuse, and alcoholism. All of these issues and many others deeply affect acculturation.

In short, stereotyping any one individual is not only discriminatory; it is also naïve!

Verbal Behavior

As noted in the chapter introduction, counseling and interviewing theory and practice have an almost infinite array of verbal frameworks within which to examine the interview. The theoretical background and beliefs of counselors deeply affect the way they listen to client stories. Four additional useful concepts for interview analysis are presented here: selective attention, key words, concreteness versus abstractions, and "I" statements and "other" statements. We will also discuss some key multicultural issues connected with verbal behavior.

Selective Attention

What are your patterns of selective attention? Clients tend to talk about what counselors are interested in and willing to hear. For example, behavioral counselors tend to encourage clients to talk about specific concrete situations, existential therapists encourage clients to talk about the meaning of life, whereas career counselors focus on life decisions and careers. Your life experience and your "theory of choice" are very real determining factors in how you listen to others.

Allen, as part of his training, "enjoyed" psychoanalytic therapy with a classical analyst, whose task was to sit and listen to "free" associations. One day, Allen started to talk about a serious interpersonal issue, but the analyst just sat there blankly. Allen tried several times to talk about his issue, but the "nonjudgmental" and "neutral" therapist did what he was supposed to do—specifically, say little or nothing and sit there. Discouraged by the lack of response, Allen mentioned a recent dream. The therapist responded with a head nod and sat up straight. Needless to say, the rest of the session focused on the dream. Allen had to figure out what to do with his interpersonal problem on his own. So much for "analytic neutrality."

A famous training film has three eminent counselors (Albert Ellis, Fritz Perls, and Carl Rogers) presenting their interviewing styles by all counseling the same client, Gloria (Shostrom, 1966). Gloria changes the way she talks about and thinks about her relationship issues, responding very differently as she works with each counselor. Further, research on verbal behavior in the film revealed that Gloria tended to match the language of the varying counselors (Meara et al., 1979, 1981).

Should clients match the language of the interviewer, or should you, the interviewer, learn to match your language and style with that of the client? Most likely, both approaches are relevant, but certainly at the beginning, we want to draw out clients' stories and issues from their own language perspective, not from ours. And we always need to consider how the client talks and makes meaning.

Relating this to your own experience, think about what you consider most important in the interview. Are there topics you are less comfortable with? Some interviewers are excellent in helping clients talk about career issues but shy away from matters such as interpersonal conflict and sexuality. Others may find their clients constantly talking about interpersonal issues, leaving little time to deal with other critical and practical issues such as getting a job or choosing a course of study.

Your personal style and theory will deeply affect how others respond in the session and talk about issues. How will you influence?

Key Words

If you listen carefully to clients, you will find that certain words appear again and again in their descriptions of situations. Noting their key words and helping them explore the facts, feelings, and meanings underlying those words may be useful. Key descriptive words are often the constructs by which a client organizes the world; these words may reveal underlying meanings. Verbal underlining through vocal emphasis is another helpful clue in determining what is most important to a client. Through intonation and volume, clients tend to stress the single words or phrases that are most important to them.

Joining clients by using their key words facilitates your understanding and communication with them. If their words are negative and self-demeaning, reflect those perceptions early in the interview but later help them use more positive descriptions of the same situations or events. Help the client change from "no, I can't" to "yes, I can."

Many clients will demonstrate problems of verbal tracking and selective attention. They may either stay on a single topic to the exclusion of other important issues or change the topic, either subtly or abruptly, when they want to avoid talking about a difficult issue. Perhaps the most difficult task of the beginning counselor or interviewer is to help the client stay on the topic without being overly controlling. Observing client topic changes is particularly important. At times it may be helpful to comment—for instance, "A few minutes ago we were talking about X." Another possibility is to follow up that observation by asking how the client might explain the shift in topic.

Concreteness Versus Abstraction

Where is the client on the "abstraction ladder"? Two major client styles of communication—concrete and abstract—are important for the counselor to observe and to be prepared for in conversation. Clients who talk with a concrete/situational style are skilled at providing specifics and examples of their concerns and problems. The language of these clients forms the foundation and "the bottom" of the abstraction ladder. Such clients may have difficulty reflecting on themselves and their situations and may have difficulties seeing patterns in their lives. (See Box 5-3.)

BOX 5-3   THE ABSTRACTION LADDER

Abstract/formal operational	Clients here tend to talk in a more reflective fashion, analyzing their thoughts and behaviors. They are often good at self-analysis. Clients high on the abstraction ladder may not easily provide concrete examples of their issues. They may prefer to analyze rather than to act. Self-oriented, abstract theories, such as person-centered or psychodynamic, are often useful with this style.
Concrete/situational	Clients who talk using this style tend to provide specific examples and stories, often with considerable detail. You'll hear what they see, hear, and feel. Helping some clients to reflect on their situations and issues may be difficult. In general, they will look to the counselor for specific actions that they can follow. Concrete behavioral theories may be preferred. More specific examples and extensions of these concepts can be found in *Developmental Counseling and Therapy: Promoting Wellness Over the Lifespan* (Ivey, Ivey, Myers, & Sweeney, 2005).

Clients who are more abstract and formal-operational, on the other hand, have strengths in self-analysis and are often skilled at reflecting on their issues. They are at "the top" of the abstraction ladder, but you will find that getting specific details from them on the concretes and specifics of what is actually going on may be difficult. Of course, most adult and many adolescent clients will talk at both levels. Children, however, can be expected to be primarily concrete in their talk—and so are many adolescents and adults.

The strength and value of these details is that you know somewhat precisely what happened, at least from their point of view. However, they often will have difficulties in seeing the point of view of others. Some with a concrete style may tell you, for example, what happened to them when they went to the hospital from start to finish with every detail of the operation and how the hospital functions. Or ask a 10-year-old to tell you about a movie, and you will practically get a complete script. Concretely oriented clients may have a difficult interpersonal relationship and discuss the situation through a series of endless stories full of specific facts and "He said . . . and then I said . . ." If asked to reflect on the meaning of their story or what they have said, they may appear puzzled.

Here are some examples of concrete/situational statements:

Child, age 5:   Jonnie hit me in the arm—right here!

Child, age 10:   He hit me when we were playing soccer. I had just scored a goal, and it made him mad. He snuck up behind me, grabbed my leg, and then punched me when I fell down! Do you know what else he did, well he . . .

Man, age 45:   I was down to Myrtle Beach—we drove there on 95, and the traffic was really terrible. Well, we drove into town, and the first thing we did was to find a motel to stay in, you know. But we found one for only $127, and it had a swimming pool. Well, we signed in and then . . .

Woman, age 27:   You asked for an example of how my ex-husband interferes with my life? Well, a friend and I were sitting quietly in the cafeteria, just drinking coffee. Suddenly, he came up behind me, he grabbed my arm (but didn't hurt me this time), then he smiled and walked out. I was scared to death. If he had said something, it might not have been so frightening.

The details are important, but clients who use a primarily concrete style in their conversation and thinking may have real difficulty in reflecting on themselves and seeing patterns in situations.

Abstract/formal operational clients are good at making sense of the world and reflecting on themselves and their situations. But some clients will talk in such abstract, broad generalities that it is hard to understand what they are really saying. They may be able to see patterns in their lives and be good at discussing and analyzing themselves, but you may have difficulty finding out specifically what is going on in their lives. They may prefer to reflect rather than to act on their issues.

Example abstract/formal operational statements include these:

Child, age 12:   He does it to me all the time. It never stops. It's what he does to everyone all the time.

Man, age 20:   As I think about myself, I see a person who responds to others and cares deeply, but somehow I feel that they don't respond to me.

Woman, age 68:   As I reflect back on my life, I see a pattern of selfishness that makes me uncomfortable. I think a lot about myself.

Many interviewers tend to be more abstract/formal operational themselves and may be drawn to clients with the analytical and self-reflective style. They may conduct entire interviews focusing totally on analysis, and the observer might wonder what the client and counselor are talking about.

In each of the conversational styles the strength is also possibly a weakness. You will want to help abstract/formal clients to become more concrete ("Could you give me an example?"). If you persist, most of these clients will be able to provide the needed specifics.

You will also want to help concrete clients become more abstract and pattern oriented. This is best effected by a conscious effort to listen to their sometimes lengthy stories very carefully. Paraphrasing and summarizing what they have said can be helpful (see Chapter 6). Just asking them to reflect on their story may not work ("Could you tell me what the story means to you?" "Can you reflect on that story and what it says about you as a person?"). More direct questions may be needed to help concrete clients step back and reflect on their stories. A series of questions such as these might help: "What one thing do you remember most about this story?" "What did you like best about what happened?" "What least?" "What could you have done differently to change the ending of the story?" Questions like these that narrow the focus can help children and clients with a concrete orientation move from self-report to self-examination.

Essentially we all need to match our own style and language to the uniqueness of the client. If you have a concrete style, abstract clients may challenge or even puzzle you. If you are more abstract, you may not be able to understand and reach those with a concrete orientation (and they are the majority of our clients). Abstract interviewers often are bored and impatient with concrete clients. If the client tends to be concrete, listen to the specifics and enter that client's world as he or she presents it. If the client is abstract, listen and join that client where he or she is. Consider the possibility of helping the client look at the concern from the other perspective.

"I" Statements and "Other" Statements

Clients' ownership and responsibility for issues will often be shown in their "I" and "other" statements. Consider the following:

"I'm working hard to get along with my partner. I've tried to change and meet her/him halfway."
versus
"It's her/his fault. No change is happening."

"I'm not studying enough. I should work harder."
versus
"The racist insults we get on this campus make it nearly impossible to study."

"I feel terrible. If only I could do more to help. I try so hard."
versus
"Dad's an alcoholic. Everyone suffers."

"I'm at fault. I shouldn't have worn that dress. It may have been too sexy."
versus
"No, women should be free to wear whatever they wish."

"I believe in a personal God who guides all actions. God is central to my life."
versus
"Our church provides a lot of support and helps us understand spirituality more deeply."

We need to be aware of what is happening in and for our clients as well as learn what occurs in their relationships with others and in their families and communities. There is a need to balance internal and external responsibility for issues.

Review the five pairs of statements above. Some of them represent positive "I" and "other" statements; some are negative. Some clients attribute their difficulties solely to themselves; others see the outside world as the issue. A woman may be sexually harassed and see clearly that others and the environment are at fault; another woman will feel that somehow she provoked the incident. Counselors need to help individuals look at their issues but also to help them consider how these concerns relate to others and the surrounding environment.

The statement about the alcoholic dad may serve as an example. Some children of alcoholics see themselves as somehow responsible for a family member's drinking. Their "I" statements may be unrealistic and ultimately "enable" the alcoholic to drink even more. In such cases, the task of the interviewer is to help the client learn to attribute family difficulties to alcohol and the alcoholic. In work with alcoholics themselves who may deny their problem, one goal is often to help move them to that critical "I" statement, "I am an alcoholic." Part of recovery from alcoholism, of course, is recognizing others and showing esteem for others. Thus, a balance of "I" and "other" statements is a useful goal.

You can also observe "I" statements as a person progresses through a series of interviews. For example, the client at the beginning of counseling may give many negative self-statements— "It's my fault." "I did a bad thing; I'm a bad person." "I don't respect myself." "I don't like myself."

If your sessions are effective, expect such statements to change to "I'm still responsible, but I now know that it wasn't all my fault." "Calling myself 'bad' is self-defeating. I now realize that I did my best." "I can respect myself more." "I'm beginning to like myself."

Multicultural differences in the use of "I" are important. We should remember that English is one of the very few languages that capitalizes the word "I." A Vietnamese immigrant comments:

> There is no such thing as "I" in Vietnamese. . . . We define ourselves in relationships. . . . If I talk to my mother, the "I" for me is "daughter," the "you" is for "mother." Our language speaks to relationships rather than to individuals.

Discrepancies and Conflict

The variety of discrepancies clients may manifest is perhaps best illustrated by the following statements:

"My son is perfect, but he just doesn't respect me." (child/parent conflict)
"I really love my brother." (said in a quiet tone with averted eyes)
"I can't get along with Charlie." (discrepancy between the client and another person)
"I deserve to pass the course." (from a student who has done no homework and just failed the final examination)
"That question doesn't bother me." (said with a flushed face and a closed fist)

Once the client is relatively comfortable and some beginning steps have been taken toward rapport and understanding, a major task of the counselor or interviewer is to identify basic discrepancies, mixed messages, conflicts, or incongruities in the client's behavior and life. A common goal in most interviews, counseling, and therapy is to assist clients in working through discrepancies and conflict, but first these have to be identified clearly.

Examples of Conflict Internal to the Client
Discrepancies in nonverbal behaviors. A client may be describing an apparently trivial event, smiling and seemingly relaxed, when you notice tears rolling down his or her face. In

such cases you can point out what you observe and respectfully help the client explain the meaning of the behavior.

Discrepancies in verbal statements. In a single sentence a client may express two completely contradictory ideas ("My son is perfect, but he just doesn't respect me" or "This is a lovely office you have; it's too bad that it's in this neighborhood"). Most of us have mixed feelings toward our loved ones, our work, and other situations. It is helpful to aid others in understanding their ambivalences.

Choking, hesitation, stammer. When facing up to difficult issues or saying something that is uncomfortable, the client's throat may tighten, and a hoarse, almost gasping, vocal tone will follow. This is often complemented by a surprise stammer or unusual vocal pause. People do "choke on their words."

Discrepancies between what one says and what one does. A parent may talk of love for a child but be guilty of child abuse. A student may say that he or she deserved a higher grade than the actual time spent studying suggests. The client may verbalize a support for multicultural, women's, or ecology causes, but fail to "walk the talk."

Discrepancies between statements and nonverbal behavior. "That question doesn't bother me" (said with a flushed face and a closed fist). Watch for the timing of "lint picking." The client may be talking of a desire to repair a troubled relationship while simultaneously picking at his or her clothes. Many clients make small or large physical movements away from the interviewer when they are confronted with a troubling issue and feel inadequately supported by the interviewer.

Examples of Conflict Between the Client and the External World
Discrepancies between people. Conflict can be described as a discrepancy between people. Noting interpersonal conflict is a key task of the interviewer, counselor, or therapist. One of the predominant issues you face in interviewing and counseling is discord and arguments among people. Mediation, in particular, focuses on this particular type of discrepancy.

Discrepancies between a client and a situation. "I want to be admitted to medical school, but I didn't make it." "I just found out I have early symptoms of arthritis." "I can't find a job." In such situations the client's ideal world is often incongruent with what really is. The counselor's task is to work through these issues in terms of behaviors or attitudes. Many gays, women, and persons with disabilities find themselves in a contextual situation that makes life difficult for them. Discrimination, heterosexism, sexism, and ableism represent situational discrepancies.

Nose wipes and lint picking. The client or friend completes a statement and then reaches up to wipe the nose, usually with the second finger. This is a strong sign of discomfort or a feeling of vulnerability—for example, "I am sure that I can do that" or just "sure" or "yes" followed by a quick nose wipe. The person is telling you clearly that he or she is not as sure or comfortable as the words say. Notice yourself here and watch for your own nose wipes. You'll often find that your nose actually itches at that exact moment when you are uncomfortable. Neuroscience tells us that the nose is the part of us most directly and immediately linked to the brain, similar to animals.

Picking real or imaginary lint on one's clothes can be a clear sign of conflict and discrepancies. A client may say, "I really love my spouse," while simultaneously picking lint. If the counselor then provides unwelcome advice, the client would speak in agreement, again picking lint.

Discrepancies in Goals

Goal setting is an important part of the *relationship—story and strengths—goals—restory—action* model. As part of clear goal setting and establishing the purposes of the interview or counseling, you will often find that clients seek incompatible goals. For example, a young client may want the approval of his friends, who see little value in education, and his parents, who want him to do well in school. To win acceptance from his peers, he may allow his academic performance to suffer; and if he pleases his parents by studying hard, his friends say he has "sold out." Or a client may want to take a vacation when work demands that he or she stay on the job.

Examples of Discrepancies Between You and the Client

One of the more challenging issues occurs when you and the client are not in synchrony. And this can occur on any of the dimensions above. Your nonverbal communication may be misread by the client. The client may avoid really facing issues. You may be saying one thing, the client another. A conflict in values or goals in counseling may be directly apparent or a quiet, unsaid thing, which still impacts the client—and either can destroy your relationship.

When you get too close to the truth, clients may wipe their nose. If you provide a clear interpretation/reframe, the client may smile slightly, perhaps laugh a bit, and look down.

Note when arms or legs are crossed or move to a more open posture. Does the client turn toward you or turn away? Hand movements and looking away at key times often indicate far more than words. Parties are a good place to note joining and rejection nonverbals. Note what happens to you at a party when you join a group. Do the others draw you in by recognition nonverbally, or are you invisible?

At times, you may carefully introduce discussion of this discrepancy into the session. In such cases, you may interpret or reframe the situation differently from how the client has been presenting it. Or you may face differences squarely and openly, discussing your mutual lack of communication at that moment. Again, missing discrepancies can undermine the session(s).

SUMMARY: OBSERVATION SKILLS

The interviewer seeks to observe client verbal and nonverbal behavior with an eye to identifying discrepancies, mixed messages, incongruity, and conflict. Counseling and therapy in particular, but also interviewing, frequently focus on problems and their resolution. A discrepancy is often a problem. At the same time, discrepancies in many forms are part of life and may even be enjoyed. Humor, for example, is based on conflict and discrepancies. In addition, observation will help you develop recovery skills you can use when you are lost or confused in the interview. Even the most advanced professional doesn't always know what is happening. When you don't know what to do, attend!

Key points of the chapter are presented below.

■ KEY POINTS

Importance of observation skills	The self-aware interviewer is constantly aware of the client and of the here-and-now interaction in the session. Clients tell us about their world by nonverbal and verbal means. Observation skills are a critical tool in determining how the client interprets the world. Simple, careful observation of the interview is basic. What can you see, hear, and feel from the client's world? Note your impact on the client: How does what you say change or relate to the client's behavior? Use these data to adjust your microskill or interviewing technique.
Three key items to observe	Observation skills focus on three areas:

1. *Nonverbal behavior.* Your own and client eye-contact patterns, body language, and vocal qualities are, of course, important. Shifts and changes in these may be indicative of client interest or discomfort. A client may lean forward, indicating excitement about an idea, or cross his or her arms to close it off. Facial clues (brow furrowing, lip tightening or loosening, flushing, pulse rate visible at temples) are especially important. Larger scale body movements may indicate shifts in reactions, thoughts, or the topic.
2. *Verbal behavior.* Noting patterns of verbal tracking for both you and the client is particularly important. At what point does the topic change and who initiates the change? Where is the client on the abstraction ladder? If the client is concrete, are you matching her or his language? Is the client making "I" statements or "other" statements? Do the client's negative statements become more positive as counseling progresses? Clients tend to use certain key words to describe their behavior and situations; noting these descriptive words and repetitive themes is helpful.
3. *Conflict, discrepancies.* Incongruities, mixed messages, and contradiction are manifest in many and perhaps all interviews. The effective interviewer is able to identify these discrepancies, to name them appropriately, and sometimes, to feed them back to the client. These discrepancies may be between nonverbal behaviors, between two statements, between what clients say and what they do, between incompatible goals, between statements and nonverbal behavior. They may also represent a conflict between people or between a client and a situation. And your own behaviors may be positively or negatively discrepant.

Multicultural issues	Note individual and cultural differences in verbal and nonverbal behavior. Always remember that some individuals and some cultures may have a meaning for a movement or use of language that is different from your own personal meaning. Use caution in your interpretation of nonverbal behavior.
Movement harmonics	Movement harmonics are particularly interesting and provide a basic concept that explains much verbal and nonverbal communication. When two people are talking together and communicating well, they often exhibit movement synchrony or movement complementarity in that their bodies move in a harmonious fashion. When people are not communicating clearly, movement dissynchrony will appear: body shifts, jerks, and pulling away are readily apparent.

COMPETENCY PRACTICE EXERCISES AND PORTFOLIO OF COMPETENCE

Many concepts have been presented in this chapter; it will take time to master them and make them a useful part of your interviewing. Therefore, the exercises here should be considered introductory. Further, it is suggested that you continue to work on these concepts throughout the time that you read this book. If you keep practicing them throughout your study of the book, material that might now seem confusing will gradually be clarified and become part of your natural style.

Individual Practice

Exercise 1: Observation of Nonverbal Patterns

Observe 10 minutes of a counseling interview, a television interview, or any two people talking. Videotape so that repeated viewing is possible.

Visual/eye contact patterns. Do people maintain eye contact more while talking or while listening? Does the "client" break eye contact more often while discussing certain subjects than others? Can you observe changes in pupil dilation as an expression of interest?

Vocal qualities. Note speech rate and changes in intonation or volume. Give special attention to speech "hitches" or hesitations.

Attentive body language. Note gestures, shifts of posture, leaning, patterns of breathing, and use of space. Give special attention to facial expressions such as changes in skin color, flushing, and lip movements. Note appropriate and inappropriate smiling, furrowing of the brow, and so on.

Movement harmonics. Note places where movement synchrony occurred. Did you observe examples of movement dissynchrony?

Where possible, observe your own videotape so that you can view the interview several times. One useful approach is to observe 5 to 10 minutes of interaction several times. Be sure to separate behavioral observations from impressions on the Feedback Form (Box 5-4).

1. Present context of observation. Briefly summarize what is happening verbally at the time of the observation. Number each observation.
2. Observe the interview for the following and describe what you see as precisely and concretely as possible: visual/eye contact patterns, vocal qualities, attentive body language, and movement harmonics.
3. Record your impressions. What interpretations of the observation do you make? How do you make sense of each observation unit? And—most important—are you cautious in drawing conclusions from what you have seen and noticed?

Exercise 2: Observation of Verbal Behavior and Discrepancies

Observe the same interview again, but this time pay special attention to verbal dimensions and discrepancies (which, of course, will include nonverbal dimensions). Consider the following issues and provide concrete evidence for each of your decisions as to what is occurring in the session. Again, separate observations from your interpretation and impressions.

Verbal tracking and selective attention. Pay special attention to topic jumps or shifts. Who initiates them? Do you see any pattern of special topic interest and/or avoidance? What does the listener seem to want to hear?

BOX 5-4   FEEDBACK FORM: OBSERVATION

_____ (Date)

_____   _____
(Name of Interviewer)   (Name of Person Completing Form)

Instructions: Observe the client or counselor carefully during the role-play session and immediately afterward complete the nonverbal feedback portion of the form. As you view the video-tape or listen to the audiotape, give special attention to verbal behavior and note discrepancies. If no recording equipment is available, one observer should note nonverbal behavior and the other verbal behavior.

Nonverbal behavior checklist

1. *Visuals.* At what points did eye contact breaks occur? Staring? Did the individual maintain eye contact more when talking or when listening? Changes in pupil dilation?

2. *Vocals.* When did speech hesitations occur? Changes in tone and volume? What single words or short phrases were emphasized?

3. *Body language.* General style and changes in position of hands and arms, trunk, legs? Open or closed gestures? Tight fist? Playing with hands or objects? Physical tension: relaxed or tight? Body oriented toward or away from the other? Sudden body shifts? Twitching? Distance? Breathing changes? At what points did changes in facial expression occur? Changes in skin color, flushing, swelling or contracting of lips? Appropriate or inappropriate smiling? Head nods? Brow furrowing?

4. *Movement harmonics.* Examples of movement complementarity, synchrony, or dissynchrony? At what times did these occur?

(continued)

BOX 5-4 (continued)

5. *Nonverbal discrepancies.* Did one part of the body say something different from another? With what topics did this occur?

Verbal behavior checklist

1. *Verbal tracking and selective attention.* At what points did the client or counselor fail to stay on the topic? To what topics did each give most attention? List here the most important key words used by the client; these are important for deeper analysis.

2. *Abstract or concrete?* Which word represents the client? How did the counselor work with this dimension? Was the counselor abstract or concrete?

3. *"I" statements and "other" statements.* While discussing the conflict or confusing situation, what self- and other statements did the client make? Which might be desirable to change, given a longer interview?

4. *Verbal discrepancies.* Write here observations of verbal discrepancies in either client or counselor.

Key words. What are the key words of each person in the communication?

Abstract or concrete conversation. Is this conversation about patterns or about specifics? Are the people involved approaching this issue in a similar fashion?

"I" statements. Consider each person. What is he or she trying to say from an "I" statement framework? Are any "I" statements present?

"Other" statements. How aware is this individual of connections with others? How accurate are those perceptions?

Discrepancies. What incongruities do you note in the behavior in either person you observe? Do you locate any discrepancies between the two? What issues of conflict might be important?

Exercise 3: Examining Your Own Verbal and Nonverbal Styles

Videotape yourself with another person in a real interview or conversation for at least 20 minutes. Do not make this a role-play. Then view your own verbal and nonverbal behavior and that of the person you are talking with in the same detail as in Exercises 1 and 2. What do you learn about yourself?

Exercise 4: Classifying Statements as Concrete or Abstract

Following are examples of client statements. Classify each statement as primarily concrete or primarily abstract. You will gain considerably more practice and thus have more suggestions for interventions in later chapters. (Answers to this exercise may be found at the conclusion of this chapter.) Circle C (concrete) or A (abstract) below:

C	A	1. I cry all day long. I didn't sleep last night. I can't eat.
C	A	2. I feel rotten about myself lately.
C	A	3. I feel very guilty.
C	A	4. Sorry I'm late for the session. Traffic was very heavy.
C	A	5. I feel really awkward on dates. I'm a social dud.
C	A	6. Last night my date said that I wasn't much fun. Then I started to cry.
C	A	7. My father is tall, has red hair, and yells a lot.
C	A	8. My father is very hard to get along with. He's difficult.
C	A	9. My family is very loving. We have a pattern of sharing.
C	A	10. My mom just sent me a box of cookies.

Group Practice

Exercise 5: Systematic Group Practice on Observation Skills

Many observation concepts have been discussed in this chapter. It is obviously not possible to observe all these in one single role-play interview. However, practice can serve as a foundation for elaboration at a later time. This exercise has been selected to summarize the central ideas of the chapter.

Step 1: Divide into practice groups. Triads or groups of four are most appropriate.

Step 2: Select a group leader.

Step 3: Assign roles for the first practice session.

▲ Client, who responds naturally and is talkative.

▲ Interviewer, who will seek to demonstrate a natural, authentic style.

▲ Observer 1, who observes client communication, using the Feedback Form: Observation, Box 5-4.

▲ Observer 2, who observes counselor communication, using the feedback form. Here the consultative microsupervision process usually focuses on helping the interviewer understand and utilize nonverbal communication more effectively. Ideally you have a videotape available for precise feedback.

Step 4: Plan. State the goals of the session. As the central task is observation, the interviewer should give primary attention to attending and open questions. Use other skills as you wish. After the role-play is over, the interviewer should report personal observation of the client made during that time and demonstrate basic or active mastery skills. The client will report on observations of the counselor.

The suggested topic for the practice role-play is "Something or someone with whom I have a present conflict or have had a past conflict." Alternative topics include the following:

My positive and negative feelings toward my parents or other significant persons
The mixed blessings of my work, home community, or present living situation

The two observers may use this session as an opportunity both for providing feedback to the interviewer and for sharpening their own observation skills.

Step 5: Conduct a 6-minute practice session. As much as possible, both the interviewer and the client will behave as naturally as possible discussing a real situation.

Step 6: Review the practice session and provide feedback for 14 minutes. Remember to stop the audiotape or videotape periodically and listen to or view key items several times for increased clarity. Observers should give special attention to careful completion of the feedback form throughout the session, and the client can give important feedback via the Client Feedback Form from Chapter 1.

Step 7: Rotate roles.

Portfolio of Competence

Determining your own style and theory can be best accomplished on a base of competence. Each chapter closes with a reflective exercise asking your thoughts and feelings about what has been discussed. By the time you finish this book, you will have a substantial record of your competencies and a good written record as you move toward determining your own style and theory.

Use the following as a checklist to evaluate your present level or mastery. Check those dimensions that you currently feel able to do. Those that remain unchecked can serve as future goals. Do not expect to attain intentional competence on every dimension as you work through this book. You will find, however, that you will improve your competencies with repetition and practice.

Check below the competencies that you have met to date.

Level 1: Identification and classification.

❑ Note attending nonverbal behaviors, particularly changes in behavior in visuals—eye contact, vocal tone, and body language.
❑ Note movement harmonics.

❑ Note verbal tracking and selective attention.
❑ Note key words used by the client and yourself.
❑ Note distinctions between concrete/situational and abstract/formal operational conversation.
❑ Note discrepancies in verbal and nonverbal behavior.
❑ Note discrepancies in the client.
❑ Note discrepancies in yourself.
❑ Note discrepancies between yourself and the client.

Level 2: Basic competence. Nonverbal and verbal observation skills are things that you can work on and improve over a lifetime. Therefore, use the intentional competence list below for your self-assessments.

Level 3: Intentional competence. You will be able to note client verbal and nonverbal behaviors in the interview and use these observations at times to facilitate interview conversation. You will be able to match your behavior to the client's. When necessary, you will be able to mismatch behaviors to promote client movement. For example, if you first join the negative body language of a depressed client and then take a more positive position, the client may follow and adopt a more assertive posture. You will be able to note your own verbal and nonverbal responses to the client. You will be able to note discrepancies between yourself and the client and work to resolve those discrepancies.

Developing mastery of the following areas will take time. Come back to this list later as you have practiced other skills in this book. For the first stages of basic and intentional mastery, the following competencies are suggested as most important:

❑ Mirror nonverbal patterns of the client. The interviewer mirrors body position, eye contact patterns, facial expression, and vocal qualities.
❑ Identify client patterns of selective attention and use those patterns either to bring talk back to the original topic or to move knowingly to the new topic provided by the client.
❑ Match clients' concrete/situation or abstract/formal operational language and help them to expand their stories in their own style.
❑ Identify key client "I" and "other" statements and feed them back to the client accurately, thus enabling the client to describe and define what is meant more fully
❑ Note discrepancies and feed them back to the client accurately. (Note that this is an important part of the skill of confrontation and conflict management, discussed in detail in Chapter 9.) The client in turn will be able to accept and use the feedback for further effective self-exploration.
❑ Note discrepancies in yourself and act to change them appropriately.

Level 4: Psychoeducational teaching competence. You will demonstrate your ability to teach observation skills to others. Your achievement of this level can be determined by how well your students can be rated on the basic competencies of this self-assessment form. Certain of your clients in counseling may be quite insensitive to obvious patterns of nonverbal and verbal communication. Teaching them beginning methods of observing others can be most helpful to them. Do not introduce more than one or two concepts to a client per interview, however!

❑ Teach clients in a helping session the social skills of nonverbal and verbal observation and the ability to note discrepancies.
❑ Teach small groups the above skills.

DETERMINING YOUR OWN STYLE AND THEORY: CRITICAL SELF-REFLECTION ON OBSERVATION SKILLS

This chapter has focused on the importance of verbal and nonverbal observation skills and you have experienced a variety of exercises designed to enhance your awareness in this area.

Again, what single idea stood out for you among all those presented in this chapter, in class, or through informal learning? What stands out for you is likely to be important as a guide toward your next steps. What are your thoughts on multicultural differences? What other points in this chapter struck you as important? How might you use ideas in this chapter to begin the process of establishing your own style and theory?

What are your thoughts?

_____

_____

HOW ALLEN RESPONDED TO THE COURTROOM SITUATION

As part of his testimony on whether Horace could take on some form of employment, Allen pointed out to the judge that many traditional African Americans from the South had a different pattern of eye contact from middle-class White standards. Looking away in some African Americans is actually a sign of respect. It also goes back to a history when Blacks had learned it was unsafe to have direct eye contact with White people in power. Also, cultural traditions of respect for elders and those in authority can lead to less eye contact. Expect this behavior in some international students and recent immigrants.

CORRECT RESPONSES FOR EXERCISE 4

1 — C	5 — A	8 — A
2 — A	6 — C	9 — A
3 — A	7 — C	10 — C
4 — C		

CHAPTER 6

Encouraging, Paraphrasing, and Summarizing: Key Skills of Active Listening

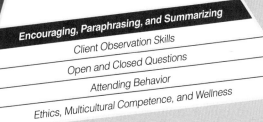

Encouraging, Paraphrasing, and Summarizing

Client Observation Skills

Open and Closed Questions

Attending Behavior

Ethics, Multicultural Competence, and Wellness

The possibility is only one sentence away. . . . Our goal is to make the eyes shine!

—**Andrew Zander**

How can these skills help you and your clients?

Chapter Goals   The goal of this chapter is to examine the basics and importance of encouragers and restatements, paraphrases, and summarizations, which are basic to helping a client feel understood. Clients need to know that the interviewer has heard what they have been saying, seen their point of view, and felt their world as they experience it. Once clients' stories have been truly heard, the clients can be much more open to and ready for change.

Competency Objectives   Awareness, knowledge, and skills in active listening will enable you to

- ▲ Help the client clarify what he or she is trying to say.
- ▲ Clarify for the interviewer what the client has said. By feeding back what you have heard, you can check on the accuracy of your listening.
- ▲ Facilitate clients talking in more detail about issues of concern.
- ▲ Show the client that you indeed have heard.
- ▲ Help an overly talkative client stop repeating the same facts or story, thus speeding up and clarifying the interview process.
- ▲ Help clients organize the key aspects of their issues and concerns via periodic summarization.

Jennifer: (enters the room and starts talking immediately) I really need to talk to you. I don't know where to start. I just got my last exam back and it was a disaster, maybe because I haven't studied much lately. I was up late drinking at a party last night, and I almost passed out. I've been sort of going out with a guy for the last month, but that's over as of last night. . . . (pause) But what really bothers me is that my Mom and Dad called last Monday and they are going to separate. I know that they have fought a lot, but I never thought it would come to this. I'm thinking of going home, but I'm afraid to. . . .

Jennifer continues for another 3 minutes in much the same vein, repeating herself somewhat, and seems close to tears. At times the data are coming so fast that it is hard to follow her. Finally she stops and looks at you expectantly.

What might be going through your mind about Jennifer at this moment? Using the ideas of active listening, what would you say and do to help her feel that you understand her and empathize?

What would you say?

_____

_____

On page 170, you'll find what thoughts occurred to us.

INTRODUCTION: ACTIVE LISTENING

If you use encouraging, paraphrasing, and summarizing skills as defined here, you can *predict* how clients will respond.

Encouraging	Predicted Result
Encourage with short responses that help clients keep talking. They may be verbal restatements (repeating key words and short statements) or nonverbal actions (head nods and smiling).	Clients elaborate on the topic, particularly when encouragers and restatements are used in a questioning tone of voice.
Paraphrasing (also known as reflection of content)	**Predicted Result**
Shorten and clarify the essence of what has just been said, but be sure to use the client's main words when you paraphrase. Paraphrases are often fed back to the client in a questioning tone of voice.	Clients will feel heard. They tend to give more detail without repeating the exact same story. If a paraphrase is inaccurate, the client has an opportunity to correct the interviewer.
Summarizing	**Predicted Result**
Summarize client comments and integrate thoughts, emotions, and behaviors. This technique is similar to paraphrase but used over a longer time span.	Clients will feel heard and often learn how the many parts of important stories are integrated. The summary tends to facilitate a more centered and focused discussion. The summary also provides a more coherent transition from one topic to the next or as a way to begin and end a full session.

Listening is not a passive process. Whether using attending skills, encouraging, paraphrasing, or summarizing, you are actively involved in the interview. You do not just sit and listen to a story. Active listening demands that you participate fully by helping the client clarify, enlarge, and enrich the story. It requires that you be able to hear small changes in thoughts, feelings, and behaviors. It asks that you walk in the other person's shoes. Active listening demands serious attention to empathy—truly being with and understanding the client as fully as possible.

Accurate empathic listening is not as common as we might wish nor is it as easy, but its effects are profound. Ask your friend or family member to tell you a story (e.g., a conflict, a positive experience, a current challenge). Simply sit and listen to what is said, perhaps asking a few questions to enrich and enlarge the story. Then say back to the volunteer client, as accurately as possible, what you have heard. Ask your friend how accurate your summary was and how it felt to be listened to. Think and then write down the other person's reactions.

What was your experience?

_____

_____

Encouraging, paraphrasing, and summarizing are basic to empathic understanding and enable you to communicate to clients that they have been heard. In these accurate listening skills you do not mix your own ideas with what clients have been saying. You say back to the clients what you have heard, using their key words. You help clients by distilling, shortening, and clarifying what has been said.

Encouragers are a variety of verbal and nonverbal means that the counselor or therapist can use to prompt clients to continue talking. They include head nods, openhanded gestures, phrases such as "Uh-huh," and the simple repetition of key words the client has uttered. Restatements are extended encouragers, the repetition of two or more words exactly as used by the client. In addition, appropriate smiling and interpersonal warmth are major encouragers that help clients feel comfortable and keep talking in the interview.

Paraphrasing, sometimes called reflection of content, feeds back to the client the essence of what has just been said. The listener shortens and clarifies the client's comments. Paraphrasing is not parroting; it is using some of your own words plus the important main words of the client.

Summarizations are similar to paraphrases but are used to clarify and distill what the client has said over a longer time span. Summarizations may be used to begin or end an interview, to move to a new topic, or to clarify complex issues. Most important, summarization helps both the client and you organize thinking about what is happening in the interview.

Counseling Children

All clients have an equal need to know they have been heard. Counseling children is much like counseling adolescents and adults. You will use the same microskills, but there tends to be more emphasis on encouraging, paraphrasing, and summarization skills. You will note that many effective elementary teachers constantly say back to students what the students have just said. These skills reinforce the conversation and help the children keep talking from their own frame of reference. Telling your story to someone who hears you accurately is clarifying, comforting, and reassuring. Box 6-1 presents key points in the use of listening with children.

BOX 6-1   LISTENING SKILLS AND CHILDREN

The listening and observing skills are just as important to use with children. Children too often go through life being told what to do. If we listen to them and their singular constructions of the world, we can reinforce their unique qualities and help them develop belief in themselves and in their own value.

Here are a few key comments on the listening skills and children:

Attending

Avoid looking down at children; whenever possible, talk to them at their level. This may mean sitting on the floor or in small chairs. Their energy is such that it helps if they have something to do with their hands; perhaps you can allow them to draw or play with clay as they talk to you. Be prepared for more topic jumps than with adults, but use attending skills later to bring them back to critical issues that need to be discussed.

Questions and Concreteness

Use short sentences, simple words, and a concrete style of language. Avoid abstractions. Children may have difficulty with a broad open question such as "Could you tell me generally what happened?" Break down such abstract questions into concrete and situational language using a mix of closed and open questions such as, "Where were you when the fight occurred?" "What was going on just before the fight?" "Then what happened?" "How did he feel?" "Was she angry?" "What happened next?" "What happened afterward?" In questioning children on touchy issues, be especially careful of closed, leading questions. Seek to get their perspective, not yours.

Encouraging, Paraphrasing, and Summarizing

Effective elementary teachers use these skills constantly, especially paraphrasing and encouraging. Seek out a competent teacher and observe for yourself. These skills, coupled with good attending and questioning, are very important in helping children get out their stories.

Other Issues

Provide an atmosphere that is suitable for children by using small chairs and interesting objects. Warmth and an actual liking for children are essential. Use names rather than pronouns, as children often get confused when under stress (as do many adults). Smiling, humor, and an active style will help.

EXAMPLE INTERVIEW: THEY ARE TEASING ME ABOUT MY SHOES

The following sample interview is an edited version of a videotaped interview conducted by Mary Bradford Ivey with Damaris, a child actor, role-playing the problem. Damaris is an 11-year-old sixth grader. The session below presents a child's problem, but all of us, regardless of age, experienced nasty teasing and put-downs, often in our closest relationships. Mary first draws out the child's story about teasing and then Damaris's thoughts and feelings about the teasing. Mary follows with a focus on the child's strengths, an example of the wellness approach.

Mary uses many encouragers and restatements. A review of the entire video transcript reveals nine minimal encouragers ("oh . . . ," "uh-huh," and single word utterances), two positive encouragers ("that's great," "nice"), four additional brief restatements, and numerous smiles and head nods. Children demand constant involvement, and showing your interest and good humor is even more essential. Active listening is especially important with children, as they tend to respond more briefly than adults.

Interviewer and Client Conversation	Process Comments
1. *Mary:* Damaris, how're you doing?	The relationship between Mary and Damaris is already established; they know each other through school activities.
2. *Damaris:* Good.	She smiles and sits down.
3. *Mary:* I'm glad you could come down. You can use these markers if you want to doodle or draw something while we're talking. I know—you sort of indicated that you wanted to talk to me a little bit.	Mary welcomes the child and offers her something to do with her hands. Many children get restless just talking. Damaris starts to draw almost immediately. You may do better with an active male teen by taking him to the basketball court while discussing issues. It can also help to have things available for adults to do with their hands.
4. *Damaris:* In school, in my class, there's this group of girls that keep making fun of my shoes, just 'cause I don't have Nikes.	Damaris looks down and appears a bit sad. She stops drawing. Children, particularly the "have-nots," are well aware of their economic circumstances. Some children must wear used sneakers; Damaris, at least, has newer ones.
5. *Mary:* They "keep making fun of your shoes"?	Encourage in the form of a restatement using Damaris's exact key words.
6. *Damaris:* Well, they're not the best; I mean— they're not Nikes, like everyone else has.	Damaris has a slight angry tone mixed with her sadness. She starts to draw again.
7. *Mary:* Yeah, they're nice shoes, though. You know?	It is sometimes tempting to comfort clients rather than just listening. Mary offers reassurance; a simple "uh-huh" would have been more effective. However, reassurance later in the interview may be a very important intervention.
8. *Damaris:* Yeah. But my family's not that rich, you know. Those girls are rich.	Clients, especially children, hesitate to contradict the counselor. Notice that Damaris says "But . . ." When clients say "Yes, but . . . ," interviewers are off track and need to change their style.
9. *Mary:* I see. And the others can afford Nike shoes, and you have nice shoes, but your shoes are just not like the shoes the others have, and they tease you about it?	Mary backs off her reassurances and paraphrases the essence of what Damaris has been saying using her key words.
10. *Damaris:* Yeah. . . . Well, sometimes they make fun of me and call me names, and I feel sad. I try to ignore them, but still, the feeling inside me just hurts.	If you paraphrase or summarize accurately, a client will usually respond with *yeah* or *yes* and continue to elaborate the story.

(continued)

Interviewer and Client Conversation	Process Comments
11. *Mary:* It makes you feel hurt inside that they should tease you about shoes.	Mary reflects Damaris's feelings. The reflection of feeling is close to a paraphrase and is elaborated in the following chapter.
12. *Damaris:* Mmm-hmm. (pause) It's not fair.	Damaris thinks about Mary's statement and looks up expectantly as if to see what happens next. She thinks back on the basic unfairness of the whole situation.
13. *Mary:* So far, Damaris, I've heard how the kids tease you about not having Nikes and that it really hurts. It's not fair. You know, I think of you, though, and I think of all the things that you do well. I get . . . you know . . . it makes me sad to hear this part because I think of all the talents you have, and all the things that you like to do and— and the strengths that you have.	Mary's brief summary covers most of the important points, and Mary also discloses some of her own feelings. Sparingly used, self-disclosure can be helpful. Mary begins the positive asset search by reminding Damaris that she has strengths to draw from.
14. *Damaris:* Right. Yeah.	Damaris smiles slightly and relaxes a bit.
15. *Mary:* What comes to mind when you think about all the positive things you are and have to offer?	An open question encourages Damaris to think about her strengths and positives.
16. *Damaris:* Well, in school, the teacher says I'm a good writer, and I want to be a journalist when I grow up. The teacher wants me to put the last story I wrote in the school paper.	Damaris talks a bit more rapidly and smiles.
17. *Mary:* You want to be a journalist, 'cause you can write well? Wow!	Mary enthusiastically paraphrases positive comments using Damaris's own key words.
18. *Damaris:* Mmm-hmm. And I play soccer on our team. I'm one of the people that plays a lot, so I'm like the leader, almost, but . . . (Damaris stops in mid-sentence.)	Damaris has many things to feel good about; she is smiling for the first time in the session.
19. *Mary:* So, you are a scholar, a leader, and an athlete. Other people look up to you. Is that right? So how does it feel when you're a leader in soccer?	Mary has added *scholar* and *athlete* for clarification and elaboration of the positive asset search. She knows from observation on the playground that other children do look up to Damaris. Counselors may add related words to expand the meaning. Mary wisely avoids leading Damaris and uses the check-out, "Is that right?" Mary also asks an open question about feelings. And we note that Damaris used that important word "but." Do you think that Mary should have followed up on that, or should she continue with her search for strengths?

(continued)

Interviewer and Client Conversation	Process Comments
20. *Damaris:* (small giggle, looking down briefly) Yeah. It feels good.	Looking down is not always sadness! The spontaneous movement of looking down briefly is termed the "recognition response." It most often happens when clients learn something new and true about themselves. Damaris has internalized the good feelings.
21. *Mary:* So you're a good student, and you are good at soccer and a leader, and it makes you feel good inside.	Mary summarizes the positive asset search using both facts and feelings. The summary of feeling *good inside* contrasts with the earlier feelings of *hurt inside.*
22. *Damaris:* Yeah, it makes me feel good inside. I do my homework and everything, (pause and the sad look returns) but then when I come to school, they just have to spoil it for me.	Again, Damaris agrees with the paraphrase. She feels support from Mary and is now prepared to deal from a stronger position with the teasing. Here we see what lies behind the "but" in 18 above. We believe Mary did the right thing in ignoring the "but" the first time. But it is also obvious that the negative feelings need to be addressed. When strengths are clear, Mary can better address those negative feelings from Damaris's wellness strengths.
23. *Mary:* They just spoil it. So you've got these good feelings inside, good that you're strong in academics, good that you're, you know, good at soccer and a leader. Now, I'm just wondering how we can use those good feelings that you feel as a student who's going to be a journalist someday and a soccer player who's a leader. Now the big question is how you can take the good, strong feelings and deal with the kids who are teasing. Let's look at ways to solve your problem now.	Mary restates Damaris's last words and again summarizes the many good things that Damaris does well. Mary changes pace and is ready to move to the problem-solving portion of the interview.

Mary had a good relationship and was able to draw out Damaris's story fairly quickly. She moved to story and strengths and wellness assets that make it easier to address client problems and challenges.

Using these same skills with an adult, you could expect to follow a similar interviewing structure. However, most adults will provide longer verbal responses. And you would usually not need to use as many verbal and nonverbal encouragers.

Damaris was stressed because of a seemingly "small" event, being teased about her shoes. You may wonder when these kinds of events become traumatic. Box 6-2 reviews this issue.

BOX 6-2   ACCUMULATIVE STRESS: WHEN DO "SMALL" EVENTS BECOME TRAUMATIC?

At one level, being teased about the shoes one wears doesn't sound all that serious—children will be children! However, some poor children go through their entire school life wearing clothes that others tease them about and laugh at, either directly or indirectly. At a high school reunion, Allen talked with a classmate who clearly was disturbed emotionally. During the talk, it became clear that teasing and bullying during schools days were still immediate and painful memories for him.

Small slights become big hurts if repeated again and again over the years. Athletes and "popular" students may talk arrogantly and dismissively about the "nerds," "townies," "rurals," or other outgroup. Teachers, coaches, and even counselors sometimes join in the laughter. Over time, these slights mount inside the child or adolescent. Some people internalize their issues in psychological distress; others may act them out in a dramatic fashion—witness the episodes of shootings throughout the country in both high schools and colleges.

Microaggressions

Discrimination and prejudice are other examples of accumulative stress and trauma. They are also known as microaggressions (Sue, 2007). One of Mary's interns, a young African American woman, spoke with her of a recent racial insult. She was sitting in a restaurant and overheard two White people talking loudly about the "good old days" of segregation. Perhaps the remark was not directed at her, but still, it hurt. The young woman went on to speak of how common racial insults were in her life, directly or indirectly. She could tell how bad things were racially by the size of her phone bill. When an incident occurred that troubled her, she needed to talk to her sister or parents and seek support. Out of the continuing indignities of microaggressions come feelings of underlying insecurity about one's place in the world (internalized oppression and self-blame) and/or tension and rage about unfairness (externalized awareness of oppression). Either way, the person who is ignored or insulted feels tension in the body, the pulse and heart rate increase, and—over time—hypertension and high blood pressure may result. The psychological becomes physical, and accumulative stress becomes traumatic.

Soldiers at war, women who suffer sexual harassment, those who are overweight or short in height, the physically disfigured through birth or accident, gays and lesbians, and many others are all at risk for accumulative stress building to real trauma. They all suffer the dangers of posttraumatic stress.

As an interviewer, you will want to be alert for signs of accumulative stress and microaggressions in your clients. Are they internalizing the stressors by blaming themselves? Or are they externalizing and building a pattern of rage and anger inside that may explode? All these people have important stories to tell, and at first, these stories may sound routine. Posttraumatic stress responses in later life may be alleviated or prevented by your careful listening and support.

Finally, social work's position on social justice is that the interviewer has the responsibility to act and intervene, where possible, to combat oppression and injustice. The counseling and psychology position on social action are becoming clearer. Where do you stand?

Informed Consent and Working With Children

When you work with children, the ethical issues around informed consent become especially important. Depending on state laws and practices, it is often necessary to obtain written parental permission before interviewing a child. And you must obtain permission when you are sharing information about the interview with others. The child and family should know exactly how the information is to be shared, and interviewing records should be available to them for their comments and evaluation. An important part of informed consent is stating that they have the right to withdraw their permission at any point.

INSTRUCTIONAL READING: THE ACTIVE LISTENING SKILLS OF ENCOURAGING, PARAPHRASING, AND SUMMARIZING

Important in this section and in all interviewing, counseling, and psychotherapy is a non-judgmental attitude by which you simply hear and accept what the client is saying. All interviewing behavior can possibly convey judgmental and negative attitudes. A real challenge is to listen nonjudgmentally when you have inner feelings discrepant from those of the client. Recall that your client is often very able to catch small facial expressions that reveal your judgments. How can you deal with this? Basically, focus your attention fully on the client in the here and now; try to enter that person's world as he or she sees it. Later, you can separate yourself from this world and more ably help your client.

Encouraging

Encouragers include head nods, open gestures, and positive facial expressions that encourage the client to keep talking. Minimal verbal utterances such as "Ummm" and "Uh-huh" have the same effect. Silence, accompanied by appropriate nonverbal communication, can be another type of encourager. All these encouragers have minimum effect on the direction of client talk; clients are simply encouraged to keep talking. Restatements, repetition of clients' key words, or brief statements are also encouragers that more directly influence what clients talk about.

Let's imagine that Jennifer, the client with multiple issues presented at the beginning of this chapter, focuses on her parents' separating as the major immediate issue.

> I feel like my life is falling apart. I've always been close to both my folks, and since they told me that they were breaking up, nothing has been right. When I sit down to study, I can't concentrate. And my roommate says that I get angry too easily. I guess everything upsets me. I was doing okay in my classes until this came along. I'm hurting so much for my Mom.

Jennifer still has a lot going on in her life. Eventually, we will want to focus on some of her strengths, but at the moment, she clearly needs to vent and explore her thoughts and feelings around her parents in more depth. There are several key words and ideas in this statement, and the repetition of any of them is likely to lead Jennifer to expand on current issues. As a counselor, we'd tend to recommend repeating the exact key words, "You're hurting." This provides an opening for her to discuss her feelings or thoughts about Mom, herself, and, if she chooses, even other issues.

"You're hurting for your Mom" would focus more narrowly, but likely would be another good choice. "Falling apart," "close to your folks," "can't concentrate," and "you get angry easily" are other possibilities. All of these will help Jennifer continue to talk, but do lead in varying directions.

Key word encouragers contain one, two, or three words, while restatements are longer. Both focus on staying very close to the client's language, most typically changing only "I" to "you." (Jennifer: "I'm hurting so much for my Mom." Counselor: "You're hurting.")

It may be helpful if you reread the paragraphs above, saying aloud the suggested encouragers and restatements. Use different vocal tones and note how your verbal style can facilitate others' talking or stop them cold.

All types of encouragers facilitate client talk unless they are overused or used badly. Excessive head nodding or gestures and too much parroting can be annoying and frustrating to the client. From the observation of many interviewers, we know that use of too many encouragers can seem wooden and unexpressive. However, too few encouragers may suggest to clients that you are not interested or involved. Well-placed encouragers help to maintain flow and continually communicate that the client is being listened to.

Paraphrasing (Reflection of Content)

At first glance, paraphrasing appears to be a simple skill, only slightly more complex than encouraging. However, if you are able to give an accurate paraphrase to a client, you are likely to be rewarded with a "That's right" or "Yes . . . ," and the client will go on to explore the issue in more depth. The goal of paraphrasing is facilitating client exploration and clarification of issues. The tone of your voice and your body language while paraphrasing will indicate whether you are interested in listening or wish for the client to move on.

Accurate paraphrasing can help clients complete their storytelling. A client who has been through a trauma may need to tell the story several times. Our goal is not to stop this talk, but paraphrasing can help work through the trauma because each time you repeat what the client has said, the client's story has been told again and *heard.* Friends who have been through a difficult hospital operation need to tell their story several times. Rather than becoming bored and saying, "I've heard that before," give full attention and say back or paraphrase what you have heard.

How do you paraphrase? Client observation skills are important in accurate paraphrasing. You need to hear the client's important key words and use them in your paraphrase much as the client does. Other aspects of the paraphrase may be in your own words, but the main ideas and concepts should reflect the client's view of the world, not yours!

An accurate paraphrase, then, usually consists of four dimensions:

1. A *sentence stem,* sometimes using the client's name. Names help personalize the session. Examples would be, "Damaris, I hear you saying . . . ," "Luciano, sounds like . . . ," and "Looks like the situation is . . ." A stem is not always necessary and, if overused, can make your comments seem like parroting. Clients have been known to say in frustration, "That's what I just said; why do you ask?"

2. The *key words* used by the client to describe the situation or person. Again, drawing on client observation skills, the effort is to include key words and main ideas that come from clients. This repetition can be confused with the encouraging restatement. A restatement, however, is almost entirely in the client's own words and covers only limited amounts of material.

3. The *essence of what the client has said* in briefer and more clear form. Here the interviewer's skill in transforming the client's sometimes confused statements into succinct, meaningful, and clarifying statements is most valuable to smoothing the interviewing process. The counselor has the difficult task of keeping true to the client's ideas but not repeating them exactly.

4. A *check-out* for accuracy. The check-out is a brief question at the end of the paraphrase, asking the client for feedback on whether the paraphrase (or summary or other microskill) was relatively correct and useful. Some example check-outs include "Am I hearing you correctly?" "Is that close?" "Have I got it right?" It is also possible to paraphrase with an implied check-out by raising your voice at the end of the sentence as if the paraphrase were a question.

Here is a client statement followed by sample key-word encouragers, restatements, and a paraphrase:

> I'm really concerned about my wife. She has this feeling that she has to get out of the house, see the world, and get a job. I'm the breadwinner, and I think I have a good income. The children view Yolanda as a perfect mother, and I do too. But last night, we really saw the problem differently and had a terrible argument.

▲ Key-word encouragers: "Breadwinner?" "Terrible argument?" "Perfect mother?"
▲ Restatement encouragers: "You're really concerned about your wife." "You see yourself as the breadwinner." "You had a terrible argument."
▲ Paraphrase: "You're concerned about your picture-perfect wife who wants to work even though you have a good income, and you've had a terrible argument. Is that how you see it?"

The key-word encourager operates like selective attention. Note that the encouragers above lead the client in very different directions for what is appropriate conversation. "Breadwinner" leads to talk about the job and possibly responsibility. "Terrible argument" may lead to the details of the argument, whereas "perfect mother" may lead to comments on his wife's behavior.

This example shows that the key-word encourager, the restatement, and the paraphrase are all different points on a continuum. In each case the emphasis is on hearing the client and feeding back what has been said. Both short paraphrases and longer key-word encouragers will resemble restatements. A long paraphrase is close to a summary. All can be helpful in an interview, or they can be overdone.

Summarizing

Summarizing encompasses a longer period of conversation; at times it may cover an entire interview or even issues discussed by the client over several interviews. In summarizing, the interviewer attends to verbal and nonverbal comments from the client over a period of time and then selectively attends to key concepts and dimensions, restating them for the client as accurately as possible. The facts, thoughts, and emotions are included in the summary. A check-out at the end for accuracy is an important part of the summarization. The following are examples of summarizations.

To begin a session: Let's see, last time we talked about your angry feelings toward your mother-in-law, and we discussed the argument you had with her around the time the new baby arrived. You saw yourself as terribly anxious at the time and perhaps even out of control. Since then you two haven't gotten along too well. We also suggested homework and ideas of what to do next. How did that go?

Midway in the interview: So far, I've seen that the ideas you came up with didn't work too well. You felt a bit guilty and worried, thinking you were getting too manipulative, and another argument almost started. "Almost" is better than a "blow up." Yet one idea did work. You were able to talk with her about her garden, and it was the first time you had been able to talk about anything without an argument. You visualize the possibility of following up with new ideas next week. Is that about it so far?

At the end of the session: In this interview we've reviewed your feelings toward your mother-in-law in more detail. Some of the following things seem to stand out: First, our plan didn't work completely, but

BOX 6-3   RESEARCH EVIDENCE THAT YOU CAN USE

Active Listening

Microskills training enables counselors to respond in a more culturally appropriate fashion (Nwachuku & Ivey, 1991). Research on encouraging, paraphrasing, and summarizing are often treated as part of a larger whole—empathic listening. Some of the most carefully designed work in this area has been done by Bensing (1999) on physician-patient relationships. She found that physicians who established a solid relationship and listened carefully tended to be rated more highly and, actually, to be more likely to have patients follow their suggestions and directives. She also found that talk-time of patients markedly increased with physicians who listened to them. Nine studies on microcounseling with nurses found that they were rated more highly on empathy, focused more on the client, and made fewer therapeutic errors. Similar findings exist among counselors and therapists (Daniels & Ivey, 2007).

Smiling is one of the most encouraging things you can do. Many researchers have found that smiling "works" and is a primary way to communicate warmth and openness (Restak, 2003). Although the following is not the result of formal research, it is based on observation: Allen Ivey is a person who does not physically show a lot of emotion—he can smile, but is a bit shy and reserved. On the other hand, Mary Bradford Ivey is known for her smile and sunny disposition. Guess which person other people talk to when Allen and Mary are together? If you are a naturally warm person, this characteristic communicates itself to your client. If you are more like Allen, it may take you a bit longer, but you can still be a very effective listener!

Active Listening and Neuropsychology

Carter (1999, p. 87) reports:

> Expressions can . . . transmit emotions to others—the sight of a person showing intense disgust turns

on in the observer's brain areas that are associated with the feeling of disgust. Similarly, if you smile, the world does indeed smile with you (up to a point). Experiments in which tiny sensors were attached to the "smile" muscles of people looking at faces show the sight of another person smiling triggers automatic mimicry—albeit so slight that it may not be visible. . . . the brain concludes that something good is happening out there and creates a feeling of pleasure.

This is a variant of movement synchrony as described in the chapter on nonverbal communication. It is also possible for you to pick up the depressed mood and style of your client and recommunicate the sadness back to the client, thus reinforcing a cycle of negativity. And, if you listen with energy and interest, and this is communicated effectively, expect your client to receive that affect as a positive resource in itself. Active listening is a key aspect in developing a relationship and drawing out the client's story and strengths.

Talking about strengths and resources impacts the brain in useful ways. For example, the neurotransmitter dopamine is released when situations are pleasant and positive, preparing the brain for new learning and development of new neural networks. New learning will not occur unless the amygdala has enough stimuli that it *energizes* the brain for opening to new information and ideas. (See Appendix II.) At the same time, we need to recall that the amygdala is a basic seat of negative emotion—and certain types of stimuli will work against learning and change. Extreme external stimulation (war, rape, home break-in) can prompt too much stimulation, can blow the "fuse," and result in loss of neural connections.

you were able to talk about one thing without yelling. As we talked, we identified some behaviors on your part that could be changed. They include better eye contact, relaxing more, and changing the topic when you see yourself starting to get angry. I also liked your idea at the end of talking with her about the fact that you really want to forgive and be forgiven. Does that sum it up? Well, we have some specifics for next week. Let's see how it goes.

Diversity and the Listening Skills

Periodic encouraging, paraphrasing, and summarizing are basic skills that seem to have wide cross-cultural acceptance. Virtually all your clients like to be listened to accurately. It may take more time to establish a relationship with a client who is culturally different from you. But again, never generalize or stereotype.

Women tend to use paraphrasing and related listening skills more than men, whereas men tend to use questions more frequently. You may notice in your own classes and workshops that men tend to raise their hands faster and interrupt more often. But there are so many exceptions to this "rule" that it should not be relied on. Nonetheless, differences in gender do exist, and it is important to be aware of them. Gender differences need to be addressed directly in the interview by both men and women.

Some Asian (Cambodian, Chinese, Japanese, Indian) clients from traditional backgrounds may be seeking direction and advice. They are likely to be willing to share their stories, but you may need to tell them why you want to hear more about their issues before the two of you come up with answers. To establish credibility, you may have to commit yourself and provide advice earlier than you wish. In such a case, be assured and confident, but let them know you want to learn more with them and that your thoughts may change as you get to know them better. Ultimately, you want to work to get them to decide on their own. Boxes 6-3 and 6-4 review evidence that you can use and applications of the listening skills from an international perspective, respectively.

BOX 6-4   NATIONAL AND INTERNATIONAL PERSPECTIVES ON COUNSELING SKILLS

Developing Skills to Help the Bilingual Client

Azara Santiago-Rivera, Past President, National Latina/o Psychological Association, and University of Wisconsin, Milwaukee

It wasn't that long ago that counselors considered bilingualism a "disadvantage." We now know that a new perspective is needed. Let's start with two important assumptions: The person who speaks two languages is able to work and communicate in two cultures and, actually, is advantaged. The monolingual person is the one at a disadvantage! Research actually shows that bilingual children have more fully developed capacities and a broader intelligence (Power & Lopez, 1985).

If your client was raised in a Spanish-speaking home, for example, he or she is likely to think in Spanish at times, even though having considerable English skills. We tend to experience the world nonverbally before we add words to describe what we see, feel, or hear. For example, Salvadorans who experienced war or other forms of oppression felt that situation in their own language.

You are very likely to work with clients in your community who come from one or more language backgrounds. Your first task is to understand some of the history and experience of these immigrant groups. Then, we suggest that you learn some key words and phrases in their original language. Why? Experiences that occur in a particular language are typically encoded in memory in that language. So certain memories containing powerful emotions may not be accessible in a person's second language (English) because they were originally encoded in the first language (for example, Spanish). And if the client is talking about something that was experienced in Spanish, Khmer, or Russian, the key words are not English; they are in the original language.

(continued)

BOX 6-4 (continued)

Here is an example of how you might use these ideas in the session:

Social worker: Could you tell me what happened for you when you lost your job?

Maria (Spanish-speaking client): It was hard; I really don't know what to say.

Social worker: It might help us if you would say what happened in Spanish, and then you could translate it for me.

Maria: Es tan injusto! Yo pensé que perdí el trabajo porque no hablo el ingles muy bien. Me da mucho coraje cuando me hacen esto. Me siento herida.

Social worker: Thanks. I can see that it really affected you. Could you tell me what you said now in English?

Maria: (More emotionally) I said, "It all seemed so unfair. I thought I lost my job because I couldn't speak English well enough for them. It makes me really angry when they do that to me. It hurts."

Social worker: I understand better now. Thanks for sharing that in your own language. I hear you saying that *injusto* hurts, and you are very angry. Let's continue to work on this and, from time to time, let's have you talk about the really important things in Spanish, okay?

The above brief example provides a start. The next step is to develop a vocabulary of key words in the language of your client. This cannot happen all at once, but you can gradually increase your skills. Here are some Spanish key words that might be useful with many clients:

Respeto: Was the client treated with respect? For example, the social worker might say, "Your employer failed to give you *respeto*."

Familismo: Family is very important to many Spanish-speaking people. You might say, "How are things with your *familia?*"

Emotions (see next chapter) are often experienced in the original language. When reflecting feeling, you could learn and use these key words with clients:

Aguantar: endure
Amor: love
Cariño: like
Coraje: anger
Miedo: fear
Orgullo: proud
Sentir: feel

We also recommend learning key sayings, metaphors, and proverbs in the language(s) of your community. *Dichos* are Spanish proverbs, like the following examples:

Al que mucho se le da, mucho se le demanda.
The more people give you, the greater the expectations of you.

Vale mas tarde que nunca.
Better late than never.

No hay peor sordo que el no quiere oir.
There is no worse deaf person than someone who doesn't want to listen.

En la unión está la fuerza.
Strength is found in unity.

Consider developing a list like this, learn to pronounce the words correctly, and you will find them useful in counseling Spanish-speaking clients. Indeed, you are giving them *respeto*. You may wish to learn key words in several languages.

Carlos Zalaquett comments: We have produced a Spanish version of the *Basic Attending Skills, Las Habilidades Atencionales Básicas: Pilares Fundamentales de la Comunicación Efectiva* (Zalaquett, Ivey, Gluckstern-Packard, & Ivey, 2008), to help both monolingual and bilingual helpers. The attending skills are illustrated with examples provided by Latina/o professionals from different Latin American countries. Using the information and exercises included in the book you can sharpen your interviewing, counseling, and psychotherapeutic tools to provide effective services to those who speak Spanish.

SUMMARY: PRACTICE, PRACTICE, AND PRACTICE

We have stressed the importance of three major listening skills in this chapter—encouraging, paraphrasing, and summarizing. Regardless of your theory of choice and however you integrate the microskills into your own natural style, these skills will remain central if you are to be effective.

Basic competence comes when you use the skills in an interview and expect that to be helpful to your clients. Every client needs to be heard; demonstrating that you are listening carefully often makes a real difference, and intentional competence in these skills requires practice.

At this point in this book, we want to share the following story, as it really drives home the importance of continuing to practice the skills. You can pass exams without practice, but if you are serious about helping, learning these skills to full mastery is the most critical issue.

Amanda Russo was a student in a counseling course at Western Kentucky University taught by Dr. Neresa Minatrea. Amanda shared with us how she practiced the skills and gave us her permission to pass this on to you. As you read her comments, ask yourself if you are willing to go as far as she did to ensure expertise.

> For my final project I selected a practice exercise entitled The Positive Asset Search: Building Empathy on Strengths. I chose this exercise early in the book because I do not have much experience with counseling, and I wanted to try a fairly simple exercise to start out. I performed the same exercise on five different people to see if I would get the same results.
>
> The exercise consists of asking the client what some of their areas of strength are, getting them to share a story regarding that strength, and then for the counselor to observe the client's gestures and be aware of any changes. The first person I tried this exercise on was Raphael, a dormitory proctor. Some of his strengths were family, friends, working out, and that he had a good inner circle/support group. As he talked about his support group and how they reminded him of the positives, he started to sit in a less tense position. He seemed very relaxed, yet excited about his topic of discussion, and I noticed a lot of hand gestures. In a matter of seconds I saw him change from tense and unsure to relaxed and enthusiastic about what he was saying.
>
> The next person I practiced this exercise on was my roommate Karol. She was a bit nervous when we started and had a difficult time thinking of strengths. Once I asked her to share a story with me she became very animated. As she spoke, I could see a sparkle in her eyes. Her voice became stronger, and her hands were moving every which way. She feels strongly about doing well at work, giving advice, working out, playing music, and finishing the song she is currently writing.
>
> Once she gave me a couple of strengths, they started her wheels turning, and she was coming up with more and more. She felt very good about jazz practice earlier that day. She introduced a new song to the group, and they really enjoyed it. She also shared a story with me about a huge accomplishment at work that day. Karol was definitely the person whose mood/persona changed the most in this exercise. (Courtesy of Amanda Russo)

Amanda went on to interview three more people to practice her skills and reported in detail on each one. If you seek to reach intentional competence, the best route toward this is systematic practice. For some of us, one practice session may be enough. For most of us, it will take more time. What commitments are you willing to make?

The following section presents key points of this chapter.

■ **KEY POINTS**

Active listening	Clients need to know that their story has been heard. Attending, questioning, and other skills help the client open up, but accurate listening through the skills of encouraging, paraphrasing, and summarizing is needed to communicate that you have indeed heard the other person fully. All of these skills involve active listening, encouraging others to talk freely. They communicate your interest and help clarify the world of the client for both you and the client. Active listening skills are some of the most difficult in the microtraining framework.
Active listening skills	Three skills of accurate listening help communicate your ability to listen:

1. *Encouragers* are a variety of verbal and nonverbal means the counselor or interviewer can use to encourage others to continue talking. They include head nods, an open palm, "Uh-huh," and the repetition of key words the client has uttered. Selective attention to various client key words can have a profound impact on client direction. Restatements are extended encouragers using the exact words of the client and are less likely to determine what the client might say next.
2. *Paraphrases* feed back to the client the essence of what has just been said by shortening and clarifying client comments. Paraphrasing is not parroting; it is using some of your own words plus the important main words of the client.
3. *Summarizations* are similar to paraphrases except that a longer time and more information are involved. Attention is also given to emotions and feelings as they are expressed by the client. Summarizations may be used to begin an interview, for transition to a new topic, to provide clarity in lengthy and complex client stories, and, of course, to end the session. It can be wise to ask clients to summarize the interview and the important points that they observed.

The "how" of active listening	Paraphrasing and summarizing usually involve four dimensions:

1. *A sentence stem.* Often you will want to use the client's name—"Jamilla, I hear you saying . . . ," "Carlos, sounds like. . . . "
2. *Key words.* The clients' exact own key words that they use to describe their situation.
3. *The essence of what the client has said in distilled form.* Here you use the client's key words in a brief clarification of what the client has said. Summaries are longer paraphrases that often include emotional dimensions as well.
4. *A check-out.* Implicitly or explicitly, check with the client to see that what you have fed back to him or her is accurate. "Have I heard you correctly?"

A caution	These skills are useful with virtually any client. However, if you do not do well or if you seem mechanical, clients will find repetition tiresome and may ask, "Didn't I just say that?" Consequently, when you use the skill you should also employ your client observation skills. As you listen to clients, seek to maintain a nonjudgmental, accepting attitude. Even the most accurate paraphrasing or summarization can be negated if you lack supporting nonverbal behaviors.

COMPETENCY PRACTICE EXERCISES AND PORTFOLIO OF COMPETENCE

The three skills of encouraging, paraphrasing, and summarizing are much less controversial than questions. Virtually all interviewing theories recommend and endorse these key skills of active listening.

Individual Practice

Exercise 1: Identifying Skills

Below is a client statement followed by several alternative interviewer responses. Identify encouragers (E), restatements (R), paraphrases (P), and summaries (S).

Client: The visit went well. I've pretty much decided to go back and finish college. But how will I pay for it? I've got a good job now, but I'll have to move to part time, and I'm not sure that they will keep me. Getting all this financed will be difficult. It is kind of scary.

_____   "Uh-huh."

_____   Silence with facilitative body language.

_____   "Scary?"

_____   "You're not sure that they will keep you."

_____   "Sounds like you've made up your mind to finish college, but financing it will be a major challenge."

_____   "In the last interview, we talked about your going back to visit the school, and so far it sounds as if it went well, and you really want to do it. But at the same time, now, it is a little scary when you think of all the financial issues. Have I heard what's been happening correctly?"

Exercise 2: Generating Written Encouragers, Paraphrases, and Summarizations

A. "Chen and I have separated. I couldn't take his drinking any longer. It was great when he was sober, but it wasn't that often he was. Yet that leaves me alone. I don't know what I'm going to do about money, the kids, or even where to start looking for work."

Write three different types of key-word encouragers for this client statement:

_____

Write a restatement/encourager:

_____

Write a paraphrase (include a check-out):

_____

Write a summarization (generate data by imagining previous interviews):

_____

B. "We worry about having a child. We've been trying for months now, but with no luck. We're thinking about going to a doctor, but we don't have medical insurance."

Write three different types of key-word encouragers for this client statement:

_____

Write a restatement:

_____

Write a paraphrase (include a check-out):

_____

Write a summarization (generate data by imagining previous interviews):

_____

Exercise 3: Practice of Skills in Other Settings

Encouraging. During conversations with friends or in your own interviews, deliberately use single-word encouragers and brief restatements. Note their impact on your friends' participation and interest. You may find that the flow of conversation changes in response to your brief encouragers. Write down your observations.

_____

Group Practice

Exercise 4: Practice of Active Listening in a Group

Experience has shown that the skills of this chapter are often difficult to master. It is easy to try to feed back what another person has said, but to do it accurately, so that the client feels truly heard, is another matter.

Step 1: Divide into practice groups. Include triads as a possibility.

Step 2: Select a group leader.

Step 3: Assign roles for the first practice session.

▲ Client
▲ Interviewer
▲ Observer 1, who uses Feedback Form (Box 6-5).
▲ Observer 2, who uses Feedback Form.

Step 4: Plan. Establish and state clear goals for the practice session. For real mastery, seek to use only the three skills in this chapter and use questions only as a last resort. The interviewer should plan a role-play in which open questions are used to elicit the client's concern. Once this is done, use encouragers to help bring out more details and deeper meanings. Use open and closed questions as appropriate, but give primary attention to the paraphrase and the encourager. End the interview with a summary (this is often forgotten). Check the accuracy of your summary with a check-out ("Am I hearing you correctly?").

The suggested topic for this practice session is the story of a past or present stressful experience that may relate to the ideas of accumulative trauma. Examples include teasing; bullying; being made the butt of a joke; an incident when you were seriously misunderstood or misjudged; an unfair experience with a school teacher, coach, or counselor; or a time you experienced prejudice or oppression of some type.

Emotions may appear as the story unfolds. Feel free to paraphrase or summarize these emotions, but this time focus first on the story and the event or situation itself. We suggest that you repeat this story again when you practice the skill of reflection of feeling in the next chapter.

All of the above incidents or topics will provide observers with the opportunity to observe nonverbal behaviors and discrepancies, incongruity, and conflict. Does the client internalize or externalize responsibility and blame? Internal attribution occurs when clients see themselves as being at fault. External attribution occurs when "they" or external matters are seen as the cause. Most clients will demonstrate some balance of internal and external attribution—self-blame versus other blame.

Step 5: Conduct a 3-minute practice session.

Step 6: Review the practice session and provide feedback to the interviewer for 12 minutes. Be sure to use the Feedback Form (Box 6-5) to ensure that the interviewer's statements are available for discussion. This form provides a helpful log of the session, which greatly facilitates discussion. And give special attention to feedback from the client, perhaps using the Client Feedback Form of Chapter 1. If you have an audio- or videotape, start and stop the tape periodically and rewind it to hear and observe important points in the interview. Did the interviewer achieve his or her goals? What mastery level was demonstrated?

Step 7: Rotate roles.

Some general reminders. It is important that clients talk freely in the role-plays. As you become more confident in the practice session, you may want your clients to become more "difficult" so you can test your skills in more stressful situations. You'll find that difficult clients are often easier to work with after they feel they have been heard.

Portfolio of Competence

Active listening is one of the core competencies of intentional interviewing and counseling. Please take a moment to review where you stand and where you plan to go in the future.

Use the following as a checklist to evaluate your present level of competency. Check those dimensions that you currently feel able to do. Those that remain unchecked can serve as future goals. Do not expect to attain intentional competence on every dimension as you work through this book. You will find, however, that you will improve your competencies with repetition and practice.

Highlight the competencies that you have met to date:

Level 1: Identification and classification.

❏ Identify and classify encouragers, paraphrases, and summaries.
❏ Discuss issues in diversity that occur in relation to these skills.
❏ Write encouragers, paraphrases, and summaries that might predict what a client will say next.

BOX 6-5   FEEDBACK FORM: ENCOURAGING, PARAPHRASING, AND SUMMARIZING

_____ (Date)

_____   _____
(Name of Interviewer)   (Name of Person Completing Form)

Instructions: Write below as much as you can of each counselor statement. Then classify the statement as a question, an encourager, a paraphrase, a summarization, or other. Rate each of the last three skills on a scale of 1 (low) to 5 (high) for its accuracy.

Counselor statement	Open question	Closed question	Encourager	Paraphrase	Summarization	Other	Accuracy rating
1. _____							
2. _____							
3. _____							
4. _____							
5. _____							
6. _____							
7. _____							
8. _____							
9. _____							
10. _____							
11. _____							
12. _____							
13. _____							
14. _____							

1. What were the key discrepancies demonstrated by the client?

2. General interview observations. Was responsibility for the concern placed internally or externally, or with some balance between the two?

Level 2: Basic competence. Aim for this level of competence before moving on to the next skill area.

- ❑ Use encouragers, paraphrases, and summaries in a role-played interview.
- ❑ Encourage clients to keep talking through use of nonverbals and through the use of silence, minimal encouragers ("uh-huh"), and the repetition of key words.
- ❑ Discuss cultural differences with the client early in the interview, as appropriate to the individual.

Level 3: Intentional competence.

- ❑ Use encouragers, paraphrases, and summaries accurately to facilitate client conversation.
- ❑ Use encouragers, paraphrases, and summaries accurately to keep clients from repeating their stories unnecessarily.
- ❑ Use key word encouragers to direct client conversation toward important topics and central ideas.
- ❑ Summarize accurately longer periods of client utterances—for example, an entire interview or the main themes of several interviews.
- ❑ Communicate with bilingual clients using some of the key words and phrases in their primary language.

Level 4: Psychoeducational teaching competence. Teaching competence in these skills is best planned for a later time, but a client who has particular difficulty in listening to others may indeed benefit by careful training in paraphrasing. There are some individuals who often fail to hear accurately and distort what others have said to them.

- ❑ Teach clients in a helping session the social skills of encouraging, paraphrasing, and summarizing.
- ❑ Teach small groups the above skills.

DETERMINING YOUR OWN STYLE AND THEORY: CRITICAL SELF-REFLECTION ON THE ACTIVE LISTENING SKILLS

This chapter has focused on encouraging/restatement, paraphrasing, and summarization as critical to obtaining a solid understanding of what clients want and need. Active listening is central, and these three skills are key.

What single idea stood out for you among all those presented in this chapter, in class, or through informal learning? What stands out for you is likely to be important as a guide toward your next steps. What are your thoughts on race/ethnicity? What other points in this chapter struck you as important? How might you use ideas in this chapter to begin the process of establishing your own style and theory? If you are keeping a journal, what trends do you see as you progress this far?

What are your thoughts?

_____

_____

OUR THOUGHTS ABOUT JENNIFER

Active listening requires actions and decisions on our part. What we listen to (selective attention) will have a profound influence on how clients talk about their concerns. When a client comes in full of information and talks rapidly, we often find ourselves confused and, we admit, a bit overwhelmed. It takes a lot of active listening to hear this type of client accurately and fully. If our work was personal counseling, we would most likely focus on her parents' separation and use an encourager by restating some of her key words, thus helping her focus on what may be the most central issue at the moment (e.g., "Your Mom and Dad called earlier this week and are separating"). We'd likely get a more focused story and could learn more about what's happening. As we understand this issue more fully, we could later move to discussing some of the other problems.

Another possibility would be to summarize the main things that Jennifer was saying as succinctly and accurately as possible. We'd do this by catching the essence of her several points and saying them back to her. Most likely, we'd use what is called the "check-out" to see if we have been reasonably close to what she thinks and feels (e.g., "Have I heard you correctly so far?"). This would then be followed by our asking her, "You've talked about many things. Where would you like to start today?"

And if we were academic counselors, not engaging in personal issues, we'd likely selectively attend to the area of our expertise (study issues) and refer Jennifer to an outside source for personal counseling.

How does this compare to what you would have done?

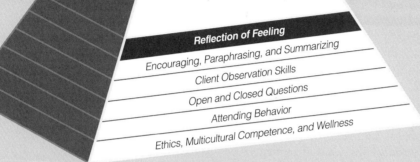

CHAPTER 7

Observing and Reflecting Feelings: A Foundation of Client Experience

The artistic counselor catches the feelings and emotions of the client. Our emotional side often guides our thoughts and actions, even without our conscious awareness.

—Allen Ivey

How can reflection of feeling help you and your clients?

Chapter Goals

The purpose of reflection of feeling is to make emotional life more explicit and clear to the client—discovering the "heart of the matter." Underlying clients' words, thoughts, and behaviors are feelings and emotions that motivate and drive action. This chapter is essential to your being fully with the client, and we shall go into depth with these skills. Closely allied to exploring emotions is empathy. You may want to read pages 203–209 now to add further depth to your understanding of this chapter.

Competency Objectives

Awareness, knowledge, and skill in reflection of feeling will enable you to

▲ Bring out the richness of the client's emotional world.
▲ Note that many clients have conflicting, discrepant, or mixed feelings about necessary decisions, thoughts about their significant others, and what direction their lives should take. You can use the skill to help clients sort out these complex feelings and thoughts.
▲ Ground the counselor and client from time to time in basic sensorimotor experience, where emotions are revealed most clearly. There is a tendency for abstract formal-operational interviewers to intellectualize and shy away from deeper feelings.

171

Thomas: My Dad drank a lot when I was growing up, but it didn't bother me so much until now. (pause) But I was just home, and it really hurts to see what Dad's starting to do to my Mum—she's awful quiet, you know. (looks down with brows furrowed and tense) Why she takes so much, I don't figure out. (looks at you with a puzzled expression) But, like I was saying, Mum and I were sitting there one night drinking tea, and he came in, stumbled over the doorstep, and then he got angry. He started to hit my mother, and I moved in and stopped him. I almost hit him myself, I was so angry. (anger flashes in his eyes) I worry about Mum. (A slight tinge of fear seems to mix with the anger in the eyes and you notice that his body is tensing.)

Paraphrasing, as presented in Chapter 6, is concerned with feeding back the key points of what a client has said. Reflection of feeling, in contrast, involves observing emotions, naming them, and repeating them back to the client. The two are closely related and often will be found together in the same statement, but the important distinction is emphasis on content (paraphrase) and emotion (reflection of feeling).

To clarify the distinction, write in the lines following a paraphrase of the client's comments above with an emphasis on content; then write a reflection of feeling, focusing on emotion. You have not yet read about reflection of feeling, so use your intuition and note the main Thomas's feeling words as he talks about his family. above. Two possible sentence stems for your consideration are presented.

Paraphrase: Thomas, I hear you saying . . .

What are your thoughts?

_____

_____

Reflection of feeling: Thomas, you are feeling . . .

What are your feelings?

_____

_____

You may want to compare your response with our thoughts presented in the following introductory section.

INTRODUCTION: REFLECTION OF FEELING

The definition of reflection of feelings is presented below. If you use this listening skill, you can make the following prediction.

Reflection of Feeling	Predicted Result
Identify the key emotions of a client and feed them back to clarify affective experience. With some clients, the brief acknowledgment of feeling may be more appropriate. Often combined with paraphrasing and summarizing.	Clients will experience and understand their emotional state more fully and talk in more depth about feelings. They may correct the interviewer's reflection with a more accurate descriptor.

Comparing Paraphrasing and Reflection of Feeling

Paraphrasing client statements focuses on the content and clarifies what has been communicated. In the case discussed here, the content includes the father's drinking history, Mum being quiet and taking it, and of course, the actual situation when the client was last home. The paraphrase will indicate to the client that you have heard what has been said and encourage him to move further to the discussion.

Paraphrase: Thomas, your father has been drinking a long time, and your mom takes a lot. But now he's started to be violent, and you've been tempted to hit him yourself. Have I heard you right?

The key content issue is escalation of violence and the need to protect Thomas's mom. It will not help the situation if Thomas becomes part of the violence. At this point, the issue is to listen, learn more about the situation, and plan with Thomas actions for the future. In this example, we are focusing on what is happening and seeking to understand the total situation. Later in the session, we can focus on emotion in depth and help him work through issues.

Reflection of feelings: The first task in eliciting and reflecting feelings is to recognize the key emotional words expressed by the client. In the above example with Thomas, you may have used the words *really hurts, anger,* and *worry.* You can know with some certainty that the client has these feelings, as they have been made explicit. The most basic reflections of feeling would be: "It really hurt," "You felt angry," and "You are worried." These reflections of feelings use the client's exact main words.

But there are also many unspoken feelings expressed in the statement—and the client may or may not be fully aware of them. If you saw Thomas on video, you would likely notice unspoken feelings expressed in his body language and vocal tone. For example, Thomas looked down with brows furrowed and body tense; anger and fear flashed in his eyes as he was talking about hitting. All this presents a powerful nonverbal portrait. While we do not have the *anger, fear, tension* said in words, it is often useful to reflect these feelings as well, but a solid, trusting relationship is necessary before reflecting or commenting on client nonverbal behavior.

Thomas says that his father's drinking didn't bother him until recently. But this seems unlikely, and it will be important at a later point to explore his early family life and how it affected him. At this point, however, the main issue is drawing out the story and noting the central emotions associated with the story.

While we believe that focusing first on the potential violence in the situation is critical, combining the paraphrase with feelings by repeating the client's stated and nonverbal key feeling words is likely also appropriate. For example, "You're really hurting with it all right now," "You're angry because your dad hit your mom," "You're worried that your dad's drinking is getting worse." Combining the feeling with the paraphrase acknowledges the client's emotions and may encourage a fuller telling of the story.

Later in the session, after the story is told more completely through your listening, you may help the client discuss and sort through the many and often conflicting emotions. There are numerous possibilities for reflecting implicit emotions that seem to be there. In each of the following we suggest a check-out so that the accuracy of your observations can be tested with the client. Here are three for your consideration:

Right now, you're hurting about the situation. I also see some anger. Is that part of what you feel too? (Uses a key word that reflects the implicit feeling.)

Stopping your dad from hitting your mom brought out a lot of emotion—I see some anger, perhaps even a little fear about what's going on. Am I close to what you're feeling? (The focus here is on unspoken feelings, seen more nonverbally than verbally. The check-out is particularly important here.)

I hear that your dad has been drinking for many years. (paraphrase) And I hear many different feelings—anger, sadness, confusion—and I also hear that you care a lot both for your mom and dad. Am I close to what you are feeling about the total situation? (This is a broader reflection of feeling that summarizes several implicit feelings and encourages the client to think more broadly.)

Which of the several possibilities (and more) presented above for a reflection of feeling is "right"? If you are intentionally attuned to where your client is at the moment and demonstrate empathy and good listening skills, any of them could be suitable. The general recommendation is to start with explicit feelings and use the client's actual emotional words. Later, you can move to an exploration of unspoken feelings.

THE LANGUAGE OF EMOTION

People are constantly expressing emotions verbally and nonverbally. General social conversation usually ignores feelings unless they are especially prominent. Thus, many of us are trained not to focus on the other person's emotional experience, and we may even be unaware of what is happening before our eyes.

It is often helpful to start work on eliciting and reflecting feelings by establishing your own vocabulary or list of emotions. We suggest that you take a moment now and develop your list using your own experience and intuitions.

An easy way to brainstorm about emotional words is to focus on four basic feelings—sad, mad, glad, and scared. These four words are listed with room for you to write related emotional words. Think particularly of different intensities of the same emotion. For example, "mad" might lead you to think of annoyed, angry, and furious.

Sad	*Mad*	*Glad*	*Scared*
depressed	angry	happy	afraid
upset	frustrated	excited	petrified
	pissed	content	

These four emotions plus surprise and disgust are termed the *primary emotions,* and their commonality has been validated throughout the world in all cultures (Ekman, 2007) in terms of people's facial expressions and language. Each of these key emotions is located in a different place in the brain, and there are some early findings that will eventually impact our counseling and clinical practice (see Appendix II).

Now, that you have completed your list, please turn to page 198 and see a list of feeling words.

More on Emotions

Feelings are also layered, like an onion. As feelings layers become more complex, clients may talk about conflicting and incongruent emotional tones such as feeling confused, lost, or frustrated; or they may be direct and forthright with a single clear emotion—fear, anger, sadness. Feeling *confused, lost, or frustrated* are more abstract layers of the onion. If we start with the "confused," reflect this feeling, often the client will go deeper, and we can define the underlying emotions more easily. For example, clients may say that they are confused in their relationship with a partner. This reflection may lead to more concrete specifics underlying confusion. This could include *anger* at lack of attention, *fear* of being alone if the relationship ends, and residual *deep caring and love* for the partner. You may discover that the partner is trying to control your client. As the client becomes aware of this issue, further emotions may be *deep anger* at the partner's controlling behavior. Later, the client may experience feelings of *relief* and anticipation of better times if the relationship ends. And in the middle of all this, the *hurt* will likely remain important.

Clients also express emotions in ways that are less clear. The client above appeared puzzled. He appeared to have mixed and conflicting emotions of caring, anger, and fear—and probably even more. A client going through a difficult time such as a divorce may express feelings of love toward the spouse at one moment and extreme anger the next. Words such as "puzzlement," "sympathy," "embarrassment," "guilt," "pride," "jealousy," "gratitude," "admiration," "indignation," and "contempt" are social emotions made up from primary emotions and learned in the cultural/environmental context.

You can be especially helpful to clients as they sort out social emotions. Think of how the word "guilt" combines anger toward oneself, sadness, and perhaps even some fear. The feeling of guilt has been learned through social interaction in the family and culture. The negative primary emotions are located deep in the brain, whereas the social emotions and happiness are found in higher areas, combining experience with underlying primary emotional reactions (Damasio, 2003). Feelings triggered by multicultural issues are presented in Box 7-1 on page 176. (Also see Appendix II for comments on emotions and social justice.)

EXAMPLE INTERVIEW: MY MOTHER HAS CANCER, MY BROTHERS DON'T HELP

Difficult life situations bring with them many emotions. Whether you are dealing with clients who experience physical illness, interpersonal conflict, alcohol or drug abuse, or challenges in the work or school setting, learning the way they feel about the situation is vital. The intentional interviewer or counselor is always alert to emotions underlying all situations and knows how to bring them out.

BOX 7-1 NATIONAL AND INTERNATIONAL PERSPECTIVES ON COUNSELING SKILLS

The Invisible Whiteness of Being
Derald Wing Sue

Multicultural issues remain a "hot button" topic and often bring out emotions that may surprise us, whether client or interviewer. I posed this open question to people on the street in San Francisco—"What does being White mean?" Here are some responses. Note the feelings that this question brought out.

42-Year-Old White Businessman

Q: What does it mean to be White?
A: Frankly, I don't know what you're talking about.

Q: Aren't you White?
A: Yes, but I come from Italian heritage. I'm Italian, not White.

Q: Well then, what does it mean to be Italian?
A: Pasta, good food, love of wine. (obviously agitated) This is getting ridiculous.

Theme: This person denies the color of his skin and speaks superficially about his Italian heritage. He shows feelings of anger around the issue.

26-Year-Old Female College Student

Q: What does it mean to be White?
A: Is this a trick question? . . . I've never thought about it . . . Well, I know that lots of Black people see us as prejudiced and all that stuff. I wish people would just forget about race differences and see one another as human beings. People are people and we should be proud to be Americans.

Theme: She never thinks of being White and shows feelings of defensiveness and confusion. She focuses on "people are people."

29-Year-Old Latina Administrative Assistant

Q: What does it mean to be White?
A: I'm not White, I'm Latina!

Q: Are you upset with me?
A: No . . . it's just that I'm light, so people always think I'm White. It's only when I speak that they realize I'm Hispanic.

Q: Well, what does it mean to be White?
A: Do you really want to know? . . . OK, it means you're always right. It means that you never have to explain yourself or apologize. . . . You know that movie Love is never having to say you're sorry? Well, being White is never having to say you're sorry. It means you think you're better than us.

Theme: Anger at being misidentified and anger and frustration with what she sees as dominant Whites who feel superior.

39-Year-Old African American Salesman

Q: What does it mean to be White?
A: If you're White, you're right. If you're Black, step back.

Q: What does that mean?
A: White folks are always thinking they know all the answers. A Black man's word is worth less than a White man's. When White customers come into our dealership and see me standing next to cars, I become invisible to them. Actually, I think they see me as a well-dressed janitor. They seek out White salesmen. I talked to my boss about this, but he says I'm oversensitive. That's what being White means. It's having the authority or power to tell me what's really happening even though I know it's not. Being White means you can fool yourself into thinking that you're not prejudiced, when you are. That's what it means to be White.

Themes: Feelings of frustration and anger around Whites who view minorities as less competent and capable; also, Whites have the power to define the world for others.

These four examples illustrate the deep-seated emotions underlying issues of race. The Latina and African American comments illustrate feelings that many People of Color have toward White people. Whites, on the other hand, are often oblivious to how color affects their lives. Being White brings with it a privilege that European Americans are often oblivious of and unaware of having.

This box by Derald Wing Sue is partially adapted from *Overcoming Our Racism: The Journey to Liberation*, Jossey-Bass, 2003. Reprinted by permission.

The discovery of cancer, AIDS, or other major physical illness brings with it an immense emotional load. Busy physicians and nurses sometimes may fail to deal with emotions in their patients. And certainly, they will have little time to help family members of those who are ill. Illness can be a frightening experience, and family, friends, and neighbors as well as professionals may have trouble dealing with it. Thus, expect to help your clients deal with both their own illness and that of their close ones.

The following transcript illustrates reflection of feeling in action. This is the second session, and Jennifer has just welcomed the client, Stephanie, into the room. They had a brief personal exchange of greetings, and it was clear nonverbally that the client was ready to start immediately.

Interviewer and Client Conversation	Process Comments
1. *Jennifer:* So, Stephanie, how are things going with your mother?	Jennifer knows what the main issue is likely to be, so she introduces it with her first open question.
2. *Stephanie:* Well, the tests came back, and the last set looks pretty good. But, I'm upset. With cancer, you never can tell. It's hard . . . (pause)	Stephanie speaks quietly, and as she talks, she speaks in an even softer tone of voice. At the word "cancer," she looks down.
3. *Jennifer:* You're really upset and worried right now.	Jennifer uses the client's emotional word ("upset") but adds the unspoken emotion of worry. With "right now," she brings the feelings to here-and-now immediacy. She did not use a check-out. Was that wise?
4. *Stephanie:* That's right. Since she had her first bout with cancer . . . (pause), I've been really concerned and worried. She just doesn't look as well as she used to; she needs a lot more rest. Colon cancer is so scary.	Often if you help clients name their unspoken feelings, they will say "That's right" or something similar, or they may nod their head. Naming and acknowledging emotions helps clarify them.
5. *Jennifer:* Scary?	Repeating the key emotional words used by clients may help them elaborate on issues in more depth.
6. *Stephanie:* Yes, I'm scared for her, and I'm scared for me. They are saying that it can be genetic. She had Stage 2 cancer, and we have really got to watch things carefully.	The intentional prediction comes true. Stephanie elaborates on the scary feelings and where they come from. She has a frightened look on her face, and she looks physically exhausted.
7. *Jennifer:* So, we've got two things here. You've just gone through your mother's operation, and that was scary. You said earlier that they got the entire tumor, but your mom really had trouble with the anesthesia, and that was frightening for a while. You had to do all the caregiving because	At this point, Jennifer decides to summarize what has been said. She repeats some key feelings identified in the first interview as well as in this session. She uses a new word, *overwhelmed,* which comes from her observations of the total situation and how very tired Stephanie appears.

(continued)

Interviewer and Client Conversation	Process Comments
the rest of the family is far away, and you felt pretty lonely. That is scary enough. And the possibility of your inheriting the genes is pretty terrifying. Putting it all together, you feel overwhelmed. Is that the right word to use—overwhelmed?	The interviewer took a chance with the word *overwhelmed*. It might produce too much emotion in the client at the moment.
8. *Stephanie:* (immediately) Yes, I'm overwhelmed, I'm so tired, I'm scared, and I'm furious with myself. (pause) But I can't be angry; my mother needs me. It makes me feel guilty that I can't do more. (starts to sob)	This reflection of feeling seems to have brought out more emotion than the interviewer expected this early in the session. Stephanie is now talking about her issues using a sensorimotor style (see page 188). This is probably okay as the relationship is solid, Stephanie has not cried in the interview before, and she likely needs to allow herself to cry and let the emotions out. Caregivers such as Stephanie often burn out and need care themselves. The primary focus is on the person with illness, and often little attention is given to the person who suddenly finds herself with all the responsibilities.
9. *Jennifer:* (sits silently for a moment) Stephanie, you've faced a lot, and you've done it alone. Allow yourself to pay attention to you for a moment and experience the hurt. (As Stephanie cries, Jennifer comments.) Let it out . . . that's okay.	Stephanie has held it all in and needs to experience what she is feeling. If you personally are comfortable with emotional experience, this ventilation of feelings can be helpful. At the same time, there is a need at some point to return to discussion of Stephanie's situation from a less emotional frame of reference.
10. *Stephanie:* (continues to cry, but the sobbing lessens)	See Box 7-2 for ideas in helping clients deal with emotional experience.
11. *Jennifer:* Stephanie, I really sense your hurt and aloneness. I admire your ability to feel—it shows that you care. Could you sit up now and take a breath?	The client sits up, the crying almost stops, and she looks at the interviewer a bit cautiously. She wipes her nose with a tissue and takes a deep breath. Jennifer did three things here: (a) she reflected Stephanie's here-and-now emotions; (b) she identified a positive asset and strength in those emotions; and (c) she suggested that Stephanie take a breath. Conscious breathing often helps clients bring themselves together.

(continued)

Interviewer and Client Conversation	Process Comments
12. *Stephanie:* I'm okay. (pause)	She wipes her eyes and continues to breathe. She seems more relaxed now that she has let out some of her emotions.
13. *Jennifer:* You've been holding that inside for a long time. That's the first time I've seen you cry. You had to have a lot of strength and power to do what you did. Your caring and strength really show. It's also strong to cry.	Jennifer provides feedback to Stephanie on her observations and outlines some positive strengths that she has seen. She provides a reframe (see Chapter 11), pointing out the positives inherent in Stephanie's caring attitude.
14. *Stephanie:* Thanks, but I still feel so guilty.	Stephanie is now back in control of herself.
15. *Jennifer:* You feel guilty?	This is the most basic reflection of feeling, and it appears in the form of a restatement. The prediction is that Stephanie will elaborate on the meaning of the feeling.
16. *Stephanie:* Who am I to cry? My mother is the one with the pain and Stage 2 cancer. I just wish I had been able to talk my brothers into coming home to see her at least.	The prediction is confirmed, and we see Stephanie elaborating on her guilt. We are beginning to see indications of Stephanie being an "overfunctioning" individual who takes on more responsibility than she needs to.
17. *Jennifer:* Your mother has gone through cancer, but you also have pain and fear, although in a different way. Do I hear you saying that you feel guilty because your brothers didn't come home?	First, Jennifer reflects Stephanie's feelings of pain and fear. She separates out the guilt, however, with a reflection of feeling in the form of a closed question. In this case, the question serves as a check-out.
18. *Stephanie:* Well, I called them daily and told them what was happening. They were fine on the phone, but they simply wouldn't come.	Stephanie talks a little faster, and her fists tighten.
19. *Jennifer:* I sense a little anger at them. Is that close?	Jennifer draws on the nonverbal observations for this reflection of unspoken feelings. Wisely, she includes a check-out, as it is possible that Stephanie will deny the anger.
20. *Stephanie:* I feel so guilty that I couldn't talk them into coming. (thoughtful pause) No, that's not right. They should have come. (angrily) I know they have jobs, and it's hard to get away, but *this* is their mother. (pause) And it's not just this time. They hardly ever call. They just seem to be in their own world. I wonder if they'll even show up *this* year for the holidays. They didn't last year.	Could it be that Stephanie has taken the caregiver role for the entire family? As she explores her feelings, she is beginning to make new discoveries.

(continued)

Interviewer and Client Conversation	Process Comments
21. *Jennifer:* Stephanie, right now I hear that you're really angry with them because they don't help and aren't involved.	A classic reflection of feeling involves 1. A sentence stem (*I hear you*), usually using the client's name or the pronoun *you.* 2. The naming of the feeling *(angry)* or feelings. 3. The underlying facts or reasons behind the feeling. 4. Bringing the emotion to the here and now of the immediate moment (*right now*).
22. *Stephanie:* And you know what else they did? (continues with another story)	The reflection unleashes Stephanie to share some long-held-back stories of frustration and anger with her brothers.
23–30. Omitted.	In this exchange, Stephanie explores her feelings toward her brothers.
31. *Jennifer:* So, Stephanie, we've talked about your anger and disappointment with your brothers. You seem very much in touch with something you weren't really aware of before. You seem to be saying, also, that feeling guilty about not getting them to shape up and take their share doesn't make sense anymore. At the same time, I sense that you still have hopes and want to involve them more. Before we go further, I wonder if we can change focus. So far, we've been talking about your concerns and difficulties. At the same time, you've managed to do a lot. You care a lot, and you've managed this past month. Could you share with me some of the things and the strengths that have enabled you to manage during this past month?	The intervening discussion is summarized. Stephanie's feelings of guilt and anger are better understood. This should free her for more open discussion in the future. Note how summaries help punctuate the interview and aid the transition to other stages and issues. Jennifer has listened to problems and challenges for the entire first session and all of this interview thus far. She decides it is time for the positive asset search. If Jennifer can help the client identify some positive feelings, thoughts, and behaviors associated with this situation, then Stephanie will be better prepared and stronger to deal with the many issues, challenges, and problems that she faces.

You may have noted that the major skill used by Jennifer throughout this session was reflection of feelings, accompanied by a few questions to help bring out emotions. This skill is important in all theories of counseling and therapy, as human change and development is often rooted in emotional experience.

You are most likely beginning your work and starting to discover the importance of reflecting feelings. It will probably take you some time before you are fully comfortable using this skill. This is so because it is less a part of daily communication than the other skills of this book, but reflecting feelings is central to every helping professional.

We would suggest that you start practice by first simply noting emotions and then reflecting them back through short acknowledging reflections. As you gain confidence and skill, you will eventually decide the extent and place of this skill area in your helping repertoire.

BOX 7-2   HELPING CLIENTS INCREASE OR DECREASE EMOTIONAL EXPRESSIVENESS

Observe Nonverbals

Breath directly reflects emotional content. Rapid or frozen breath signals contact with intense emotion. Also note facial flushing, pupil contraction/dilation, body tension, and changes in vocal tone; note especially speech hesitations. You may also find apparent absence of emotion when discussing a difficult issue. This might be a clue that the client is avoiding dealing with feelings or that the expression of emotion is culturally inappropriate for this client.

Pace Clients

You can pace clients and then lead them to more expression and awareness of affect. Many people get right to the edge of a feeling and then back away with a joke, change of subject, or intellectual analysis.

Some of the Things You Can Do

▲ Say to the client that she looked as though she was close to something important. "Would you like to go back and try again?"
▲ Discuss some positive aspect of the situation. This can free the client up to face the negative. You as counselor also represent a positive asset yourself.
▲ Consider asking questions. Used carefully, questions may help some clients explore emotions.
▲ Use here-and-now sensorimotor techniques, especially in the present tense: "What are you feeling right now—at this moment?" "What's occurring in your body as you talk about this?" Use Gestalt exercises or anything to enable a client to become more aware of body feeling. Use the word "do" if you find yourself uncomfortable with emotion: "What do you feel?" or "What did you feel then?" starts to move the client away from here-and-now experiencing.

When Tears, Rage, Despair, Joy, or Exhilaration Come Up

Your comfort level with your own emotions and feelings will affect how your client faces emotion. If you aren't comfortable with a particular emotion, your client will likely avoid it also, and you may not be effec-

tive in helping the client work through feeling issues. A balance between, on the one hand, being very present with your own breathing and showing culturally appropriate and supportive eye contact, and on the other, still allowing room to sob, yell, or shake is important.

You can also use phrases such as these:

▲ I'm here. I've been there too.
▲ I'm standing right with you in this. Let it out . . . that's okay.
▲ These feelings are just right. I hear you.
▲ I see you. Breathe with it.

Sometimes it is helpful to keep emotion expression within a fixed time; 2 minutes is a long time when you are crying. Afterward, helping the person reorient is important.

Tools for reorienting the interview include these:

▲ Slowed, rhythmic breathing
▲ Counselor and client discussion of positive strengths inherent in the client and situation
▲ Discussion of direct, empowering, self-protective steps that the client can take in response to the feelings expressed
▲ Standing and walking or centering the pelvis and torso in a seated position
▲ Positive reframing of the emotional experience
▲ Commenting that the story needs to be told many times, and each time helps

A Caution

As you work with emotion, there is always the possibility of reawakening issues in a client who has a history of painful trauma. This is an area where the beginning interviewer often needs to refer to a more experienced professional. Even the 2-minute expression of emotion suggested here may be too long. In such situations, seek supervision and consultation.

This box is adapted from a presentation by Leslie Brain, a student in Allen's counseling skills course at the University of Massachusetts. It is used here with her permission.

Before moving further, it is important for you to reflect on yourself and your own personal style. How comfortable are you with emotional expression? If discussing feelings was not common in your experience, this skill may be difficult for you. Thus, as you work through the ideas presented here, reflect on your own personal history and ability to deal with emotion. If you find this area uncomfortable, you may have difficulty in helping clients explore their issues in depth. The exercises here and throughout the book may help you gain greater access to your own experiential and emotional world. In your own practice with reflection of feeling, attempt to use the skill as frequently as possible. In the early stages of mastery it is wise to combine the skill with questioning, encouraging, and paraphrasing.

INSTRUCTIONAL READING: BECOMING AWARE OF AND SKILLED WITH EMOTIONAL EXPERIENCE

Many authorities argue that our thoughts and actions are only extensions of our basic feelings and emotional experience. The skill of reflecting feeling is aimed at assisting others to sense and experience the most basic part of themselves—how they really feel about another person or life event.

At the most elementary level, the brief encounters we have with people throughout the day involve our emotions. Some are pleasant; others can be fraught with tension and conflict even though the interaction may be only with a telemarketer desperate to make a sale, with a hurried clerk in a store, or with the police as they stop you for speeding. Feelings undergird these situations just as the more complex feelings we have toward significant others underlie more intimate relationships.

At another level, our work and social relationships and the decisions we make are often based on emotional experience. In an interviewing situation it is often helpful to assist the client in identifying feelings clearly. For example, an employee making a move to a new location may have positive feelings of satisfaction, joy, and accomplishment about the opportunity but simultaneously feel worried, anxious, and hesitant about new possibilities. The effective interviewer notes both dimensions and recognizes them as a valid part of life experience.

A basic feeling we have toward our parents, family, and best friends is love and caring. This is a deep-seated emotion in most individuals. At the same time, over years of intimate contact, negative feelings about the same people may also appear, possibly overwhelming and hiding positive feelings; or negative feelings may be buried. A common task of counselors is to help clients sort out mixed feelings toward significant people in their lives. Many people want a simple resolution and want to run away from complex mixed emotions. However, ideally the counselor should help the client discover and sort out both positive and negative feelings.

Trust between counselor and client is necessary for full emotional exploration, but in some cultures expression of feelings is discouraged (see Box 7-3). You will find that a brief acknowledgment of feeling is at times more appropriate than a deep exploration of feelings. In acknowledging feelings, you state the feeling briefly ("You seem to be sad about that," or "It makes you happy") and then move on with the interview. However, do not use individual or cultural uniqueness as an excuse to avoid talking about emotion in the interview. All clients have vital emotional lives, whether they are aware of them or not.

With children, you will find acknowledgment of feelings may be especially helpful, particularly when they themselves are unaware of what they are feeling. At the same time, many children respond well to the classic reflection of feeling, "You feel (sad, mad, glad, scared) because . . ."

When addressing client feelings and emotions in an interview, the following specifics are important to keep in mind.

BOX 7-3   NATIONAL AND INTERNATIONAL PERSPECTIVES ON COUNSELING SKILLS

Does He Have Any Feelings?
Weijun Zhang

Illustration: A student from China comes in for counseling, referred by his American roommate. According to the roommate, the client quite often calls his wife's name out loud while dreaming, which usually wakes the others in the apartment, and he was seen several times doing nothing but gazing at his wife's picture. Throughout the session the client is quite cooperative in letting the counselor know all the facts concerning his marriage and why his wife is not able to join him. But each time the counselor tries to identify or elicit his feelings toward his wife, the client diverts these efforts by talking about something else. He remains perfectly polite and expressionless until the end of the session.

No sooner had the practicing counselor in my practicum class stopped the videotape machine than I heard comments such as "inscrutable," and "He has no feelings!" escape from the mouths of my European American classmates. I do not blame them, for the Chinese student did behave strangely, judged from their frame of reference.

"How do you feel about this?" "What feelings are you experiencing when you think of this?" How many times have we heard, or used, questions such as these? The problem with these questions is that they stem from a European American counseling tradition, which is not always appropriate.

For example, in much of Asia, the cultural rationale is that the social order doesn't need extensive consideration of personal, inner feelings. We make sense of ourselves in terms of our society and the roles we are given within the society. In this light, in China, individual feelings are ordinarily seen as lacking social significance.

For thousands of years, our ancestors have stressed how one behaves in public, not how one feels inside. We do not believe that feelings have to be consistent with actions. Against such a cultural background, one might understand why the Chinese student was resistant when the counselor showed interest in his feelings and addressed that issue directly. But I am not suggesting here that Asians are devoid of feelings or strong emotions. We are just not supposed to telegraph them as do people from the West. Indeed, if feelings are seen as an insignificant part of an individual and regarded as irrelevant in terms of social importance, why should one send out emotional messages to casual acquaintances or outsiders (the counselor being one of them)?

What is more, most Asian men still have traditional beliefs that showing affection toward one's wife, even verbally, while others are around is a sign of being a sissy, being unmanly, or weak. I can still vividly remember when my child was four years old, my wife and I once received some serious lecturing on parental influence and social morality from both our parents and grandparents, simply because our son reported to them that he saw "Dad give Mom a kiss." You can imagine how shocking it must be for most Chinese husbands, who do not dare even touch their wives' hands in public, to see on television that American presidential candidates display such intimacy with their spouses on the stage! But the other side of the coin is that not many Chinese husbands watch television sports programs while their wives are busy with household chores after a full day's work. They show their affection by sharing the housework!

Observing Client Verbal and Nonverbal Feelings

When a client says "I feel sad"—or "glad" or "frightened"—and supports this statement with appropriate nonverbal behavior, identifying emotions is easy. However, many clients present subtle or discrepant messages, for often they are not sure how they feel about a person or situation. In such cases the counselor may have to identify and label the implicit feelings. However, with skilled listening you can often help such clients label their own emotions.

The most obvious technique for identifying client feelings is simply to ask the client an open question ("How do you feel about that?" "Could you explore any emotions that come to mind about your parents?" "What feelings come to mind when you talk about the loss?").

With some quieter clients, a closed question in which the counselor supplies the missing feeling word may be helpful ("Does that feel hurtful to you?" "Could it be that you feel angry at them?" "Are you glad?").

At other times the counselor will want to infer, or even guess, the client's feelings through observation of verbal and nonverbal cues such as discrepancies between what the client says about a person and his or her actions, or a slight body movement contradicting the client's words. As many clients have mixed feelings about the most significant events and people in their lives, inference of unstated feelings becomes one of the important observational skills of the counselor. A client may be talking about caring for and loving parents while holding his or her fist closed. The mixed emotions may be obvious to the observer though not to the client.

Stephanie, the client in the example interview, used the following feeling words during the session: upset, worried, concerned, scared, tired, furious, guilty, anger, disappointed, and hope. Jennifer, the counselor, reflected those words but also brought in some of her own observations. We saw Stephanie's reaction to the word "overwhelmed." But the interviewer also sought to emphasize some positive emotions such as caring and the strengths that the client demonstrated. As the interview moves on, we can anticipate that Stephanie will express fewer difficult emotions and will move to more awareness of positive feelings and strengths. At the moment, however, she needs to be encouraged to talk through the worries and problems.

The intentional counselor does not necessarily respond to every emotion, congruent or discrepant, that has been noted; reflection of feeling must be timed to meet the needs of the individual client. Sometimes it is best simply to note the emotion and keep it in mind for possible comment later.

The Techniques of Reflecting Feelings

The classic reflection of feeling consists of the following dimensions:

1. A sentence stem using, insofar as possible, the client's mode of receiving information (auditory, visual, or kinesthetic) often begins the reflection of feeling ("I hear you are feeling . . . ," "It looks like you feel . . . ," "Sounds like . . ."). Unfortunately, these sentence stems have been used so often that they can almost sound like comical stereotypes. As you practice, you will want to vary sentence stems and sometimes omit them completely. Using the client's name and the pronoun "you" helps soften and personalize the sentence stem.
2. A feeling label or emotional word is added to the stem ("Jonathan, you seem to feel bad about . . ." "Looks like you're happy," "Sounds like you're discouraged today; you look like you feel really down"). For mixed feelings, more than one emotional word may be used ("Maya, you appear both glad and sad . . .").
3. A context or brief paraphrase may be added to broaden the reflection of feeling. (To use the examples above, "Jonathan, you seem to feel bad about all the things that have happened in the past 2 weeks"; "Maya, you appear both glad and sad because you're leaving home.") The words "about," "when," and "because" are only three of many that add context to a reflection of feeling.
4. The tense of the reflection may be important. Reflections in the present tense ("Right now, you are angry") tend to be more useful than those in the past ("You felt angry then"). Some clients will have difficulty with the present tense and talking in the "here

and now." But occasionally, a "there and then" review of past feelings can be quite helpful. You'll find that experiencing feelings and thinking back on them in the past are often quite different.

5. A check-out may be used to see whether the reflection is accurate. This is especially helpful if the feeling is unspoken ("You feel angry today—am I hearing you correctly?").

Individual and cultural diversity in the way one expresses feelings needs to be respected. The student from China discussed in Box 7-3 is an example of cultural emotional control. Emotions are obviously still there, but they are expressed differently. Do not expect all Chinese or Asians to be emotionally reserved, however. Their style of emotional expression will depend on their individual upbringing, their acculturation, and other factors. Many New England Yankees may be fully as reserved in emotional expression as the Chinese student described by Weijun Zhang. But again, it would be unwise to stereotype all New Englanders in this fashion.

Emotions are very personal. Regardless of cultural background, each client will be unique. Build trust and relationship first. When you work with clients who are emotionally reticent, you might want to try the suggestions for helping them express their feelings presented in Box 7-2. But be sure the clients know what is going to happen and express a willingness to move into their feelings more deeply with your help.

The Place of Positive Emotions in Reflecting Feelings

Positive emotions, whether joyful or merely contented, are likely to color the ways people respond to others and their environments. Research shows that positive emotions broaden the scope of people's visual attention, expand their repertoires for action, and increase their capacities to cope in a crisis. Research also suggests that positive emotions produce patterns of thought that are flexible, creative, integrative, and open to information (Gergen & Gergen, 2005). "Sad, mad, glad, scared"—this is one way to organize the language of emotion. But perhaps we need more attention to glad words such as "pleased," "happy," "contented," "together," "excited," "delighted," "pleasured," and the like. If you were to take just a moment now and think of specific situations when you experienced each of the positive emotions listed in the previous sentence, it is very likely that you would smile, your body tension would be reduced, and even your blood pressure might change in a more positive direction.

When you experience emotion, your brain signals bodily changes. Thus, when you feel sad or angry, a set of chemicals floods your body, and usually these changes will show nonverbally. Emotions change the way your body functions and thus are a foundation for all our thinking experience (Damasio, 2003). As you help your clients experience more positive emotions, you are also facilitating wellness and a healthier body. The route toward health, of course, often entails confronting negative emotions.

Research examining the life of nuns found that those who had expressed the most positive emotions in early life lived longer than those who expressed a less positive past (Danner, Snowdon, & Friesen, 2001). Stress reaction to the 9/11 disaster found that students who had access to the most positive emotions showed fewer signs of depression (Fredrickson, Tugade, Waugh, & Larkin, 2003). A resilient affective style with a fast recovery style when stressed results in lower cortisol levels and even control of cortisol. This aspect of well-being and wellness is located predominantly in the prefrontal cortex with lower levels of activation in the amygdala (Davidson, 2004, p. 1395). Drawing on long-term memory for positive experiences is one route to well-being and stress reduction. Additional research is presented in Box 7-4, and the figures in Appendix II will be helpful.

BOX 7-4   RESEARCH EVIDENCE THAT YOU CAN USE

Reflection of Feeling

Emotional processing—the working through of emotional states and the ability to examine feelings and body states—has been found fundamental in effective experiential counseling and therapy. For example, gains in treatment of depressed clients was found to be highly related to emotional processing skills (Pos, Greenberg, Goldman, & Korman, 2003). As you work with all clients, your skill in reflecting feelings can be a basic factor in helping them take more control of their lives.

Dealing with emotion is not only a central aspect of interviewing, counseling, and psychotherapy; it is also key to high-quality interviewing (Bensing, 1999b; Daniels & Ivey, 2007). Working with emotion requires attention to nonverbal dimensions. Head nodding, eye contact, and especially smiling are important. Clearly, warmth, interest, and caring are communicated nonverbally as much as or more than through the use of the skill. Moreover, Hill and O'Brien (1999), and Tamase and Kato (1990) found that using questions oriented toward affect increased client expression of emotion. Nonetheless, once a client has expressed emotion, continued use of questions may be too intrusive, and the more reflective approach will be more useful.

"Several studies have shown that between 30 and 60 percent of patients in general practice present health problems for which no firm diagnosis can be made" (Bensing, 1999a). Recognizing emotional complexity can be important. Older persons tend to manifest more mixed feelings than others (Carstensen et al., 2000). Perhaps this is because life experience has taught them that things are more multifaceted than they once thought. Helping younger clients become aware of emotional complexity may also be a goal of some interviewing sessions. Tamase, Otsuka, and Otani (1990), through their work in Japan, have provided clear indication that the reflection of feelings is useful cross-culturally. Hill (2001) reported on a series of studies in this area and noted the facilitating impact of reflective responses. She notes that clients are usually not aware when helpers are using good restatement and reflections. Effective listening facilitates exploration.

Reflection of Feeling and Neuropsychology

Neuropsychology has found that counseling's traditional categories of emotion as presented in this chapter also appear in brain imaging (Kolb & Wishaw,

2003). Most important are mirror neurons that are basic to empathy—one cannot reflect feelings without basic empathic understanding (see pages 203–209 for more information and research). Among other functions, mirror neurons fire when we see or hear what the client is experiencing. These brain functions are basic if we are to understand and be with another person. You may view areas discussed here in Appendix II, Figures AII-1 and AII-2. While emotional experience is important throughout the entire brain structure, the limbic system is the center, which can be described as the place where primitive needs meet with demands of the outside world (Cozolino, 2002, p. 70). Feelings of fear and anger are located in the amygdala, which also reacts to intensity of emotion. In fact, some emotional intensity is critical if learning is to occur in the memory center, the hippocampus. Those who experience feelings of depression have a less active frontal cortex. Selective attention to events and thoughts that bring about positive emotions can help calm the more turbulent parts of the limbic system and activate the frontal cortex.

More complex and mixed emotions are also defined and organized in the prefrontal cortex (Kolb & Wishaw, 2003). For example, you may receive a gift from a person who is very important to you, but the gift isn't to your liking. You may feel irritation about the gift (mild anger directed outward), positive appreciation for the thoughtfulness it represents, and guilt about not responding more positively to the giver (anger directed inward). You may even feel mild bodily fear from the amygdala that your lack of enthusiasm shows too prominently.

In the middle of the conflicting emotions above, the executive left brain makes the decision on how to behave and respond. All this happens within milliseconds. Usually the response will be an appropriate but somewhat muted "thank you." But imagine that you have had a bad day and are tired from the stress and many difficult decisions you've had to make. Amir (2008) has found that "making decisions tires the brain." He suggests that the brain is like a muscle. Yes, it needs exercise to develop and make neural networks, but it also needs rest. Without rest, the brain becomes less effective. Multitasking—doing too many things at once—is a classic way to tire the brain and

(continued)

BOX 7-4 (continued)

prepare it for an outburst and release of damaging cortisol.

To carry this a bit further, real damage to neuronal functioning often occurs during a single traumatic event (war, rape, hostage situation, flood). This can produce posttraumatic stress and even destroy memory centers. Think of serious stress as either one "big bang"

or a series of small continuous acts of harassment, insults, or teasing/bullying, which are called *microaggressions*. Microaggressions on a daily basis damage neural functioning and can result in serious outbursts such as the rash of killings in universities, or they may beat down the individual, leading to "learned helplessness," depression, and inactivity.

Searching for wellness and positive assets will likely be helpful. As part of a wellness assessment, be sure that you reflect the positive feelings associated with aspects of wellness. For example, your client may feel safety and strength in the spiritual self, pride in gender and/or cultural identity, caring and warmth from past and/or present friendships, and the intimacy and caring of a love relationship. It would be possible to anchor these emotions early in the interview and draw on these positive emotions during more stressful moments. Out of a wellness inventory can come a "backpack" of positive emotions and experiences that are always there and can be drawn on as needed.

When reflecting feelings, observe your client and elicit strengths. Make this part of your reflection of feeling strategy. For example, the client may be going through the difficult part of a relationship breakup and crying and wondering what to do. We don't suggest that you should interrupt the emotional flow, but with appropriate timing, reflecting back the positive feelings that you have observed can be helpful.

Couples with relationship difficulties can be helped if they focus more on the areas where things are going well—what remains good about the relationship. Many couples focus on the 5% where they disagree and fail to note the 95% where they have been successful or enjoyed each other. Some couples respond well when asked to focus on the reasons they got together in the first place. These positive strengths can help them deal with very difficult issues.

When providing your clients with homework assignments, have them engage daily in activities associated with positive emotions. For example, it is difficult to be sad and depressed when running or walking at a brisk pace. Meditation and yoga are often useful in generating more positive emotions and calmness. Seeing a good movie when one is down can be useful, as can going out with friends for a meal. In short, help clients remember that they have access to joy, even when things are at their most difficult.

Service to others often helps people feel good about themselves. When one is discouraged and feeling inadequate, volunteering for a church work group, working on a Habitat for Humanity home, or giving time to work with animal shelters can all be helpful in developing a more positive sense of self.

Important caution: But please do not use the above paragraphs as a way to tell your clients that "everything will be okay." Some interviewers and counselors are so afraid of negative emotions that they never allow their clients to express what they really feel. ***Do not minimize difficult emotions by too quickly focusing on the positive.***

Noting Emotional Intensity: A Developmental Skill

Clients have varying styles and levels of intensity with which they describe emotional experiences. You may recall that some emotional styles are more resilient and positive than others.

All of the styles here have both positive and negative possibilities. You will find that some clients are overwhelmed by emotion and feelings while others are more remote and may use thinking and cognition to avoid really examining how they feel.

In developmental counseling and therapy (DCT), key observational skills have been identified to help organize the depth of client emotional experiencing (see Ivey, Ivey, Myers, & Sweeney, 2005, for elaboration of ideas presented here). Feelings are not just "feelings." They vary in intensity and in the ways they are expressed. It is important that you be able to determine how a client reacts emotionally. Once you have awareness of a client's style of emotional expressiveness, you will have a better idea of how you can help the client explore the complex world of affect and feeling.

DCT's four emotional styles are described in the following paragraphs. Spend some time noticing these important differences as practice is critical for developing competence. The effort will pay off.

Sensorimotor emotional style. These clients *are* their emotions. They experience emotions rather than naming them or reflecting on them. The body is fully involved. They may cry, they may laugh, but emotional experience is primary, and there is only limited separation of thought and feeling. The positives of this style of emotional involvement are access to the real and immediate experiences of being sad, mad, glad, or scared in the moment. On the negative side, these clients may be overwhelmed by too much emotion.

If you are to help clients experience their issues, you will often want to encourage full sensorimotor experiencing of emotion. A question that may enhance sensorimotor, deeper emotions is "What are you feeling/experiencing *at this moment*?" You may also suggest that the client develop a visual, auditory, or kinesthetic image: "Can you develop an image of that experience—what are you seeing/hearing/feeling at this moment?" This imagery exercise can be powerful. Use with care and a clear sense of ethics.

Concrete emotional style. The skills of reflecting feeling as presented in this chapter are primarily focused on a concrete emotional style. See Jennifer 3, 5, 15, 21 for four examples. Feelings are named (early concrete) with statements such as "You seem to feel sad." Late concrete emotion is emphasized when we add causation statements to the reflection— "You feel sad because. . . ." Many of your clients will come to you with only a vague sense of the emotions underlying their concerns. Concrete exploration of emotions helps clarify issues. The resulting increased awareness can form a foundation for moving to more in-depth emotional exploration. Note that as we make emotions concrete or examine their patterns, we move away from direct here-and-now sensorimotor emotional expression.

Most of this chapter is focused on concrete emotional experience. The first task is concretely naming or labeling the emotion—"What are you feeling?" If necessary, ask, "What name would you give to that feeling?" The simplest, most direct concrete reflection of feeling is "You are feeling (or felt) X?" For example, "You feel sad" (or glad, mad, or scared). This naming of feelings itself can be very therapeutic to some clients, as they may be very out of touch with their emotions.

Abstract formal-operational emotional style. The client becomes less concrete and more abstract and reflects on emotions, but he or she may avoid *experiencing* them at the sensory level. Formal clients may be quite effective at seeing repeating patterns of emotion, but they may also have difficulty being concrete and specific. Stephanie at

22 starts processing her emotions. You may find that some clients are very good at reflecting abstractly on their feelings, but they never allow themselves to experience emotion with the full sensorimotor style. Reflection is very useful in helping clients understand their patterns of emotional reaction. "As you look back on your situation/ yourself, what types of feelings do you notice?" "As you reflect on that feeling you just talked about, what do you think?" "What are your patterns of emotion?" "Do you feel that way a lot?" At this point emotion is more abstract. Clients think back rather than directly experiencing their feelings. Jennifer at 31 seeks to help Stephanie develop a more reflective style.

Abstract dialectic/systemic emotional style. Clients using this style are very effective at analyzing their emotions, and their emotionality will change with the context. A client may say, "I am terribly sad to lose my lover through AIDS, but I am proud of how he/she lived effectively despite it. In a way, I am joyful at the triumph over adversity my lover has demonstrated." Note that this is a more analytic and multiperspective view of emotionality, which moves even further away from direct, here-and-now experiencing. Jennifer does not illustrate this type of emotion as yet.

Dialectic/systemic emotions are complex, and they change in context. The style tends to be theoretical, and emotions are analyzed more than experienced. "How do your emotions change when you take another perspective on your issue(s)?" "Where do you think you learned that pattern of emotional expression—in your family or elsewhere?" These are only two of many possible questions that lead to multiperspective thought on emotional experience.

What can the DCT view of emotions do for you? If you assess clients' emotional styles, you'll be better able to empathize and understand how they are experiencing a situation. For example, with clients who are predominantly abstract in style, you can join them in analyzing and reflecting. It may be helpful to aid them in experiencing emotion more immediately within the concrete or sensorimotor styles. A person experiencing significant loss (divorce, death, job loss, illness) will usually benefit from talking about her or his issues using all four emotional styles. With concretely oriented clients, a long-term interviewing goal may be to help them get more distance on their emotions and think more abstractly. With reflective clients, help them become more here and now. Emotion is holistic, and no one way of experiencing emotion is best.

Box 7-2 presents techniques for working with emotion in the session. Note that specific ways are suggested to help clients get in touch with their feelings. At other times, you may want to assist clients in slowing down the emotional flow.

SUMMARY: A CAUTION ABOUT REFLECTION OF FEELINGS IN THE INTERVIEW

Reflection has been described as a basic feature of the counseling process, yet it can be overdone. Many times a short and accurate reflection may be the most helpful. With friends, family, and fellow employees, a quick acknowledgment of feelings ("If I were you, I'd feel angry about that. . . ." or "You must be tired today") followed by continued normal conversational flow may be most helpful in developing better relationships. In an interaction with a harried waiter or salesperson, an acknowledgment of feelings may change the whole tone of a meal or business interchange. Similarly, with many clients a brief reflection of feeling may be more useful. Identifying unspoken feelings can be helpful, and as clients move toward complex issues, the sorting out of mixed feelings may be a central ingredient of successful counseling, be it school counseling, vocational interviewing, personal decision making, or in-depth individual counseling and therapy.

Nevertheless, it is important to remember that not all clients will appreciate or welcome your commenting on their feelings. Clients tend to disclose feelings only after rapport and trust have been developed. Less verbal clients may find reflection puzzling at times or may say, for instance, "Of course I'm angry; why did you say that?" With some cultural groups, reflection of feeling may be inappropriate and represent cultural insensitivity. Some men, for example, may believe that expression of feelings is "unmanly," yet a brief acknowledgment may be helpful to them. Be aware that an empathic reflection can sometimes have a confrontational quality that causes clients to look at themselves from a different perspective; it may therefore seem intrusive to some clients. Though noting feelings in the interview is essential, acting on your observations may not always be in the best interests of the client. Timing is particularly important with this skill.

Key points of the chapter are presented below.

■ KEY POINTS

Emotions undergird life experience

Emotions are the source of many of our thoughts and actions. If we can identify and sort out clients' feelings, we have a foundation for further action.

Identifying emotions

Emotions and feelings may be identified through labeling client behavior with affective words such as "sad," "mad," "glad," and "scared." The counselor will want to develop an array of ways to note and label client emotions. In naming client feelings, it is important to note the following:

1. Emotional words used by the client
2. Implicit emotional words not actually spoken
3. Nonverbally expressed emotions discovered through the observation of body movement
4. Mixed verbal and nonverbal emotional cues, which may represent a variety of discrepancies

Reflecting feelings

Emotions may be observed directly, drawn out through questions ("How do you feel about that?" "Do you feel angry?"), and then reflected through the following steps:

1. Begin with a sentence stem such as "You feel . . ." or "Sounds like you feel . . ." or "Could it be you feel . . . ?" Use the client's name.
2. Feeling word(s) may be added (sad, happy, glad).
3. The context may be added through a paraphrase or a repetition of key content ("Looks like you feel happy about the excellent rating").
4. In many cases a present-tense reflection is more powerful than one in the past or future tense. "You feel happy right now" rather than "you felt" or "you will feel."
5. Following identification of an unspoken feeling, the check-out may be most useful. "Am I hearing you correctly?" "Is that close?" This lets the client correct you if you are either incorrect or uncomfortably close to a truth that he or she is not yet ready to admit.

(continued)

Key Points (continued)

	Brief reflections of feeling may be particularly helpful with friends, family, and people met during the day. Deeper reflections and a stronger emphasis on this skill may be appropriate in many counseling situations, but they require a relatively verbal client. The skill may be inappropriate with clients of certain cultural backgrounds. The acknowledgment of feeling puts less pressure on clients to examine their feelings and may be especially helpful in the early stages of interviewing a client who is culturally different from you. Later, as trust develops, you can explore emotion in more depth.
Developmental skills facilitating exploration of emotion	Clients present their feelings within four styles. Your ability to identify their emotional approach will enable you to match exploration of feelings with where clients are in the here and now. Later you can mismatch and help clients explore other possible styles for dealing with emotional experience.
	Sensorimotor style. Clients are embedded in emotion and do not separate self from the experience. The strength is that core emotions are reached. The challenge is that these clients may have difficulty in separating the self from emotions, and for them, reflecting on their feelings will be most difficult.
	Concrete style. Clients can name and label emotions, although they sometimes oversimplify: "I feel sad." A causal statement would be "I feel sad because . . ." Their strength is identifying clearly what is happening; their weakness is lack of reflection and inability to understand and deal with complex mixed feelings.
	Formal-operational style. Clients can reflect on and work with more complex emotions. The concept "reflection of feeling" is most associated with this style. The challenge of clients with this style is that they may be emotionally removed from actually experiencing deeper feelings as is possible within the sensorimotor style.
	Dialectic/systemic style. Multiperspective thought is characteristic of this emotional style. Clients can intellectually understand, name, and work with mixed emotions and see emotional experience broadly in context. They have easier access to different interpretations. But they can become so intellectually involved that they do not experience their emotions and lose touch with the concrete style as well.

COMPETENCY PRACTICE EXERCISES AND PORTFOLIO OF COMPETENCE

We observe feelings in many daily interactions, but we usually ignore them. In counseling and helping situations, however, they can be central to the process of understanding another person. Further, you will find that increased attention to feelings and emotions may enrich your daily life and bring you to a closer understanding of those with whom you live and work.

Individual Practice

Exercise 1: Increasing Your Feeling Vocabulary
Return to the list of affective words you generated at the beginning of this chapter. Take some more time to add to that list. One way to lengthen the list is to consider two categories of feeling words that would provide additional ideas about how the client feels about the world.

The first category is words that represent mixed or ambivalent feelings. In such cases the feelings are often very unclear, and your task is to help the client sort out the deeper emotions underlying the surface, expressed word. Write a list of words that represent confused or vague feelings (for instance, "confused," "anxious," "ambivalent," "torn," "ripped," "mixed").

_____

A common mistake is to assume that these words represent the root feelings. Most often they cover deeper feelings. The word "anxiety" is especially important to consider in this context: It is sometimes a vague indicator of mixed feelings. If you accept client anxiety as a basic feeling, counseling may proceed slowly. An important task of the interviewer when noting mixed-feeling words is to use questions and reflection of feeling to help the client discover the deeper feelings underlying the surface ambivalence. Underlying anxiety or confusion, for example, you may find anger, hurt, love, or other feelings.

Now take two of the words from your mixed-feelings list and lay out more basic words that might underlie affectively oriented words such as confused or frustrated. Again, the basic words "mad," "sad," "glad," or "scared" may be helpful in this process. See page 198 for a list of emotion-laden words.

_____

As a second area, feelings are often presented through metaphors and similes, concrete examples, and body language. It is often more descriptive of your emotions to say that you feel like a limp dishrag than to say you are tired and exhausted. Other examples might include "down in the pits," "high as a kite," "crashed worse than a computer," or "proud as a peacock." Because metaphors are often masks for more complex feelings, you may want to search for the underlying feelings through careful listening and perhaps some questions. After you have developed a list of at least five metaphors, you may wish to generate a list of basic feeling words underlying the metaphors.

_____

Exercise 2: Distinguishing a Reflection of Feeling From a Paraphrase

The key feature that distinguishes a reflection of feeling from a paraphrase is the affective word. Many paraphrases contain reflection of feeling; such counselor statements are classified as both. Consider the two following examples. In the first example, you are to indicate which of the leads is an encourager (E), which a paraphrase (P), and which a reflection of feeling (RF).

> "I am really discouraged. I can't find anywhere to live. I've looked at so many apartments, but they are all so expensive. I'm tired, and I don't know where to turn."

Mark the following counselor responses with an E, P, RF, or combination if more than one skill is used.

_____ "Where to turn?"

_____ "Tired . . ."

_____ "You feel very tired and discouraged."

_____   "Searching for an apartment simply hasn't been successful; they're all so expensive."

_____   "You look tired and discouraged; you've looked hard but haven't been able to find an apartment you can afford."

For the next example, write an encourager, paraphrase, reflection of feeling, and a combination paraphrase/reflection of feeling in response to the client.

"Right, I do feel tired and frustrated. In fact, I'm really angry. At one place they treated me like dirt!"

_____

Exercise 3: Acknowledgment of Feeling

We have seen that the brief reflection of feeling (or acknowledgment of feeling) may be useful in your interactions with busy and harried people during the day. At least once a day, deliberately tune in to a server/waitstaff person, teacher, service station attendant, telephone operator, or friend, and give a brief acknowledgment of feeling ("You seem terribly busy and pushed"). Follow this with a brief self-statement ("Can I help?" "Should I come back?" "I've been pushed today myself, as well") and note what happens in your journal.

Exercise 4: Developmental Skills Area 1—Recognizing Varying Styles Toward Emotional Expression (see pages 188–189 for definitions)

Classify the following emotions as either sensorimotor (S), concrete (C), formal-operational (FO), or dialectic/systemic (D/S).

A client discusses arguments he (she) has with his (her) parents that occur just before leaving home.

____S____   (tears) "I'm overwhelmed."

____C____   "I feel really sad because of the argument I had with my parents last week."

____FO____   "As I think about it, I feel bad because we have so many arguments. It seems to be a pattern, and we argue every time I am about to leave home for school."

____D/S____   "I suppose we could look at it from several perspectives. First, it really hurts to have these arguments, but I know I have to find my own space, and perhaps it is part of my becoming a separate person. I know my parents care for me; perhaps that's why we argue just before I leave home."

A friend discusses reactions she (he) has to anxiety about an examination.

____C____   "It's maddening, and it made me angry when the professor didn't bring the exam to class today."

____FO____   "I suppose I can see the professor's frame of reference. After all, she has 40 papers to look over, and I know she looks awfully hassled. But it sure does make it difficult for me to know where I am. A lot of students are really angry."

_____ S   "I'm scared. I can't eat. My stomach hurts. I'm confused."

_____ FO "Professor Jones is often late. It's typical of me to feel angry and upset when I have to wait. It's an emotional pattern for me."

Exercise 5: Developmental Skills Area 2—Facilitating Clients' Exploration of Emotion Using Varying Styles

Assume that you are working with one of the preceding clients who is overwhelmed by emotion. How would you help this person move away from direct, here-and-now sensorimotor experiencing?

_____

Assume you are working with one of the preceding clients who avoids really looking at here-and-now emotional experiencing. How would you help this client increase affect and feeling?

_____

Group Practice

Exercise 6: Practice Reflection of Feeling in a Group

One of the most challenging skills is reflection of feeling. Mastering this skill, however, is critical to effective counseling and interviewing.

Step 1: Divide into practice groups.
Step 2: Select a group leader.
Step 3: Assign roles for the first practice session.

▲ Client
▲ Interviewer
▲ Observer 1, who gives special attention to noting client feelings, using the Feedback Form in Box 7-5. The focus of microsupervision needs to be on the ability of the interviewer to bring out and deal with emotions.
▲ Observer 2, who gives special attention to interviewer behavior and writes down each specific interviewer lead.

Step 4: Plan. We suggest that you examine a past or present story of a stressful experience (bullying, teasing, being seriously misunderstood, an unfair situation with school personnel, going through a hurricane or flood, or a time when you experienced prejudice or oppression of some type).

Most of us experienced real stress and some form of trauma as we watched planes fly into the World Trade Center. You very likely have a story to tell about where you were when you saw or heard the news. What were you seeing, hearing, and feeling? How did it affect you both then and now?

Another possibility is to explore accumulative trauma (see Box 6-2). How have repeating events in one's life added up to real trauma?

Give some attention to observing styles of emotional expression. With what style does the client present the story: sensorimotor, concrete, abstract formal-operational, or abstract

BOX 7-5   FEEDBACK FORM: OBSERVING AND REFLECTING FEELINGS

_____ (Date)

_____   _____

(Name of Interviewer)   (Name of Person Completing Form)

Instructions: Observer 1 will give special attention to client feelings via notations of verbal and nonverbal behavior below. On a separate sheet, Observer 2 will write down the wording of interviewer reflections of feeling as closely as possible and comment on their accuracy and value.

1. Verbal feelings expressed by the client. List here all words that relate to emotions.

2. Nonverbal indications of feeling states in the client. Facial flush? Body movements? Others?

3. Implicit feelings not actually spoken by the client. Check these out with the client later for validity.

4. Reflections of feelings used by the interviewer. As closely as possible, use the exact words of the interviewer and record them on a separate sheet of paper.

5. Orientation to emotional expression. Within what developmental style or styles was this client telling the story: here-and-now sensorimotor, detailed concrete description, abstract formal-operational reflection, or abstract dialectic/systemic with multiple perspectives?

6. Comments on the reflections of feeling. What sentence stems were typically used? Were the feeling words reflected by the interviewer implicitly or explicitly expressed by the client? Was the interviewer's use of the skill accurate and valid? Was the check-out used?

dialectic/systemic? You may also wish to explore specific questions to facilitate emotional discussion using different styles of experience (see pages 188–189).

Establish clear goals for the session. You can use questioning, paraphrasing, and encouraging to help bring out data. Periodically, the interviewer reflects feelings. This may be facilitated by one-word encouragers that focus on feeling words and by open questions ("How did you feel when that happened?"). The practice session should end with a summarization of both the feelings and the facts of the situation. Examine the basic and active mastery goals in the "Portfolio of Competence" section to determine your personal objectives for the interview.

The observers should use this time to examine the feedback forms and to plan their own sessions.

Step 5: Conduct a 5-minute practice session using this skill.

Step 6: Review the practice session and provide feedback to the interviewer for 10 minutes.
How well did the interviewer achieve goals and mastery objectives, and what feedback does the client provide verbally and perhaps through the Client Feedback Form of Chapter 1? As skills and client role-plays become more complex, you'll find that this time is not sufficient for in-depth practice sessions, and you'll want to contract for practice time outside the session with your group. Again, it is particularly important that the observers and the interviewer note the level of mastery achieved by the interviewer. Was the interviewer able to achieve specific objectives with a specific impact on the client?

Step 7: Rotate roles.

A reminder. The client may be "difficult" if he or she wishes, but must be talkative. Remember that this is a practice session, and unless affective issues are discussed, the interviewer will have no opportunity to practice the skill.

Portfolio of Competence

Skill in reflection of feeling rests in your ability to observe client verbal and nonverbal emotions. Reflections of feeling then can vary from brief acknowledgment to exploration of deeper emotions. You may find this a central skill as you determine your own style and theory.

Use the following as a checklist to evaluate your present level of mastery. As you review the items below, ask yourself, "Can I do this?" Check those dimensions that you currently feel able to do. Those that remain unchecked can serve as future goals. Do not expect to attain intentional competence on every dimension as you work through this book. You will find, however, that you will improve your competencies with repetition and practice.

Highlight competencies that you have met to date.

Level 1: Identification and classification.

❏ Generate an extensive list of affective words.
❏ Distinguish a reflection of feeling from a paraphrase.
❏ Identify and classify reflections of feeling.
❏ Discuss, in a preliminary fashion, issues in diversity that occur in relation to this skill.
❏ Write reflections of feeling that might encourage clients to explore their emotions.
❏ Recognize the developmental styles of emotion: sensorimotor, concrete, formal-operational, and dialectic/systemic.

Level 2: Basic competence. Aim for this level of competence before moving on to the next skill area.

❑ Acknowledge feelings briefly in daily interactions with people outside of interviewing situations (restaurants, grocery stores, with friends, and the like).
❑ Use reflection of feeling in a role-played interview.
❑ Use the skill in a real interview.

Level 3: Intentional competence. Review the following skills, all related to predictability and evaluation of the effectiveness of your abilities in working with emotion. These are skills that may take some time to achieve. Be patient with yourself as you gain mastery and understanding.

❑ Facilitate client exploration of emotions. When you observe clients' emotions and reflect them, do clients increase their exploration of feeling states?
❑ Reflect feelings so that clients feel their emotions are clarified. They may often say, "That's right . . . and . . ." They then continue to explore their emotions.
❑ Help clients move out of overly emotional states to a period of calm.
❑ Facilitate client exploration of multiple emotions one might have toward an important relationship (confused, mixed, positive, and negative feelings).
❑ Recognize and facilitate client exploration within the four styles of emotional expression—sensorimotor, concrete, formal-operational, and dialectic/systemic.

Level 4: Pychoeducational teaching competence. Teaching competence in these skills is best planned for a later time, but a client who has particular difficulty in listening to others may indeed benefit by training in observing emotions. Many individuals fail to see the emotions occurring all around them. Empathic understanding is rooted in awareness of the emotions of others. All of us, including clients, can benefit from bringing this skill area into use in our daily lives. There is clear evidence that people diagnosed with antisocial personality disorder have real difficulty in recognizing and being empathic with the feelings of others. You will also find this problem in some acting-out children. Here psychoeducation on recognizing the other person may be a critical treatment. A good place to start is with acknowledgment of feelings.

❑ Teach clients in a helping session how to observe emotions in those around them.
❑ Teach clients how to acknowledge emotions and, at times, to reflect the feelings of those around them.
❑ Teach small groups the skills of observing and reflecting feelings.

DETERMINING YOUR OWN STYLE AND THEORY: CRITICAL SELF-REFLECTION ON REFLECTION OF FEELING

This chapter has focused on emotion and the importance of grounding both yourself and your client. Special attention was given to identifying four varying styles of emotional expression as well as how you might help clients express more or less emotion, as appropriate to their situations.

What single idea stood out for you among all those presented in this chapter, in class, or through informal learning? What stands out to you is likely to be important as a guide toward your next steps. What are your thoughts on diversity? What other points in this chapter struck you as important? How might you use ideas in this chapter to begin the process of establishing your own style and theory?

What are your thoughts?

_____

_____

LIST OF FEELING WORDS

Sad—unhappy, depressed, tearful, uninterested, blue, bored, cheerless, dismal, dispirited, dull, gloomy, grief, grieving, miserable, anguished, sorrow, regret, sorry, guilty, deplorable, devastated, devalued, pitiful, derided, joyless, melancholy, melancholic, dejected, desolate, heavy-hearted, low, spiritless, rejected, woebegone, falling apart, wistful, wretched.

Mad—angry, annoyed, insulted, irritated, indignant, irate, hostile, offended, ripped, displeased, aggressive, furious, ferocious, rabid, stormy, inflamed, infuriated, hatred, strongly opposed, antagonistic, uncompromising, dislike, animosity, distaste, threatening, dissatisfied, undesirable, unfair, unreasonable, rude, insensitive.

Fear—scared, fretting, fright, frightened, threatened, anxious, anxiety, apprehension, dangerous, concerned, worried, worrisome, agitated, alarmed, dread, horror, panic, terror, trepidation, distressed, troubled, tormented, angst, disquieted, unease, nervous, brooding, moping.

Glad—happy, relaxed, safe, comfortable, calm, at ease, pleased, feeling of "wholeness," valued, accepted, "together," interesting, excited, confident, cheerful, spirited, joy, joyful, heartfelt, appreciative, grateful, cheery, pleasure, bright, contented, satisfied, delight, delighted, enjoy, thankful, relieved, fascinating, lovely, light, cared for, caring, pleasing, eager, compliant, festive, tickled, merry, fortunate, lucky, chipper, lighthearted, esteemed, respected, honored, cherished, welcomed.

CHAPTER 8

Integrating Listening Skills: How to Conduct a Well-Formed Interview

The Five-Stage Interview Structure

Reflection of Feeling

Encouraging, Paraphrasing, and Summarizing

Client Observation Skills

Open and Closed Questions

Attending Behavior

Ethics, Multicultural Competence, and Wellness

The secret of joy in work is contained in one word—excellence. To know how to do something well is to enjoy it.

—Pearl S. Buck

The five stages of the interview allow you and your client to integrate what has been said into a meaningful whole. Can you complete a full session using only listening skills? How can the skills and concepts of listening be used to help you and your clients?

Chapter Goals

The central aim of this chapter is to move you toward intentional competence. You will be able to engage in a full, well-formed interview using only listening skills. *The relationship—story and strengths—goals—restory—action* model is elaborated here and will enable you to structure a full session and will serve as a foundation for effective listening and success in the interview. This decisional model can be used with multiple theories of counseling.

Competency Objectives

Awareness, knowledge, and skills in listening will enable you to

▲ Understand the *basic listening sequence* (BLS) that integrates the listening skills you have learned thus far. You will also see how the BLS is used in multiple settings—from counseling to management to medicine.

▲ Identify the specific dimensions of empathy and understand how the listening skills are vital for empathetic interviews.

▲ Develop further awareness of the impact of interviewing, counseling, and psychotherapy on the brain, particularly as it relates to empathic understanding.

▲ Conduct a complete interview using only the listening skills within the *relationship—story and strengths—goals—restory—action* model.

What is your story?

Most chapters begin with a case study. As we start here, let's change the focus to you and your story. Or perhaps you may prefer to work with a partner. Please select one of the following topics that relates to you at this moment. We'd like you to think about the present story or script that you are enacting. Describe your life in relationship to one of the following possibilities: leisure/ play, physical activities, relationships, spiritual matters.

What is your present *story*? How are you living your life in one of these areas? Name at least two specific strengths that you observe in your story or life that might enable you to improve on where you are now.

Can you identify a goal for doing something better? Make it as specific, concrete, and measurable as possible. Begin with a general statement such as "I want a better relationship" or "I need to exercise more." Than make it more specific and doable within the coming week. Instead of reaching for a 10-point change, go for one or two smaller steps toward the larger objective.

Now it is time to restory. Take your present story and your goal. Note the difference or discrepancy between where your story is and where you would like it to be. Start the restorying process by imagining an ideal solution in which the present story is transformed. Note how you might feel to have that more desirable world. Then go to your specific, concrete, and more doable goal for the coming week. Do your two strengths help in this process? Imagine how actually meeting that smaller goal would result in a new paragraph or two in the larger story, perhaps even serving as a wedge for greater change later on.

Can you live into your new story?

Helping clients take action and live into new stories, narratives, and ways of being is what this chapter is about. You will use only empathy and listening skills to help clients find their inner strengths, wisdom, and abilities to restory and meet their goals.

INTRODUCTION: A REVIEW OF CULTURAL INTENTIONALITY AND INTENTIONAL COMPETENCE

A critical issue in interviewing is that the same skills may have different effects on people with varying individual and cultural backgrounds. Diversity will always characterize the mainstream of interviewing, counseling, and psychotherapy. In each interview you have, you will encounter people with varying life experiences. To this must be added the many issues of multiculturalism (e.g., ethnicity/race, people with disabilities, sexual orientation, spirituality/ religion). Remember, all interviewing contains multicultural dimensions.

To this diversity must be added individual differences. One Arab American client cannot be expected to behave the same as the next. Many are Christians of varying denominations, and the Islamic religions are widely diverse in beliefs. Veterans from the Iraq war are not all the same. The individual client will behave differently from one interview to the next. Almost as soon as you think you "fully understand" a client, a new side of personality and style will appear.

The microskills offer us a way to have some predictability in our work with individuals, almost regardless of multicultural background. But individual men and women and those of varying groups may have surprising differences in response, despite our best efforts at prediction. We must always be aware that our wide-ranging clientele will constantly vary. What "works" as expected one time may not the next. Intentional competence requires flexibility and the ability to move and change in the moment with constantly shifting client needs.

Let us examine the basic listening sequence and its critical importance in more detail.

The Basic Listening Sequence: Its Real Promise to Make a Difference

Observations of interviews in counseling and therapy as well as in management, medicine, and other settings reveal a common thread of skill usage. After developing a relationship, successful interviewers often begin their sessions with an open question followed by closed questions for diagnosis and clarification. The paraphrase checks out the content of what the client is saying, and the reflection of feeling (usually brief in the early stages) notes key emotions. These skills are followed by a summary of the concern or issue before continuing to resolution, setting goals, and restorying. Encouragers may be used throughout the interview to highlight important points.

Though these skills can be used in many different situations, they need not be used in any rigid sequence. Each counselor or interviewer adapts these skills to meet the needs of the client and the situation. The effective interviewer uses client observation skills to note client reactions and intentionally flexes at that moment to provide the support the client needs. As you learn other skills and observe individual and cultural differences, you may find it appropriate to begin some interviews with a self-disclosure or even a directive instead of the frequent open question.

In using the basic listening sequence (BLS), you have a three-part goal. Whenever you are working with a client on any topic, you will want to elicit the following:

1. *The key facts and thoughts around the situation.* These are obtained through "what" questions, encouragers, and paraphrases. This includes client thoughts about what happened.
2. *The central emotions and feelings.* You elicit emotions through questions (such as "Could you share your feelings about that issue?" "How does that feel?"), reflection of feeling, and encouragers that focus on emotional words.
3. *An overall summary of the issue.* At the close of a section of the interview, you may want to summarize the client's main facts and feelings. Sometimes it is useful to start a session with a summary.

When you use the basic listening sequence (BLS) shown below, you will be able to predict how clients may respond.

Basic Listening Sequence	Predicted Result
Select and practice all elements of the basic listening sequence, open and closed questions, encouraging, paraphrasing, reflection of feeling, and summarization. These are supplemented by attending behavior and client observation skills.	Clients will discuss their stories, problems, or concerns, including the key facts, thoughts, feelings, and behaviors. Clients will feel that their stories have been heard.

Table 8-1 gives examples of the basic listening sequence in counseling, management, and medical interviewing. A special advantage of mastering this sequence is that once a basic set of skills is defined, it can be used in many different situations. It is not unusual for a person skilled in the concepts of intentional interviewing to be conducting career counseling at a college in the morning, training parents in communication skills in the afternoon, and working as a management consultant in the evening. The microskills approach can be applied in many settings. In each case the BLS has the objective of bringing out client data—behaviors, key facts, thoughts, and feelings—for later interviewer and client action.

TABLE 8-1   Three examples of the basic listening sequence

Skill	Counseling	Management	Medicine
Open question	"Could you tell me what you'd like to talk to me about?"	"Could you tell me what you'd like to talk to me about?"	"Could you tell me what you'd like to talk to me about?"
Closed question	"Did you graduate from high school?" "What specific careers have you looked at?"	"Who was involved with the production line problem?" "Did you check the main belt?"	"Is the headache on the left side or on the right?" "How long have you had it?"
Encouragers	Repetition of key words and restatement of longer phrases.		
Paraphrases	"So you're considering returning to college."	"Sounds like you've consulted with almost everyone."	"It looks like you feel it's on the left side and may be a result of the car accident."
Reflection of feeling	"You feel confident of your ability but worry about getting in."	"I sense you're troubled by the supervisor's reaction."	"It appears you've been feeling very anxious and tense lately."
Summarization	In each case the effective counselor, manager, or physician summarizes the story from the client's point of view *before* bringing in the interviewer's framework.		

What is critical here is your awareness that the basic listening sequence will appear in some form in all theories of counseling. Different theories will focus on and bring out different types of stories from clients, but all will use listening skills to understand what is happening. Please return to Figure 1.1 on page 24 where the skills used with different theories are presented.

Theories of counseling and therapy can be considered *stories* or *narratives* with different focuses and goals. Cognitive-behavioral counselors' narratives focus on behaviors and thoughts, with a fair number of questions as they draw out the story. Person-centered counselors focus on the client and often let the interview flow to whatever happens, using reflective listening skills. However, this narrative does seek to bring the *real person* more in synchrony with the *ideal person.* Stories in the psychodynamic model often develop around unconscious issues, while the decision-oriented counselor focuses the client's story on decisions the client must make. Brief counseling seeks to draw out stories around client goals while motivational interviewing is very similar to the five-stage models presented in this chapter. The narratives and stories in counseling focus on the client, those of management focus on problem solving, and the physician emphasizes cure and wellness.

Counseling, interviewing, and psychotherapy can be difficult experiences for some clients. They have come to discuss their problems and resolve conflicts, so the session can rapidly become a depressing litany of failures and fears. Again, people grow from their strengths. Taking a wellness approach and using the basic listening sequence to draw out the client's strengths and resources, and then reflecting them back, ensures a more optimistic and directed interview. (For the research base of microskills, see Box 8-1.)

BOX 8-1   RESEARCH EVIDENCE THAT YOU CAN USE

Overview of Microskills Research

Daniels (2007) has assembled and reviewed 450 data-based studies on microskills. This review is available in the accompanying CD-ROM. Among his central observations and conclusions that have direct relevance to interviewing practice are the following:

▲ The microskills listed in the hierarchy do exist and can be classified with excellent reliability. However, the basic listening sequence skills have more research validation than the influencing skills discussed later in this text. Microcounseling skill training appears to be as effective or more effective than alternative systems.

▲ Maintenance of skills requires practice and use. Research has found that if the skills are not used in actual practice, they may disappear over time. You may have seen practicing professionals so sure of themselves that they fail to listen to the client. Continued practice is necessary if these skills are to be transferred to your own interviewing sessions. Understanding does not automatically indicate ability to perform.

▲ Research in the Netherlands and Japan clearly indicates the cross-cultural validity of the skills. Studies with medical practitioners in the Netherlands reveal the importance of establishing a solid relationship and then using listening skills and show that this leads to better compliance in using needed medications and patient satisfaction. Microskills (often

termed "microcounseling") have been translated into at least 20 languages. For example, the skills approach has been used in Africa by UNESCO staff to train peer counselors who work with AIDS patients and with refugees suffering from war trauma. Consultants in Sweden, Indonesia, and Japan use it to train top-level managers. Aboriginal social workers in Australia and Canadian Inuit community organizers have used this framework.

▲ A number of studies have shown that intentional skill usage in counseling results in predictable client impact and satisfaction with the session. Hill and O'Brien's (2004) review of parallel research on counseling skills produced similar findings on this important issue. What you say and do in the interview affects what clients say and do.

▲ The basic interviewing skills of this book all can be used effectively with children, particularly if you use concrete language and match their cognitive/emotional style (Ivey, Ivey, Myers, & Sweeney, 2005; Van Velsor, 2004).

▲ Teaching clients skills of communication can be effective. Research has shown that clinical practice with clients who may demonstrate avoidant personality style (shyness), depression, schizophrenia, and other diagnoses has been particularly successful. Individual, group, and family communication skills can be taught. Clearly, more research needs to be conducted in this important area.

INSTRUCTIONAL READING 1: EMPATHY AND MICROSKILLS

Carl Rogers (1957, 1961) brought the importance of empathy to our attention. He made it clear that it is vital to listen carefully, enter the world of the client, and communicate that we understand the client's world as the client sees and experiences it. Putting yourself "into another person's shoes" and "viewing the world through someone else's eyes and ears" are other ways to describe empathy. The following quotation has been used by Rogers himself to define empathy:

> This is not laying trips on people. . . . You only listen and say back the other person's thing, step by step, just as that person seems to have it at that moment. You never mix into it any of your own things or ideas, never lay on the other person anything that the person did not express. . . . To show that you understand exactly, make a sentence or two which gets exactly at the personal meaning the person wanted to put across. This might

be in your own words, usually, but use that person's own words for the touchy main things. (Gendlin & Hendricks, *Rap Manual,* undated)

The basic listening sequence is deeply involved with empathy. Of special importance is reflection of feelings. Again, please recall the importance of empathy to the relationship, the "working alliance"; it is central to the 30% of *common factors* that make for successful interviewing, counseling, and psychotherapy (Hubble, Duncan, & Miller, 1999). When you provide an empathic response, you can predict how clients may respond. Note below another description of empathy and the predictions that you can make.

Empathic Response	Predicted Result
Experiencing the client's world as if you were the client; understanding his or her key issues and feeding them back to clarify experience. This requires attending skills and using the important key words of the client, but distilling and shortening the main ideas. In additive empathy the interviewer may add meaning and feelings beyond those originally expressed by the client. If done ineffectively, it may subtract from the client's experience.	Clients will feel understood and engage in more depth in exploring their issues. Empathy is best assessed by clients' reaction to a statement.

Many others have followed and elaborated on Rogers's influential definition of empathy (cf. Carkhuff, 2000; Egan, 2007; Ivey, D'Andrea, Ivey, & Simek-Morgan, 2006). A common practice is to describe three types of empathic understanding:

▲ *Basic empathy:* Interviewer responses are roughly interchangeable with those of the client. The interviewer is able to say back accurately what the client has said. Skilled intentional competence with the basic listening sequence demonstrates basic empathy.

▲ *Additive empathy:* Interviewer responses add something beyond what the client has said. This may be adding a link to something the client has said earlier, or it may even be a congruent idea or frame of reference that helps the client see a new perspective. Skilled use of listening skills and/or influencing skills (see the later chapters of this book) enable an interviewer to become additive.

▲ *Subtractive empathy:* Interviewer responses sometimes give back to the client less than what the client says and perhaps even distort what has been said. In this case, the listening or influencing skills are used inappropriately.

The three anchor points above are often expanded to classify and rate the quality of empathy shown in the interview. A 5-point scale for examining the level of a counselor's empathic response and its related constructs is presented below, followed by examples of the varying degrees of empathic responses.

Level	1	2	3	4	5
	Subtractive	Somewhat subtractive	Interchangeable	Somewhat additive	Additive

Client:	I don't know what to do. I've gone over this problem again and again. My husband just doesn't seem to understand that I don't really care any longer. He just keeps trying in the same boring way—but it doesn't seem worth bothering with him any more.
Level-1 Counselor:	(subtractive) That's not a very good way to talk. I think you ought to consider his feelings, too.
Level-2 Counselor:	(slightly subtractive) Seems like you've just about given up on him. You don't want to try any more.
Level-3 Counselor:	(basic empathy or interchangeable response) You're discouraged and confused. You've worked over the issues with your husband, but he just doesn't seem to understand. At the moment, you feel he's not worth bothering with. You don't really care.
Level-4 Counselor:	(slightly additive) You've gone over the problem with him again and again to the point that you don't really care right now. You've tried hard. What does this mean to you?
Level-5 Counselor:	(additive) I sense your hurt and confusion and that right now you really don't care any more. Given what you've told me, your thoughts and feelings make a lot of sense to me. At the same time, you've had a reason for trying so hard. You've talked about some deep feelings of caring for him in the past. How do you put that together right now with what you are feeling? (Or if you truly sense that the relationship is hopeless, it may be wise to help the client acknowledge that fact.)

Box 8-2 presents an interesting perspective on empathy. A Level-3 interchangeable response is fairly safe and direct. However, to be empathic is to take risks, which in this case means risk of error. When counselors strive for higher level, additive responses, they may sometimes be completely off the mark and actually be subtractive, in which case the client will respond negatively. Use unpredicted and surprising client responses as an opportunity to understand the client more fully. *It's not the errors you make; it is your ability to repair them and move on that counts!*

Empathy has been further refined with specific ways to enhance the quality of the interviewing relationship—positive regard, respect and warmth, concreteness, nonjudgmental attitude, and authenticity or congruence. All of the following empathic dimensions can be rated on a 5-point scale.

Positive Regard

Positive regard, closely related to the search for strengths and positive assets, is responding to the client as a worthy human being. More concretely, it may be defined as selecting positive aspects of clients' stories and selectively attending to positive aspects of client statements.

In a subtractive response, the counselor focuses too much on the negative, forgetting possibility. In an interchangeable response, the counselor notes or reflects accurately what the client has talked about. This is important as it is the usual foundation of effective empathy—hearing what the client has said. But we need to move on to additive responses where the counselor points out that, even in the most difficult situation, the client is doing something positive. For example, the counselor may say the following to a client suffering from depression and talking about many problems: "John, I respect your ability to talk about and analyze your issues and problems so well. I can understand at least some of your feelings as you have described them. Clearly, you've got good insight into what is happening. Let's change the focus just a bit. When isn't the problem occurring? Tell me about a time in the past when you were able to feel better about yourself and others." Search constantly for positives and strengths in your clients.

BOX 8-2   NATIONAL AND INTERNATIONAL PERSPECTIVES ON COUNSELING SKILLS

Is Empathy Always Possible?

Kathryn Quirk, Student in Counseling Program at Cambridge College

As beginning students in counseling, one of the first concepts we run into is empathy—experiencing and understanding the world of the client. Certainly this is core to the helping interview.

But, as I read my text (not this one), I felt increasingly uncomfortable. Was this almost magical concept really possible? I'll tell you why. First the happy ending. I am now the delighted mother of a lovely child, the darling of my life. But Ryan did not come easily, and my husband and I needed the help of two fertility clinics.

The "simpler" strategies of getting pregnant failed for three years. Those years were agonizing, but only a sample of the trauma we were to face (yes, dear reader, going through fertility procedures meets the full definition of trauma). We then moved to complicated in vitro procedures involving petri dishes and surgery. The first three procedures failed, and the fourth resulted in a pregnancy that ended when twins died after three months. I don't like the word "fetus"—and grieving for lost babies was horrible. We moved to a new clinic, and our fifth try was fantastically successful.

How does all this relate to empathy? I recall a pleasant and expert nurse who counseled a group of us experiencing primary infertility. She was helpful and had good suggestions, but when things got emotional and we cried, she simply didn't get it. She would say that she understood and knew what we were going through. But let's face it; she hadn't been there herself. How could she truly understand the physical pain or the feelings of failure, shame, and hopelessness? She didn't understand our loneliness and, perhaps worst of all, the crushed hopes. How can she understand what we were really feeling? I resented it when she said she understood when she clearly did not and could not. Fortunately, those in the group who had "been there" supplied the needed empathy and support.

Does this mean that if you haven't experienced the inner world and actual experience of the client that you can't be empathic? At first I thought that understanding of my experience was impossible except for those who had experienced what I had gone through. However, I've softened my thoughts somewhat as I learn about

and think about good counseling. I still feel that being there is what serves as a foundation for the deepest empathy. But the nurse could have provided a deeper empathy than she did if she had admitted openly that she understood our feelings and experience only partially. She did, after all, have more experience listening to people with pregnancy challenges than we did. She did have something to offer.

By failing to discuss and admit that she was different from us suggested to me and to others in the group that she did not understand. What could she have done? First, I think she should have said early in her work with us that she herself had not experienced the difficulties that we went through and, as such, she could have admitted that her understanding and empathy were only partial. But she could have pointed out that she understood pain and loss and perhaps even shared some of her own difficult experiences. Saying this and also outlining her expertise and knowledge would have developed more trust and given us all a deeper feeling that she was an empathic person.

As I've gone through my counseling program, I increasingly become aware that I too will have problems with being truly empathic and communicating the understanding I do have. When I meet clients who are different from me multiculturally (e.g., race, sexual orientation, religious commitment), I now know that I need to discuss these issues up front. And I have the obligation to learn as much about the cultural background of these clients as I can. To maximize my empathic potential, I need to read, get out in the community where these people live, and participate with them when I can.

This also holds true for me when I work with alcoholics, cancer survivors, and those who have been raped. I haven't been there, but I have a responsibility to learn more about those whose life experience is different from mine.

Empathy is clearly important, but it is not learned just from classes and books. We all need to examine the human experience and become more fully aware of the life of those around us.

Source: Kathryn Quirk, counseling graduate student, Cambridge College.

Respect and Warmth

Respect and warmth helps the client feel safe and encourages openness. It may be most easily seen in the body and nonverbal behavior. You show respect and warmth by your open posture, your smile, and your vocal qualities. Your ability to be congruent with your body language is an indicator of respect and warmth. Capture yourself on videotape and obtain feedback on how much respect and warmth you show the client.

Concreteness

Concreteness has been and will be stressed throughout this book. It is important to seek specifics rather than vague generalities. As interviewers, we are most often interested in specific feelings, specific thoughts, and specific examples of actions. One of the most useful of all open questions here is "Could you give me a specific example of . . . ?" Concreteness makes the interview live and real. Likewise, communication from the interviewer—the directive, the feedback skill, and interpretation—needs to be highly specific or it may become lost in the busy world of the client.

Immediacy

Immediacy is often described as being in the moment with the client. What are you and the client experiencing *here and now*? It is most easily described in terms of language. You may respond in three tenses to a client who is angry: "You were angry . . . you are angry right now . . . you will be angry." We tend to respond to others in the same tense in which they are speaking. Some clients always talk in the past tense; they may profit from present-tense discussion. Other clients are always thinking of the future; still others are constantly in the present. Responses stated in present tense are frequently useful. However, responses that include all three tenses tend to be the most effective.

Another way to view immediacy is in terms of the relationship between the counselor and the client (Egan, 2007). The more personal the relationship is, the more immediate it is. "At this moment, what might be your thoughts about how the two of us are doing?" "What are you feeling and thinking at this very moment?"

As interviews move more to here-and-now immediacy, your presence in the interview will become more powerful and important. Clients will start responding to you as they responded to significant people from their past. They have *transferred* past experiences to the here-and-now relationship. A client's reaction to you is a clue to the past reenacted in the present.

This may help you understand possible issues with other significant persons in the client's life. It is in such instances that the immediacy of "I–you" talk, the skill of artful self-disclosure and feedback (Chapter 12), between counselor and client becomes especially important.

Nonjudgmental Attitude

A nonjudgmental attitude is difficult to describe. Closely related to positive regard and respect, a nonjudgmental attitude requires that you suspend your own opinions and attitudes and assume value neutrality in relation to your client. Many clients have attitudes toward

BOX 8-3   RESEARCH EVIDENCE THAT YOU CAN USE

Empathy, Neuroscience, and Mirror Neurons

Empathy is clearly central to all interviewing, counseling, and psychotherapy, regardless of the theory used. Empathy, however, is not just an abstract idea; empathy is identifiable and measurable in the physical brain. Fascinating research on brain activity validates what the helping field has been saying for years. "The basic building blocks [of empathy] are hardwired into the brain and await development through interaction with others. . . . [E]mpathy [is] an intentional capacity" (Decety & Jackson, 2004, pp. 71, 93).

A lot is said in the research quote above. Let us "unpack" the meaning and implications. *Mirror neurons* fire when we see actions or act ourselves. Mirror neurons enable you to sense and understand what the client is saying and feeling. These neurons fire and even impact your internal bodily responses when you are truly experiencing the world of the client. This is a natural talent that you can encourage and develop by increasing your awareness of the client and noting what happens inside your own body.

And you are awakening the mirror neurons in the client and facilitating their development of new connections in thoughts, feelings, and action. This awakening shows in verbal behavior of clients and the action that they take as a result of the interview. And as the clients restory their issues, new neural connections are born. Your empathic behavior and the relationship are central to change. Needless to say, this emphasizes the importance of a positive approach to change. If we listen and selectively attend only to problems, we reinforce negative patterns in the brain, and this will make the change process slow and clumsy.

What we learn here is that the empathic person's brain responds to another person's experience, even though he or she does not actually experience that person's world. Many studies over the years back up this central point. For example, children around their second year indicate concern for others cognitively, emotionally, and behaviorally by comprehending others' difficulties and trying to help (Zhan-Waxler et al., 1992). You also may have seen two young children playing together. One falls and starts crying. Even though the second child has not been hurt, he or she also cries. This ability to observe the feelings of others could be considered the developmental roots of empathic understanding.

Consider the following pain experiment with volunteer couples. A painful electric shock was administered to one member of each couple, and brain scans revealed brain activity throughout the full pain-processing network of the shocked individual. *Important for our understanding of empathy,* when the partner only watched the shock given to a loved one, the affective section of the observer's brain was activated, but not the sensory component where the physical pain is felt (Singer et al., 2004). In effect, we see here that brain patterns related to pain appear in the person who does not directly experience pain.

Research, cited by Decety and Jackson (2004), has shown that the antisocial, criminal personality has a reduced ability to appreciate the emotions of others. There is less firing of mirror neurons in the prefrontal cortex, and this deficit also appears to be a dysfunction of the energizing amygdala and hippocampus (long-term memory) (Blair, 2001, 2003; Blair & Cipoletti, 2000). Visit Appendix II for visuals and elaboration.

What does this mean for you when you work with difficult clients? First, remember the importance of a nonjudgmental and positive approach. Even so, the relationship with acting-out children or adults requires that you also be firm and clear as well. These clients almost always have a history of neglect and abuse. Tony Crespi, one of Allen's students, studied the developmental history of adolescents on death row, waiting to be killed. These were not pleasant individuals, but Tony found that they had experienced an average of 10 identifiable events of abuse, incest, trauma, and neglect as they grew up. Small wonder that they lack understanding and respect for others! Crespi's findings suggest that life events have produced their lack of empathy, with resultant less firing of mirror neurons. Some believe that genetics are the "cause" of antisocial and acting-out behavior. But it is also clear that we need to understand and consider the likely developmental history of individuals exhibiting such behavior.

their issues and concerns that may be counter to your own beliefs and values. People who are working through difficulties and issues do not need to be judged or evaluated, and your neutrality is important if you want to maintain the relationship.

A nonjudgmental attitude is expressed through vocal qualities and body language and by statements that indicate neither approval nor disapproval. However, as with all qualities and skills, there are times when your judgment may facilitate client exploration. There are no absolutes in counseling and interviewing.

For a moment, stop and think of a client whose behavior troubles you personally. It may be someone whom you regard as dishonest, a person who is a perpetrator of violence, or one who shows clear sexism and racism in the interview. These are challenging moments for the nonjudgmental attitude. You do not have to give up your personal beliefs to present a nonjudgmental attitude; rather you need to suspend your private thoughts and feelings. If you are to help these people change and become more intentional, presenting yourself as nonjudgmental is critical. You do not have to agree with or approve of the thoughts and behaviors of the client to be nonjudgmental. But you may have an obligation to educate the client and help her or him move to new understandings and new stories. Nonetheless, you will still be nonjudgmental in expressing yourself, as change requires a basis of trust and honesty.

Authenticity and Congruence

Are you personally real? Authenticity and congruence are the reverse of discrepancies and mixed messages. The hope is that the counselor or interviewer can be congruent and genuine and not display many discrepancies. Needless to say, however, life is full of discrepancies and paradoxes, and your ability to be flexible in response to the client may be the most basic demonstration of your authenticity.

Box 8-3 provides evidence that empathy is identifiable and measurable in the brain.

INSTRUCTIONAL READING 2: THE FIVE STAGES/DIMENSIONS OF THE WELL-FORMED INTERVIEW

When you use the five-stage interview structure, you can predict how clients may respond. Note below the brief description of the interview and the general prediction that you can make.

The Five-Stage Interview Structure	General Prediction
Relationship—story and strengths—goals—restory—action	The client will establish a positive relationship with the interviewer, will tell the story, set realistic goals, develop a new story or way of viewing issues, and transfer new learning to daily life.

Relationship—story and strengths—goals—restory—action is a basic framework that you can use in virtually all types of interviewing, counseling, and therapy. Combine these five interview stages with the basic listening sequence, and you will be able to complete a full interview using only listening skills. They are basic to what is called decisional counseling.

The *relationship—story and strengths—goals—restory—action* model is based on a classic problem-solving model that you have likely encountered before:

1. Defining the problem. (Story)
2. Defining goals. (Goals)
3. Generating alternative solutions and selecting a more effective approach. (Restory)

To this basic triad, we add developing a *relationship* with the client as you start the interview. And after a solution has been selected through the underlying three-part model of problem solving, you then need to make sure that the client takes *action* and does something after the interview to follow up and use the decisions that have been made. Later in this book, you will see how this structure can be used in many approaches to counseling and therapy, even though the theories appear to be very different from one another (Ivey & Matthews, 1984; Ivey, Pedersen, & Ivey, 2001).

Each interview will be different from all others, and your relationship with each client needs to be intentional. You must have the ability to work with different styles. All interviewers use microskills and strategies and tend to follow a sequence of stages from the beginning to the end. *Each stage requires using the basic listening sequence to accomplish the aims of the stage.* As you move from stage to stage, continue listening. The basic decisional model is:

1. *Relationship*—Initiating the session. Rapport, trust building, and structuring.
2. *Story and Strengths*—Gathering data. Drawing out stories, concerns, problems, or issues.
3. *Goals*—Mutual goal setting. What does the client want to happen?
4. *Restory*—Working. Exploring alternatives, confronting client incongruities and conflict, restorying.
5. *Action*—Terminating. Generalizing and acting on new stories.

These five stages, however, do not always need to be completed in order. Expect to recycle back to an earlier stage as you discover new information. For example, you may be working on restorying in Stage 4, but new issues come up that are really part of Stage 2; you now see that the original story changes, perhaps requiring new goals to be set. Throughout, the relationship remains central, and you may need to recyle and refocus on relationship building at any stage of the session. The five stages of the interview are summarized in Table 8-2.

The Circle of Interviewing Stages (Figure 8-1) is a graphic depiction of the interview process, another way of visualizing it. Note that the words "relationship, positive assets, and wellness" form the hub of the circle, continuing in all stages. Relationship and the working alliance are so important that we include them twice.

A circle has no beginning or end and can sometimes be used as the symbol of an egalitarian relationship in which interviewer and client work together on concerns. Some clients prefer that you take charge, particularly in the early stages as you gain credibility as a helper. However, eventually you will want to work *with* clients on a mutual basis in helping them reach their goals.

TABLE 8-2 The five stages of the microskills interview

Recall that the core of the five-stage structure is the relationship and the strength-based wellness approach. The basic listening skills are crucial for all stages. Decisional counseling and all other theories can be related to this model.

Stage	Function and Purpose	Commonly Used Skills	Predicted Result
1. **Relationship:** Initiate the session. Develop rapport and structuring. "Hello, what would you like to talk about today?"	Build a working alliance and enable the client to feel comfortable with the interviewing process. Explain what is likely to happen in the session or series of interviews, including informed consent and ethical issues.	*Attending, observation skills, information giving* to help *structure* the interview. If the client asks you questions, you may use *brief self-disclosure.*	The client will feel at ease with an understanding of the key ethical issues and the purpose of the interview. The client may also come to know you more completely as a person and professional.
2. **Story and strengths:** Gather data. Draw out client stories, concerns, problems, or issues. "What's your concern?" "What are your strengths and resources?"	Discover and clarify why the client has come to the interview and listen to the client stories and issues. Identify strengths and resources as part of a wellness approach.	*Attending* and *observation* skills, especially *the basic listening sequence* and the *positive asset search.*	The client will share thoughts, feelings, and behaviors and tell the story in detail as well as present strengths and resources.
3. **Goals:** Mutual goal setting. "What do you want to happen?"	Provides purpose and direction for the session. Without clear goals, the interview will drift.	*Attending skills,* especially the *basic listening sequence,* certain *influencing skills,* especially *confrontation* (Chapter 9), may be useful.	The client will discuss directions in which he or she might want to go, new ways of thinking, desired feeling states, and behaviors that might be changed. The client might also seek to learn how to live more effectively with situations or events that cannot be changed at this point (rape, death, an accident, an illness). A more ideal story ending might be defined.
4. **Restory:** Working. Exploring alternatives, confronting client incongruities and conflict, restorying. "What are we going to do about it?" "Can we generate new ways of thinking, feeling, and behaving?"	Generate at least *three* alternatives that may resolve the client's issues. Creativity is useful here as you seek to find at least three alternatives so that the client has a choice. One choice at times may be to do nothing but accept things as they are.	*Summary* of major discrepancies with a supportive *confrontation.* More extensive use of *influencing skills,* depending on theoretical orientation (e.g., *interpretation, reflection of meaning, feedback*). But this is also possible using only *listening skills.* Use *creativity* to solve problems.	The client may reexamine individual goals in new ways, solve problems from at least those alternatives, and start the move toward new stories and actions.
5. **Action:** Terminating. Generalizing and acting on new stories. Conclude with plan for generalizing interview learning to "real life" and eventual termination of the interview or series of sessions. "Will you do it?"	Generalize new learning and facilitate client changes in thoughts, feelings, and behaviors in daily life. Commit the client to homework and action. As appropriate, plan for termination of sessions.	*Influencing skills,* such as *directives* and *information/ explanation,* plus *attending* and *observation skills* and the *basic listening sequence.*	The client will demonstrate changes in behavior, thoughts, and feelings in daily life outside of the interview.

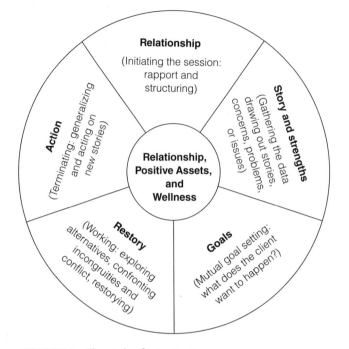

FIGURE 8-1   The circle of interviewing stages.

The importance of Stage 1 relationship and Stage 3 goal setting cannot be overemphasized. If you and your client agree to mutual goals for Stage 3, you have a solid basis for trust and actually reaching those goals.

Failure to treat is one of the most common causes of malpractice suits. Clients and counselors who agree on goals in a clear fashion have a direction and contract for the sessions. Jointly revise goals as necessary. Clients who participate in goal setting and understand the reasons for your helping interventions may be more likely to participate in the process and are more open to change. Keeping a joint written record of goals and progress toward them will help clarify the purpose and direction of counseling and provide accountability standards for insurance companies. Following is an outline of the five-stage decisional model.

Stage 1. *Relationship*—Initiating the Session: Rapport, Trust Building, and Structuring ("Hello")

Building Rapport

"Hello, _____." A prime rule for establishing rapport is to use the client's name and repeat it periodically through the session. Some interviewers give extensive attention to the relationship (rapport) stage, whereas others simply assume rapport and start immediately. Introducing the interview and building rapport obviously are most important in the first interview with a client. In some cases relationship building will be lengthy and blend into treatment.

In much of Western counseling and interviewing theory and practice, the relationship stage is quite short. After a brief "Hello," the interview begins with rapport being assumed, right or wrong. In reality, Allen found that it could take more than one inteview for him to develop sufficient trust to build a working relationship with Australian Aboriginal people. With a history of oppression by White people—such as seeing their children stolen and used as servants and themselves suffering continued acts of violence—this group starts with an automatic distrust of Whites. If you are White, expect to work through issues of distrust as you establish a relationship. You may encounter similar areas of distrust as you work to establish a relationship with Canadian Dene, Native American Indians, African Americans, gays/lesbians, people with disabilities, and many other groups who may be different from you. And if you are are a counselor from any of these groups, you can encounter difficulties in establishing a real trusting relationship with some Whites, who automatically distrust you.

As you begin the session, focus on developing rapport and trust (and then continue that awareness throughout the session). Sometimes new clients want to start immediately, and you can assume that rapport with them is satisfactory. Observe client attending behavior as a clue to the nature of the relationship—eye contact, body language, and vocal tone often indicate the real nature of what the client is saying.

Structuring

Structuring is the second part of this first stage. Informed consent and ethical issues should be discussed with the client. Talk with your client about what is going to happen, be explicit, and ask if he or she understands. Explain the purpose of the interview and what he or she can or cannot do. Some clients need to have the interview explained for them. This may be their first interview, and they may not know how to behave.

Interviewers conducting a welfare intake interview, for example, find that they can better assist clients if they indicate very early in the session what they can and can't do and the limits and goals of the session. If the client has needs that you know you can't fulfill, consider immediate referral. But don't leave clients hanging. Stay with them until they are situated with someone who can help.

Your ability to observe clients will tell you when it is appropriate to gather data (Stage 2). One of the best clues for time to start is when the client begins talking spontaneously about concerns. Many—perhaps most—interviews include in Stages 1 or 2 some variation on "What would you like to talk about today?" "I'd like to know what your story is and what your goals are."

Multicultural Considerations

In addition to the comments above, maintain awareness that different cultural groups develop relationships in varying ways. In Latin cultures, the concepts of respect and dignity are particularly important and may require a more formal approach. The more experienced your client is in English-speaking cultures, the more likely he or she will be to understand and accept traditional counseling and interviewing theory and approaches.

Authorities are in increasing agreement that cultural and ethnic differences need to be addressed in a straightforward manner relatively early in counseling, often in the first interview (for example, see Cheatham & Stewart, 1990; Kim et al., 2003). Your ability to recognize and respect differences is essential for the success of the interview or counseling.

Stage 2. *Story and Strengths*—Gathering Data: Drawing Out Stories, Concerns, Problems, or Issues ("What is your concern?" "What are your strengths and resources?")

Listening to the Client's Story

"What's your concern?" The second major task of the interviewer is to find out why the client has come for counseling and what the client wants. Coupled with that is gathering necessary information about the client. Some clients may confuse you right at the beginning as they present you with a long list of issues and concerns. We have discovered that the last item in the client "laundry list" is often the central issue. Never forget, however, that the entire list is important. When there are several issues, you need to summarize the list as well as you can and then ask "What have I missed?" It may be helpful for you and the client to write down the several concerns and then prioritize them, identifying the ones that need to be addressed first.

The central task in this stage is drawing out the story and clarifying client issues. Use the basic listening sequence to elicit facts, thoughts, and feelings. The newspaper reporter's checklist of *who, what, where, when, how,* and *why* questions helps to ensure that you have covered the most important issues and concerns underlying client challenges and problems.

In your attempts to define the central client concerns, always ask yourself, what is the real world of the client? What problem seeks resolution or what opportunity needs to be actualized? Failure to clearly answer these questions often results in an interview that wanders and lacks purpose.

Positive Asset Search and Wellness

Finding strengths as you initiate the session can make an important difference; use the BLS in a positive asset search for strengths that later will help with problem resolution. During the working and termination phases, frequent reference to these positives produces a "can do" attitude in clients that enables them to look at serious challenges in their lives. Checking out where the client stands on wellness issues discussed in Chapter 2 enables the client to discover *what he or she is doing right.*

If we help *center* clients with existing strengths, they are better able to delve in and discuss deeper problems and issues. The power of the positive will enable clients to deal honestly and bravely with even the most complex issues.

Multicultural Considerations

Recall that the word "problem" can be a problem for some clients. Consider using *story, issue,* or *challenge* as alternatives. Keep in mind that the problem-focused language of much of counseling and therapy may be inappropriate for some clients. Furthermore, the second stage of the interview, for many clients from all cultural backgrounds, may be better placed in the third stage, which focuses on client goals. If you establish joint goals with your client, you empower that client, and the two of you may then return to concerns oriented toward helping that client reach her or his goals. In fact, in brief solution-focused counseling, the interview often starts with a focus on client goals (Chapter 14).

Stage 3. *Goals*—Mutual Goal Setting ("What do you want to happen?")

If you can't tell me what you'd like to be happening . . . you don't have a problem yet. You're just complaining. A problem only exists if there is a difference between what is actually happening and what you desire to be happening.

— Kenneth Blanchard

Where does the client want to go? Many counselors and interviewers summarize the problem or challenge, then ask a question such as "What would you imagine the ideal solution to be?" "Where do you want to go with this?" "Could you take a moment to sit back and develop a fantasy of what you would like to happen?" Expect clients to discuss general, nonspecific goals. You must help them see the importance of concrete, attainable goals. Use the basic listening sequence and help clients clarify what they really want.

The word "mutual" is important in goal setting. Your active involvement in client goal setting is important. You will have clients who constantly talk about their problems but when asked, "What might you want to do about it? What is your goal?" they ignore you and return to their issues. *If you and the client don't know where the interview is going, you may end up somewhere else!*

A maxim for the confused interview—be it a difficulty in school, career, partner relationships, depression or anxiety—is this:

> Define a goal, make the goal explicit and concrete, search for strengths, resources, and assets that are likely to facilitate goal attainment. Only then move on to restorying and and action.

The question of determining outcomes is of interest from a theoretical as well as a practical perspective. Rogers (1957, 1961) talks about many clients who have incongruencies between the real self and the ideal self. Behavioral psychologists often talk about present behavior compared to desired behavioral goals. Reality therapists talk about fulfilling unmet needs, and decisional and brief solution-focused counselors tend to focus on the whatever concrete goal the client has. Most theoretical orientations ask, "Where is the client, and where does she or he want to go?" "What is the difference between the real world and the desired world?"

Consider the following model goal summary sentences from different theoretical orientations:

▲ *Decisional counseling:* "On the one hand, your concern may be summarized as . . . , and on the other hand, your desired outcome is . . . , and you have the following assets and strengths to help you reach your goals. . . ."
▲ *Rogerian client-centered:* "Your real self, as you describe yourself, is. . . . Yet, you see your ideal self as . . . , and you have several positive qualities, such as. . . ."
▲ *Cognitive-behavioral:* "Your present behavior is . . . , but you would like to behave differently. For example, you would really like to achieve the change goal of. . . ." "As we've talked, I've noticed some strengths you already have to meet this challenge."

> ▲ *Marital counseling:* "Your present relationship is described as . . . , but you would like to see it change as follows. . . . As a couple you seem to have several strengths, such as . . . , that will be helpful in resolving the conflict."
> ▲ *Career counseling* (confused client): "You are searching for a college major (or life career) and aren't really sure of your possibilities. Yet, you've described your short- and long-term goals rather clearly. . . . You've had several positive work experiences in the past. . . . How do you put the the challenges you face with all those strengths?"

Note that all these sentences aim toward concreteness and point out the discrepancy between the problem definition and the desired outcome. Many clients can use this summary as a springboard to action. With your support, they will resolve the discrepancy on their own. Clients who are extremely emotionally involved with their problems may require more influencing skills and active direction on your part. Observing your client's verbal and nonverbal reaction to this summary of the key discrepancies will help you determine what to do in Stage 4 restorying.

Stage 4. *Restory*—Working: Exploring Alternatives, Confronting Client Incongruities and Conflict, Restorying ("What are we going to do about it?")

"What are we going to do about it?" The purpose of this stage is to resolve concerns and find relief for the client. You want to find workable alternatives for the client's daily life and long-term living. The client at this stage may be stuck and unable to come up with productive alternatives. The task of the counselor or interviewer is to help explore possibilities and to assist the client in finding new ways to act more intentionally in the world. This can be done through common sense and brainstorming or with the aid of a specific theoretical approach.

Let us consider a school counselor talking with an acting-out teen who just had a major showdown with the principal. With the five-stage interview structure, the first task is to establish relationship (rapport). The teen may be expected to challenge you, particularly as he or she expects you to support the school administration. Your task is not to judge but to gather data from the teen's point of view. If this is done effectively, the teen will feel that he or she has been heard. Follow by finding what the teen would like to have happen. Often, if you hear the client's story first, her or his goals become more workable and realistic. It does little good, of course, to work with unrealistic goals. Work with the teen to find a way to "save face" and move on. You may have to become an ally of teens if you are to be effective in important conflicts in the future. If things are going reasonably well, teens in such situations may be able to describe the conflict from the principal's point of view.

The model summarizations described in the discussion of Stage 3 can be helpful. "On one hand, you see the situation as . . . , and your goal is. . . . But on the other hand, the principal tells a different story, and as you say, the principal's goal is likely to be. . . ." Given that you, the counselor, have some specific goals, what do you think you can do to reach this client and find a solution that works? If you have listened well and developed rapport, many teens at this point are able to generate ideas to help resolve the situation.

The basic listening sequence and skilled questioning are useful in facilitating client exploration of answers and solutions. Here are some useful questions to assist client problem solving:

"What other alternatives can you think of?"
"Can you brainstorm ideas—just anything that occurs to you?"
"What has worked for you before?"
"What part of the problem is workable if you can't solve it all right now?"
"Which of the ideas that you have generated appeals to you most?"
"What would be the consequence of your taking that alternative?"

In effect, all of these are oriented toward opening clients' thought, leading to new solutions. Encouraging skills are useful in helping clients stop and explore possibilities. Repeat key words that might lead the client to new alternatives for action.

Counseling and long-term therapy both try to resolve issues in clients' lives in a similar fashion. The counselor needs to establish rapport, define the problem, and establish certain desired client outcomes. The distinction between the problem and the desired outcome is the major incongruity the therapist seeks to resolve. This incongruity or discrepancy may be resolved in three basic ways. The counselor can use attending skills to clarify the client's frame of reference and then feed back a summary of client concerns and the goal. Often clients will generate their own synthesis and resolve their challenges. If clients do not generate their own answers, then the therapist can add interpretation, self-disclosure, and other influencing skills in an attempt to resolve the discrepancy. In that case, the counselor would be working from a personal frame of reference or theory. Finally, in systematic problem solving the counselor and client might generate or brainstorm alternatives for action and set priorities for the most effective possibilities.

During Stage 4, it is particularly important to keep the concrete issue, or challenge, in view while generating alternatives for a solution. However, a decision for action is not enough. You also need to plan to make sure that feelings, thoughts, and behaviors generalize beyond the interview itself. Stage 5 of the interview speaks to this task.

Stage 5. *Action*—Terminating: Generalizing and Acting on New Stories ("Will you do it?")

"Will you do it?" The complexities of the world are such that taking a new behavior back to the home setting is difficult.

Some therapies work on the assumption that behavior and attitude change will come out of new unconscious learning; they "trust" that clients will change spontaneously. This can happen, but there is increasing evidence that planning and actually contracting in written form for specific change increases the likelihood of transfer of learning to the "real world" and actually changing things.

Consider the situation again with the teen in conflict with the principal. You may have generated some good ideas together, but unless the teen follows up on them, soon he or she will be back in your office again, and the principal may be wondering why you aren't more effective. (And if the problem is not with the teen but with an insensitive teacher and/or principal, then your challenge is that much greater.) Your task is to find something "that works" and changes the repeating story.

Change does not come easily, and maintaining any change in thoughts, feelings, or behavior is even more difficult. Behavioral psychology has given considerable thought to the transfer of training and has developed an array of techniques for transfer; even so, clients still revert to earlier, less intentional behaviors. At this point, we suggest that you examine how to maintain change in behavior, thoughts, and feelings (see Maintaining Change and Relapse Prevention Worksheet on page 430). More and more interviewers, therapists, and counselors are using some variation of this formal way to ensure that the hard work done in the session has relevance and impact in the outside world.

Here are examples that different theoretical schools have used to facilitate the transfer of learning from the interview. Many of these examples are elaborated on later in this book.

Homework. The interviewer may suggest specific tasks for the client to try during the week as a follow-up to the interview. If you have developed concrete goals for the interview or series of counseling sessions, a review of progress will lead to specific homework activities.

Role-playing. The client can practice the new behavior in a role-play with the counselor or interviewer. This emphasizes the specifics of learning and increases the likelihood that the client will recognize the need for the new behavior after the session is over.

Imagery. Ask the client to imagine the future event and what he or she will specifically need to do to manage the situation more effectively. Suggest that this image be brought to attention when the difficulties arise again.

Behavioral charting and journaling. The client may keep a record of the number of times certain behaviors occur and report back to the counselor. With other clients a journal of reactions to the sessions will be helpful.

Follow-up and support. Ask the client to return periodically, so you can check on the maintenance of behavior. Alternatively, you can use a telephone call or e-mail exchange. The counselor can also provide social and emotional support through difficult periods. This may be especially helpful in implementing a wellness plan. Phone calls and e-mails to clients can focus on whether they actually are exercising, are eating and sleeping better, or are engaging in positive leisure activities to help them break computer addiction or stop smoking.

Here are some questions you can use to help clients plan their own generalization from the interview:

"What one thing from the interview stands out for you right now that you might take home?"
"You've generated several ideas and selected one to try. How are we going to know if you actually do it?"
"What comes to your mind to try as homework for next week that we can look at when we get together?"

Each of these can be coupled with the basic listening sequence to draw out the generalization plan in more detail. You may want to ask your client at the close of the interview "Will you do it?" If you have a written contract, review the goals and actions needed.

These are just a few of the many possibilities to help develop and maintain client change. Each individual will respond differently to these techniques, and client observation skills are called for to determine which technique or set of techniques is most likely to be helpful to a particular person.

EXAMPLE DECISIONAL COUNSELING INTERVIEW: USING LISTENING SKILLS TO HELP CLIENTS WITH INTERPERSONAL CONFLICT

"I can't get along with my boss." This interview illustrates how listening skills can be used to help the client understand and cope with interpersonal conflict. The interview has been edited to identify how the five- stage model works in the interview (*relationship—story and strengths—goal—restory—action*). When you conduct your own interview and develop a transcript indicating your own ability to use listening skills, you may want to arrange your transcript in a similar fashion to that presented here.

The client in this case is a relatively verbal 20-year-old part-time student who is in conflict with his boss at work. A verbal, cooperative client is required for a counselor to work through a complete interview using only listening skills.

This is a very abbreviated and edited example interview. Our goal here is to show you how the five stages can be identified in practice. The interview also illustrates how you can set up and analyze your own interview, focusing on listening skills.

Stage 1. *Relationship*—Initiate the Session

Develop rapport and structuring. "Hello, what would you like to talk about today?"

Interviewer and Client Conversation	Process Comments
Machiko: Robert, do you mind if we tape this interview? It's for a class exercise in interviewing. I'll be making a transcript of the session, which the professor will read. Okay? We can turn the recorder off at any time. I'll show you the transcript if you are interested. I won't use the material if you decide later you don't want me to use it.	Machiko opens with a closed question followed by structuring information. It is critical to obtain client permission and offer the client control over the material before taping. As a student, you cannot legally control confidentiality, but it is your responsibility to protect your client.
Robert: Sounds fine; I do have something to talk about.	Robert seems at ease and relaxed. As the taping was presented casually, he is not concerned about the use of the recorder. Rapport was obviously established as they knew each other. There are some clients who will start the session as quickly as Robert, but don't expect it.
Machiko: What would you like to share?	The open question, almost social in nature, is designed to give maximum personal space to the client. But Machiko might have been wise if she had spent more time telling him what to expect in this interview.
Robert: My boss. He's pretty awful.	Robert indicates clearly through his nonverbal behavior that he is ready to go. We also see body movements indicating distress. However, Machiko observes that he is comfortable with her, and she will deal with the stress later. She decides to move immediately to Stage 2 and gather data on the client's story. With some clients, several interviews may be required to reach this level of rapport and relationship.

Stage 2. *Story and Strengths*—Gather Data

Draw out client stories, concerns, problems, or issues. "What's your concern?" "What are your strengths and resources?"

Interviewer and Client Conversation	Process Comments
Machiko: Could you tell me about it?	This open question is oriented toward obtaining a general outline of the concern the client brings to the session.
Robert: Well, he's impossible.	Instead of the expected general outline of the issues, Robert gives a brief answer. The predicted consequence didn't happen. This is where your cultural intentionality and ability to flex in the here and now of the session become so important.
Machiko: Impossible?	Encourager. Intentional competence requires you to be ready with another response if the client does not elaborate. This time the encourager started things.
Robert: Yeah, really impossible. It seems that no matter what I do, he is on me, always looking over my shoulder. I don't think he trusts me.	This time the predicted result occurs, and Machiko begins to obtain a picture of what is going on with Robert.
Machiko: Could you give me a more specific example of what he is doing to indicate he doesn't trust you?	Robert is a bit vague in his discussion. Machiko asks an open question eliciting concreteness.
Robert: Well, maybe it isn't trust. Like last week, I had this customer lip off to me. He had a complaint about a shirt he bought. I don't like customers yelling at me when it isn't my fault, so I started talking back. No one can do *that* to me! And of course the boss didn't like it and chewed me out. It wasn't fair.	As events become more concrete through specific examples, we understand more fully what is going on in the client's life and mind.
Machiko: As I hear it, Robert, it sounds as though this guy gave you a bad time, and it made you angry, and then the boss came in. It wasn't fair.	Machiko's response is relatively similar to what Robert said. Her paraphrase and reflection of feeling represent Level 3 basic interchangeable empathy. This is a good example of the last item mentioned by the client being the important one. Fairness is an important construct for Robert.
Robert: Exactly! It really made me angry. I have never liked anyone telling me what to do. I left my last job because the boss was doing the same thing.	Accurate listening often results in the client's saying "exactly" or something similar.
Machiko: So your last boss wasn't fair either?	Machiko's vocal tone and body language communicate nonjudgmental warmth and respect. She brings back Robert's key word *fair* by paraphrasing with a questioning tone of voice, which represents an implied check-out. This is an interchangeable empathic response (Level 3).

The interview continues to explore Robert's conflict with customers, his boss, and past supervisors. There appears to be a pattern of conflict with authority figures over the past several years. This is a common among young males in their early careers. After a detailed discussion of the specific conflict situation and several other examples of the pattern, Machiko decides to conduct a positive asset search.

Machiko: Robert, we've been talking for a while about difficulties at work. I'd like to know some things that have gone well for you there. Could you tell me about something you feel good about at work?	Paraphrase, structuring, open question, and beginning positive asset search.
Robert: Yeah; I work hard. They always say I'm a good worker. I feel good about that.	Robert's increasingly tense body language starts to relax with the introduction of the positive asset search. He talks more slowly.
Machiko: Sounds like it makes you feel good about yourself to work hard. Could give me a specific time you felt good?	Reflection of feeling, emphasis on positive regard, Level 3 interchangeable empathy. Note emphasis on concreteness.
Robert: Yeah. For example . . .	We start hearing Robert identify strengths. This will help center him and also help him deal with issues in his life. We deal with problems best if we have a sense of our personal worth.

Robert continues to talk about his accomplishments. Machiko has used the basic listening sequence to help Robert feel better about himself. She also learns that Robert has several important assets to help him resolve his own problems—among them, determination and willingness to work hard.

Stage 3. *Goals*—Mutual Goal Setting

"What do you want to happen?"

Interviewer and Client Conversation	Process Comments
Machiko: Robert, given all the things you've talked about, could you describe an ideal solution? How would you like things to be?	Open question. The addition of a new possibility for the client represents additive empathy (Level 3). It enables Robert to think of something new. Here we see the discrepancy that we hope will be at least partially resolved in this session. Specifically, we have a good sense that Robert doesn't like the work environment, but how would he like it to be?
Robert: Gee, I guess I'd like things to be smoother, easier, with less conflict. I come home so tired and angry.	

(continued)

Interviewer and Client Conversation	Process Comments
Machiko: I hear that. It's taking a lot out of you. Tell me more specifically how things might be better.	Paraphrase, open question oriented toward concreteness.
Robert: I'd just like less hassle. I know what I'm doing, but somehow that isn't helping. I'd just like to be able to resolve these conflicts without always having to give in.	Robert is not as concrete and specific as anticipated. But he brings in a new aspect of the conflict—giving in, which turns out to be an important key word in Robert's meaning system.
Machiko: Give in?	Encourager. This led to Robert sharing again the importance of fairness. Machiko also saw the words "give in" as one of Robert's important meaning issues. From a meaning perspective, he wants fairness and won't give in on it. This will present an interesting challenge to Machiko.

Again, keep in mind that this session is abbreviated and edited. In the goal-setting process, you will often find yourself changing the problem definition. Here Machiko learns another dimension of Robert's conflict with others. Subsequent use of the basic listening sequence brings out this pattern with several customers and employees. It seems we have both an immediate problem with the boss and a possible longer term issue in Robert's relationships with authority.

Machiko: So, Robert, I hear two things in terms of goals. One that you'd like less hassle, but another, equally important, is that you don't like to give in. Have I heard you correctly?	Machiko uses a summary to help Robert clarify his problem, even though no resolution is yet in sight. She checks out the accuracy of her hearing. Level 4 additive empathy because we are moving toward more integration.
Robert: You're right on, but what am I going to do about it?	Robert hears this and is ready to move on.

Normally, the goal-setting process is *not* completed in such a short exchange. Actually, there were 10 interactions before a concrete goal was clearly established by Machiko and Robert.

Stage 4. *Restory*—Working: Exploring Alternatives, Confronting Client Incongruities and Conflict, Restorying

"What are we going to do about it?" "Can we generate new ways of thinking, feeling, and behaving?"

Interviewer and Client Conversation	Process Comments
Machiko: So, Robert, on the one hand I heard you have a long-term pattern of conflict with supervisors and customers who give you a bad time. On the other hand, I also heard just as loud and clear your desire to have less hassle and not give in to others. We also know that you are a good worker and like to do a good job. Given all this, what do you think you can do about it?	Machiko remains nonjudgmental and appears to be very congruent with the client in terms of both words and body language. In this major empathic Level 4 summary, she distills and clarifies what the client said. This adds to and integrates what has been said so far and leads to the future.
Robert: Well, I'm a good worker, but I've been fighting too much. I let the boss and the customers control me too much. I think the next time a customer complains, I'll keep quiet and fill out the refund certificate. Why should I take on the world?	Robert talks more rapidly. He, too, leans forward. However, his brow is furrowed, indicating some tension. He is "working hard."
Machiko: So one thing you can do is keep quiet. You could maintain control in your own way, and you would not be giving in.	Paraphrase, Level 4 additive empathy. Machiko is using Robert's key words and feelings from earlier in the interview to reinforce his present thinking, but she waits for Robert's response.
Robert: Yeah, that's what I'll do, keep quiet.	He sits back, his arms folded. This suggests that the "good" response above was in some way actually subtractive. There is more work to do.
Machiko: Sounds like a good beginning, but I'm sure you can think of other things as well, especially when you simply can't be quiet? Can you brainstorm more ideas?	Machiko gives Robert brief feedback. Her open question is a Level 4 response adding to the interview. She is aware that his closed nonverbals suggest that more work is needed on relationships, specificity, and generating new possibilities for action.

Clients are often too willing to seize the first idea that comes up. It may not be the best thing for them. It is helpful to use a variety of questions and listening skills to draw out the client further. With Robert, some ideas came easily but others more slowly. Eventually, he was able to generate two other useful suggestions: (1) to talk frankly with his boss about the continuing problem and seek his advice, and (2) to plan an exercise program after work to help blow off steam and energy. In addition, Robert began to realize that his problem with his boss was but one example of a continuing problem. He and Machiko discussed the possibility of talking more or for him to visit a professional therapist. Robert decided he'd like to talk with Machiko a bit more. A contract was made: If the situation did not improve within 2 weeks, Robert would seek professional help.

Stage 5. *Action*—Terminating: Generalizing and Acting on New Stories

Conclude with plan for generalizing interview learning to "real life" and eventual termination of the interview or series of sessions. "Will you do it?"

Interviewer and Client Conversation	Process Comments
Machiko: So you've decided that the most useful approach is to talk with your boss. But the big question is "Will you do it?"	Paraphrase, open question.
Robert: Sure, I'll do it. The first time the boss seems relaxed.	
Machiko: As you've described him, Robert, that may be a long wait. Could you set up a specific plan, so we can talk about it the next time we meet?	Paraphrase, open question. To generalize from the interview, it is important to encourage specific and concrete action in your client. Level 4, as it challenges Robert to take action.
Robert: I suppose you're right. Okay, occasionally he and I drink coffee in the late afternoon at Rooster's. I'll bring it up with him tomorrow.	Robert concretizes his plan, and this is a start toward his goal.
Machiko: What, specifically, are you going to say?	Open question, again eliciting even more concreteness. This is a time that a role-play might be useful. Machiko could play the boss's role, and Robert could practice what to say.
Robert: I could tell him that I like working there, but I'm concerned about how to handle difficult customers. I'll ask his advice on how he does it. In some ways, it worries me a little; I don't want to give in to the boss . . . but maybe he'll be willing to give some.	Robert is able to plan something that might work. With other clients, you might role-play, give advice, or actually assign homework. You will also note that Robert is still concerned about "giving in." There is the possibility that he is exhibiting a rigid style that requires more examination by a professional counselor.
Machiko: Would you like to talk about giving in more the next time we meet? Maybe through your talk with the boss we can figure out how to deal with that. Sounds like a good contract. Robert, you'll talk with your boss, and we'll meet later this week or next week.	Open question, structuring. Level 3 interchangeable. Machiko stops too soon here. Many counselors and clients are tired at this point and do not give enough attention to generalization and maintainance of new ideas.

Machiko presented an especially important response during Stage 5 when she asked what Robert was going to do specifically. Again, you'll find that concreteness is very important in assisting clients to make and act on decisions. It would have been wise for Machiko to specify the follow-up contract even more precisely, but this would most likely have entailed the development of clearly defined homework. By not working harder on action, Machiko and many other counselors undermine their own work. Please give special attention to follow-up and ensuring that change does happen.

Note Taking

A frequently asked question is whether one should take notes during an interview. As might be expected, the answer is dependent on you, the client, and the situation. If you personally are relaxed about note taking, it will seldom become an issue in the interview. If you are worried about taking notes, it likely will be a problem. When working with a new client, obtain permission early in the session. We find it useful to say something like the following, according to your own natural style:

> I'd like to take a few notes while you talk. Would that be okay? It often helps us refer back to important thoughts. I also like to write down your exact words for the important points. As you know, all notes in your file are open to you at any time.

Clearly, detailed note taking where the writing takes precedence over listening is to be avoided. Beyond that one major warning, you and your client can usually work out an arrangement suitable for both of you. We also suggest that any case notes be made available to clients as well, and we recommend that you share with your volunteer clients any notes or transcripts of interviews you complete with practice sessions in this book.

There is nothing wrong with not taking notes in the session, but it is essential to have a comprehensive and accurate summary afterward. Most counselors probably write their notes after the session, but in doing so they may miss key items, as no one's recall is perfect. In-session note taking is often most helpful in the initial portions of interviewing and counseling, and it will become less important as you get to know the client better.

Audiotaping and videotaping the session follow the same guidelines. If you are relaxed and share on an equal basis with your client, making this type of record of the interview generally goes smoothly. Some clients find it helpful to take audio recordings of the session home and listen to them, thus enhancing learning from the interview. We have found it very effective to videotape sessions occasionally and then review them together with the client. Of course, videos for supervision are immensely helpful.

SUMMARY: CONDUCTING A WELL-FORMED INTERVIEW

The five-stage structure of the microskills decisional counseling model has been demonstrated in the preceding example, showing that it is possible to integrate all the microskills and concepts presented into a meaningful, well-formed session. It may be challenging to work through the *relationship—story and strengths—goals—restory—action* interview and not use advice and influencing skills, yet it can be done. It is a useful format for individuals who are verbal and anxious to resolve their own issues—and with resistant clients who want to make their own decisions. By acting as a mirror and asking questions, we can encourage many clients to find their own direction.

Theoretically and philosophically, decisional counseling using only listening skills is related to Carl Rogers's person-centered therapy (Rogers, 1957). Rogers developed guidelines for the "necessary and sufficient conditions of therapeutic personality change," and the empathic constructs described in this chapter are derived from his thinking. Rogers originally was opposed to the use of questions but in later life modified his position. Implicit in your ability to conduct an interview without using information, advice, and influencing skills is a respect for the person's ability to find her or his own unique direction. In conducting an interview using only attending, observation, and the basic listening sequence, you are using a very person-centered approach to counseling and interviewing.

Key points of this chapter are presented below.

■ KEY POINTS

Basic listening sequence	Draw out the client's story and strengths via questioning, encouraging, paraphrasing, reflection of feeling, and summarization. This sequence is used in multiple settings, not just in interviewing and counseling.
Empathy	Seeing the world through the clients' eyes, hearing them as they have been heard and want to be heard, imagining what it would be like to be in their shoes. Experiencing them in the moment. At the same time, maintaining your own self and not mixing in "your own things" in understanding the client. Mirror neurons represent the physical basis of our ability to sense the world of client. But recall that the client also has mirror neurons!
Additive empathy	Clarifying and adding meaning and feelings beyond those originally expressed by the client. If done ineffectively, it may subtract from the client's experience. Empathy is best assessed by the client's reaction to a statement, not by a simple rating of the interviewer's comments.
Subtractive empathy	Hearing clients inaccurately and feeding back what you have heard and experienced in a way that is less than what they shared with you.
Positive regard	Selecting positive aspects of client experience and selectively attending to positive aspects of client statements.
Respect and warmth	Respect and warmth are attitudinal dimensions usually shown through nonverbal means—smiling, touching, and a respectful tone of voice—even when differences in values are apparent between interviewer and client.
Concreteness	Being specific rather than vague in interviewing statements constitutes concreteness.
Immediacy	Interviewer statement in the present, past, or future tense. Here-and-now present-tense statements tend to be the most powerful. Immediacy is also viewed as the immediate "I–you" talk between interviewer and client.
Nonjudgmental attitude	Suspend your own opinions and attitudes and assume a value neutrality with regard to your clients.
Authenticity and congruence	These are the opposite of incongruity and discrepancy. The interviewer is congruent with the client and is authentic in their relationship.
Five stages of the interview	Stage 1: *Relationship:* Rapport and structuring ("Hello.") Stage 2: *Story and Strengths:* Gathering information and defining issues ("What's your concern?" "What are your strengths?") Stage 3: *Goals:* Determining outcomes ("What do you want to happen?") Stage 4: *Restory:* Exploring alternatives and client incongruities ("What are we going to do about it?") *Stage 5: Action:* Generalization and transfer of learning ("Will you do it?")
Circle of decision making	The five stages of the interview need not always follow the five steps in order. Think of the stages as dimensions that need to be considered in each session. Also, give continuous attention to relationship, positive assets, and wellness at the hub of the circle.

COMPETENCY PRACTICE EXERCISES AND PORTFOLIO OF COMPETENCE

Mastery of the skills of this chapter is a complex process that you will want to work on over an extended period of time. Some basic exercises for the individual and systematic group practice in each skill area follow.

Individual Practice: Basic Listening Sequence

Exercise 1: Illustrating How the Basic Listening Sequence Functions in Different Settings

Write counseling leads as they might be used to help a client solve the problem "I don't have a job for the summer." Imagine a full statement of issues, and write responses that represent the BLS.

Open question _____

Closed question _____

Encourager _____

Paraphrase _____

Reflection of feeling _____

Summary _____

Now imagine you are talking with a client who has just been told that her or his parents are getting a divorce after more than 25 years of marriage. Your task is to use the BLS to find out how the client is thinking, feeling, and behaving in reaction to this news.

Open question _____

Closed question _____

Encourager _____

Paraphrase _____

Reflection of feeling _____

Summary _____

Finally, how would you use these skills in talking with an elementary school student who has come to you crying because no one will play with her or him?

Open question _____

Closed question _____

Encourager _____

Paraphrase _____

Reflection of feeling _____

Summary _____

Exercise 2: The Positive Asset Search and the Basic Listening Sequence

Imagine that you are role-playing a counseling interview.

> In a career interview, the client says, "Yes, I am really confused about my future. One side of me wants to continue a major in psychology, while the other—thinking about the future—wants to change to business." Use the BLS to draw out this client's positive assets. In some cases, you will have to imagine client responses to your first question.

Open question _____

Closed question _____

Encourager _____

Paraphrase _____

Reflection of feeling _____

Summary _____

> You are counseling a couple considering divorce. The husband says, "Somehow the magic seems to be lost. I still care for Chantell, but we argue and argue—even over small things." Use the positive asset search to bring out strengths and resources on which they may draw to find a positive resolution to their problems. In marriage counseling, in particular, many counselors err by failing to note the strengths and positives that originally brought the couple together.

Open question _____

Closed question _____

Encourager _____

Paraphrase _____

Reflection of feeling _____

Summary _____

Exercise 3: Writing Helping Statements Representing the Five Levels of Empathy

> I'm having trouble here at the community college. I'm the first one in my family who has ever even attempted college. The work doesn't seem all that hard, but when I turn papers in, my grades seem so low. It's hard to make friends. I have to work, and I don't have as much money as the other kids seem to.

Write statements here that represent the five levels of empathic response to this client's concern. These statements can employ any skill, but paraphrasing and reflection of feeling are perhaps the clearest and easiest statements to write.

Level 1 (subtractive) _____

Level 2 (slightly subtractive) _____

Level 3 (interchangeable response) _____

Level 4 (slightly additive) _____

Level 5 (additive) _____

> Carlena says (near tears), "Alexander and I just broke up. I don't know what to do. We've been living together for almost a year. I don't have any place to live, and I'm so confused."

Again, write statements representing the five levels of empathic responding.

Level 1 (subtractive) _____

Level 2 (slightly subtractive) _____

Level 3 (interchangeable response) _____

Level 4 (slightly additive) _____

Level 5 (additive) _____

Exercise 4: Rating Interview Behavior Using Empathic Dimensions

Use any systematic group practice exercise from this chapter or the whole book and rate the interviewer's empathic response. Alternatively, you may wish to use an interview on audio-tape, videotape, in transcript form, or a live interview. Provide specific and behavioral evidence for your conclusions using the Feedback Form (Box 8-4).

Group Practice: A Decisional Counseling Interview Using Only Attending and Listening Skills

Exercise 5: Practice the BLS

The goal here is a challenging one—can you go through a full interview using only attending behavior and the microskills of the basic listening sequence? You may even wish to try to complete a full session without any questions or with minimal use of questions. You are not fully prepared to encounter the influencing skills of later chapters until you demonstrate that you can hear the client's story through listening carefully—and this is best shown through your ability to conduct a full session using only listening skills.

We suggest that you find a volunteer client who is relatively verbal and willing to talk about something of real interest. Negotiate the topic before you start. Let the client know what you plan to do by sharing the interview stages with him or her beforehand. It is okay to enlist client cooperation both in practice and real situations. This can result in a more egalitarian interviewing or counseling relationship.

After the session is over, ask the client to complete the Client Feedback Form of Chapter 1. Also ask the client to give you immediate feedback on what was helpful and things that might have been missed. Ask the key questions, "What did we miss discussing today?" and "What else?" Review the audio- or videotape with your client and start and stop at various points to obtain her or his impressions.

Under ideal circumstances, you would conduct this session in a room equipped with a one-way mirror through which an observer could see the interview directly. Ask a classmate to serve as observer, and he or she can complete the observation form by listening to your

BOX 8-4 FEEDBACK FORM: EMPATHY

_____ (Date)

_____   _____
(Name of Interviewer)   (Name of Person Completing Form)

Instructions: Observers are to (1) view an interview or segment of an interview, rating the empathic responding on a 5-point scale, and (2) provide specific behavioral evidence for their decisions.

	Level 1 (subtractive)	Level 2	Level 3 (interchangeable)	Level 4	Level 5 (additive)
1. Overall empathy rating					
2. Positive regard					
3. Respect and warmth					
4. Concreteness					
5. Immediacy					
6. Nonjudgmental attitude					
7. Authenticity and congruence					
8. Other observations					

(continued)

BOX 8-4 (continued)

Provide specific behavioral evidence in the space provided to justify your rating of empathic behaviors on the chart.

1. _____

_____

2. _____

_____

3. _____

_____

4. _____

_____

5. _____

_____

6. _____

_____

7. _____

_____

8. _____

_____

audiotape or observing your videotape. See the Feedback Form in Box 8-5. Also, the Client Feedback Form of Chapter 1 is important to help you obtain needed feedback, particularly if you do not have an external observer.

Following is an outline of things to consider as you work through the five stages of the well-formed interview.

1. *Relationship* (Initiating the session; rapport and structuring). Develop rapport with your client. Structure the interview by informing the client about video or audio recording and how the tape will be used. You may wish to say something like "What we are going to do today is discuss the issues around your career or your balance of work and play. A choice of specific career is a good topic (or we agreed to talk about . . .). Then we'll search out some of your strengths, examine goals, and look at possibilities. Finally, we will talk about how to take home some of the things we've talked about today. Is that okay?" Include informed consent and ethical issues.
2. *Story and Strengths* (Gathering data; drawing out stories, concerns, problems, or issues). Use the basic listening sequence to draw out the client's story. Be sure to obtain at least one positive asset that may later be helpful in facilitating decision making.

BOX 8-5   FEEDBACK FORM: DECISIONAL INTERVIEW USING ONLY THE BASIC LISTENING SEQUENCE

_____ (Date)

_____   _____
(Name of Interviewer)   (Name of Person Completing Form)

Instructions—   Conduct a brief five-stage session using only the skills of the basic listening sequence. We
Interviewer: suggest sharing the steps beforehand with your volunteer client. Suggested topics are making
a career decision, finding a balance between work and play, or working on a wellness issue
from Chapter 2. You and your client could take any current life issue that is not too complex.

Instructions—   Please provide feedback and commentary to the interviewer. Was this interviewer able to
Observer: conduct a session using only listening skills?

1. RELATIONSHIP. Initiating the session. Nature of rapport? Was enough established before the interview
   continued to the next stage? Did the interviewer provide structuring? Was rapport maintained throughout
   the session? Observations on BLS?

2. STORY. Gathering data, defining concerns, and identifying assets. Did the interviewer draw out the story from
   the client using only listening skills? Was at least one positive supportive asset of the client examined?
   Observations on BLS?

3. GOALS. Mutual goal setting. Was a specific outcome or goal outlined for the client through use of listening skills?
   Was it concrete and doable? Observations on BLS?

(continued)

BOX 8-5 (continued)

4. RESTORYING. Working. Was the interviewer able to assist the client in generating new ideas through the use of listening skills only? Did the session move toward achieving the goal, or was it too broad for the client to accomplish? Observations on BLS?

5. ACTION. Terminating and generalizing. Were specific plans made and contracted for with the client for taking ideas home? Is there a systematic plan of action for follow-up and maintenance? Observations on BLS?

6. INTENTIONAL COMPETENCE. *What did the interviewer do right?* Did use of the basic listening sequence result in predicted outcomes? When the expected result did not occur, was the interviewer able to flex intentionally and use a different listening skill?

3. *Goals* (Mutual goal setting; what does the client want to happen?). Using the basic listening sequence, work with the client to find objectives and a satisfactory solution. With some clients, goal setting may precede Stage 2. In fact, brief solution-focused counseling often uses goal setting as one of the first agenda items.

4. *Restory* (Working; exploring alternatives, confronting incongruities and conflict, restorying). Summarize the real and ideal world from 2 and 3 above—this may appear as a summary confrontation ("On one hand, your story/concern may be summarized as . . . On the other hand, your goals seem to be . . . How can we put these together in some new ways?"). With the client, brainstorm and discover new ways of thinking, feeling, and behaving. Discuss and examine discrepancies and incongruities. Help the client rewrite old stories into new narratives.

5. *Action* (Terminating; generalizing and acting on new stories). Review and summarize the session and work specifically on things that the client can do concretely to follow up on what happened in this session.

Portfolio of Competence

A lifetime can be spent increasing one's understanding and competence in the ideas and skills from this chapter. You are asked here to learn and even master the basic ideas of predictability from skill usage in the session, several empathic concepts, and the five stages of the well-formed interview. We have learned that student mastery of these concepts is possible, but for most of us (including the authors) we find that reaching beginning competence levels makes us aware that we face a lifetime of practice and learning.

You should feel good if you can conduct an interview using only listening skills. Focus on that accomplishment and use it as a building block toward the future. As you do, you are even better prepared for developing your own style and theory.

Use the following as a checklist to evaluate your present level of mastery. As you review the items below, ask yourself, "Can I do this?" Check those dimensions that you currently feel able to do. Those that remain unchecked can serve as future goals. Do not expect to attain intentional competence on every dimension as you work through this book. You will improve your competencies with repetition and practice. Highlight competencies that you have met to date:

Level 1: Identification and classification.

❏ Identify and classify the microskills of listening.
❏ Identify and define empathy and its accompanying dimensions.
❏ Identify and classify the five stages of the structure of the interview.
❏ Discuss, in a preliminary fashion, issues in diversity that occur in relation to these ideas.

Level 2: Basic competence. Aim for this level of competence before moving on to the next skill area.

❏ Use the microskills of listening in a real or role-played interview.
❏ Demonstrate the empathic dimensions in a real or role-played interview.
❏ Demonstrate five dimensions of a well-formed interview in a real or role-played session.

Level 3: Intentional competence. Review the following skills, all related to predictability and evaluation of your effectiveness in working with emotion. These are skills that may take some time to achieve. Be patient with yourself as you gain mastery and understanding.

❏ Anticipate predicted results (Ivey Taxonomy) in clients using the listening microskills.
❏ Facilitate client comfort, ease, and emotional expression by being empathic.
❏ Enable clients to reach the objectives of the five-stage interview process—(a) *relationship:* develop rapport and feel that the interview is structured; (b) *story and strengths:* share data about the concern and also positive strengths to facilitate problem resolution; (c) *goals:* identify and even change the goals of the interview; (d) *restory:* work toward problem resolution; (e) *action:* generalize ideas from the interview to their daily lives.

Level 4: Psychoeducational teaching competence. Teaching competence in these skills is best planned for a later time, but those who run meetings or do systematic planning can

profit from learning the five-stage interview process. It serves as a checklist to ensure that all-important points are covered in a meeting or planning session.

❑ Teach clients the five stages of the interview.
❑ Teach small groups this skill.

DETERMINING YOUR OWN STYLE AND THEORY: CRITICAL SELF-REFLECTION ON INTEGRATING LISTENING SKILLS

You are now at the stage to initiate construction of your own interviewing process. You certainly cannot be expected to agree with everything we say. You likely have found that some skills work better for you than others, and your values and history deeply affect the way you conduct an interview. Some skills you'd like to keep, and some you might like to change.

We encourage you to look back on these first eight chapters as you consider the following basic questions leading toward your own style and theory.

What single idea stood out for you among all those presented in this text, in class, or through informal learning? Allow yourself time to really think through the one key idea or concept—and it may be something you discovered yourself. What stands out for you is likely important as a guide toward your next steps.

Continue your development of your own style and theory through writing.

What are your thoughts?

_____

_____

Helping Clients Generate New Stories That Lead to Action: Influencing Skills and Strategies

The listening skills and the five-stage interview structure presented in previous chapters provide the foundation for the more action-oriented skills and strategies presented here. The influencing skills and strategies will help you actively intervene to facilitate the process of change in your clients. In addition, you will find many of these skills and strategies useful in psychoeducational treatment, which supplements the interview.

Relationship—story and strengths—goals—restory—action is a general model of interviewing, counseling, and psychotherapy. Clients share their stories and problems and can do something about them. Moreover, your mastery of these basics will enable you to develop competence in multiple theories of helping.

Look for the following in this section.

Chapter 9. The Skills of Confrontation: Supporting While Challenging Clients   Supportive confrontation is considered by some the most important stimulus enabling client change and development. Confrontation builds on your present ability to listen empathically and observe client conflict. Skilled supportive confrontation enables resolution of incongruity, which leads to new behaviors, thoughts, feelings, and meanings.

Chapter 10. Focusing the Interview: Exploring the Story From Multiple Perspectives   Focusing, extends the concept of confrontation and illustrates how to ensure that you have made a comprehensive examination of clients' stories. This skill will enable clients to clarify how other people, situations, and the total environment relate to their lives and stories. Skilled focusing facilitates clients in reframing their stories in new ways without you supplying the answers for them.

Chapter 11. Reflection of Meaning and Interpretation/Reframing: Helping Clients Restory Their Lives   Here you will examine the relationship between and among behaviors, thoughts, feelings, and their underlying meaning structure. These skills help you gain a deeper understanding of each client's issues and history. Clients will gain valuable new perspectives on their problems and stories.

Chapter 12. Influencing Skills: Five Strategies for Change   Logical consequences, self-disclosure, feedback, information/psychoeducation, and directives are explored with specific suggestions for facilitating client restorying and action.

The substance of this section is to suggest ways in which you can take a more active stance with clients, helping them move on to new ways of thinking, feeling, behaving, and finding meaning in their lives. Yet influencing without careful listening can take too much responsibility away from the client. Use these skills empathically.

As you develop competence in influencing skills, you may accomplish the following:

1. Master the art of supportive confrontation and the ability to assess your client's developmental change in response to your interventions.
2. Demonstrate the ability to change focus in the interview and to facilitate client exploration of the full complexities of the story.
3. Use the skills of reflection of meaning and interpretation/reframing to help clients move to deeper levels of self-exploration and self-understanding.
4. Use an array of influencing skills and strategies to assist client developmental progress, particularly when the more reflective listening skills fail to produce change and understanding.

With this solid base, you can gradually move to the predictability associated with intentional competence. The effective interviewer is always in process—growing and changing in response to new challenges.

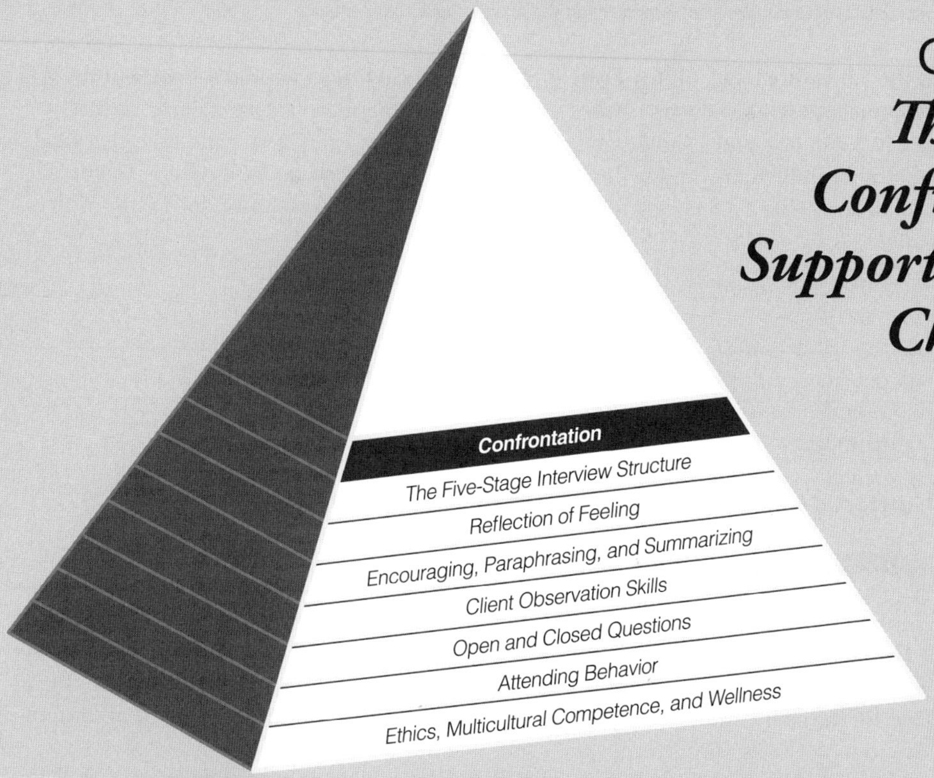
Confrontation
The Five-Stage Interview Structure
Reflection of Feeling
Encouraging, Paraphrasing, and Summarizing
Client Observation Skills
Open and Closed Questions
Attending Behavior
Ethics, Multicultural Competence, and Wellness

CHAPTER 9
The Skills of Confrontation: Supporting While Challenging Clients

The creation of the New is a powerful, poetic way to talk about change and restorying. What is really New is the discovery of stories leading to long-term change. Change includes the creation of new neural networks made possible by your client's brain plasticity and capacities for long-term memory. You can make a lifetime difference through the New. A supportive confrontation can be basic in the process of growth in Newness.

—Allen Ivey

How can confrontation help you and your clients?

Chapter Goals

This chapter defines and explains the value of confronting client issues directly through clarification and supportive challenge. Skill in confrontation will help minimize resistance and facilitate client development in new areas. Facing challenges is basic to client restorying.

Competency Objectives

Awareness, knowledge, and skills in confrontation will enable you to

- ▲ Identify incongruity, conflict, discrepancies, or mixed messages in behavior, thought, feelings, or meanings through effective use of listening and observation skills.
- ▲ Clarify issues, concerns, and problems with a view toward explanation and/or resolution of conflict and discrepancies.
- ▲ Identify client change in the here and now of the interview and throughout treatment, using confrontation or other skills.
- ▲ Use confrontation skills and the five-stage interview as part of mediation and conflict resolution.

Chris: I'm really between a rock and hard place. Here I am, ready to go to grad school, but my dad just had a heart attack and won't be back at work any time soon. He says that it doesn't make any difference, and I should go. My mom insists that I stay home and take over Dad's garage. I've been planning and saving for school for 2 years, and now everything might totally fall apart. But I'd feel so guilty if I didn't stay here. What should I do? I've been sitting on this for 2 weeks.

Chris is obviously stuck over this decision. How would you respond?

What would you say?   _____

_____

You can compare your responses with ours at the end of this chapter.

INTRODUCTION: HELPING CLIENTS MOVE FROM INACTION TO ACTION

Creativity is the root of change. From the very first chapter, we have stressed that you as interviewer, counselor, or therapist are changing the client, and the client is changing you—this is the "creation of the *New*" as defined by the famed theologian Paul Tillich. "The *New* appears in three aspects, as creation, as restoration, as fulfillment" (Tillich, 1964, pp. 163–164). In that sense, there is a spiritual dimension to the helping process, both for you and for the client. There is even evidence that a location in the brain is related to meaning and spirituality (Ratey, 2008).

What does this say about creativity and client change? Basically, counselors need to access client resources and listen. Draw out the story and/or the conflict, and clearly summarize the issues (including an emotional base). Then an empathic supportive confrontation leads to the *New,* change, and client growth.

Cultural intentionality is not just a goal for interviewers; it is also a goal for clients. A client comes to an interview "stuck"—having either no alternatives for solving a problem or a limited range of possibilities. The task of the interviewer is to eliminate "stuckness" and substitute intentionality and the creative *New. Stuckness* is an inelegant but highly descriptive term coined by Fritz Perls of Gestalt therapy to describe the opposite of intentionality. Other words that represent stuckness include immobility, blocks, repetition compulsion, lack of understanding, limited behavioral repertoire, limited life script, impasse, and lack of motivation. Stuckness may also be defined as an inability to reconcile discrepancies and incongruity. In short, clients often come to the interview because they are stuck and seek intentionality—they need a new story.

A basic and common form of stuckness occurs when there is a discrepancy between clients' stories and their goals for change. Confronting this incongruence directly often leads to a new story, resolution of the concern, and a plan for action.

Uncontrollable conflict and discrepancy bring stress to the client. Neurons may be actively inhibited and damaging glucocorticoids and cortisol are released (see Box 9-1 and Appendix II). Long-term conflict and stress or trauma can result in memory loss (Grawe, 2007, p. 24). The importance and reparative value of effective counseling and therapy should be evident. Development in the interview may be described as the resolution of conflict and incongruity, the working through of an impasse or developmental delay. This process of transformation and change leads clients to manage their lives more competently and joyfully. The perspective of the *New* leads to active restorying and change.

Out of pain can come growth. The stress of conflict, however, is not always destructive. We need a certain amount of stress to energize the client and facilitate cognitive and emotional growth. An appropriate stress level of confrontation can lead clients to gain new information and thus reduce stress in the long term. Too much stress is damaging, but appropriate stress and challenge is key to development and the creative *New*. With your support, clients can enter into their pain and problems, clarify what is going on, and start the process of healing and growth.

Later in this chapter, we discuss the process of positive change as we draw from observations of clients working through loss, working their way out of alcoholism, and working through the change process in general. But first we consider the skills of confrontation.

INSTRUCTIONAL READING: CHALLENGING CLIENTS IN A SUPPORTIVE FASHION

Clients usually come for counseling because they have some degree of stress associated with conflict. They may be stuck in their behaviors, thoughts, feelings, and value decisions. There is a real need to move these fixed problems and stories to the *New*, changing these issues to challenges and opportunities for change. Working on the discrepancy between present issues and goals will supply the client with creative strength to move and become "unstuck."

The person-centered counselor seeks to resolve the discrepancy between the real and ideal self; the cognitive-behavioral therapist aims to resolve behavioral and thought inconsistencies; and the decisional counselor facilitates resolving conflicting wishes and desires. Different theories focus on and confront different aspects of client conversation. In effect, they see different routes to problem resolution. However, all have the goal of supporting the client and resolving conflict.

Although all counseling skills are concerned with facilitating change, it is the clarification and confrontation of discrepancies that acts as a lever for the activation of human potential. Most clients come to an interview seeking some sort of movement or change in their lives. Yet they may resist your efforts to bring about the transformation they seek. Your task is to help them move beyond their issues and problems to realize their full potential as human beings. An understanding of confrontation is basic to helping clients restory their lives.

When client discrepancies, mixed messages, and conflicts are confronted skillfully and nonjudgmentally, clients are encouraged to talk in more detail and to resolve their problems and issues. Confrontation can be defined in this way:

> Confrontation is not a direct, harsh challenge. Think of it, rather, as a more gentle skill that involves listening to the client carefully and respectfully; and, then, seeking to help the client examine self or situation more fully. Confrontation is not "going against" the client; it is "going with" the client, seeking clarification and the possibility of a creative *New*, which enables resolution of difficulties.

But while the gentle, empathic approach to confrontation is basic, there are many clients who will need and even prefer a more direct challenge. For example, if you are working with an acting-out or antisocial client, a firm and more solid confrontation may be necessary. The client may sneer at and manipulate "nice" helpers, but may be more likely to respect and work with a counselor

BOX 9-1   RESEARCH EVIDENCE THAT YOU CAN USE

Confront, but Also Support

Attending and listening skills are used frequently in the session, but you'll find that confrontation strategies are used only occasionally. A review of research found that confrontations account for only 1% to 5% of interviewer statements (Hill & O'Brien, 1999). The reviewers noted that confrontations are useful, but that they also often make clients uncomfortable. Especially important, they noted that clients frequently become defensive and may not deal fully with feelings and issues following a confrontation. They also note that confrontations that are intentionally modified to meet individual needs were more effective. Our point of listening and using supportive challenge becomes all the more important.

Counselor eye contact affected client perception of rapport in the session. Specifically, less direct eye contact early in the session when discussing sensitive matters was helpful, and at this point the clients appreciated a non-confrontational approach. As the interview progressed, more eye contact and more confrontation were acceptable (Sharpley & Sagris, 1995).

Clinical sessions can lead to new, researchable ideas. Allen Ivey (1973) identified confrontation as a microskill when he was conducting therapy with inpatient veterans of the Vietnam war. He videotaped his interviews with hospitalized psychiatric patients. Allen and the patients reviewed the video, and Allen asked them what they observed. This helped the former soldiers see their verbal and nonverbal discrepancies. Often they would identify the discrepancy themselves. At other times, Allen had to provide the confrontation within a supportive, empathic, and trusting atmosphere. The first task in supportive confrontation was and remains to establish a solid base of relationship, best demonstrated by your ability to listen.

Confrontation and Neuroscience

Interviewing and counseling are very much concerned with helping clients create the *New* and discover pathways to growth. As part of this, fresh neural networks are developed. To facilitate significant change, seek an appropriate balance of stress while supporting the client. Too much stress is damaging, but too little stress likely won't lead to change. Important in this is your awareness that ". . . released adrenaline *(resulting from stress)* influences almost all regions of the brain—the entire cortex, the hypothalamus, the hind brain, and the brain stem" (Grawe, 2007, p. 220).

Too much stress can flood the brain with damaging cortisol and fix negative memories in the mind (such as in posttraumatic stress). There are a few unethical and charismatic "therapists" who encourage clients to reach strong emotions. They use this here-and-now base to reach back to so-called "long forgotten and repressed" memories of trauma. Unfortunately, this can result in permanently imprinting false memories that do not exist (Loftus, 2003). Clients come to believe that things happened that never did. This type of "therapy" introduces new damaging neural networks in the brain.

We suggest that counseling or therapy can also be quite damaging when the counselor repeatedly focuses only on client difficulties and negative stories. Some cynics even suggest that classical long-term orthodox psychoanalytic therapy takes years because so much time is spent on negative ruminating. We need to listen to negative stories and client problems, but the prime focus is the *here and now,* and the objective is action in the future to implement positive change.

What we call "creativity" may be located in the connections between the holistic right brain and the linear left brain as well as the participation of the mainly unconscious limbic system (Carter, 1999). However, this idea is not yet fully proven. Nonetheless, research continues to suggest that new learning occurs when the left and right brains synchronize their activity through the connecting corpus callosum (Gazzaniga, 2000; Goodwin & Sherrard, 2005).

who listens and offers respect but takes no "garbage" from the client. Nonetheless, empathic listening remains central if you are going to establish any type of working relationship.

Box 9-1 presents research information on confrontation as well as a clinical example of how confrontation can be used with war veterans. Making videos of your counseling sessions and reviewing them with clients can be valuable. Video feedback adds to the clarity and depth of the interview.

Confrontation involves three major steps. First: *Listen and identify conflict in clients' mixed messages, discrepancies, and incongruity.* These are discussed briefly in Chapter 5 on observation and repeated here for emphasis.

Second: *Clarify and clearly point out issues to clients and help them work through conflict to resolution.* Questions, observation, reflective listening, and feedback are effective in confrontation. Relationship is critical here as we make the conflict clearer to the client.

Finally: *Listen, observe, and evaluate the effectiveness of your intervention on client change and growth.* The effectiveness of a confrontation can be measured by how the client responds and whether something *New* has been created. The Client Change Scale (CCS) (see page 250) is a systematic way to evaluate the effectiveness of confrontations and whether clients are moving to new ways of thinking, feeling, and behaving. Keep your attention on goal attainment, and use microskills flexibly to help the client move toward change.

If you use confrontation skills as structured above, you can *predict* how clients will respond.

Confrontation	Predicted Result
Supportively challenge the client: 1. Listen, observe, and note client conflict, mixed messages, and discrepancies in verbal and nonverbal behavior. 2. Point out and clarify internal and external discrepancies by feeding them back to the client, usually through the listening skills. 3. Evaluate how the client responds and whether the confrontation leads to client movement or change. If the client does not change, the interviewer flexes intentionally and tries another skill.	Clients will respond to the confrontation of discrepancies and conflict with new ideas, thoughts, feelings, and behaviors, and these will be measurable on the 5-point Client Change Scale. Again, if no change occurs, *listen.* Then try an alternative style of confrontation.

The following brief session illustrates the fundamentals of confrontation—supporting while challenging. In this example, the client keeps moving toward a new resolution. However, many clients stay stuck in the same place and repeat the same discussion of their concerns over and over again. With these clients, listen more carefully, draw out issues more clearly, and summarize the conflict.

Step 1: Identify the Conflict; Note Mixed Messages and Incongruity

Interviewer and Client Conversation	Process Comment
Client: (shows excitement through body language) I've found this great friend on the Internet. He sounds wonderful, and we're doing e-mail at least four times a day. It feels great. I think I'd like to meet him. (Her body language becomes more hesitant, and she breaks eye contact.) But it means I may have to go out of town. I wonder what my partner would think if he found out. It makes me a bit anxious, but I really want to meet this guy.	Instances of incongruity or discrepancies include desire and excitement about meeting the Internet friend, but with internal anxiety and hesitation: the inevitable external conflict of being involved with two people at once. Which discrepancy might you discuss first? In addition, clients who discuss these mixed feelings and conflicts usually also show them in their body.

(continued)

Interviewer and Client Conversation	Process Comment
Counselor: You really want to meet him. On the other hand, you're a bit anxious.	The paraphrase and reflection of feeling confront the mixed feelings in the client. This catches both verbal and nonverbal observations.
Client: Yes, but what would happen if my partner found out? It scares me. I've got so much involved with him over the past 2 years. But, wow, this guy on the Internet . . .	The client responds, but turns her focus to the discrepancy between her and the partner. There are clearly two issues at least in this situation.

Step 2: Clarify Issues of Incongruity and Work to Resolve Them

Interviewer and Client Conversation	Process Comments
Counselor: Could I review where we've been this far? I know you have been having some difficulties with your partner, and you've detailed them over the last two sessions. I also hear that you want to work things out despite your present anger at him. You have a lot of positive history together that you'd hate to give up. But on the other hand, you've found this man on the Internet, and he doesn't live that far away. You seem really excited about the possibility of meeting him. In the middle of all this, I sense you feel pretty conflicted. Have I got the issues right?	This summary indicates the counselor has been listening. Both verbally and nonverbally, the counselor communicates respect and a nonjudgmental attitude. We support by listening and searching for strengths (in this case, the positive history). The counselor summarizes the major discrepancies that led to internal and external conflict and checks out with the client to see if the listening has been accurate.
Client: Yes, I think you've got it. As I hear you, it makes me think that I've got to work a bit harder on the present relationship, but—wow—I sure would like to meet that guy.	Through having her thoughts and feelings said back to her, the client starts some movement. Resolution of conflict and discrepancy best occurs after the situation is understood fully.

The conversation continues over the next 10 minutes, and the client's thoughts and feelings evolve to a new perspective.

Step 3: Evaluate the Change

The effectiveness of a confrontation is measured by how the client responds. In the example on the next page, the client keeps moving toward a new resolution. However, there are many clients who stay stuck in the same place and repeat the same discussion of their concerns over and over again. Later in this chapter, the Client Change Scale (CCS) is presented. This is a systematic way to evaluate the effectiveness of confrontations and whether clients are moving to new ways of thinking, feeling, and behaving.

Interviewer and Client Conversation	Process Comments
Client: (said with conviction) It's beginning to make sense to me. I've got so much time invested in my partner, and I've really got to try harder. (Her nonverbals again show hesitancy.) But how am I going to work this out with my Internet friend?	The conflict is moving, and the client is starting to show evidence of new ways of thinking that weren't there in the first two sessions. Nonetheless, conflict remains.
Counselor: (solid supportive body language and vocal tone) It looks like you really want to work it out with your partner. You sounded and looked very sure of yourself. But let us explore a bit more what the possibilities are with your Internet friend.	You can confront and help clients face discrepancies, incongruity, and conflict if you are able to listen and be fully supportive.

With this brief introduction, let us turn to a more detailed examination of supporting while challenging.

Step 1: Identify Conflict by Observing Mixed Messages, Discrepancies, and Incongruity

Your ability to observe incongruities and mixed messages in the interview is fundamental to effective confrontation. (The words "conflict," "discrepancy," "incongruity," and "mixed message" are used interchangeably in this chapter.) The major types of discrepancies are summarized briefly below, and you are encouraged to return to Chapter 5 to review the observation skills, which present them from another frame of reference.

What Are the Internal Conflicts?

Discrepancies internal to the client include mixed messages in nonverbal behavior, incongruities in verbal statements, discrepancies between what the person says and what he or she does, and discrepancies between verbal statements and nonverbal behavior. The client discussing relationship difficulties with the partner in the preceding section presents a mixed message in nonverbal behavior, such as smiling inappropriately while talking about her mixed feelings. The client's statements balancing her present partner against the excitement of the new Internet friend represent an internal conflict. The client has discrepancies within herself that need to be resolved so that she can make a decision.

What Are the External Conflicts?

Discrepancies between the client and the external world highlight conflict between people, and between clients and the situation in which they find themselves. In the example of the client with the conflict regarding the Internet decision, she likely has other conflicts with her partner. The external conflict with the partner, of course, leads to internal conflict. We need to examine discrepancies between the client and the partner as part of the counseling process. Beyond that, the client may have difficulties at work, problems with her parents, discrepancies between you and the client, or other situational issues that relate to external discrepancies.

Resistance can be an important dimension of incongruity between the interviewer and the client. Some theorists talk of "breaking down" client resistance and see client hesitancy as inappropriate. Our view is that many clients need to defend themselves from what they see as external threat, and simply sharing important information can be frightening and highly confrontative to some clients. In this sense, the so-called resistance is actually a good thing for the client. It protects the internal self from dealing with obvious contradiction. See resistance as an opportunity, not as a problem. When you see resistance, listen and understand from whence it comes.

However, the protection offered by resistance can be faulty. Once you have observed resistance and listened for the reasons it may be there, you are better prepared for confronting the client in a respectful, supportive fashion.

Step 2: Point Out and Clarify Issues of Incongruity and Work to Resolve Them

As noted earlier, simply labeling the incongruity through a nonjudgmental confrontation may be enough to resolve a situation. More likely, however, incongruity will remain a problem to be resolved. Remember the importance of focusing on the elements of incongruity rather than on the person. Confrontation is too often thought of as blaming a person for her or his faults; rather, the issue is facing the incongruity squarely through such measures as the following:

1. Clearly identify the incongruity or conflict in the story or comment. Using reflective listening skills, summarize the inconsistency for the client. Often the simple question "How do you put these two together?" will lead a client to self-confrontation and resolution.

2. Draw out and clarify the specifics of the conflict or mixed messages with the basic listening sequence. One at a time, give attention to each part of the mixed message, contradiction, or conflict. If two people are involved, attempt to have the client examine both points of view. It is important at this stage to be nonjudgmental and nonevaluative and to reflect this in your tone of voice and body language—aim for facts, but expect that you will have to deal with emotions and feelings as well, particularly mixed emotions.

3. Periodically summarize the several dimensions of the incongruity. The model confrontation statement, "On the one hand . . . , but on the other hand . . . ," appears to be particularly useful. Variations include "You say . . . , but you do . . . ," "I see . . . at one time, and at another time I see . . . ," and "Your words say . . . , but your actions say . . ." Follow this with a check-out (for example, "How does that sound to you?"). When you use these examples to point out incongruities, the client is confronted with facts. Be nonjudgmental and include the check-out as a final part of the clarifying confrontation.

4. Use the positive asset search with special attention to wellness issues. When we challenge clients, we often place them off balance. One of the values of confronting is indeed disturbing client complacency and comfort. We all seek our "center," and effective confrontation builds new strengths in the client. You may simply comment that the client is describing the problem clearly—for example, "I like the way you are dealing with some very real challenges." Your personal support can help clients build new behaviors, thoughts, and meanings.

5. If necessary, provide feedback giving your opinions and observations about the discrepancies. You may wish to use influencing skills such as directives, logical consequences, and other skills to facilitate resolution (see later chapters).

If the incongruity is not resolved by this process, it may be necessary to say to your client, "You see it that way; the other person sees it another way. We'll have to go at it again." Don't give up on positions you believe are correct, but allow your point of view to be modified by input from the client. Many clients are unaware of their mixed messages and discrepancies; pointing these out gently but firmly can be very beneficial to them. Finally, a wide variety of attending and influencing skills may be used to follow up and elaborate on confrontations.

Individual and Multicultural Cautions

Confrontation of discrepancies can be highly challenging to any client. The following options will help in your work with clients.

Questioning and elaboration: Rather than challenging the client immediately, select various dimensions of the discrepancy or conflict and sort out the stories carefully. For example, "Tell me more about your feelings when you come home from work so tired. What's going on for you when you walk in the door?" After having heard that story and the accompanying emotions, ask about the partner: "What is your partner's typical story of one day like? What thoughts and feelings do you notice in her or him?" In these situations, you are helping the client to explore the discrepancy, and new alternatives for resolution may result.

Direct challenge: The direct challenge may take the form of sharing your own thoughts with the client ("It might be helpful for your relationship if you considered how your partner experiences those first few moments when you both arrive home from work. Homecoming, the end of the day, is often difficult for tired, busy couples"). Influencing skills such as logical consequences (what happens if you don't change), the interpretation/reframe (telling a new story about the situation), and specific directives, are all direct challenges to the status quo.

It may be helpful to take "time out" from confronting discrepancies and focus on positive stories and wellness assets. For example, if clients are encouraged to talk about past successes, they may then generate the emotional courage to look at more difficult issues. Or in working with relationship issues, a particularly useful method is to have the individual or couple go back to when their connection began—ask them to tell you a story of what brought them together or to tell you about the last time things were going well between them. Working with the two of them in a wellness assessment can be a useful strategy.

Not confronting is one final issue for you to consider. With a forward-moving client, simply listening to the story may be sufficient. Pushing for change at times can get in the client's way. And sometimes life is such that we must accept and learn to live with discrepancies that simply cannot be resolved.

Confrontation of discrepancies can be particularly challenging when you work with clients who are culturally different from you (race, ethnicity, economic or social status, etc.). The confrontation process will be more successful if you take time to establish a solid relationship of trust and rapport. Mistrustful clients are usually not ready for you to confront them.

You will find that direct, aggressive confrontations are not necessary if the client contradiction is stated kindly, with a sense of warmth and caring. Direct confrontations are likely to be especially culturally inappropriate for Asian, Latina/ Latino, and Native American clients. But even here, if good rapport and understanding exist between you and the client, confrontations can be most helpful.

McMinn (1996) has outlined several approaches that may be useful in broadening your thinking beyond the empathic, supportive confrontation presented in this chapter. Silence can be a powerful confrontation—you simply listen to client discussion and sit quietly while the client struggles with internal or external contradiction. Silence is a value within much of the Native American, Dene, and Inuit traditions. You may sit silently with these clients at times for several minutes as they sort out issues. How comfortable are you with silence? Each mode of confrontation must be personally authentic and meaningful for the client, or it is likely to fail.

Sometimes disagreements have become entrenched conflict that plays out on a large stage. As an example, Northern Ireland has long been full of such serious conflict. Confrontation in Northern Ireland has not been a supportive challenge; it has meant hatred, building walls between Protestants and Catholics, and a constant state of near-war (see Box 9-2). The counseling role demands justice for the individual and society. How do

BOX 9-2   NATIONAL AND INTERNATIONAL PERSPECTIVES ON COUNSELING SKILLS

Counseling in the Real World—Children in Northern Ireland
Owen Hargie, Professor, University of Ulster, Northern Ireland

Conflict looms large in Northern Ireland, as it does in other international "hot spots," such as between Israel and other countries in the Middle East and between India and Pakistan. Conflicts centered on religion and race/ethnicity may appear in counseling sessions no matter where you work. I'd like to share an example from my country and discuss implications for counseling.

The Context of the Catholic/Protestant Conflict

After centuries of often problematic relations between Great Britain and Ireland, the predominantly Catholic Republic (south) of Ireland was established as an independent nation in 1922. Simultaneously, Northern Ireland continued as part of Great Britain. Originally, the south was 90% Catholic and 10% Protestant, but over the years the Protestant minority has dropped to some 2%. Northern Ireland currently is deeply divided between 40% Catholics and 60% Protestants, and in recent years it has been one of the most violent places in the world. The ongoing conflict, or "Troubles," has culminated in a mortality toll of over 3,700 people—the pro rata equivalent of 600,000 deaths in the United States. Not surprisingly, this violence has impacted almost every aspect of the lives of the population. The two groups are almost totally segregated, and despite a current downturn in violence, indications are that they are separating further.

Not least of the problems in Northern Ireland has been that of diametrically opposite political aspirations. The Protestant/Unionist community wishes to remain part of Great Britain, while the Catholic/Nationalist community seeks unification with the Republic of Ireland. The entrenched divisions emanating from these politico-religious differences have led to overt physical combat.

What Happens to Children in This Context?

The tension and violence are likely to continue, and even young children are indoctrinated into an ideology of hate and mistrust. Here 10-year-olds Kevin and Liam (both Catholic) report on a visit to a Protestant school. The visit was designed to promote understanding.

Interviewer: Why did you not like the Protestant school?

Kevin: 'Cos they're Protestants. (They giggle and look at the interviewer in amazement as if it is obvious.)

Liam: 'Cos they're slabbering [calling names] to us and all, and we were like messing about.

Kevin: They were slabbering to me so I . . .

Interviewer: Why did they slabber at you?

Liam: 'Cos they don't like us.

Kevin: See the wee Protestant, ahm, I was beside, he said to me (whispers), "I'm going to block you out," and I said (leans forward as though he didn't hear him), "Wha?" And he was going to dig [hit] me.

Research has shown that children even as young as 3 have these attitudes. Hate clearly is engendered at an early age.

Working Toward Understanding

Counseling theories are not enough. The skills of constructive challenge/confrontation and mediation are obviously important parts of building toward better understanding and community. My experience and research suggest the following:

▲ There is a need for actual contact and communication between combatants. You can't establish a relationship through a closed door.

▲ This contact needs to be positive and rewarding to both sides of the conflict. A win-win attitude is needed as both parties expect a win-lose result.

▲ Intimacy and getting to know the other person as a person are essential. There is need to search for common goals and areas where they might agree.

▲ Power needs to be equal between the two individuals or groups.

▲ Conflict cannot easily be resolved alone. Two individuals or two groups need support from the larger community. Key people beyond the counselor, educator, or mediator need to support resolution. This is a situation where the person is truly a person-in-community.

(continued)

BOX 9-2 (continued)

> With severe conflict such as this, I question the need or wisdom of drawing out the detailed story underlying the conflict. While I think in-depth sharing is important and vital in counseling, when we move to severe conflict we are going to have to focus on positive assets and strengths and use a wellness approach. Without extensive community and governmental support and increased contact among groups, we are going to see ever more "recreational rioting" in which young people riot because it becomes the stimulating and fun thing to do. As counselors and therapists, we can become part of the solution to these complex issues by first learning how to listen effectively and then working to resolve contradictions and conflict in our individual client, family, and group work. With this experience, we can then move to the real challenges of broadly and deeply based misunderstandings. It is a large task, but it is one we must undertake.

we resolve real problems of deep conflict between peoples? What is our role as counselors in such conflict prevention and social justice? Skilled confrontation skills are vital in such counseling.

Step 3: Evaluate the Change

The effectiveness of a confrontation is measured by how the client responds. If you observe closely in the *here and now of the session,* you can rate how effective your interventions have been. You also can assess whether your client has changed as a result of your interviewing and counseling skills. The method of evaluating cognitive/emotional change described here is based on an adaptation of work by Elisabeth Kübler-Ross (1969), who revolutionized our thinking about death and dying.

Kübler-Ross identified five stages of cognitive and emotional change as people faced death and dying (see Box 9-3). Her well-known theory has implications for change in areas ranging from interviewing to addiction and career choice. When people face death, they tend to display one of five reactions ranging from denial to acceptance to transcendence. The framework has five levels, and many people move through them in order; some people stay in one stage and never change, while others may "bounce" among the stages. Change is not always as linear and progressive as Kübler-Ross suggests.

Apply the Kübler-Ross framework to interviewing, counseling, and therapy using the *restory* framework. The client tells us a story and we, of course, listen. If the client is in the denial stage, the story may be distorted, others could be blamed unfairly, and the client's part in the story may be denied. In effect, the client in *denial* does not deal with reality. When the client is confronted effectively, his or her story changes to discussing inconsistencies and incongruity and we see *bargaining and partial acceptance*—the story is changing. At *acceptance* (Level 3), the reality of the story is acknowledged, and storytelling is more accurate and complete; it is possible to move to *new solutions* and *transcendence* (Levels 4 and 5). When changes in thoughts, feelings, and behaviors are integrated into a new story, we see the client move into major new ways of thinking accompanied by action after the session is completed.

Virtually any problem a client presents may be assessed at one of the five levels. The five levels may be seen as a general way to view the change process in interviewing, counseling, and therapy. Each confrontation or other interview intervention in the *here and now* may lead to identifiable changes in client awareness.

BOX 9-3   KÜBLER-ROSS'S FIVE STAGES OF DEATH AND DYING COMPARED WITH FIVE LEVELS OF CHANGE IN THE INTERVIEW

To help see parallels between death and dying theory and the interviewing process, think about counseling an alcoholic. In counseling, breaking out of firmly rooted denial (Stage 1) may take some time, and clearly confrontation of many issues will be important. Many alcoholics go through a stage of bargaining (Stage 2) in which they begin to explore what is really going on in their lives. Stage 3 represents the first real breakthrough when the alcoholic admits to the problem. But many alcoholics at this stage continue drinking even though they are aware of their alcoholism. True change occurs when the drinking ceases (Stage 4). Some alcoholics are able to transcend and make major life changes, including mentoring others, as they work the "steps" of Alcoholics Anonymous.

Following are Kübler-Ross's five stages of death and dying:

Stage 1: Denial. The patient cannot accept and denies the reality that he or she will die. The inevitable is ignored. "It won't happen—not to me!"

The client denies that inconsistency, incongruity, or conflict exists. The story may show considerable confusion and/or key information may not be disclosed.

Stage 2: Partial acceptance of reality: bargaining and anger. "Bargaining" occurs when patients engage in magical thinking. "If I lead a better life, then God will let me live." Another type of partial acceptance is anger, when the affected person may be angry at God ("Why me? I've lived a good life") or at the overall unfairness of it all ("It isn't right"). Often anger is a cover-up for more basic issues, not only in death and dying but in other situations as well.

The client partially recognizes and understands the conflict, but does not understand it fully. The client has moved beyond active denial.

Stage 3: Acceptance and recognition. Dying is acknowledged by the individual, and along with this comes a marked shift of emotion. The underlying emotions of sadness, fear, and grief reactions surface. Loss is faced rationally, but with appropriate emotions.

The client recognizes the situation as it is and has appropriate emotions, but no significant change occurs.

Stage 4: Generation of a new solution—early transcendence. Death may be reframed as an opportunity to forgive and forget old family arguments or an opportunity to meet God and loved ones in heaven. The patient may decide to donate organs. A new meaning has been given to death that allows both acceptance and some degree of transcendence.

The client generates new thoughts, feelings, and behaviors.

Stage 5: Development of new, larger, and more inclusive constructs, patterns, or behaviors—transcendence. The person may appear to have made a major change in consciousness. Sadness may continue, but the individual becomes peaceful and serene—transcendence.

The client makes major changes in thoughts, feelings, and behaviors. This is coupled with a new sense of being—transcendence of past issues.

Smaller changes in the interview will result in larger client change over a session or series of sessions. Not only can you measure these changes over time, but you can also list specific goals and contract with the client in a partnership to resolve conflict, integrate discrepancies, and work through issues and problems.

The Client Change Scale (CCS):* The Creation of the *New*

The creation of the *New* is a creative act. You will want to access client resources and search for strengths that can be useful in confrontion. For example, "You've said that your

---

* A paper-and-pencil measure of the Client Change Scale was developed by Heesacker and Pritchard and was later replicated by Rigazio-DiGilio (cited in Ivey et al., 2005). Factor analytic study of over 500 students and a second study of 1,200 revealed that the five CCS levels are identifiable and measurable.

roommate is totally impossible, and you don't know what to do. On the other hand, you've just told me how you worked through a similar situation in high school. How might those skills be helpful now?" Search for strengths, draw out the story and/or the conflict, and clearly summarize the issues (including an emotional base). Bring in the strengths as needed. Be aware of goals, and the client and you likely can move to creative change and something *New.*

You can assess the direct impact of confrontation or other microskill on the client in the *here and now* by using the Client Change Scale (CCS). This scale will give you a frame of reference for determining how well clients have responded to your use of the skill of confrontation (Ivey, 2000; Ivey, Ivey, Myers, & Sweeney, 2005). When you identify where your clients are in the change process, you can adapt and change skills and strategies to help them move further along. One confrontation may be helpful, but do not expect the use of a single skill to resolve issues.

Maintain awareness that major change in areas such as addiction, lifestyle, and other complex issues may take considerable time. Movement on the scale can occur in one interview, or it may take a year or more to help a person move out of denial or partial acceptance—witness how difficult it is for the alcoholic to interpret life experiences via the creative *New.*

Client Change Scale (CCS)	Predicted Result
The CCS helps you evaluate where the client is in the change process. Level 1. Denial. Level 2. Partial examination. Level 3. Acceptance and recognition, but no change. Level 4. Generation of a new solution. Level 5. Transcendence.	You will be able to determine the impact of your use of skills and the creation of the *New.* Suggest new ways that you might try to clarify and support the change process though more confrontation or the use of another skill that might facilitate growth and development.

The CCS is not only used to assess the effectiveness of confrontation. *It is also equally important to use with all other skills as you observe clients react to your interventions.* Assessing client progress throughout the interview will help you select appropriate skills and strategies. The CCS can also be used to assess larger client movement over several interviews. This is especially so if you and the client have established clear, concrete, and accountable goals for your meetings.

The following example shows five different reactions to a divorce. Any time clients are working through change, they talk about their issues with varying levels of awareness. The client may be in denial one moment, talking the next as if he or she accepts the problem, and then returning to bargaining to avoid change.

Level 1. Denial. The individual may deny or fail to hear that an incongruity or mixed message exists. "I'm not angry about the divorce. These things happen. I do feel sad and hurt, but definitely not angry."

BOX 9-4   CONFLICT RESOLUTION AND MEDIATION:
AN IMPORTANT PSYCHOEDUCATIONAL STRATEGY

You will find that confrontation skills are important in the mediation process. In conflict resolution and mediation, whether between children, adolescents, or adults, the following steps are useful.

Relationship

Develop rapport and outline the structure of your session. Pay equal, neutral attention to each participant. Four useful rules for children are "(a) agree to solve the problem, (b) no name-calling or put-downs, (c) be as honest as you can, and (d) do not interrupt" (Lane & McWhirter, 1992). Agreeing to some variation of these with adults is important to obtain commitment to the process of mediation.

Story and Strengths

Define the problem (concern). Use the basic listening sequence to clearly and *concretely* draw out the point of view of each person involved in the dispute. To avoid emotional outbursts, acknowledgment of feeling rather than reflection of feeling is recommended. Clearly summarize each person's frame of reference and carefully check out your accuracy with each one of them. You may ask each disputant to state the opponent's point of view. Outline and summarize the points of agreement and disagreement, perhaps in written form, if the conflict is complex. In this process, include an attitude of respect and spend time on the strengths of each participant.

Goals

Use the basic listening sequence to draw out each person's wants and desires for satisfactory problem solution. Focus primarily on concrete facts rather than emotions and abstract intangibles. This is the beginning of the negotiation process in which problems and concerns may be redefined and clarified. Summarize the goals for each person, with attention to possible joint goals and points of agreement.

Restory

Begin negotiation in earnest. Rely on your listening skills to see whether the parties can generate their own satisfactory solutions. When a level of concreteness and clarity has been achieved (Steps 2 and 3), the parties involved may come close to agreement. If the parties are very conflicted, meet each one separately as you brainstorm alternative solutions. With touchy issues, summarize them in writing. Many of the influencing skills discussed later in this section will be useful in the process of negotiation.

Action

Contract and generalize. Use the basic listening sequence; summarize the agreed upon solution (or parts of the solution if negotiations are still in progress). Make the solution as concrete as possible and write down touchy main issues to make sure each party understands the agreement. Obtain agreement about subsequent steps. With children, congratulate them on their hard work and ask each child to tell a friend about the resolution.

The Martin Luther King Jr. Center (1989) summarizes six steps for nonviolent change that are closely related to the mediation model above: (1) information gathering, (2) education, (3) personal commitment, (4) negotiations, (5) direct action, and (6) reconciliation. When you work on complex issues of institutional or community change, a review of Dr. King's model may help you in thinking through your approach to major challenges.

Level 2. Partial examination. The individual may work on a part of the discrepancy but fail to consider the other dimensions of the mixed message. "Yes, I hurt, and perhaps I should be angry, but I can't really feel it."

Level 3. Acceptance and recognition, but no change. The client may engage in the confrontation but make no resolution. Until the client can examine incongruity, stuckness, and mixed messages accurately, real change in thoughts, feelings, and behavior is difficult. "I guess I do have mixed feelings about it. I hurt about the marriage, but now I realize how angry I am." Coming to terms with anger or other emotion is an important breakthrough.

Level 4. Generation of a new solution. The client moves beyond recognition of the incongruity or conflict and puts things together in a new and productive way. "Yes, I've been avoiding my anger, and I think it's getting in my way. If I'm going to move on, I will have to experience anger as part of the total situation."

Level 5. Development of new, larger, and more inclusive constructs, patterns, or behaviors— transcendence. A confrontation is most successful when the client recognizes the discrepancy and generates new thought patterns or behaviors to cope with and resolve the incongruity. "You helped me see that mixed feelings and thoughts are part of every relationship. I've been expecting too much. If I expressed both my hurt and anger more effectively, perhaps I wouldn't be facing a divorce."

Moving from confrontation to resolution will often involve mediation, whether between two individuals or in a much larger context. Box 9-4 illustrates the importance of conflict resolution skills in mediation processes.

As you examine mediation, consider how you might help children, adolescents, and adults become more tolerant and accepting of each other when serious differences loom in the future. Mediation is an important psychoeducational tool in conflict resolution, and the skills of mediation can be used effectively to further understanding and promote closer community ties.

EXAMPLE INTERVIEW: BALANCING FAMILY RESPONSIBILITIES

Three skills are reviewed simultaneously in this example interview: (1) listening skills are used to obtain client data, (2) confrontations of client discrepancies are noted, and (3) the effectiveness of the confrontations is considered using the Client Change Scale (CCS). You may want to read the segment several times and study it carefully. With more experience and practice, these concepts will be useful and important in your interviewing practice.

The following interview presents a conflict that is common to many working couples—balancing home tasks. Male attitudes and behaviors are changing, but many working women are still burdened with the responsibility for most home tasks. In this example we have both internal and external incongruity. Dominic is struggling internally with the discovery that things have changed and externally with the realization that his wife is behaving differently. Arguments may be particularly intense as two tired people come home from a hard day's work to face needy children and undone housework. Couples may blame each other rather than attributing a major portion of their issues to external causes at work. An exhausted partner or spouse often has little energy left to deal with the concrete issues at home.

Interviewer and Client Conversation	Process Comments
1. *Dominic:* I'm having a terrible time with my wife right now. She's working for the first time, and we're having lots of arguments. She isn't fixing meals like she used to or watching the kids. I don't know what to do.	On the CCS, this client response is rated Level 2; he is partially aware of the problem. At the same time, he denies (Level 1) his role in the problem and is attributing the difficulties in the home to husband/wife issues, failing to see how demands at work play into the system.

(continued)

Interviewer and Client Conversation	Process Comments
2. *Ryan:* (holds out one hand to the right) So, on one hand, your wife is working outside the home, but (holds out the left hand), on the other hand you expect her to continue with all the housework, too. You don't like what's going on right now.	Confrontation presented as paraphrase and reflection of feeling. The use of hands with the words helps strengthen and clarify the conflict.
3. *Dominic:* You damn betcha she's expected to do what she's always done—I'm not confused about that.	CCS Level 1—denial. This needs more clarification of what is going on in the relationship.
4. *Ryan:* I see. You're not confused; you really don't like what's going on. Could you give me a specific example of what's happening—something that goes on between the two of you when you both get home?	Paraphrase, open question oriented to concreteness.
5. *Dominic:* (sighs, pauses) Yeah, that's right, I don't like what's going on. Like last night, Sara was so tired that she didn't get around to fixing dinner 'til half an hour late. I was hungry and tired myself. We had a big argument. This type of thing has been going on for 3 weeks.	While Dominic is able to identify and talk about the conflict (CCS Level 2), he lacks awareness of how his wife feels. He seems insensitive to the fact that his wife is working, and he expects her to do everything she did in the past for him (CCS Level 1).
6. *Ryan:* I hear you, Dominic; you're pretty angry. Let's change focus for a minute. When dealing with conflict, it helps to concentrate on positive things. Could we search for some positive things that have worked for you and Sara in the past?	This summary introduces an incongruity between the difficult present situation and a positive past through the positive asset search.
7. *Dominic:* Yes, I am angry and discouraged. Sara and I were doing pretty well until the babies came. Somehow things just got off kilter. (5-second pause and silence) . . . Well, let me try it your way. We sure had fun times together over the 3 years we've been together. We both like outdoor activity and doing things together. We never seem to have time for that now.	In couples work it is useful to remind the partners of positive stories from the past. This interview has been edited for brevity. It is important to explore positives in the relationship in much more depth than presented here.
8. *Ryan:* It's important to remember that you have a good history and have enjoyed each other. I'm wondering if part of the solution isn't finding time just to be together doing fun things. Let's make that part of our discussion later. But for the moment, let's go back to your main issues. Could you give me a specific example of what goes on between you when she gets home?	Paraphrase, suggestion, open question oriented to concreteness. Getting specifics helps clarify the situation. When counseling around couples issues, search for strengths. What brought the couple together and what maintains them now? This helps clients center themselves in positives as they struggle with the difficult negatives in a relationship.

(continued)

Interviewer and Client Conversation	Process Comments
9. *Dominic:* (pause) . . . Well, lately, I've had a lot of pressure at work. They're downsizing, and morale is bad. I worry I'll be next. I try really hard, but when I get home, I just want to sit. Sara's got the same thing. Her new boss just wants more all the time. I guess when she comes home, she's about as exhausted and confused as I am.	It is very common for partners to get angry with each other when external stressors hit one or both individuals in the relationship. One can't argue with the boss or colleagues easily, so the partner is the scapegoat. Imagine how much more difficult this would be if one of them lost their job.
10. *Ryan:* I see. You both work all day and come home exhausted. Dominic, do you really think Sara can work that hard and still take care of you like she used to?	Paraphrase, closed question. Ryan's implicit confrontation is now more concrete.
11. *Dominic:* I guess I hadn't thought of it that way before. If Sara is working, she isn't going to be physically able to do what she did. But where does that leave me?	For the first time, Dominic is able to see that Sara can't continue as she has in the past (CCS Level 3). Note that he is still thinking primarily of himself. To move to higher levels on the CCS, Dominic would have to be able to take Sara's perspective and articulate what she likely thinks and feels when she gets home.
12. *Ryan:* Yes, where does that leave you? Dominic, let me tell you about my experience. My wife started working, and I, too, expected her to continue to do the housework, take care of the kids, do the shopping, and fix the meals. She went to work because we needed the money to make a down payment on a house. Well, what I found was that my wife couldn't work unless I helped around the house. I had to decide which was more important—getting a house or maintaining our traditional roles. I share some of the household work with her. I don't like it, but it seems like it's got to be done. I shop, and I pick the kids up at day care, too. How does that sound to you? Do you think you want to continue to expect Sara to do it all?	Self-disclosure followed by a check-out. Ryan is operating like a coach. He is speaking up directly with his ideas, but he is also allowing the client to react to them. Clearly, Ryan is trying to get Dominic's thinking and behavior to move. His last statement contains an important implicit confrontation: "On the one hand, your wife is working, and if she continues, she'll need some help; on the other hand, perhaps you don't want her to work—what is your reaction to this?"
13. *Dominic:* Uhhh . . . we need the money. We've missed the last car payment. It was my idea that Sara go to work. But isn't housework "women's work"?	Dominic is now facing up to the contradiction he is posing (CCS Level 3). He has not yet synthesized his desire for his wife to work with the need for his sharing the workload at home, but at least he is moving toward a more open attitude. As he acknowledges his part in the situation and the need for more money, he is beginning to come to a new understanding, or synthesis, of the problem—he is less incongruent and is taking beginning steps toward resolving his discrepancies.

For each developmental task completed in the interviewing process, it often seems that a new problem arises. Just as the counselor is beginning to facilitate client movement, a new obstacle ("women's work") arises. To make the progress shown took half the session. Changing the concept from "women's work" to "work in the home that must be shared if the car payments are to be met" took the rest of the session. Achieving a new and likely more lasting level of male/female cultural differences would require a major transformation in thinking and behavior (CCS Level 5). Major change may require several interviews, group sessions, and time for Dominic, or any other client, to internalize new ideas.

SUMMARY: CONFRONTATION AND CHANGE

We have covered three steps of confrontation and change:

1. Identify conflict via observing incongruities, discrepancies, and mixed messages.
2. Point out issues of incongruity and work toward resolution.
3. Evaluate the change process via the Client Change Scale.

Confrontation itself is a not a distinct skill; it is a set of skills that may be used in different ways. The most common confrontation uses the paraphrase, reflection of feeling, and summarization of discrepancies observed in the client or between the client and her or his situation. However, questions and influencing skills and strategies can also lead to client change. The CCS can be used throughout the session with all skills. It is also an informal assessment tool for the success of your interviews.

The creative *New* reminds us that each individual works through change in her or his own way. Some clients will move rapidly through all five levels in one session. Most clients will move more slowly. If you work with a major grief reaction around divorce or a highly significant change such as stopping drinking, do not expect clients to respond to your confrontations very rapidly. Creating the *New* and then acting on a new story may take time and patience.

Key points of the chapter are presented below.

■ KEY POINTS	
Confrontation	Clients come to us stuck and immobilized in their developmental processes. Through the use of microskills—confrontation in particular—we facilitate change, movement, and transformation—restorying and action. Confrontation has been defined as a supportive challenge in which you note incongruities and discrepancies and then feed back or paraphrase those discrepancies to the client. Our task is then to work through the resolution of the discrepancy.
The creative *New*	This is a more poetic and positive way to describe the human growth process. Change is a creative action. Tillich's concept of the *New* helps make confrontation and change more fluid and strength producing.

(continued)

Key Points (continued)

Confrontation and change strategies	An explicit confrontation can be recognized by the model sentences, "On one hand . . . , but on the other hand. . . . How do you put those two together?" In addition, many interviewer statements contain implicit confrontations that can be helpful in promoting client growth and developmental movement. For example, you may summarize client conversation, pointing out discrepancies, or use an influencing skill such as the interpretation/reframe, feedback, or other strategies to produce change.
Death and dying as a change model	We seek to help our clients change their ways of thinking and behaving. We have examined the ideas of Kübler-Ross on death and dying and the notion that clients often work through five identifiable stages as they change their thoughts, feelings, and behaviors: (1) denial, (2) partial acceptance of reality/bargaining/anger, (3) acceptance and recognition, (4) generation of a new solution, and (5) development of new, larger, and more inclusive constructs, patterns, or behaviors.
The Client Change Scale	The Client Change Scale is a tool to examine the effect microskills and confrontation have on client verbalizations immediately in the interview. At the lowest level clients deny their incongruities; at middle levels they acknowledge them; at higher levels they transform or integrate incongruity into new stories and action.
Mediation	Mediation and conflict resolution can be facilitated by using listening skills, confrontation, and the five-stage interview model.
Multicultural and individual issues	Confrontation is believed to be relevant to all clients, but it must be worded to meet individual and cultural needs. A narcissistic or self-centered client may resist confrontation, and with this client the microskills of interpretation and feedback may be more helpful. Clients from more direct and outspoken cultures such as European Americans and African Americans may respond well to appropriate confrontations. Cultures that place more emphasis on subtlety and an indirect approach such as Asian groups may prefer gentler, more polite confrontations. Modification in style to accommodate various individuals will be necessary. Do not expect individuals in any cultural group always to follow one pattern; avoid stereotyping.

COMPETENCY PRACTICE EXERCISES AND PORTFOLIO OF COMPETENCE

This chapter is designed to help you construct a view of helping oriented toward change. If you master the cognitive concepts of the reading material and the exercises that follow, you will be able to promote client change and assess the effectiveness of your interventions. Again, this is an area that takes practice and experience. Apply the ideas here throughout the rest of your work with this book.

Individual Practice

Exercise 1: Identifying Discrepancies, Incongruity and Mixed Messages, and Strengths Leading Toward Resolution

Exercises 1–3 of Chapter 5 on observation skills are basic to confrontation. Please review them again. Viewing videotapes of interviews, especially your own, is the best way to practice identification skills. The following discussion will also be useful as it encourages self-examination. Unless you can identify incongruity in your own self, seeing it in others may be difficult or even inappropriate.

Discrepancies internal to the self. Can you identify specific times in which your nonverbal behavior contradicted your verbal statements and gave you away? Are there times when you say two things at once, and your verbal statements are incongruous? Have you done one thing while saying another?

Discrepancies between you and the external world. Part of life is living with contradictions. Many of these are unresolvable, but they can give considerable pain. What are some of the discrepancies between you and other individuals? What are some of the mixed messages, contradictions, and incongruities you face in your world of schooling or work?

Discrepancies between you and the client. You may have already experienced this and can easily summarize times when you felt out of tune with and discrepant from the client. Or if you have not interviewed extensively, it may be helpful to think of situations when you had major differences with someone else. Often we have typical situations that "push our buttons" and move us toward actions that are too quick. Self-awareness in this area can be most helpful.

Specific strengths. Resolution of conflict and discrepancy is often made from a positive frame of reference. Can you identify personal strengths and wellness assets that can help you resolve internal and external differences? What strengths do you admire in others that you might like to add to your repertoire?

Exercise 2: Practicing Confrontation of Incongruity

Write confrontation statements for the following situations. Using the model sentence "On the one hand . . . , but on the other hand . . ." provides a standard and useful format for the actual confrontation. Of course, you may also use variations such as "You say . . . , but you do . . . ," and remember to follow up the confrontation with a check-out.

A client breaks eye contact, speaks slowly, and slumps in the chair while saying, "Yes, I really like the idea of getting to the library and getting the career information you suggest. Ah . . . I know it would be helpful for me."

"Yes, my family is really important to me. I like to spend a lot of time with them. When I get this big project done, I'll stop working so much and start doing what I should. Not to worry."

"My partner is good to me most of the time—this is only the second time he's hit me. I don't think we should make a big thing out of it."

"My daughter and I don't get along well. I feel that I am really trying, but she doesn't respond. Only last week I bought her a present, but she just ignored it."

Exercise 3: Practicing With the Client Change Scale

Here are some statements made by clients. Identify which of the five levels each client statement represents.

1. Denial
2. Partial examination

3. Acceptance and recognition
4. Generation of a new solution
5. Development of new, larger, and more inclusive constructs, patterns, behaviors—transcendence

Health issues. Look for movement from denial to new ways of taking care of one's body.

_____ I can't have a heart attack. It will never happen to me. I need to eat real food.

_____ Oh, I suppose I am overweight, but if I cut down a bit on butter and perhaps no more milk shakes, I'll be okay.

_____ I guess I can see that I need to balance my diet, but the busy life I lead won't really allow that to happen.

_____ I'm now able to cut out fats. At least that's taken care of.

_____ I've completely changed my way of doing things. I eat right—not much fat—I exercise, and I'm even getting to like relaxation and stress management.

Career planning. Look for a movement from inaction or randomness to action.

_____ Okay, I guess I see your point. I've been released from two work-study programs because I didn't show up on time. But those were the bosses' fault. They should have made what they wanted clearer.

_____ The teacher referred me to you. Everyone has to have a job plan, but I see no need to worry about it so much. I'll be okay.

_____ Yes, I need a job plan. I can see now that is necessary. I'll write one and bring it to you tomorrow.

_____ I've got a job! The plan worked, and I interviewed well, and now I'm on my way.

_____ The plan has been helpful. I think I see now how to interview more effectively and present myself better.

Awareness of racism, sexism, heterosexism. Look for movement from denial that these issues exist to awareness and action.

_____ I feel committed. I've started action at home and at work, and I'm really going to concentrate on a more active approach to this issue. It's so important.

_____ Well, some people do discriminate, but I think that many people are just exaggerating.

_____ I don't really believe there is such a thing as racism or sexism. It's just people complaining.

_____ I've starting working with my family and children on being more tolerant, fair, and understanding of people different from us.

_____ There is a fair amount of prejudice, racism, and sexism everywhere.

Exercise 4: Writing Model Confrontation Statements

Review the Client Change Scale described. Then read the following confrontations.

Dominic: How can she expect me to work around the house? That's women's work!

Ryan: On one hand, I hear you wanting that second income she brings in. On the other hand, you seem to want her to keep up her housework as she did in the past without any help. How do you put that together?

Dominic could respond to that confrontation level by using denial, or he could work toward new ways of thinking. Can you write below example statements for Dominic representing the five levels of the Client Change Scale?

Level 1 (Denial): _____

_____

Level 2 (Partial examination): _____

_____

Level 3 (Acceptance and recognition): _____

_____

Level 4 (Generation of a new solution): _____

_____

Level 5 (Transcendence: Development of new, larger, and more inclusive constructs, patterns, or behaviors): _____

_____

Client: I'm getting tired of talking with you. You always seem to think I'm taking the easy way out.

Counselor: Sounds as if you're telling me that, on the one hand, you want to change—that's why you started counseling—but on the other, now that change is getting close, you want to leave. That seems similar to the way you handle your relationships with the opposite sex: When someone gets close, you leave. How do you respond to that?

Write statements that would represent each level of the Client Change Scale.

Group Practice

Exercise 5: Practice Confrontation Skills in a Group
Step 1: Divide into groups.

Step 2: Select a group leader.

Step 3: Assign roles for the first practice session.

- ▲ Client.
- ▲ Interviewer.
- ▲ Observer 1, who will rate each client statement using the Feedback Form (Box 9-5) and, during a replay of an audio- or videorecording, will stop the tape after each client statement and rate it carefully.

BOX 9-5   FEEDBACK FORM: CONFRONTATION USING THE CLIENT CHANGE SCALE

_____ (Date)

_____   _____
(Name of Interviewer)   (Name of Person Completing Form)

Instructions: Video and/or audio recording will be necessary for the best type of feedback. Otherwise, it will be best for the observer(s) to stop the session shortly after a confrontation has occurred and then discuss what was observed. Rate how the client responded to the confrontation on the five-point Client Change Scale. 1 = Denial, 2 = Partial examination, 3 = Acceptance and recognition, 4 = Generation of a new solution, 5 = Transcendence and the creation of the New.

Interviewer Confrontation (Write key words to help recollection and discussion.)	Client Response (Write key words to help recollection and discussion.)	CCS Rating

▲ Observer 2, who will record the key words of each interviewer statement on a separate sheet of paper, thus making it possible to construct a picture of the interviewer as well. Pay special attention to the microskill leads of the interviewer.

Step 4: Plan. State the goal of the session. The interviewer's task is to use the basic listening sequence to draw out a conflict in the client and then to confront this conflict or incongruity. Your ability to observe and note discrepancies on the spot during the session and to feed them back to the client will be important.

A useful topic for this practice session is any issue on which the volunteer client feels conflicted within or without. Internal conflict often shows around a difficult decision, past or present. External conflict most often appears when one has difficulty in dealing with a family member, a friend, or someone at work. Usually you will find both internal and external conflict in the client. Potentially useful topics include these:

▲ An important purchase.
▲ Choosing between two equally attractive majors in college.
▲ A career choice between a larger income and work that would be enjoyed more fully.
▲ Moral decisions ranging from telling the truth when one has held it back, to differences of opinion on abortion, divorce, or making a commitment, to issues of diversity or the role of spirituality in one's family.
▲ Debating whether you should go to the best party of the year or study for the next day's test.
▲ Deciding to tell your roommate that he or she needs to clean up his or her mess.
▲ Deciding how to tell your fiancé that you don't love him any more.
▲ Confronting a friend you saw stealing a book from the library.
▲ Virtually any other type of interpersonal conflict.

Step 5: Conduct a 5-minute practice session using confrontation skills as part of your listening and observation demonstration.

Step 6: Review the practice session using confrontation skills.

Step 7: Rotate roles.

Some general reminders. The volunteer client may be asked to complete the Client Feedback Form of Chapter 1. This exercise is an attempt to integrate many of the skills and concepts used thus far in this book. Allow sufficient time for thinking through and planning this practice session, and recall the potential value of the positive asset search, coupled with full awareness of wellness potential.

Portfolio of Competence

Skill in confrontation depends on your ability to listen first and then to take an active role in the helping process. This needs to be done in a nonjudgmental fashion with respect for differences. As you work through this list of competencies, think ahead to how you would include confrontation skills in your own Portfolio of Competence.

Use the following as a checklist to evaluate your present level of mastery. Check those dimensions that you currently feel able to do. Those that remain unchecked can serve as future goals. Do not expect to attain intentional competence on every dimension as you work through this book. You will find, however, that you will improve your competencies with repetition and practice.

Level 1: Identification and classification.

☐ Identify discrepancies and incongruities manifested by a client in the interview.
☐ Classify and write counselor statements indicating the presence or absence of elements of confrontation.
☐ Identify client change processes through observation on the Client Change Scale.

Level 2: Basic competence.

❑ Demonstrate confrontation skills in a real or role-played interview.
❑ In the here and now of the interview, observe and identify client responses on the five levels of the Client Change Scale.
❑ Utilize wellness and the positive asset search to help clients find strengths that might help them move forward toward positive change when confronted.

Level 3: Intentional competence. You will be able to use confrontational skills in such a manner that clients improve their thinking and behaving as reflected on the CCS.

❑ Help clients change their manner of talking about a problem as a result of confrontation. This may be measured formally by the CCS or by others' observations.
❑ Move clients from a discussion of issues at the lower levels of the CCS, when beginning discussion of a problem, to discussion at higher developmental levels at the end of the interview, or when the topic has been fully explored.
❑ Identify client responses inferred from the CCS on the spot in the interview and change counseling interventions to meet those responses.

Level 4: Psychoeducational teaching competence. Are you able to teach change and confrontation concepts to clients and to others?

The basic dimensions of confrontation are really designed more for counselors and interviewers than for clients. But there are some very specific ways that psychoeducation will be important in your practice. First, those going through the stages of grief associated with death may find it helpful to have the change stages identified for them, thus enabling them to understand their feelings and thoughts more fully. These stages of change will also be helpful in understanding reactions to serious illness, accidents, alcoholism, and traumatic incidents. Second, you can set up change goals with your clients and work with them to discover how far they have progressed in meeting goals and making life changes. Finally, the mediation process is obviously an important psychoeducational tool.

DETERMINING YOUR OWN STYLE AND THEORY: CRITICAL SELF-REFLECTION ON CONFRONTATION

Confrontation is based primarily on listening skills, but it does require you to move more actively in the session through highlighting discrepancies and conflict. The Client Change Scale (CCS) was presented to show that you can assess the influence of your interventions in the here and now of the session. The creative *New* provides a more philosophical dimension to confrontation and change.

What single idea stood out for you among all those presented in this chapter, in class, or through informal learning? What stands out that is likely to be important as a guide toward your next steps? How might confrontation relate to diversity issues? What other points in this chapter struck you as important? How might you use ideas in this chapter to begin the process of establishing your own style and theory? How would you use mediation as a psychoeducational treatment program?

*What are your
thoughts?*

_____

_____

OUR THOUGHTS ABOUT CHRIS

Every day in the counseling process we encounter clients facing difficult decisions. Our first task is to identify the key discrepancies, conflicts, or contradictions faced by the client. In this case, Chris has several obvious conflicting issues that he needs to explore before making a decision:

▲ The central issue, at the moment, is attending graduate school versus staying at home and caring for his parents.
▲ He is dealing with mixed family messages: that he continue with his studies or that he stay home and help with the family business.
▲ We have incomplete data on his finances. If he has been saving up for 2 years for school, why is he worried that grad school will disappear if he doesn't go now?
▲ There is also the important issue of mixed, conflicting emotions. He has a desire to continue his education but guilt about the possibility of leaving home.
▲ Multicultural issues come in as well. Within certain cultures, the idea of leaving home after a family crisis may not even occur to the client. In a very individualistic culture, there may not even be a question—Chris should go on to graduate school. Thus, staying or leaving can also represent a cultural conflict to the client.

Our first goal would be to hear Chris's story more fully, and at first we'd likely focus on the central issue, recognizing that later what appears to be the central dimension of decision may change as we learn more.

"Chris, I really hear the pressure you feel you are under. On one hand, you do want to care for your parents, but your dad suggests that they can take care of themselves. On the other hand, it leaves you feeling guilty, and your mom adds to that conflict. Tell me more."

We think this lead catches the essence of the conflict, although we have left out (for the moment) his desire for grad school. We call this type of response a confrontation because it focuses on conflicting, contradictory possibilities.

As our work with Chris evolves and we hear his story more completely, we can explore the other issues listed above—and likely, there will be new issues to encounter. One may be a cultural tradition of responsibility to family—or the cultural tradition may be that the individual needs to make the decision totally on his own. Resolution of the contradictions occurs when Chris is able to find a solution that satisfies him.

CHAPTER 10

Focusing the Interview: Exploring the Story From Multiple Perspectives

Focusing

Confrontation

The Five-Stage Interview Structure

Reflection of Feeling

Encouraging, Paraphrasing, and Summarizing

Client Observation Skills

Open and Closed Questions

Attending Behavior

Ethics, Multicultural Competence, and Wellness

One very important aspect of motivation is the willingness to stop and to look at things that no one else has bothered to look at. This simple process of focusing on things that are normally taken for granted is a powerful source of creativity.

—Edward de Bono

How can focusing help you and your clients?

Chapter Goals

Focusing is a skill that enables multiple tellings of the story and will help you and clients think of creative new possibilities for restorying. Client issues are often complex, and the systematic framework of focusing can help in reframing and reconstructing problems, concerns, issues, and challenges. Focusing is a skill you may not see in other books. We developed the skill of focusing because it remains the clearest way to (1) stress the importance of the *individual,* and (2) expand awareness of how individual clients develop in social context—especially community and family.

Competency Objectives

Awareness, knowledge, and skill in focusing will enable you to

▲ Increase client cognitive and emotional complexity. Too often issues are considered from only one frame of reference.

▲ Better understand the viewpoints of others with an accompanying increase in empathic understanding and cultural intentionality.

▲ Incorporate family and cultural issues, particularly through family and community genograms.

▲ Be aware of the role of advocacy and social change as part of the focus of your interviewing practice.

Vanessa walks swiftly into the office and starts talking even before she sits down:

> I'm really glad to see you. I need help. My sister and I just had an argument. She won't come home for the holidays and help me with Mom's illness. My last set of exams was a mess, and I can't study. I just broke up with the guy I was going with 3 years. And now I'm not even sure where I'm going to live next term. And my car wouldn't start this morning. . . . (She continues with her list of issues and begins repeating stories almost randomly, but always with energy and considerable emotion.)

How would you respond to Vanessa and help her focus on key issues? Compare your response with ours on page 290.

INTRODUCTION TO FOCUSING

There are many clients like Vanessa, who have numerous issues in their lives. You may feel overwhelmed when you get 5 or more minutes of problematic stories in which the client rapidly jumps from topic to topic. Frequently there is an insistence to do "something" immediately and start solving the issues before you have heard even one complete story. Solving problems too early results in poor listening and may result in your generating more problems and difficulties.

We should first note that Vanessa has many short- and long-term stressors acting on her, which literally could tear her apart emotionally and send damaging cortisol to her brain. She is in a high state of incongruence. Focusing will help clearly identify the major areas of conflict and discrepancy to be approached first. Listening and using supportive challenges will help her clarify her situation and move more readily to problem solution. If you use focusing skills as defined below, you can *predict* how clients respond.

Focusing	Predicted Result
Use selective attention and focus the interview on the client, problem/concern, significant others (partner/spouse, family, friends), a mutual "we" focus, the interviewer, or the cultural/environmental/contextual issues. You may also focus on what is going on in the *here and now* of the interview.	Clients will focus their conversation or story on the dimensions selected by the interviewer. As the interviewer brings in new focuses, the story is elaborated from multiple perspectives.

Selective attention (Chapter 3) is basic to focusing. Clients tend to talk about or focus on topics to which you give your primary attention. Through your attending skills (visuals, vocal tone, verbal following, and body language), you indicate to your client that you are listening. But we all tend to focus or listen in different ways. It is important to be aware of both your conscious and unconscious patterns of selective attention; clients follow your lead rather than talk about what they really want to say.

If you focus solely on individual issues, clients will talk about themselves and their frame of reference. A second important area to focus on is the client's problem, issue, or concern. If a client has gone through a breakup of a significant relationship, has study difficulties, has cancer or another serious illness, you need to hear the details, and often you need to hear a lengthy story. Just telling the story is relieving for the client. We feel better when someone seriously tries to listen and understand. However, too many beginners and even professionals fall into a voyeur state and become so interested in the problematic story that they fail to focus on the unique client before them. The following examples show how to redirect the focus.

"Which issue would you like to focus on first?"

"Sounds like your mother is the most important issue. Have I heard correctly?"

"On one hand, you are hurting over the breakup, but I also hear your strengths and your desire to keep going. You've done a lot in the middle of all this to keep yourself together."

In addition, people live in a broad context of multiple systems. The concept of *self-in-relation* may be helpful. The idea of *person-in-community* was developed from an Afrocentric frame by Ogbonnaya (1994), who points out that our family and community history live within each of us. The client brings to you many community voices that influence the view of self and the world. The community genogram introduced in this chapter will be useful in helping clients gain new perspectives on themselves and their relationships to significant others. The community genogram is a useful way to understand your client's history and a good place to identify strengths and resources.

Clearly, much individual counseling focuses on issues of conflict, incongruity, and discrepancies between the individual and family and friends. In addition, many client problems are caused by and related to issues and events in the broader context (e.g., poor schools, floods, economic conditions). If you help clients see themselves and their issues as *persons-in-community*, they can learn new ways of thinking about themselves and use existing support systems more effectively. The following list offers potential comments and questions that allow the interviewer to focus the session in a specific area:

Significant others (partner, spouse, friends, family)

"Vanessa, tell me a bit more about your mother and how she is doing."

"How are your friends helpful to you?"

"What's going on with your sister?"

"You seem to be thinking a lot about the breakup after three years of living together. Let's explore that a bit more now."

"Your grandmother was very helpful to you in the past. What would she say to you?"

Mutual focus and immediacy ("we" statements and talking about what is going on in the session here and now)

(early in the session) "Vanessa, you have a lot on your plate, but *we* will work through your issues. Right now I can almost feel your hurt."

(later sessions) "Vanessa, we've been working together for 2 weeks now. I sense at this moment that you felt angry at what I just said. I'm glad that you can openly express your feelings to me."

Interviewer focus (sharing one's own experiences and reactions)

"I felt really confused and worried when my mother had the same illness. I simply didn't know what to do. Is that close to the way you feel?"

"I can understand your frustration with the car. It happened to me last week."

Cultural/environmental/contextual and broader issues such as the impact of the economy

"Finding a new place to stay is difficult, and you think that landlords won't rent to people of color."

"What are some strengths that you gain from your spiritual orientation?"

"You feel that the college is simply not supporting you at all. Could you tell me a bit more about what they are doing to make it difficult for you?"

As an interviewer, be aware of how you focus an interview and how you can broaden the session so that clients are aware of themselves more fully in relation to others and social systems, as persons-in-relation, and as persons-in-community. In a sense you are like an orchestra conductor, selecting which instruments (ideas) to focus on, enabling a better understanding of the whole. Some of us focus exclusively on the client and the problem, failing to recognize the impact and importance of other issues. Others may fail to give sufficient attention to the client as a person and use the interview as a chance to learn interesting details—almost as a voyeur.

EXAMPLE INTERVIEW: IT'S ALL MY FAULT—HELPING THE CLIENT UNDERSTAND SELF-IN-RELATION

Carl Rogers's person-centered counseling has had an immense and lasting influence on the way we conduct helping sessions. We can only work with the individual before us, and ultimately, it is this unique person who is most important. We live in a culture that focuses on individuality, individual responsibility, and individual achievement—the "I" focus. It is only natural that counseling focuses on the immediate person in the *here and now*. At the same time, the issues of self-in-relation and the environmental/cultural/contextual setting in which the person exists also need full attention.

The following interview with Janet focuses first on her individual issue of taking virtually all the responsibility for difficulties in her relationship with Sander. This is the second interview, and during the first session Janet completed a community genogram of the home where she grew up in Eugene, Oregon. The community genogram (see Box 10-1, page 273) is used to help Janet understand how her history might affect her present behavior, feelings, and thoughts.

Interviewer and Client Conversation	Process Comments
1. *Samantha:* Nice to see you, Janet. How have things been going?	A solid relationship was established during the first session. When Samantha met Janet outside her office, Janet seemed anxious to start the session.
2. *Janet:* Well, I tried your suggestion. I think I understand things a bit better.	Last week, Samantha suggested homework that asked Janet to spend the week listening carefully to Sander to try to identify what he was really saying. She also advised continuing her own behavioral patterns so that she could note patterns of behavior more easily.
3. *Samantha:* You understand better? Could you tell me more?	Encourage. Focus on problem.

(continued)

Interviewer and Client Conversation	Process Comments
4. *Janet:* Well, I listened more carefully to Sander, so I could understand what he really wants from me. He really wants a lot. He wants me to be a better cook, he doesn't like the way I keep the house, and the more I listened, the more he wanted. I guess I haven't given him enough attention. I should be doing more and doing it better.	Self-disclosure on observations. Focus on self (the client), others (Sander), and the main theme of relationship.
5. *Samantha:* Janet, it looks like you feel sad and a little guilty for not doing more. Am I hearing you right?	Reflection of feeling, focus on Janet with secondary focus on the problem.
6. *Janet:* Yes, I do feel sad and guilty; it's my responsibility to keep things together at home. He works so hard.	The key word *responsibility* appears. Janet seems to believe that she is responsible for keeping the relationship together. The acceptance of individual responsibility (and blame) is often characteristic of, but not exclusive to, women of Northern European background. Change in the interview can be measured in the session using the Client Change Scale (CCS). At this point of the session, Janet takes almost total responsibility for the problems she is experiencing with Sander. She seems to be at Level 1 with no recognition of relevant aspects of her situation.
7. *Samantha:* Sounds like you're punishing yourself, when you're trying so hard and being so responsible. Could you tell me more about how Sander reacts?	Samantha makes an interpretation ("punishing yourself"). The focus is first on Janet and changes to Sander.
8. *Janet:* Yes, I get angry with myself for not doing better. It's hard with work, too. But Sander wants the house to run perfectly. I try and try, but I always miss something. Then Sander blows up.	Focus on self and the problem with the close relationship-in-connection. Level 3 on Client Change Scale.
9. *Samantha:* I hear you trying very hard. Could you give me a specific example when Sander blew up?	Paraphrase and open question. The focus changes from Janet to Sander in a search for concreteness.
10. *Janet:* Last night I made a steak dinner; I try to fix what he wants. He seemed so pleased, but I forgot to buy steak sauce, and he blew up. It really shook me up . . . then, when we went to bed, he wanted to make love, but I was so tense that I couldn't. He got angry all over again.	Examples help us understand the specifics of situations. Some clients talk in vague generalities and asking for specifics can make a real difference.
11. *Samantha:* What happened then?	Open question/encourage, focus on problem. Search for more concreteness.
12. *Janet:* He went to sleep. I lay there and shook. As I calmed down, I realized that I need to do a better job, and maybe he won't get so angry.	Focus on self and Sander and actions that might be taken on her part to resolve or prevent the difficulties.

(continued)

Interviewer and Client Conversation	Process Comments
13. *Samantha:* Let's see if I understand what you are saying. You are doing everything you can to make the relationship work. Sander blows up at little things, and you try harder. And then you feel the problem is your fault. Have I heard you correctly?	Samantha summarizes the situation. She recognizes it as a pattern in Janet's relationship with her husband. No matter how hard she tries, the situation escalates, and Janet accepts the blame. Focus on Janet, Sander, and the main theme/problem. Level 3 interchangeable empathy.
14. *Janet:* It is my fault, isn't it?	Self-focus. Note the acceptance of individual responsibility for the difficulties. She is still at Level 1 of the CCS.
15. *Samantha:* I'm not so sure. We've talked about your problem with Sander. Now let's talk about something that went right for you in the past. As you look at your community genogram, what do you see that reminds you of good times, when things were going well?	Focus shifted from interviewer and problem to cultural/environmental/context issues and a search for positive assets and wellness strengths. Strengths and positive behaviors are helpful in understanding present situations.
16. *Janet:* (pause) Well, I loved visiting Grandma and Grandpa. They were friendly and helped me with my problems. Mom was the same way—she kept pointing out that I could do things.	Focus on family.
17. *Samantha:* Could you tell me a story about Grandma and Grandpa or your mom when they made things better for you?	Open questions focused on positive stories in the family.
18. *Janet:* Well, Mom had to work and carry the family, and I had the most fun with Grandma and Grandpa. One time kids were teasing me at school about my braces. I thought I was funny looking because I was different. They listened to me; they told me that I was beautiful and smart. They taught me to ignore the others, just try harder, and it would all work out. I found that it does— if I try harder, it usually does get better.	We learn about a supportive family. We also hear about a problem-solving style that focuses on ignoring underlying issues and trying harder—exactly what Janet is doing with Sander.
19. *Samantha:* So you learned in your family that ignoring things and trying harder usually helped work things out. Where did your desire to try so hard come from?	We see the emphasis on individual responsibility. Paraphrase with focus on family and client. Linking of past history with present situation. Level 3 interchangeable empathy.
20. *Janet:* (pauses and looks at community genogram) Dad made impossible demands on Mom. After she talked with the minister of our church, she started standing up for herself.	Focus on community, the family, and the church. Janet is beginning to draw on resources from the past; she is becoming aware of being a person-in-community. This helps her to start thinking of herself in new ways. She's now moving into Level 2, partial examination, of the CCS.

(continued)

Interviewer and Client Conversation	Process Comments
21. *Samantha:* Connections and relationships are very important to you. There is also need to care for yourself if you are to care for others. Too many women fall into the trap of always caring for others and not caring for themselves. That seems to represent the family lesson of caring.	A brief summary followed by reframing what Janet has said with a new perspective. This is a slight extension of what Janet seemed to be already saying. Reframing and interpretation tend to be best received when they are not too far from the client's present thinking. Level 4 additive empathy.
22. *Janet:* I think so. That is the reason I wouldn't be able to change. How would I do it? I've become too much of a doormat for Sander.	Janet understands the importance of caring for herself but doesn't believe she can do something about it because she has learned to behave in this way and has become a "doormat." Confronting this incongruence directly may lead to a new story, resolution of the concern, and a plan for action
23. *Samantha:* (pauses and looks at the community genogram) On the one hand, the genogram helped you recognize your family lesson of caring mostly for others, but on the other hand, the same genogram reminded you of how your mother was able to stand up for herself.	Samantha supportively challenges the client by pointing out an apparent discrepancy in her life story. She is using the model confrontation statement, "On the one hand . . . , but on the other hand . . ."
24. *Janet:* I so appreciate your pointing this out. Yes, I learned that lesson from my mother. I also learned in church that it is important to care for others, but that I can't care for others unless I care for myself. But I worry about what Sander would do if I stood up more for myself. I wonder how my mother did it?	Janet focuses on herself as a person in a family within the community. She also realizes that she may be able to stand up for herself but worries about the consequences. Her mother's experience and her spiritual background may be helpful. She's now in Level 3. Acceptance and recognition, but no change.
25. *Samantha:* That is an interesting question. Maybe we need to focus on your mom starting to care for herself, how she made a difference in her relationship with your father and the family.	The focus turns to the mother's history that may help Janet build on the past to work through current issues.
26. *Janet:* Connections with friends and family were important. Also my mother's church minister was helpful. She learned to value herself and establish a different way to interact with my father.	Janet is identifying her mother as a self-in-relation that drew on community resources for additional strength.
27. *Samantha:* Friends, family, and her minister gave her a foundation for making positive changes. I wonder how this could make a similar difference for you now?	Open question with focus on the client and the cultural/environmental/contextual in terms of maternal model and the church. Level 3 interchangeable empathy. The question seeks to move Janet to more understanding and change.

(continued)

Interviewer and Client Conversation	Process Comments
28. *Janet*: Well, for starters I realize that my mother did it, and some of my friends did it, and their relationships with their spouses changed in positive ways. I could speak with my friends and family about positive ways of standing up for myself. Also, my minister and you can help me achieve this.	Janet focuses on herself as a person in a family within the community. The interviewing session can now move to problem solving and restorying. Her spiritual background may be helpful. She is beginning to see herself as a being-in-relation. This may serve as the foundation for moving into Levels 4 and 5 of the CCS.
29. *Samantha*: Family, friends and spirituality often give us a foundation for deciding what is best. Let us explore that in more detail in the next meetings.	The focus turns to external sources that may help Janet build on the past to work through current issues. With all of this, the creation of the *New* becomes more possible.

Samantha helped Janet examine her situation in its broader context. The community genogram can be helpful in aiding clients who place too much responsibility on themselves. This can help both clients and helping professionals see the client as a being-in-relation or a person-in-community. It is a more leading approach, but remember the importance of listening carefully to what the person has to say.

As focus shifts to various dimensions of the larger client story, the client and the counselor begin to understand more fully the true complexity of the story. Janet's problem is not just "Janet's problem." Rather, her issues interact with many aspects of her past and present situation.

Change can be measured in the session, regardless of skills used. At the beginning of this session, Janet took almost total responsibility for the problems she experienced with Sander. After review and discussion of the community genogram, she seems to have moved from denial (Level 1) on the Client Change Scale (CCS) to acceptance and recognition, but no change (Level 3). She clearly has a new way of thinking and is beginning to see herself as a being-in-relation. But real and lasting creation of the *New* will require new behavior that works successfully in the relationship. The resources of understanding her mother's later life change plus her spirituality may help Janet experience further progress in her understanding and future relationship with Sander.

INSTRUCTIONAL READING: MULTIPLE CONTEXTUAL PERSPECTIVES ON CLIENT CONCERNS

Three exercises frame this chapter. The community genogram will help you see the broad cultural/environmental/contextual aspects of your client. Then, the family genogram focuses on the family. Experience with these two systems moves us beyond the individual client sitting before us in isolation. This is followed by a challenging exercise in which we ask you to examine yourself and your biases as you might work with the complex issue of abortion.

Clients bring us many stories. Most often we tend to work with only one individual story. But stories and issues of many others (e.g., friends, family, unique factors of diversity) deeply affect the client's narrative. In addition, there are many other factors that we can focus on as well if we are to help the client deal with complexity in living and personal decisions.

Nonetheless, each person we interview or counsel is totally unique. Interviewing and counseling are for the client, and learning about and focusing on that unique human being

BOX 10-1 DEVELOPING A COMMUNITY GENOGRAM

Identifying Personal and Multicultural Strengths

The community genogram is a "free-form" activity in which clients are encouraged to present their community of origin or perhaps a present or recent community, using their own unique artistic style. Through the community genogram, you can better visualize your own developmental history and that of your clients. With a positive approach, you will identify strengths that both you and your clients can use later to help them reach goals. Clients may construct a genogram by themselves or be assisted by you through questioning while you listen to their stories and strengths.

Before you start using this strategy with your clients, it is important that you first develop your own community genogram. With this understanding, you are better prepared to engage your client.

Step 1: Develop a Visual Representation of the Community

▲ Select the community in which you were primarily raised; the community of origin is where you tend to learn the most about culture. But any other community, past or present, may be used.
▲ Represent yourself or the client with a significant symbol. Use a large poster board or flipchart paper. Place yourself or the client in that community, either at the center or at another appropriate place. Encourage clients to be innovative and represent their communities in a format that appeals to them. This could include maps or unique personal constructions of the community. Remember that the two examples presented here are only suggestions. Reward creativity in unique presentations.
▲ It is important to place family or families—adoptive, single-parent, nuclear, or extended—on the paper, represented by the symbol that is most relevant for you or the client. Different cultural groups define family in varying ways.
▲ Place important, most influential groups on the community genogram, represented by distinctive visual

symbols. School, family, neighborhood, and spiritual groups are most often selected. For teens, the peer group is often particularly important. For adults, work groups and other special groups tend to become more central.
▲ You may wish to suggest relevant aspects of the RESPECTFUL model (page 46). In this way, diversity issues can be included in the genogram.

Step 2: Search for Images and Narratives of Strengths in Your Own Life or That of the Client

▲ Post the community genogram on the wall during counseling sessions. Think through the diagram yourself or help the client provide an overall summary of what he or she sees.
▲ Then focus on one single dimension of the community. Emphasize positive stories even if the client wants to start with a negative story. Do not work with the negatives until positive strengths are solidly in mind unless the client clearly needs to tell you the difficult story. (The community genogram can later be used as a useful diagnostic framework to look at the client's past challenges as they still play out today.)
▲ Help the client share one or more positive stories relating to the community dimension selected. If you are doing your own genogram, you may want to write your story in journal form.
▲ Develop at least two positive visual, auditory, or kinesthetic images from different groups within the community. It is often useful to have one positive family image, one spiritual image, and one cultural image so that several areas of wellness and support are included. Relax, close your eyes, and get that positive experience in your mind. With a meditative approach, it is possible to actually recall and experience those positives from the past. These thoughts and body feelings can serve as resources to the client in problem solving and living more effectively.

before us is the most critical and important focus. Using the person's name and the word *you* is central to every interview. But if we are to fully discover client uniqueness, we need to understand the broader context of the client. To help in this understanding we present the community genogram (Boxes 10-1 and 10-2). The family genogram illustrates the importance

BOX 10-2   THE COMMUNITY GENOGRAM: TWO VISUAL EXAMPLES

We encourage clients to generate their own visual representations of their "community of origin" and/or their current community support network. The two examples here are only samples of the possibilities. Some clients like to draw concrete pictures, while others prefer abstract symbols.

1. *The map:* The client draws a literal or metaphoric map of the community, in this case a rural setting. Note how this view of the client's background reveals a close extended family and a relatively small experiential world. The absence of friends in the map is interesting. Church is the only outside factor noted.

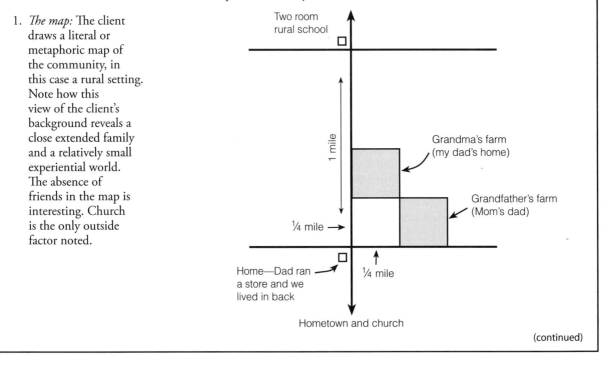

(continued)

of family (Box 10-3). Multiple focusing and these two genograms provide a framework for broader understanding and action, which takes issues beyond the individual into full recognition that challenges, issues, and problems exist in relation to others.

Let's start the process of focusing more broadly by completing a community genogram. You will gain a real understanding of the influence of your community of origin by following the suggestions of Box 10-1 and the examples in Box 10-2. With this type of personal understanding, you will be better prepared to help your clients consider themselves as persons-in-community. In addition, the community genogram provides a snapshot of the culture from which you and your clients come.

We urge you to use the community genogram as a strength and positive asset. Rather than discuss the many difficulties the client had in the original community, focus on positives and identify client strengths and resources.

We can bring broader understanding and multiple perspectives to the session by what we focus on in the client's life and social context. Part of what leads us to focus on certain issues is our own social context. Your developmental past and present issues can affect the interview. You can consciously or unconsciously avoid talking about certain

BOX 10-2 (continued)

2. *The star:* Janet's world during elementary school tells us a good bit about a difficult time in her life. Nonetheless, note the important support systems.

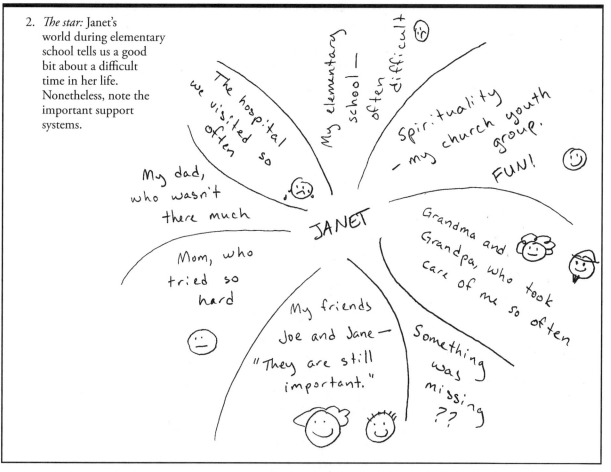

subjects that make you uncomfortable. You may do the same thing with clients. Becoming aware of your possible biases will free you to understand the uniqueness of each individual more fully.

The Family Genogram

The family genogram brings additional information about all-important family history. We frequently use both strategies with clients and often hang the family and community genograms on the wall in our office during the session, thus indicating to clients that they are not alone in the interview. Many clients find themselves comforted by our awareness of their strengths and social context.

Box 10-3 presents the major "how's" of developing a family genogram. Specific symbols and conventions have been developed that are widely accepted and help professionals communicate information to each other. The family genogram is one of the most fascinating exercises

BOX 10-3   THE FAMILY GENOGRAM

This brief overview will not make you an expert in developing or working with genograms, but it will provide a useful beginning with a helpful assessment and treatment technique. First go through this exercise using your own family; then you may want to interview another individual for practice.

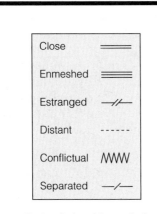

Basic relationship symbols

1. List the names of family members for at least three generations (four is preferred) with ages and dates of birth and death. List occupations, significant illnesses, and cause of death, as appropriate. Note any issues with alcoholism or drugs.
2. List important cultural/environmental/contextual issues. These may include ethnic identity, religion, economic, and social class considerations. In addition, pay special attention to significant life events such as trauma or environmental issues (e.g., divorce, economic depression, major illness).
3. Basic relationship symbols for a genogram are shown on the left, and an example of a genogram is shown below.
4. As you develop the genogram with a client, use the basic listening sequence to draw out information, thoughts, and feelings. You will find that considerable insight into one's personal life issues may be generated in this way.

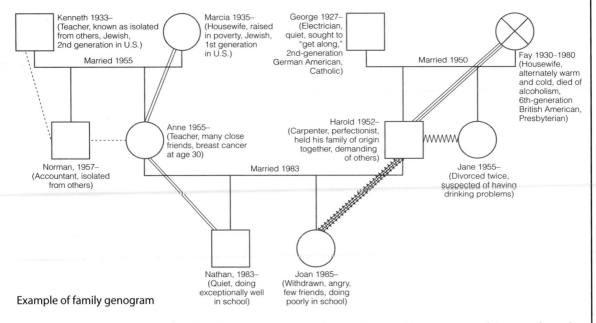

Example of family genogram

Developing a genogram with your clients and learning some of the main facts of family developmental history will often help you understand the context of individual issues. For example, what might be going on at home that results in Joan's problems at school? Why is Nathan doing so well? How might intergenerational alcoholism problems play themselves out in this family tree? What other patterns do you observe? What are the implications of the ethnic background of this family? The Jewish and Anglo backgrounds represent a bicultural history. Change the ethnic background and consider how this would impact counseling. Four-generation genograms can complicate and enrich your observations. (Note: The clients here have defined their ethnic identities as shown. Different clients will use different wording to define their ethnic identities. It is important to use the client's definitions rather than your own.)

that you can undertake. You and your client can learn much about how family history affects the way individuals behave in the here and now. The classic source for family genogram information is McGoldrick and Gerson (1985). You will find the book *Ethnicity and Family Therapy* a most valuable and enjoyable tool for helping to expand your awareness of racial/ethnic issues (McGoldrick, Giordano, & Garcia-Preto, 2005).

Many of us have important family stories that are passed down through the generations. These can be sources of strength (such as a story of a favorite grandparent or ancestor who endured hardship successfully). Family stories are real sources of pride and can be central in the positive asset search. There is a tendency among most counselors and therapists to look for problems in the family history and, of course, this is appropriate. Be sure to search for positive family stories as well as problems.

Children often enjoy the family genogram and a simple adaptation called the "family tree" makes it work for them. The children are encouraged to draw a tree and put their family members on the branches, wherever they wish.

Using Focusing to Examine Your Own Beliefs

Before you continue with your reading, take some time to think through your own thoughts on a difficult and challenging issue—abortion. It will help if you take time to write your responses to the *italicized* questions below.

As an interviewer, counselor, or psychotherapist, you will encounter controversial cases and work with clients who have made different decisions than perhaps you would. Abortion is part of what is sometimes called the "culture wars." There are deeply felt beliefs and emotions around this issue. Even the language of "pro-choice" and "pro-life" can be upsetting to some. *What is your personal position around this challenging issue?*

_____

_____

What do your family, your friends, and others close to you think about abortion? What does your community and church, both past and present, say and think? And how do your understanding of state laws and the extensive national media coverage affect your thinking? *From a more complex, contextual point of view, spend a little time thinking about what has influenced your thinking on this issue; record what you discover. Who decided? You or your family/community context?*

_____

_____

As a counselor, it is vital that you understand the situations, thoughts, and feelings of those who take varying positions around abortion or any other controversial issue, whether you agree with them or not. *Can you identify some of the thoughts and feelings of those who have a different position from your own? How do they think and feel?*

_____

_____

Counseling is not teaching clients how to live or what to believe. It is helping clients make their own decisions. Regardless of your personal position, you may find yourself using the interview to further that position. Most would agree that counselors should avoid bias in counseling. You may need to help your clients understand more than one position on abortion or recognize and deal with their conscious or unconscious sexism, racism, anti-Semitism, anti-Islamism, or other forms of intolerance. The art and mastery of effective counseling merges awareness of and respect for beliefs with unbiased probing in the interest of client self-discovery, autonomy, and growth.

Applying Focusing With a Challenging Issue

Some school systems and agencies have written policies forbidding any discussion of abortion. Further, if you are working within certain agencies (e.g., a faith-based agency or a pro- or anti-abortion counseling clinic), the agency may have specific policies regarding counseling around abortion. Ethically, clients should be made aware of specific agency beliefs before counseling begins. Again, write your answers to the questions below, but there are no necessarily "right" answers to these difficult issues.

Imagine that a client comes to you who just terminated a pregnancy. How would you help this client, who clearly needs to tell her story? Below are several issues; we ask you to think through how you would respond using different focus dimensions.

Focus on the Individual and on Significant Others

Teresa: I just had an abortion, and I feel pretty awful. The medical staff was great, and the operation went smoothly. But Cordell won't have anything to do with me, and I can't talk with my parents.

What would you say to focus on Teresa as an individual?
What could you say to focus on Cordell, the significant other?
How might you focus on the attitudes and possible supports from her friends?

_____

_____

Choosing an appropriate focus can be most challenging. Too many beginners focus only on the problem. The prompt, "Tell me more about the abortion," may result in drawing out details of the abortion, but little may be learned about the client's distinctive personal experience. An extremely important task is drawing out the client's story: "I'd like to hear *your* story" or "What do *you* want to tell me?" There are no final rules on where to focus, but generally, we want to hear the client's unique experience. Focusing on the individual is usually where to start—note where the word *you* was used in the example interviewer comments above.

Other key figures (Cordell, family, friends) are part of the larger picture. What are their stories? How do they relate to Teresa as a person-in-relation? You can more fully understand her situation when you draw out other stories or viewpoints. It is important to keep all significant others in mind in the process of problem examination and resolution.

For a full understanding of the client's experience, all pertinent relationships eventually need to be explored.

Focus on Family

Teresa: My family is quite religious, and they have always talked strongly against abortion; it makes me feel all the more guilty. I could never tell them.

How might you focus on the family in response to her statement?
How would you search for others in the family who might be helpful or supportive?

The family is where personal values and ethics are first learned. How does Teresa define "family"? There are many styles of family beyond the nuclear. African American and Hispanic clients may think of the extended family; a lesbian may see her supportive family as the gay community. Issues of single parenthood and alternative family styles continue to make the picture of the family more complex. Developing a community or family genogram may help Teresa locate resources and models that might help her. If her parents are not emotionally available, perhaps an aunt or grandmother might help.

What are your thoughts?   _____

_____

Mutuality Focus

Teresa: I feel everyone is just judging me. They all seem to be condemning me. I even feel a little frightened of you.

How would you appropriately focus on the relationship between yourself and the client?
What might you say to Teresa that focuses on the "here and now" feelings?

A mutual immediate focus often emphasizes the "we" in a here-and-now relationship. Working together in an egalitarian relationship can empower the client. Also, helping clients recognize the depth of their feelings in the here and now can be immensely valuable and powerful. "Right now at this moment, *we* have an issue." "Can *we* work together to help you?" "What are some of your thoughts and feelings about how *we* are doing?" The emphasis is on the relationship between counselor and client. Two people are working on an issue, and the interviewer accepts partial ownership of the problem.

In feminist counseling, the "we" focus may be especially appropriate: *"We* are going to solve this problem." The "we" focus provides a sharing of responsibility, which is often reassuring to the client regardless of his or her background. Many feminist counselors emphasize "we." In some counseling theories and Western cultures, emphasizing the distinction between "you" (client focus) and "me" (interviewer focus) is more common, and "we" would be considered inappropriate.

The mutual focus often includes a here-and-now dimension and brings immediacy to the session. To focus on the here and now, there are several different types of responses. "Teresa, right now you are really hurting and sad about the abortion." "I sense a lot of unsaid anger right now." There is also the classic "What are you feeling right now, at this moment?"

What are your thoughts?   _____

_____

Interviewer Focus

Teresa: What do you think about what I did? What should I do?

What would you say? An interviewer focus could be self-disclosure of feelings and thoughts or personal advice about the client or situation, "*I feel concerned and sad over what happened*"; "Right now, *I* really hurt for you, but *I* know that you have what it takes to get through this"; "*I* want to help"; or "*I*, too, had an abortion . . . *my* experience was. . . ." Opinions vary on the appropriateness of interviewer or counselor involvement, but the value and power of such statements are increasingly being recognized. They must not be overused; keep self-disclosures brief.

How might you share your own thoughts and feelings appropriately?
Would you give advice from your frame of reference? What would it be?

What are your thoughts?

_____

_____

Cultural/Environmental/Contextual Focus

Given Teresa's discussion thus far, what would you say to bring in broader cultural/environmental/contextual issues?

What are your thoughts?

_____

_____

Perhaps the most complex focus dimension is the cultural/environmental/contextual one. Some topics within these broad areas are listed here, along with possible responses to the client. A key cultural/environmental/contextual issue in discussing abortion will often be religion and spiritual orientation. Whether she is a conservative or liberal Christian, a Jew, Hindu, Muslim, or a nonbeliever, discussing the values issue from a spiritual perspective may be important.

▲ *Moral/religious issues:* "What can you draw from your spiritual background to help you?"
▲ *Legal issues:* "The topic of abortion brings up some legal issues in this state. How have you dealt with them?"
▲ *Women's issues:* "A support group for women is just starting. Would you like to attend?"
▲ *Economic issues:* "You were saying that you didn't know how to pay for the operation."
▲ *Health issues:* "How have you been eating and sleeping lately? Do you feel aftereffects?"
▲ *Educational/career issues:* "How long were you out of school/work?"
▲ *Ethnic/cultural issues:* "What is the meaning of abortion among people in your family/church/neighborhood?"

Any one of these issues, as well as many others, could be important to a client. With some clients all of these areas might need to be explored for satisfactory problem resolution. The counselor or interviewer who is able to conceptualize client issues broadly can introduce many valuable aspects of the problem or situation. Note that much of cultural/environmental/

BOX 10-4   NATIONAL AND INTERNATIONAL PERSPECTIVES ON COUNSELING SKILLS

Where to Focus: Individual, Family, or Culture?
Weijun Zhang

Case study: Carlos Reyes, a Latino student majoring in computer science, was referred to counseling by his adviser because of his recent academic difficulties and psychosomatic symptoms. The counselor was able to discern that Carlos's major concern was his increasing dislike of computer science and growing interest in literature. While he was intrigued about changing his major, he felt overwhelmed by the potential consequences for his family, in which he is the oldest of four siblings. He is also the first in his family to ever attend college. Carlos has received some limited financial support from his parents and one of his younger siblings, and the family income is barely above the poverty line. The counseling was at an impasse, for Carlos was reluctant to take any action and instead kept saying, "I don't know how to tell this to my folks. I'm sure they'll be mad at me."

During class discussion of this case, almost everyone argued that Carlos's problem is that he does not give priority to his personal career interests, that he should learn to think about what is good for his own mental health, and that he needs assertiveness training. I did not quite agree with my fellow students, who are all European Americans. I thought they were failing to see a decisive factor in the case: Carlos is Latino!

In traditional Hispanic culture, the extended family, rather than the individual, is the psychosocial unit of cooperation. The family is valued over the individual, and subordination of individual wants to the family needs is assumed. Also, traditional Hispanic families are hierarchical in form; parents are authority figures, and children are supposed to be obedient. Given this cultural background, to encourage Carlos to make a major career decision totally by himself was impossible. Any counseling effort that does not focus on the whole family is doomed to fail.

Because it is the financial support from the family that made his college education possible, Carlos may be expected to contribute to the family when he graduates. This reciprocal relationship is a lifelong expectation in Hispanic culture, and the oldest son is especially responsible in this regard. Changing his major in his junior year does not only mean he will be postponing the date when he will be able to help his family financially, but it also means he may not be able to do so at all, for we all understand how hard it is to find a job that pays well in the field of literature. When interdependence is the norm among Hispanic Americans, how can we expect Carlos to focus entirely on his personal interests without giving more weight to his family's pressing economic needs?

If I were Carlos's counselor, rather than focusing immediately on his needs, I would first support him with his family loyalty and then help him understand that there are not just two solutions: either . . . or . . . Together, we may brainstorm to generate some alternatives, such as having literature as his minor now and as his pastime after he graduates, changing his career when his younger siblings are off on their own, or exploring possibilities that may combine the two. He could, for example, design computer programs to help schoolchildren learn literature. Each of these takes into account family needs as well as those of Carlos.

The professor praised me highly for my "different and sensitive perspective," but I shrugged it off; this is just common sense to most Third World minority people and probably many Italian and Jewish Americans as well. (I remember years ago, when I was trying to make major career decisions with my parents; at least ten of my relatives were involved. And these days, I am still obligated to help anyone in my extended family who is in financial need.)

If the meaning of family in Hispanic culture is confusing to many counselors, the traditional extended family clan system of Native American Indians, Canadian Dene, or New Zealand Maori can be even more difficult for them to grasp. This family extension can include at times several households and even a whole village. Unless majority group counselors are aware of these differences in family structure, they may cause serious harm through their own ignorance.

contextual focusing requires sensitive leading and influencing from the interviewer. Box 10-4 illustrates working with diverse students.

Box 10-5 provides a summary of research related to the skill of focusing.

BOX 10-5   RESEARCH EVIDENCE THAT YOU CAN USE

Focusing

Training students to focus on cultural/environmental/contextual issues resulted in greater awareness and willingness to discuss racial and gender differences early in the session and to make these issues a consistent part of the interview (Zalaquett, 2008). Moos (2001) has reviewed much of the contextual literature and points out that the way we appraise a situation can be self-centered or environmental/contextual. Clients often come to the interview with a focus that may work against their own best interest. Too much of an "I" focus may result in self-blame and lack of awareness of context. In contrast, too much of a "they" focus may mean that clients are avoiding responsibility or their part in the conflict. As you know, there are two or more stories as people look at the same event. The extensive research on attribution theory further highlights this point (cf. Harvey & Manusov, 2001). By focusing on multiple dimensions, you can help the client find a suitable balance.

Work on training students in multicultural counseling provides an excellent research model for the future (Torres-Rivera et al., 2001). The researchers used a set of multicultural skills training videos prepared by Arredondo et al. (2000). They showed these videotapes of culture-specific counseling to their students and found that the students' multicultural effectiveness and understanding increased. Moos (2001) noted that teaching clients the context of their issues helps them understand themselves in new ways and "makes possible a transformative experience."

Focus and Neuroscience

The skill of focusing is closely related to selective attention (see Chapter 3). Client selective attention is guided by existing patterns in the mind (Allport, 1998), and focusing is an intentional interviewing skill that can open up more possibilities for client thoughts, feelings, and actions. A number of regions of the prefrontal cortex "are activated selectively during different aspects of attentional task preparation and execution" (Allport, 1998). Self-regulation and understanding of others (empathy) is also deeply affected by attentional systems in the brain (Decety & Jackson, 2004).

ADVOCACY AND SOCIAL JUSTICE

What Is the Interviewer's Role in Advocacy and Social Justice?

You are going to face situations when your best counseling efforts are insufficient to help your client resolve their issues and move on with their lives. The social context of homelessness, poverty, racism, sexism, and other contextual issues may leave your client in an impossible situation. The problem may be bullying on the playground, an unfair teacher, or an employer who refuses to follow fair employment practices. Helping clients resolve issues is much more challenging when we examine the societal stressors that they may face.

Advocacy is speaking out for your clients; working in the school, community, or larger setting to help clients; and also working for social change. What are you going to do on a daily basis to help improve the systems within which your clients live? Here are some examples showing that simply talking with clients about their issues may not be enough:

▲ As an elementary school counselor, you counsel a child who is being bullied on the playground.
▲ You are a high school counselor and work with a 10th grader who is teased and harassed about being gay, while the classroom teacher quietly watches and says nothing.
▲ As a personnel officer, you discover systematic bias against promotion for women and minorities.
▲ Working in a community agency, a client speaks of abuse in the home but fears leaving because she sees no future financial support.

▲ You are working with an African American client who has dangerous hypertension. You know that there is solid evidence that racism influences blood pressure.

What are your thoughts?

_____

_____

The elementary counselor can work with school officials to set up policies around bullying and harassment, actively changing the environment that allowed bullying to occur. The high school counselor faces an especially challenging issue as interview confidentiality may preclude immediate classroom action. If this is not possible, then the counselor can initiate school policies and awareness programs against oppression in the classroom. The passive teacher may become more aware through training you offer to all the teachers. You can help the African American client understand that hypertension is not just "his problem," but rather his blood pressure is partially related to racism in his environment, and you can work to eliminate oppression in your community.

"Whistle-blowers" who name problems that others like to avoid can face real difficulty. The company may not want to have their systematic bias exposed. On the other hand, through careful consultation and data gathering, the human relations staff may be able to help managers develop a more fair, honest, and equitable style. Again, the issue of policy becomes important. Counselors can advocate policy changes in work settings and equal pay for equal work. You can help the client who suffers racial, gender, or sexual orientation harassment. You can speak to employers about how they can employ more people with handicaps.

The counselor in the community agency knows that advocacy is the only possibility when a client is being abused. For clients in such situations, advocacy in terms of support in getting out of the home, finding new housing, and learning how to obtain a restraining order against the abusing person may be far more important than self-examination and understanding.

Counselors who care about their clients also act as advocates for them when necessary. They are willing to move out of the counseling office and seek social change. You may work with others on a specific cause or issue to facilitate general human development and wellness (e.g., prenatal care, child care, fair housing, the homeless, athletic fields for low-income areas). These efforts require you to speak out, to develop skills with the media, and to learn about legal issues. *Ethical witnessing* moves beyond working with victims of injustice to the deepest level of advocacy (Ishiyama, 2006). Counseling, social work, and human relations are inherently social justice professions. Speaking out for social concerns needs our time and attention.

SUMMARY: BEING-IN-RELATION, BECOMING A PERSON-IN-COMMUNITY

Focusing helps the client see issues and concerns in a broader setting. While the "I" focus remains central, we are also *beings-in-relation*. We must start, of course, by focusing on the special individual before us. Interviewing and counseling are for the person. However, by focusing on various dimensions, we can help clients expand their horizons. Connection and interdependence are as important for mental health as are independence and autonomy.

The community genogram places the client in connection with the family in a cultural context. Rather than focusing on the many possible negatives in our communities of origin, the genogram reveals that these are places where we also learned strengths. The imaging of stories about community and cultural strengths can be an important resource in interviewing and counseling.

We recommend that you consider developing family and community genograms with your clients and placing them on the wall throughout the counseling series. In this way, both you and the client are reminded of the self-in-relation and the need to take multiple perspectives on any issue.

It is not necessary to generate the genograms with every client. What is most important is to be aware that for any client issue, there may be multiple explanations and multiple new stories.

Key points of Chapter 10 are summarized below.

■ KEY POINTS

Draw out stories with multiple focusing	Client stories and issues have many dimensions. It is tempting to accept problems as presented and to oversimplify the complexity of life. Focusing helps interviewer and client to develop an awareness of the many factors related to an issue as well as to organize thinking. Focusing can help a confused client zero in on important dimensions. Thus, focusing can be used to either open or tighten discussion.
Seven focus dimensions	There are seven types of focuses. The one you select determines what the client is likely to talk about next, but each offers considerable room for further examination of client issues. As a counselor or interviewer, you could say many things, including the following:

    ▲ Focus on client: "Tari, you were saying last time that you are concerned about your future. . . ."

    ▲ Focus on the main theme or problem: "Tell me more about your getting fired. What happened specifically?"

    ▲ Focus on others: "So you didn't get along with the sales manager. I'd like to know a little more about him. . . ."

    ▲ Focus on family: "How supportive has your family been?"

    ▲ Focus on mutual issues or group: "We will work on this. How can you and I (our group) work together most effectively?"

    ▲ Focus on interviewer: "My experience with difficult supervisors was . . ."

    ▲ Focus on cultural/environmental/contextual issues: "It's a time of high unemployment. Given that, what issues will be important to you as a woman seeking a job?"

Focusing and other skills	Focusing can be consciously added to the basic microskills of attending, questioning, paraphrasing, and so on. Careful observation of clients will lead to the most appropriate focus. In assessment and problem definition it is often helpful to consciously and deliberately assist the client to explore issues by focusing on all dimensions, one at a time. Advocacy and social action may be necessary when you discover that the client's issues cannot be resolved through the interview alone. Counseling could be described as a social justice profession.
Multicultural issues	Focusing will be useful with all clients. With most clients the goal is often to help them focus on themselves (client focus), but with many other people, particularly those of a Southern European or African American background, the family and community focuses may at times be more appropriate. The goal of much North American counseling and therapy is individual self-actualization, whereas among other cultures it may be the development of harmony with others—self-in-relation.

    Deliberate focusing is especially helpful in problem definition and assessment, where the full complexity of the problem is brought to light. Moving from focus to

(continued)

Key Points (continued)

	focus can help increase your clients' cognitive complexity and their awareness of the many interconnecting issues in making important decisions. With some clients who may be scattered in their thinking, a single focus may be wise.
The importance of the individualistic "I" focus	Recall that counseling is for the client. Though expanding awareness of context and self-in-relation and understanding alternative stories of a situation are obviously useful, ultimately the unique client before you will be making decisions and acting. The bottom line is to assist that client in writing her or his own new story and plan of action.

COMPETENCY PRACTICE EXERCISES AND PORTFOLIO OF COMPETENCE

This chapter has presented several practice exercises within the instructional reading; therefore, the number of exercises in this section will be reduced to two, followed by the usual self-assessment section.

Individual Practice

Exercise 1: Writing Alternative Focus Statements

A 35-year-old client comes to you to talk about an impending divorce hearing. He says the following:

> I'm really lost right now. I can't get along with Elle, and I miss the kids terribly. My lawyer is demanding an arm and a leg for his fee, and I don't feel I can trust him. I resent what has happened over the years, and my work with a men's group at the church has helped, but only a bit. How can I get through the next 2 weeks?

Fill in the client's main issue as you see it; write several alternative focus statements in the spaces that follow. Be sure to brainstorm a number of cultural/environmental/contextual possibilities.

Main issue as presented:

Client focus _____

Problem/main theme focus _____

Others focus _____

Family focus _____

Mutual, group, "we" focus _____

Interviewer focus _____

Cultural/environmental/contextual focus _____

Now write alternative focus statements, as indicated:

Reflection of feeling focusing on the client _____

Open question focusing on the problem/main theme _____

Closed question focusing on others _____

Open question focusing on the family _____

Reassurance statement focusing on "we" _____

Self-disclosure statement focusing on yourself, the interviewer _____

Paraphrase focusing on a cultural/environmental/contextual issue _____

An imaginary confrontative summary (assuming a longer interview) in which you demonstrate a mixed focus, pointing out a major client discrepancy

_____

Exercise 2: Developing a Family Genogram

Using information from the chapter and the illustrations in Box 10-3, develop a family genogram with a volunteer client or classmate. After you have created the genogram, ask the client the following questions and note the impact of each. Change the wording and the sequence to fit the needs and interests of the volunteer.

What does this genogram mean to you? (individual focus)

As you view your family genogram, what main theme, problem, or set of issues stands out? (main theme, problem focus)

Who are some significant others, such as friends, neighbors, teachers, or even enemies who may have affected your own development and your family's? (others focus)

How would other members of your family interpret this genogram? (family, others focus)

What impact do your ethnicity, race, religion, and other cultural/environmental/contextual factors have on your own development and your family's? (C/E/C focus)

What I have learned as an interviewer working with you on this genogram is (state your own observations). How do you react to my observations? (interviewer focus)

Summarize in journal form some of what you've learned from this exercise. What questions did you find most helpful?

_____

Exercise 3: Developing a Community Genogram

Pages 273–275 present specific step-by-step instructions for developing a community genogram. For your first effort, we suggest that you use your own personal experience in your community of origin. As an alternative, work through the process with a class member and each of you practice drawing out the data. Present the completed genogram and briefly summarize what you learned.

Group Practice

Exercise 4: Practice Focusing the Interview in a Group

Step 1: Divide into groups.

Step 2: Select a group leader.

Step 3: Assign roles for the first practice session.

▲ Client, who has completed a family or community genogram.
▲ Interviewer.
▲ Observer 1, who will give special attention to focus of the client, using the Feedback Form, Box 10-6. The key microsupervision issue is to help the interviewer continue a central focus on the client while simultaneously developing a comprehensive picture of the client's contextual world.
▲ Observer 2, who will give special attention to focus of the interviewer, using the Feedback Form, Box 10-6.

Step 4: Plan. Establish clear goals for the session. The task of the interviewer in this case is to go through all seven types of focus, systematically outlining the client's issue. If the task is completed successfully, a broader outline of client issues should be available.

A useful topic for this role-play is a story from your family or community. Your goal here is to help the client see the issues in broader perspective.

Observers should take this time to examine the feedback form and plan their own interviews. The client may fill out the Client Feedback Form of Chapter 1.

Step 5: Conduct a 5-minute practice session using the focusing skill.

Step 6: Review the practice session and provide feedback for 10 minutes. Give special attention to the interviewer's achievement of goals and determine the mastery competencies demonstrated.

Step 7: Rotate roles.

Some general reminders. Be sure to cover all types of focus; many practice sessions explore only the first three. In some practice sessions three members of the group all talk with the same client, and each interviewer uses a different focus.

Portfolio of Competence

The history of counseling and therapy has provided the field with a primary "I" focus in which the client is considered and treated within a totally individualistic framework. The microskill of focusing is key to the future of interviewing, counseling, and psychotherapy as it broadens the way both interviewers and clients think about the world. This does not deny the importance of the "I" focus. Rather, the multiple narratives made possible by the use of microskills actually strengthen the individual, for we all live as selves-in-relation. We are not alone. The collective strengthens the individual.

At the same time, the above paragraph represents a critical theoretical point. Some might disagree with the emphasis of this chapter and argue that only the individual and problem focus are appropriate. What are your thoughts and feelings on this important point? As you work through this list of competencies, think ahead to how you would include or adapt these ideas in your own Portfolio of Competence.

BOX 10-6   FEEDBACK FORM: FOCUS

_____ (Date)

_____   _____

(Name of Interviewer)   (Name of Person Completing Form)

Instructions: Observer 1 will give special attention to the client and Observer 2 to the interviewer. Note the correspondence between interviewer and client statements. In the space provided, record the main words used. Classify each statement by checking a box.

Main words	Client							Interviewer						
	Client (self)	Concern/problem	Significant others	Family	Mutual "we"	Interviewer	Cultural/environmental/contextual	Client	Concern/problem	Significant others	Family	Mutual "we"	Interviewer (self)	Cultural/environmental/contextual
1. _____														
2. _____														
3. _____														
4. _____														
5. _____														
6. _____														
7. _____														
8. _____														
9. _____														
10. _____														
11. _____														
12. _____														
13. _____														
14. _____														

(continued)

BOX 10-6   (continued)

Observations about client verbal and nonverbal behavior:

_____

_____

Observations about inverviewer verbal and nonverbal behavior:

_____

_____

_____

Use the following as a checklist to evaluate your present level of mastery. As you review the items below, ask yourself, "Can I do this?" Check those dimensions that you currently feel able to do. Those that remain unchecked can serve as future goals. Do not expect to attain intentional competence on every dimension as you work through this book. You will find, however, that you will improve your competencies with repetition and practice.

Level 1: Identification and classification. You will be able to identify seven types of focus as interviewers and clients demonstrate them. You will note their impact on the conversational flow of the interview.

❑   Identify focus statements of the interviewer.
❑   Note the impact of focus statements in terms of client conversational flow.
❑   Write alternative focus responses to a single client statement.

Level 2: Basic competence. You will be able to use the seven focus types in a role-play interview and in your daily life.

❑   Demonstrate use of focus types in a role-play interview and draw out multiple stories.
❑   Use focusing in daily life situations.

Level 3: Intentional competence. Use the seven types of focus in the interview, and clients will change the direction of their conversation as you change focus. Maintain the same focus as your client, if you choose (that is, do not jump from topic to topic). Combine this skill with earlier skills (such as reflection of feeling and questioning) and use each skill with alternative focuses. Check those skills you have mastered and provide evidence via actual interview documentation (transcripts, tapes).

❑   Clients tell multiple stories about their issues.
❑   Maintain the same focus as my clients.
❑   During the interview, observe focus changes in the client's conversation and change the focus back to the original one if it is beneficial to the client.
❑   Combine this skill with skills learned earlier; use focusing together with confrontation to expand client development.
❑   Use multiple-focus strategies for complex issues facing a client.

Level 4: Psychoeducational teaching competence. Some clients will have a primary "I" focus as they discuss their issues. By learning the skills of multiple focus dimensions, they can discover important new ways to view their issues and become more amenable to change. Those clients who are more externally directed may avoid the "I" focus as they blame others. They also can benefit from learning their responsibility in the larger system. The impact of teaching is measured by the achievement of students, given the preceding criteria.

❑   Teach focusing to clients and small groups.

DETERMINING YOUR OWN STYLE AND THEORY: CRITICAL SELF-REFLECTION ON FOCUSING

What single idea stood out for you among all those presented in this chapter, in class, or through informal learning? What stands out for you is likely to be important as a guide toward your next steps. What are your thoughts on multicultural issues and the use of the focusing skill? What other points in this chapter struck you as important? How might you use ideas in this chapter to begin the process of establishing your own style and theory? What are your thoughts and experiences with the community and family genograms?

What are your thoughts?

_____

_____

OUR THOUGHTS ABOUT VANESSA

There are many clients like Vanessa, who have numerous issues in their lives. We, perhaps like you, often feel overwhelmed when we get 3 to 5 nonstop minutes of stories about different complex problems. Sometimes there is an insistence that we do "something" immediately and start solving the issues. We have noted that when we have fallen into solving problems for clients they often refuse to listen to us and generate still more problems and difficulties.

So what needs to be done here? Each client, of course, is unique, and there is no magic answer. But one rule really helps us settle down and start working with the client. Counseling is for the individual client. Our responsibility is to focus on what we can see and work with—specifically, the unique human being in front of us. We can't see the family, we can't see the boyfriend, and we can't study for the client.

Thus, our first efforts are to focus on the individual client. Although this chapter emphasizes a broadly based contextual perspective, we always want to recognize that counseling is most effective when we focus on the client as the core of our interviewing and treatment plan.

Most likely, with Vanessa, we would listen to her for no more than 3 to 5 minutes— and there are clients who will engage in nonstop talking for that time and longer. At some point, we would likely interrupt and say something like this:

> Vanessa, could we stop for a moment? I really hear you loud and clear. There are a lot of things happening right now. One of the best ways to approach these issues is to focus on how you are feeling and what's happening with you. Once I understand you a bit more, we can move to working on the issues that you describe.

I get the sense that you, Vanessa, are hurting a lot right now and are confused as to what to do next. Could you take a deep breath and tell me what you are feeling and thinking this very minute—right now? What are these things doing to Vanessa? What's happening with you?

In these two paragraphs, we have used the words *Vanessa* and *you* a total of 12 times. The goal here is to help Vanessa focus on herself. If you go back to the beginning of the chapter, you will see that the primary focus of conversation has been on others and problems. We have heard very little about Vanessa, the person.

Once we have a better grasp of the person before us, we can work more effectively toward problem resolution.

Another possibility is to summarize what you have heard the client say and then ask her or him to name the one thing that he or she would like to start working on first. You can comment that we can explore other issues later.

While this chapter aims to broaden the view you can take of the client through multiperspective thought and bring in many ways to enlarge the conception of an issue, counseling is still for the individual person. We need to focus our efforts on the unique individual.

CHAPTER 11

Reflection of Meaning and Interpretation/ Reframing: Helping Clients Restory Their Lives

Pyramid (bottom to top):

- Ethics, Multicultural Competence, and Wellness
- Attending Behavior
- Open and Closed Questions
- Client Observation Skills
- Encouraging, Paraphrasing, and Summarizing
- Reflection of Feeling
- The Five-Stage Interview Structure
- Confrontation
- Focusing
- **Reflection of Meaning and Interpretation/Reframe**

Then I spoke of the many opportunities of giving life a meaning. I told my comrades . . . that human life, under any circumstances has meaning. . . . I said that someone looks down on each of us in difficult hours—a friend, a wife, somebody alive or dead, or a God—and He would not expect us to disappoint him. . . . I saw the miserable figures of my friends limping toward me to thank me with tears in their eyes.

—Viktor Frankl (who helped many Jews survive and find meaning
while imprisoned at Auschwitz during the Holocaust)

How can reflection of meaning and interpretation/reframing help you and your clients?

Chapter Goals   The goal of reflection of meaning is to facilitate clients in finding deeper meanings and values that provide a guiding sense of vision and direction for their lives. The goal of interpretation/reframing is to provide a new way of restorying and understanding thoughts, feelings, and behaviors, which often results in new ways of making meaning. Clients usually generate their own meanings, whereas interpretations/reframes

---

This chapter is dedicated to Viktor Frankl. The initial stimulus for the skill of reflection of meaning came from a 2-hour meeting with him in Vienna shortly after we visited the German concentration camp at Auschwitz, where he had been imprisoned in World War II. He impressed on us the central value of meaning in counseling and therapy—a topic to which most theories give insufficient attention. It was Frankl's unusual ability to find positive meaning in the face of impossible trauma that impressed us most. His thoughts also impacted our wellness and positive strengths orientation. His theoretical and practical approach to counseling and therapy deserves far more attention than it receives. We have often recommended his powerful book, *Man's Search for Meaning* (1959) to clients who face serious life crises.

usually come from the interviewer. Interpretation often comes from a specific theoretical perspective such as psychodynamic counseling, rational-emotive behavioral therapy, or multicultural counseling and therapy. Reframing, a kind of interpretation, tends to be more in the moment and enables clients to see new ways of looking at their situations more positively. Both skills are central in restorying, creating the *New,* and developing new neural networks in the brain.

Competency Objectives Awareness, knowledge, and skill in reflection of meaning and interpretation/reframing will enable you to

▲ Understand reflection of meaning and interpretation/reframing and their similarities and differences.
▲ Assist clients, through reflection of meaning, to explore their deeper meanings and values and to discern their goals or life purpose. Important in this is the process of discerning life's mission.
▲ Realize the power of perceptions. The way you perceive things affects how you feel and behave.
▲ Help clients, through interpretation/reframing, find an alternative frame of reference or way of thinking that facilitates personal development.
▲ Understand how these skills bring about change that can be measured on the Client Change Scale.

Charlis, a workaholic 45-year-old middle manager, has a heart attack. After several days of intensive care, she is moved to the floor where you, as the hospital social worker, work with the heart attack aftercare team. Charlis is motivated; she is following physician directives and progressing as rapidly as possible. She listens carefully to diet and exercise suggestions and seems to be the ideal patient with an excellent prognosis. However, she wants to return to her high-pressure job and continue moving up through the company; you observe that she feels some fear and puzzlement about what has happened.

INTRODUCTION: DEFINING THE SKILLS OF REFLECTING MEANING AND INTERPRETATION/REFRAMING

Both interpretation/reframing and reflection of meaning seek implicit issues and meanings below the surface of client conversation. Reflecting meaning is the art of encouraging the client to find the *New* on her or his own through your in-depth listening for deeper issues and visions. Interpretation/reframing is the art of supplying the client with new perspectives, words, and ideas that can be used to create the *New.* From these two definitions, you can see the logic of presenting the two closely related skills together. Figure 11-1 illustrates the centrality of meaning. Both interpretation/reframing and the reflection of meaning may help clients find their "center of being."

Eliciting and reflecting meaning are both skills and strategies. As skills, these are fairly straightforward. To elicit meaning, ask the client some variation of the basic question,

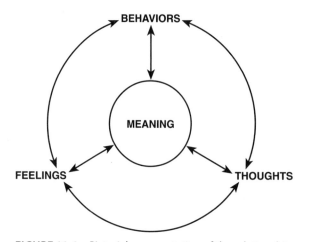

FIGURE 11-1   Pictorial representation of the relationships among behaviors, thoughts, feelings, and meanings.

"What does . . . *mean* to you, your past, or future life?" At the same time, effective exploration of meaning becomes a major strategy in which you bring out client stories, *past, present,* and *future.* You use all the listening, focusing, and confrontation skills to facilitate this self-examination, yet the focus remains on the client's meaning and finding purpose in his or her life.

Interpretations and reframes vary with theoretical orientation, and the joint term *interpretation/reframing* is used because both skills focus on providing a new way of thinking or a new frame of reference for the client, but the word "reframe" is a gentler construct that usually comes from your here-and-now observations. Keep in mind when you use influencing skills that interpretive statements are more directive than reflecting meaning. When we use interpretation/reframing we are working primarily from the interviewer's frame of reference. This is neither good nor bad; rather, it is something we need to be aware of when we use influencing skills.

If you use reflection of meaning and interpretation/reframing skills as defined here, you can *predict* how clients will respond

Reflection of Meaning	Predicted Result
Meanings are close to core experiencing. Encourage clients to explore their own meanings and values in more depth from their own perspective. Questions to elicit meaning are often a vital first step. A reflection of meaning looks very much like a paraphrase but focuses on going beyond what the client says. Often the words "meaning," "values," "vision," and "goals" appear in the discussion.	The client will discuss stories, issues, and concerns in more depth with a special emphasis on deeper meanings, values, and understandings. Clients may be enabled to discern their life goals and vision for the future.

(continued)

Interpretation/Reframe	Predicted Result
Interpretation and reframing can provide the client with a new meaning or perspective, frame of reference, or way of thinking about issues. Interpretations/reframes may come from your observations; they may be based on varying theoretical orientations to the helping field; or they may link critical ideas together.	The client will find another perspective or meaning of a story, issue, or problem. The new perspective could be generated by a theory used by the interviewer, from linking ideas or information, or by simply looking at the situation afresh.

The case of Charlis presented above will serve as a way to illustrate similarities and differences between reflection of meaning and interpretation.

Reflection of Meaning

You recognize that Charlis is reevaluating the meaning of her life. She asks questions that are hard to answer—"Why me? What is the meaning of my life? What is God saying to me? Am I on the wrong track? What should I *really* be doing?" You sense that she feels something is missing in her life, and she also wants to reevaluate where she is going and what she is doing. How might you help Charlis? What thoughts occur to you? What do you see as the key issues that relate to the meaning and purpose of her life?

To elicit meaning, we may ask Charlis some variation of a basic meaning question, "What does the heart attack *mean* to you, your past, and your future life?" We may also ask Charlis if she would like to examine the meaning of her life through the process of *discernment,* a more systematic approach to meaning and purpose defined in some detail in this chapter. If she wishes, we could share the specific questions of discernment presented there and ask her which areas she'd like to explore. In addition, we'd ask her to think of questions and issues that are particularly important to her as we work to help her discern the meaning of her life, her work, her goals, and her mission. These questions often bring out emotions, and they certainly bring out meaning in the client's thoughts and cognitions. When clients explore meaning issues, the interview becomes less precise as the client struggles with defining the almost indefinable. As appropriate to the situation, questions such as the following can address the general issue of meaning in more detail:

"What has given you most satisfaction in your job?"
"What's *been missing* for you in your present life?"
"What do you *value* in your life?"
"What *sense* do you make of this heart attack and the future?"
"What things in the future will be most *meaningful* to you?"
"What is the *purpose* of your working so hard?"
"You've said that you wonder what God is saying to you with this trial. Could you share some of your thoughts?"
"What gift would you like to leave the world?"

Eliciting meaning often precedes reflection. Reflection of meaning as a skill looks very much like a reflection of feeling or paraphrase, but the key words "meaning," "sense," "deeper understanding," "purpose," "vision," or some related concept will be present explicitly or implicitly. "Charlis, I sense that the heart attack has led you to question some basic understandings in your life. Is that close? If so, tell me more." Eliciting and reflecting meaning is an *opening* for the client to explore issues for which there is not a final answer but rather a deeper awareness of the possibilities of life. Both reflecting meaning and interpretation/reframing are designed to help clients look deeper, first by careful listening and then by helping clients examine themselves from a new perspective.

Reflecting meaning involves *client* direction; the interpretation/reframe implies *interviewer* direction. The client provides the new and more comprehensive perspective in reflection of meaning, while an interpretation/reframe supplies the new way of being as suggested by the interviewer or counselor.

Comparing Reflection of Meaning and Interpretation/Reframing

Here are brief examples of how reflection of meaning and interpretation may work for Charlis as she attempts to understand some underlying issues around her heart attack.

Charlis: My job has been so challenging, and I really feel that pressure all the time, but I just ignored it. I'm wondering why I didn't figure out what was going on until I got this heart attack. But I just kept going on, no matter what.

Eliciting and Reflection of Meaning

Counselor: I hear you—you just kept going. Could you share what it feels like *to keep going on* and what it *means* to you? (encourager focusing on the key words *"keep going on"*; open question oriented to meaning)

Charlis: I was raised to keep going. My mother always prided herself on doing a good job, even in the worst of times. Grandma did the same thing.

Counselor: Charlis, I hear that keeping going and persistence have been a key family value that remains very important to you. (reflection of meaning) "Hanging in" is what you are good at. (Positive asset leading to wellness is mentioned.) Could we focus now on how that value around persistence and *keeping going on* relates to your rehabilitation? (open question that seeks to use the wellness dimensions to help her plan for the future)

Interpretation/Reframe

Counselor: It sounds very much like you have learned to do "the right thing," but right now it isn't working. But I also hear that you value the strengths from your family. How does that sound to you? (mild positive reframe followed by check out)

Charlis: Yes, I saw my mother and my grandmother do this through all kinds of hardships.

Counselor: Many of us become who we are because of family history. It sounds as if several generations have taught you to struggle and *keep going on, no matter what.* Do you want to continue that tradition? Or could you use *keeping going on* in a more positive way? (This interpretation/reframe comes more from a multicultural/family therapy theoretical perspective, closed question, open question.)

Both reflection of meaning and the reframing of the situation ended up in nearly the same place, but Charlis is more in control of the process with reflection of meaning. Whichever approach is used, we are closer to helping Charlis work on the difficult questions of the meaning and direction of her future life. If the client does not respond to reflective strategies, move to the more active reframing or a theoretical interpretation. We need to give clients power and control of the session whenever possible. They can often generate new interpretations/reframes and new ways of thinking about their issues.

Linking is an important part of interpretation, although it often appears in an effective reflection of meaning as well. In linking, two or more ideas are brought together, providing the client with a new insight. The insight comes primarily from the client in reflection of meaning, but almost all from the interviewer in interpretation/reframing. Consider the following four examples:

Reframe/interpretation 1: Charlis, what stands out to me at this moment is how able you are, and we can use your "smarts" and ability to understand situations to find new, more comfortable directions. (Reframe/interpretation with a very positive spin; positive reframes in the here and now are often the most useful. Links strengths to problem.)

Interpretation/reframe 2: Charlis, you seem to have a pattern of thinking that goes back a long way—we could call it an "automatic thought." You seem to have a bit of perfectionism there, and you keep saying to yourself (self-talk), "Keep going no matter what." (cognitive-behavioral theory; links the past to the present perfectionism)

Interpretation/reframe 3. It sounds as if you are using hard work as a way to avoid looking at yourself. The avoidance is similar to the way you avoid dealing what you think you need to change in the future to keep yourself healthier. (combines confrontation with linking to what is occurring in the interview series; this linking pushes her to take care of herself)

Interpretation/reframe 4. The heart attack almost sounds like unconscious self-punishment, as if you wanted it to happen to give you time off from the job and a chance to reassess your life. (interpretation from a psychodynamic perspective focusing first on negative issues [self-punishment], then on new opportunities that are now open [reassess your life])

In summary, a reflection of meaning looks very much like a paraphrase but focuses beyond what the client says. Often the words "meaning," "values," and "goals" will appear in the discussion. Clients are encouraged to explore their own meanings in more depth from their own perspective. Questioning and eliciting meaning are often vital as first steps.

Interpretations/reframes provide the client with a new perspective, frame of reference, or way of thinking about issues. They may come from observations of the counselor, they may be based on varying theoretical orientations to the helping field, or they may link critical ideas together.

The two skills are similar in helping clients generate a new and potentially more helpful way of looking at things. Reflection of meaning focuses on the client's worldview and seeks to understand what motivates the client; it provides more clarity on values and deeper life meanings. An interpretation/reframe results from interviewer observation and seeks new and more useful ways of thinking.

EXAMPLE INTERVIEW: TRAVIS EXPLORES THE MEANING OF A RECENT DIVORCE

In the following session, Travis is reflecting on his recent divorce. When relationships end, the thoughts, feelings, and underlying meaning of the other person and the time together often remain an unsolved mystery. Moreover, some clients are likely to repeat the same mistakes in their relationships when they meet a new person.

However, both the interpretation/reframe and reflection of meaning are central skills in helping clients take a new perspective on themselves and their world. Terrell, the interviewer, seeks to help Travis think about the word "relationship" and its meaning. Note that Travis stresses the importance of connectedness with intimacy and caring. The issue of self-in-relation to others will play itself out very differently among individuals in varying cultural contexts. Many clients will focus on their need for independence.

Interviewer and Client Conversation	Process Comments
1. *Terrell:* So, Travis, you're thinking about the divorce again. . . .	Encourager/restatement.
2. *Travis:* Yeah, that divorce has really thrown me for a loop. I really cared a lot about Ashley and . . . ah . . . we got along well together. But there was something missing.	Level 2 on Client Change Scale. Travis partially understands the issue but wants to know more.
3. *Terrell:* Uh-huh . . . something missing?	Encouragers appear to be closely related to meaning. Clients often supply the meaning of their key words if you repeat them back exactly.
4. *Travis:* Uh-huh, we just never really shared something very basic. The relationship didn't have enough depth to go anywhere. We liked each other, we amused one another, but beyond that . . . I don't know. . . .	Travis elaborates on the meaning of a closer, more significant relationship than he had with Ashley.
5. *Terrell:* You amused each other, but you wanted more depth. What sense do you make of it?	Paraphrase using Travis's key words followed by a question to elicit meaning.
6. *Travis:* Well, in a way, it seems like the relationship was shallow. When we got married, there just wasn't enough depth for a meaningful relationship. The sex was good, but after a while I even got bored with that. We just didn't talk much. I needed more. . . .	Note that Travis's personal constructs for discussing his past relationship center on the word *shallow* and the contrast *meaningful*. This polarity is probably one of Travis's significant meanings around which he organizes much of his experience.
7. *Terrell:* Mm-hmmm . . . you seem to be talking in terms of shallow versus meaningful relationships. What does a meaningful relationship feel like to you?	Reflection of meaning followed by a question designed to elicit further exploration of meaning.
8. *Travis:* Well, I guess . . . ah . . . that's a good question. I guess for me, there has to be some real, you know, some real caring beyond just on a daily basis. It has to be something that goes right to the soul. You know, you're really connected to your partner in a very powerful way.	Connection appears to be a central dimension of meaning.
9. *Terrell:* So connections, soul, and deeper aspects strike you as really important.	Reflection of meaning. Note that this reflection is also very close to a paraphrase, and Terrell uses Travis's main words. The distinction centers on issues of meaning. A reflection of meaning could be described as a special type of paraphrase.

(continued)

Interviewer and Client Conversation	Process Comments
10. *Travis:* That's right. There has to be some reason for me to really want to stay married, and I think with her . . . ah . . . those connections and that depth were missing. We liked each other, you know, but when one of us was gone, it just didn't seem to matter whether we were here or there.	Here we are beginning to see movement on the Client Change Scale. Travis appears to be thinking at Level 3. He is understanding his situation fairly well, but we see no real change.
11. *Terrell:* So there are some really good feelings about a meaningful relationship even when the other person is not there. You didn't value each other that much.	Reflection of meaning plus some reflection of feeling. Note that Terrell has added the word *values* to the discussion. In reflection of meaning it is likely that the counselor or interviewer will add words such as "meaning," "understanding," "sense," and "value." Such words lead the client to make sense of experience from the client's own frame of reference.
12. *Travis:* Uh-huh.	
13. *Terrell:* Ah . . . could you fantasize how you might play out those thoughts, feelings, and meanings in another relationship?	Open question oriented to meaning.
14. *Travis:* Well, I guess it's important for me to have some independence from a person, but when we were apart, we'd still be thinking of one another. Depth and a soul mate is what I want.	Travis's meaning and desire for a relationship are now being more fully explored. Travis moves a bit further as he defines his life goals more precisely. This could be called a "Level 3+" response.
15. *Terrell:* Mm-hmmm.	
16. *Travis:* In other words, I don't want a relationship where we always tag along together. The opposite of that is where you don't care enough whether you are together or not. That isn't intimate enough. I really want intimacy in a marriage. My fantasy is to have a very independent partner I care about and who cares about me. We can both be individuals but still have bonding and connectedness.	Connectedness is an important meaning issue for Travis. With other clients, independence and autonomy may be the issue. With still others, the meaning in a relationship may be a balance of the two. We see further progress on the Client Change Scale, but Level 4 (generation of a new solution) clearly has not been achieved yet.
17. *Terrell:* Let's see if I can put together what you're saying. The key words seem to be independence with intimacy and caring. It's these concepts that can produce bonding and connectedness, as you say, whether you are together or not.	This reflection of meaning becomes almost a summarization of meaning. Note that the key words and constructs have come from the client in response to questions about meaning and value. Level 4 change will occur when Travis truly incorporates new meanings and acts on them.

Further counseling would aim to bring behavior or action into accord with thoughts. Other past or current relationships could be explored further to see how well the client's behaviors or actions illustrate or do not illustrate expressed meaning.

As we look at this session, we see that the counselor used listening skills and key word encouragers to focus on meaning issues. The counselor's open questions oriented to values and to meaning are often effective in eliciting client talk about meaning issues. You may also note that a reflection of meaning looks very much like a paraphrase except that the focus is on implicit deeper issues, often not expressed fully in the surface language or behavior of the client.

BOX 11-1   NATIONAL AND INTERNATIONAL PERSPECTIVES ON COUNSELING SKILLS

What Can You Gain From Counseling Persons With AIDS and Serious Health Issues?
Weijun Zhang

A good friend of mine had just started working with persons with AIDS. When I asked him, "What does this mean to you?" he started to grumble: "It's being around people with serious illness who could die at any time. It also means that no one gets cured, despite my best efforts." What a bleak picture he painted. No wonder some counselors are reluctant to work with AIDS clients or those facing truly serious health issues. Can something as miserable and difficult as AIDS be meaningful? How can one work as a counselor in a kidney dialysis unit? What about working in a hospice where all are expected to die within a reasonably short time? Certainly, there is much to learn and profit from in this type of work. What my friend was missing is that there are precious rewards available.

Years ago, I happened on an article outlining the positive aspects of working with clients who face extremely difficult futures or death.

1. We can help clients appreciate each moment. Working with people who have no choice but to live in the present can change our perspective. It can help us find a deeper meaning in life and learn to watch the world in wonder and appreciation as if this were our last day.

2. It is possible to learn something from each patient or client if we are willing to listen and be with his or her experience. Having direct contact with clients who face unthinkable pain and suffering but who still strive to live fully can help us understand the great strength that is the human spirit. This courage in the face of adversity can be very contagious.

3. Doing AIDS work or counseling the seriously ill will enable us to witness a lot of truly unconditional love and help us to become bigger hearted, more loving and caring persons.

I have always liked doing work with the seriously ill because of the selfless giving needed on the part of the counselor. But I have learned that it can mean a tremendous taking, learning, and satisfaction for those who help others most in need. Though this taking should never be our primary motive in doing AIDS or related work, it is certainly rewarding to see the light when we deal with the darkness associated with major life challenges. This positive insight I wish to share with my good friends.

INSTRUCTIONAL READING 1: THE SPECIFIC SKILLS OF ELICITING AND REFLECTION OF MEANING

Meaning issues often become prominent after a person has experienced a serious illness (AIDS, cancer, heart attack, loss of sight), encountered a life-changing experience (death of a significant other, divorce, loss of a job), or gone through serious trauma (war, rape, abuse, suicide of a child). Issues of meaning are also prominent among older clients who face major changes in their lives. These situations cannot be changed; they are a permanent part of the life experience. Box 11-1 presents national and international perspectives in finding meaning in a serious illness.

Reflecting meaning can also help clients work through issues of daily life. We saw in the sample interview how Travis gained understanding of himself as he reflected on the meaning of divorce. Everyday issues of life and many of our typical concerns can be resolved if we turn to serious examination of meaning, values, and life purpose. Religion and spiritual life provide a value base and can be a continuing source of strength and clarity.

Eliciting Client Meaning

Understanding the client is the essential first step. Consider storytelling as a useful way to discover the background of a client's meaning-making. If a major life event is critical, illustrative stories can form the basis for exploration of meaning. Clients do not often volunteer meaning issues, even though these may be central to the clients' concerns. Critical life events such as illness, loss of a parent or loved one, accident, or divorce often force people to encounter deeper meaning issues. If spiritual issues come to the fore, draw out one or two concrete example stories of the client's religious heritage. Through the basic listening sequence and careful attending, you may observe the behaviors, thoughts, and feelings that express client meaning.

Fukuyama (1990, p. 9) outlined some useful questions for eliciting stories and client meaning systems. Adapted for this chapter, they include the following:

"When in your life did you have existential or meaning questions? How have you resolved these issues thus far?"

"What significant life events have shaped your beliefs about life?"

"What are your earliest childhood memories as you first identified your ethnic-cultural background? Your spirituality?"

"What are your earliest memories of church, synagogue, mosque, a higher power, or lack of religion?"

"Where are you now in your life journey? Your spiritual journey?"

Reflecting Client Meanings

Say back to clients their exact key meaning and value words. Reflect their own meaning system, not yours. Implicit meanings will become clear through your careful listening and questions designed to elicit meaning issues from the client. Using the client's key words is preferable, but occasionally you may supply the needed meaning word yourself. When you do so, carefully check that the word(s) you use feel right to the client. Simply change "You feel . . ." to "You mean . . ." A reflection of meaning is structured similarly to a paraphrase or reflection of feeling. "You value . . . ," "You care . . . ," "Your reasons are . . . ," or "Your intention was . . ." Distinguishing among a reflection of meaning, a paraphrase, and a reflection of feeling can be difficult. Often the skilled counselor will blend the three skills together. For practice, however, it is useful to separate out meaning responses and develop an understanding of their import and power in the interview. Noting the key words that relate to meaning ("meaning," "value," "reasons," "intent," "cause," and the like) will help distinguish reflection of meaning from other skills.

Reflection of meaning becomes more complicated when meanings or values conflict. Here concepts of confrontation (Chapter 9) may be useful. Conflicting values, either explicit or implicit, may underlie mixed and confused feelings expressed by the client. For instance, a client may feel forced to choose between loyalty to family and loyalty to spouse. Underlying love for both may be complicated by a value of dependence fostered by the family and the independence represented by the spouse. When clients make important decisions, helping them sort out key meaning issues may be even more important than the many other issues that affect the decision.

For example, a young person may be experiencing a value conflict over career choice. Spiritual meanings may conflict with the work setting. The facts may be paraphrased accurately, and the feelings about each choice duly noted, yet the underlying *meaning* of the choice may be most important. The counselor can ask, "What does each choice mean for you? What sense do you make of each?" The client's answers provide the opportunity for the counselor to reflect back the meaning, eventually leading to a decision that involves not only facts and feelings but also values and meaning. And, as in confrontation, you can evaluate client change in meaning systems using the Client Change Scale in Chapter 9.

Discernment: Identifying Life Mission and Goals

Listen. Listen, with intention, with love, with the "ear of the heart." Listen not only cerebrally with the intellect, but with the whole of feelings, our emotions, imaginations, and ourselves.

—Esther de Waal

Discernment is "sifting through our interior and exterior experiences to determine their origin" (Farnham, Gill, McLean, & Ward, 1991, p. 23). The word "discernment" comes from the Latin *discernere,* which means "to separate," "to determine," "to sort out." In a spiritual or religious sense, discernment means identifying when the spirit is at work in a situation—the spirit of God or some other spirit. The discernment process is important for all clients, regardless of their spiritual or religious orientation or lack thereof. Discernment has broad applications to interviewing and counseling; it describes what we do when we work with clients at deeper levels of meaning. Discernment is also a process whereby clients can focus on visioning their future as a journey into meaning.

"There is but one truly serious problem, and that is . . . judging whether or not life is worth living" (Camus, 1955, p. 3). Viktor Frankl (1978) talks about the "unheard cry for meaning." Frankl claimed that 85% of people who successfully committed suicide saw life as meaningless, and he blamed this perception on an excessive focus on self. He said that people have a need for transcendence and living beyond themselves.

The vision quest, often associated with the Native American Indian, Dene, and Australian traditions, is oriented to helping youth and others find purpose and meaning in their lives. These individuals often undertake a serious outdoor experience to find or envision their central life goals. Meditation is used in some cultures to help members find meaning and direction.

Visioning and finding meaning may often be facilitated if issues are explored with a guide, counselor, or interviewer. This can be a spiritual or religious quest for some clients, but the discernment process will be useful for all. (Review Ivey, Ivey, Myers, and Sweeney, 2005, for additional information.) In Box 11-2, the specific discernment questions lead to further examination of goals, values, and meaning. Share the list of questions and encourage the client to participate with you in deciding which questions and issues are most important.

Multicultural Issues and Reflection of Meaning

For practical multicultural interviewing and counseling, recall the concept of focus. When helping clients make meaning, focus exploration of meaning not just on the individual but also on the broader life context. In much of Western society, we tend to assume that the individual is the person who makes meaning. And with most of your clients, you will be helping them find meanings and determine their individual life goals, but cultural context remains important.

In many other cultures—for example, the traditional Muslim world, the individual will make meaning in accord with the extended family, the neighborhood, and religion. If individual meaning is not in accord with cultural beliefs, making that meaning work in daily life will present major challenges to the client. An African American or Latina/o client will often feel more comfortable if meanings are made in a broader context. In contrast, many White

BOX 11-2   QUESTIONS LEADING TOWARD DISCERNMENT OF LIFE'S PURPOSE AND MEANING

You may find it helpful to share this list with the client before you begin the discernment process and identify together the most helpful questions to explore. In addition, we encourage adding topics and questions that occur to you and the client. Discernment is a very personal exploration of meaning, and the more the client participates, the more useful it is likely to be. Questions that focus on the *here and now* and intuition may facilitate deeper discovery.

Following is a systematic approach to discernment. First, you or your client may wish to begin by thinking quietly about what might give life purpose, meaning, and vision.

Here-and-now body experience and imaging can serve as a physical foundation for intuition and discernment.

▲ Relax, explore your body, and find a positive feeling of strength that might serve as an anchor for your search. Allow yourself to build on that feeling and see where it goes.
▲ Sit quietly and allow an image (visual, auditory, kinesthetic) to build.
▲ What is your gut feeling? What are your instincts? Get in touch with your body.
▲ Discerning one's mission cannot be found solely through the intellect. What feelings occur to you at this moment?
▲ Can you recall feelings and thoughts from your childhood that might lead to a sense of direction now?
▲ What is your felt body sense of spirituality, mission, and life goal?

Concrete questions leading to telling stories can be helpful.

▲ Tell me a story about that image above, or a story about any of the *here-and-now* experiences listed there.
▲ Can you tell me a story that relates to your goals/vision/mission?
▲ Can you name the feelings you have in relation to your desires?
▲ What have you done in the past or are doing presently that feels especially satisfying and close to your mission?
▲ What are some blocks and impediments to your mission? What holds you back?
▲ Can you tell about spiritual stories that have influenced you?

For self-reflective exploration, the following are often useful.

▲ Let's go back to that original image and/or the story that goes with it. As you reflect on that experience or story, what occurs for you?
▲ Looking back on your life, what have been some of the major satisfactions? Dissatisfactions?
▲ What have you done right?
▲ What have been the peak moments and experiences of your life?
▲ What might you change if you were to face that situation again?
▲ Do you have a sense of obligation that impels you toward this vision?
▲ Most of us have multiple emotions as we face major challenges such as this. What are some of these feelings and what impact are they having on you?
▲ Are you motivated by love/zeal/a sense of morality?
▲ What are your life goals?
▲ Can you see some specific examples of these goals?
▲ What do you see as your mission in life?
▲ What does spirituality mean to you?

The following questions place the client in larger systems and relationships—the self-in-relation. They may also bring multicultural issues into the discussion of meaning.

▲ Place your previously presented experiences and images in broader context. How have various systems (family, friends, community, culture, spirituality, and significant others) related to these experiences? Think of yourself as a self-in-relation, a person-in-community.
▲ *Family.* What do you learn from your parents, grandparents, and siblings that might be helpful in your discernment process? Are they models for you that you might want to follow, or even oppose? If you now have your own family, what do you learn from them, and what is the implication of your discernment for them?
▲ *Friends.* What do you learn from friends? How important are relationships to you? Recall important developmental experiences you have had with peer groups. What do you learn from them?
▲ *Community.* What people have influenced you and perhaps serve as role models? What group activities in your community may have influenced you? What

(continued)

BOX 11-2 (continued)

would you like to do to improve your community? What important school experiences do you recall?

▲ *Cultural groupings.* What is the place of your ethnicity/race in discernment? Gender? Sexual orientation? Physical ability? Language? Socioeconomic background? Age? Life experience with trauma?

▲ *Significant other(s).* Who is your significant other? What does he or she mean to you? How does this person (or persons) relate to the discernment

process? What occurs to you as the gifts of relationship? The challenges?

▲ *Spiritual.* How might you want to serve? How committed are you? What is your relationship to spirituality and religion? What does your holy book say to you about this process?

Discernment questions from Ivey, A., Ivey, M., Myers, J., & Sweeney, T. (2005). *Developmental Counseling and Therapy: Promoting Wellness Over the Lifespan.* Boston: Lahaska/Houghton Mifflin. Reprinted by permission.

European Americans focus on individual meaning making with minimal attention to broader issues—indicative of the individualistic culture of the United States.

Cultural, ethnic, religious, and gender groups all have systems of meaning that give an individual a sense of coherence and connection with others. Muslims draw on the teachings of the Qur'an. Similarly, Jewish, Buddhist, Christian, and other religious groups will draw on their writings, scriptures, and traditions. African Americans may draw on the strengths of Malcolm X or Martin Luther King Jr. or on support they receive from Black churches as they deal with difficult situations. Women, who are often more relational than men, may make meaning out of relationships, whereas men may focus more on issues of personal autonomy.

You may counsel clients who have experienced some form of religious bias or persecution. As religion plays such an important part in many people's lives, members of dominant religions in a region or a nation may have different experiences from those who follow minority religions. For example, Schlosser (2003) talks of Christian privilege in North America where people of Jewish and other faiths may feel uncomfortable, even unwelcome, during Christian holidays. Anti-Semitism, anti-Islamism, anti-liberal Christianity, anti-evangelical Christianity are all possible results when clients experience spiritual and/or religious intolerance. We also recall that when Christians and other religious groups find themselves in countries where they are a minority, they can suffer serious religious persecution—to the point of death.

FRANKL'S LOGOTHERAPY: MAKING MEANING UNDER EXTREME STRESS

If one person were to be identified with meaning and the therapeutic process, that individual would have to be Viktor Frankl, the originator of logotherapy. Frankl (1959) has pointed out the importance of a life philosophy that enables us to transcend suffering and find meaning in our existence. He argues that our greatest human need is for a core of meaning and purpose in life. Most counseling theories give little attention to meaning.

Frankl, a survivor of the German concentration camp at Auschwitz, could not change his life situation, but he was able draw on important strengths of his Jewish tradition to change the meaning he made of it. The Jewish tradition of serving others facilitated his survival. When times were particularly bad and prisoners had been whipped and were not being given food, Frankl (1959, pp. 131–133) counseled his entire barracks, helping them reframe their terrors and difficulties, pointing out that they were developing strengths for the future:

I quoted from Nietzsche, "That which does not kill me, makes me stronger." I spoke to the future. I said that . . . the future must seem hopeless. I agreed that each of us could guess . . . how small were [our] chances for survival. . . . I estimated my chances at about one in twenty. But I also told them that, in spite of this, I had no intention losing hope and giving up. . . . I also mentioned the past; all its joys and how its light shone even in the present darkness. . . . Then I spoke of the many opportunities of giving life a meaning. I told my comrades . . . that human life, under any circumstances has meaning. . . . I said that someone looks down on each of us in difficult hours—a friend, a wife, somebody alive or dead, or a God—and He would not expect us to disappoint him. . . . I saw the miserable figures of my friends limping toward me to thank me with tears in their eyes.

Shortly after his liberation, Frankl wrote his famous book *Man's Search for Meaning* (1959) within a 3-week period. This short, emotionally impactful book has remained a consistent best seller since that time. Frankl believed that finding positive meanings in the depth of despair was vital to keeping him alive. During the darkest moments, he would focus his attention on his wife and the good things they enjoyed together; or in the middle of extreme hunger, he would meditate on a beautiful sunset.

Mary and Allen spent 2 hours with Dr. Frankl after their lecture tour to Poland, which included a visit to Auschwitz. There we saw the gas chambers and the ovens that had incinerated Jews, Gypsies, gays and lesbians, many individuals with disabilities, and Polish people. Frankl shared again the importance of positive meaning for survival. He quoted the German philosopher Nietzsche:

He who has a why will find a how.

If your clients can find a meaningful vision and life direction (the *why*), they often will bear many difficult things as they seek their way to resolve their issues and continue life. Also memorable was Frankl's comment, "The best of us did not survive." It was an incredible experience to be in the presence of the man who was the real forerunner of the cognitive-behavioral movement (Mahoney & Freeman, 1985). Frankl was fully aware that meaning is not enough in itself—we also must *act* on our meaning and value system.

We have given *Man's Search for Meaning* to many clients facing a real crisis. We recommend that you read it while studying this chapter. It will make a difference to you and to those clients who face real life challenges. As you can see, Frankl was influential in our focus on wellness and the positive asset search. The recent trends toward a positive psychology and "learned optimism" (Seligman, 1998) are other examples of how an emphasis on strengths can aid the client.

Logotherapists search for positive meaning that underlies behavior, thought, and action. Dereflection is a specific strategy that logotherapy uses when clients focus solely on negatives. It helps to uncover deeper meanings and enables clients to become more positive in outlook. Dereflection and modification of underlying attitudes are specific techniques that logotherapy uses to uncover meaning and facilitate new actions. Many clients "hyperreflect" (think about something too much) on the negative meaning of events in their lives and may overeat, drink to excess, or wallow in depression. They are constantly attributing a negative meaning to life.

The direct reflection of meaning may encourage such clients to continue these negative thoughts and behavior patterns. Dereflection, by contrast, seeks to help clients discover the values that lie deeper in themselves. This strategy is similar to positive reframing/interpretation, but the client, rather than the counselor, does much of the positive thinking. The goal is to enable clients to think of things other than the negative issue and to find alternative positive meanings in the same event. The questions listed on page 295 of the instructional reading section represent first steps in helping clients dereflect and change their attitudes. The following abbreviated example illustrates this approach.

Client: I really feel at a loss. Nothing in my life makes sense right now.

Counselor: I understand that—we've talked about the issues with your partner and how sad you are. Let's shift just a bit. Could you tell me about what has been meaningful and important to you in the past? (The client shares some key supportive religious experiences from the past. The counselor draws out the stories and listens carefully.)

Counselor: (reflecting meaning) So, you found considerable meaning and value in worship and time spent quietly. You also found worth in service in the church. You drifted away because of your partner's lack of interest. And now you feel you betrayed some of your basic values. Where does this lead you in terms of a meaningful way to handle some of your present concerns?

As you may note, the process of dereflection is a special form of the positive asset search. But rather than focusing just on the concretes (spirituality, service to others, walking in the outdoors, enjoying one's friends), the counselor explores the positive meaning of these specifics. "What does spirituality mean to you?" "What sense do you make of a person who finds such joy in walking outdoors and enjoying sunsets?" "What values do you find in service to others?"

Out of the exploration of meaning may come data for restorying one's problems and even life-transformative actions. But Frankl was interested in more than just meaning. He would also discuss specific actions that the client would take in the here and now of daily life. Meaning without implementation and action is not enough. Frankl's emphasis on action beyond thinking new thoughts in the interview was pathbreaking and innovative.

Box 11-3 presents relevant research regarding reflection of meaning.

BOX 11-3 RESEARCH EVIDENCE THAT YOU CAN USE

Reflection of Meaning and Reframing

Carl Rogers brought meaning issues to center stage in counseling and therapy, whereas Viktor Frankl provided both philosophical and practical applications of meaning in counseling. A solid relationship with your client helps give meaning to your encounter.

Classic research by Fiedler (1950) and Barrett-Lennard (1962) set the stage for the present when they found that relationship variables (closely related to the listening skills) were vital to the success of all forms of interviewing, regardless of theory. Now you will find the relationship issues are termed "common factors," and the idea of relationship as central has become almost universally accepted. Those working in the Heart and Soul of Change Project cite data suggesting that 30% or more of successful therapy is based on relationship (Miller, Duncan, & Hubble, 2005).

Luszczynska et al. (2007) studied coping profiles of patients after cancer surgery. Those who accommodated to the surgery exhibited low levels of active problem-directed strategies but high acceptance and humor; patients who coped more actively exhibited positive reframing and active strategies. Patients' partners provided most support to patients demonstrating active coping.

Research has demonstrated that "families that seek support and try to accept what happened after a traumatic injury may experience less injury-related stress and family dysfunction over time" (Wade et al., 2001, p. 412). Turning to religion was the second most used strategy among parents whose children suffered traumatic injury. Connectedness with others and the comfort of spirituality can be a most important positive

(continued)

BOX 11-3 (continued)

asset and wellness strength for many clients. Religiously oriented clients did better in cognitive-behavioral therapy when their spirituality was part of the process in a classic and often cited study by Probst (1996). Recovery from heart surgery has been found more rapid among those with religious involvement, particularly among women (Contrada et al., 2004, p. 227).

Lucas (2007/2008) examined experiences of 19 caregivers and teachers working with traumatized children. She found learning coping strategies of reframing and realistic goal setting helped them reduce emotional exhaustion and increased personal sense of accomplishment. Li and Lambert (2008) found positive reframing to be one of the best predictors of job satisfaction among 102 intensive care nurses from the People's Republic of China.

Neuroscience and Meaning

At the surface, the broad idea of meaning would appear to be beyond measurement in a physical sense. Our sense of meaning brings our thoughts, feelings, and behavior into a whole, enabling us to make sense of our experience. Surprisingly, this seemingly abstract concept of meaning can be pinpointed rather precisely:

> Meaningfulness is inextricably bound up with emotion. Depression is marked by wide-ranging symptoms, but the cardinal feature of it is the draining of meaning from life. . . . By contrast, those in a state of mania see life as a gloriously ordered, integrated whole. Everything seems to be connected and the smallest events are bathed in meaning. (Carter, 1999, p. 197)

Creation of the *New* also means that new neural networks are formed in the brain and long-term memory.

Ratey (2008) indicates that there is a key moral and spiritual dimension in the brain that we are close to identifying. Stimulation of a portion of the brain appears to evoke spiritual images for many people. Morality may be partially hardwired. Sapolsky's lectures on neuroscience support this frame of reference. *The Political Brain* (Weston, 2007) follows this logic. Weston speaks of how candidates literally reach the mirror neurons of the public, creating empathy and changing neural connections. Morality as described by neuroscientists is awareness of the Other. An interesting challenge in brain science is explaining the individualistic mind versus the collectivist mind. Gene expression is clearly part of this, but gene expressions often require environmental events before they are triggered. Some genes may lie dormant throughout a lifetime.

Ratey, a leader in neuroscience applications and research has commented:

> You have to find the right mission, you have to find something that's organic, that's growing that keeps you focused on and continues to provide meaning and growth and development for yourself.
>
> I see meaning as a big part of neuroscience. We start with neuroscience and now we're talking about transcendence. Spirituality even lights up key centers in the brain. Meaning drives the lower centers and is connected to emotions and motivation areas. It's a huge, huge, human construct that means so much to our race and our species. Obviously it involves memory and learning and remembering the good stuff, remembering what your goals are, remembering what you want to do, and so you need all those things working well to keep you on the right meaning path. If you can get *people* into a situation where they have the meaning direction provided by their mission or their job or their goal, they don't need medicine. (Ratey, 2008, p. 41)

INSTRUCTIONAL READING 2: THE SKILLS OF INTERPRETATION/REFRAMING

When you use the microskill of interpretation/reframing, you are helping the client to restory or look at the problem or concern from a new, more useful perspective. This new way of thinking is central to the restorying and action process. In the microskills hierarchy, the words "interpretation" and "reframe" are used interchangeably. Interpretation reveals new perspective and new ways of thinking beneath what a client says or does. The reframe provides another frame of reference for considering problems or issues. And eventually the client's story may be reconsidered and rewritten as well.

The basic skill of interpretation/reframing may be defined as follows:

▲ The counselor listens to the client story, issue, or problem and learns how the client makes sense of, thinks about, or interprets the story or issue.

▲ The counselor may draw from personal experience and/or observation of the client (reframe) or may use a theoretical perspective, thus providing an alternative meaning or interpretation of the narrative. This may include *linking* together information or ideas discussed earlier that relate to each other. Linking is particularly important as it integrates ideas and feelings for clients and frees them to develop new approaches to their issues.

▲ (positive reframe from personal experience) "You feel that coming out as gay led you to lose your job, and you blame yourself for not keeping quiet. Maybe you just really needed to become who you are. You seem more confident and sure of yourself. It will take time, but I see you growing through this difficult situation." Here self-blame has been reinterpreted or reframed as a positive step in the long run.

▲ (psychoanalytic interpretation with multicultural awareness) "It sounds like the guy who fired you is insecure about anyone who is different from him. He sounds as if he is projecting his own unconscious insecurities on you, rather than looking at his own heterosexism or homophobia."

Consider an example interpretation/reframe developed from the logic of the interviewer. Allen, the client, was going through a divorce and was very angry—a common reaction for those engaged in a major breakup, particularly when finances are involved. He was telling his attorney, at some length, about what he wanted and why. Attorneys use a form of interviewing involving many questions, and it sometimes involves informal counseling. After listening carefully to Allen's issues and acknowledging his strong feeling (acknowledging, not reflecting), the attorney got out from behind the desk and stood over Allen saying: "Allen, that's your story. But I can tell you that you won't get what you want. Your wife has a story as well, and what will happen is something between what you both feel you need and deserve. For your own and your children's sake, think about that." This was a rather rough and confrontative reframing of Allen's story. It also changed the focus from Allen and his problems to his wife and children. Fortunately, he heard this powerful reframe, and resolution of differences in the divorce finally began.

This story has several implications. First, even with the most effective listening, clients may still hold on to unworkable stories, ineffective thinking, and self-defeating behaviors. Clearly, they need a new perspective. Respect clients' frame of reference before interpreting or reframing their words and life in new ways. In effect, *listen before you provide your interpretation or reframe.* There will always be some clients who will need the strong, confrontative interpretation that Allen got, but recall that the attorney first listened attentively to Allen.

We may also consider the interpretation/reframe as the creation of the *New* because we and the client are building another way to think about issues—and ultimately create a more effective and happy self. The value of an interpretation or reframe depends on the client's reaction to it and how he or she changes thoughts, feelings, or behaviors. Think of the Client Change Scale (CCS)—how does the client react to each interpretation? If the client denies or ignores the interpretation, you obviously are working with denial (Level 1 on the CCS). If the client explores the interpretation/reframe and makes some gain, you have moved that client to bargaining and partial understanding (Level 2 on the CCS). Interchangeable responses and acceptance of the interpretation (Level 3) will often be an important part of the gradual growth toward a new understanding of self and situation. If the client develops useful new ways of thinking and behaving (Level 4 on CCS) movement is clearly occurring. Transcendence, perhaps the ultimate creation of the *New* (Level 5), will appear only with major breakthroughs that change the direction of interviewing, counseling, and psychotherapy. But let us recall that movement from denial (Level 1) to partial consideration of issues (Level 2) may be a major breakthrough, beginning client improvement.

The potential power of the effective interpretation/reframe can be seen in the divorce example above. Allen was in denial about what he could "win" in the divorce and refused even to bargain. But confronted by the attorney towering over him, he moved almost immediately from denial (Level 1) to a new understanding (Level 3) by accepting the attorney's reframe. The real test of change would be whether *he does change his behavior as a result of his new insights*. New solutions (Level 4) are never reached without behavior change. Transcendence (Level 5) is rarely found in complex cases of divorce!

Interpretation/Reframing and Other Microskills

Focusing, like reflection of meaning and interpretation/reframing, is another influencing skill that greatly facilitates the creation of new client perspectives. In the story of Allen and his attorney, the focus on the wife and her needs was key to the successful reframe. As another example, you may work with a male or female client who feels that he or she has been subjected to gender discrimination or sexual harassment. If you just focus on the individual, the client may blame himself or herself for the problem. By focusing on gender or other multicultural issues, you are expanding client perspectives, and these clients may generate a new perspective, meaning, or way of solving the problem on their own—again, the creation of the *New*.

Interpretation may be contrasted with the paraphrase, reflection of feeling, focusing, and reflection of meaning. In those skills the interviewer remains in the client's *own* frame of reference, and effective listening often enables creation of the *New*. In interpretation/reframing, the frame of reference comes from the counselor's personal and/or theoretical constructs. The following are examples of interpretation/reframing paired with other skills.

Annaliese: (with a low self-concept) I just feel so bad about myself. I don't feel that I'm performing at work. I think the boss is going to be down on me pretty soon.

Counselor: (reflection of feeling) Annaliese, You're really troubled and worried, perhaps even scared.

Counselor: (paraphrase/restatement) You're not doing as well as you like, and you know your boss doesn't like it.

Counselor: (eliciting meaning) Could we move in a different direction for a moment? Annaliese, what does this job really mean to you? Does it fit with your life goals?

Annaliese: No! I'm bored and frustrated. The job just doesn't make sense to me. I thought it would, but it doesn't. I need something that I care about—that is meaningful to me, so I can go home feeling that I've done something worthwhile. I'd like to care for others instead of working with numbers all the day long.

Counselor: (reflecting meaning) I hear you, Annaliese. You care. So what we have is a job that has little meaning for you, but it pays well. (Next the counselor reflects deeper meanings.) You seem to feel that you would have more value for yourself if you could help others more. (Next is elicitation of meaning for more depth.) Let's explore that vision of caring for others a bit more and see how it might lead to more meaningful work. You may be interested in discernment as a way to explore life goals more deeply.

Counselor: (positive reframe/interpretation) Let's look at this another way, Annaliese. The fact that you're bored is a sign that you've accomplished what you need to in that job. You've shown great skills, and you are ready to move on to something new where you can use your strengths more effectively.

Counselor: (interpretation/reframe—linkage) Annaliese, this seems to tie in with what you said last week about enjoying your volunteer work several years ago with children in the inner city. You derived joy from that, but your job eventually took you away. It seems that you may be ready now to be taken away to something that is more satisfying at a deeper level . . . something that brings you more fun and joy.

Interpretation has traditionally been viewed as a mystical activity in which the interviewer reaches into the depths of the client's personality to provide new insights. However, we can demystify interpretation; we consider it to be merely a new frame of reference. Interpretation reframes the situation. Viewed in this light, the depth of a given interpretation refers to the magnitude of the discrepancy between the frame of reference from which the client is operating and the frame of reference supplied by the interviewer. Gradually, reframing has become a more prominent term as it provides a more understandable view of interpretation.

To ensure mutuality and not influence the client too much, you should follow most interpretation/reframes with a check-out—"How does that sound to you?" "What meaning do you take from what I just said?"

Theories of Counseling and Interpretation/Reframing

Theoretically based interpretations can be extremely valuable as they provide the interviewer with a tested conceptual framework for thinking about the client. Each theory is itself a story—a story told about what is happening in interviewing, counseling, and therapy and what the story means. Integrative theories find that each theoretical story has some value. Most likely, as you generate your own natural style you will develop your own integrative theory, drawing from those approaches that make most sense to you.

Below are several examples of how different orientations to theory might interpret the same information. You will see a dream that Charlis had and how different theories might interpret it. Before the actual interpretation, you will see a brief theoretical paragraph that provides a background for the theory-oriented interpretation that follows.

Imagine that you have worked with Charlis over a longer period, and she came to you upset over a troubling dream. This dream recurred frequently in childhood, and after the heart attack, it returned with a vengeance, and Charlis would awaken, sweating, in the middle of the night. Different counseling theories would interpret and work with the dream differently, but each would provide a new frame of reference, a new perspective for the client. Charlis tells you her dream story.

Charlis: I dreamed that I was walking along the cliffs with the sea raging below. I felt terribly frightened. There was a path that I could have taken away from the cliffs, but I just felt so undecided about what to do. The dream just went on and on. I woke up in a cold sweat—and I've had that dream almost every night since I last saw you.

Decisional theory. A major issue in interviewing for all clients is making appropriate decisions and understanding alternatives for action. Decisions need to be made with awareness of cultural/environmental/contextual issues. Interpretation/reframing helps clients find new ways of thinking about their decisions. Linking ideas together is particularly important.

Counselor: Charlis, you're facing new challenges since the heart attack and have many key decisions to make, including what you want to do with the rest of you life. You feel almost as if you might fall off the cliff if things don't straighten out soon. The whole situation is frightening, and making decisions can make it worse. On the other hand, we already identified several strengths that will enable you to make the important decisions you have to make. (Interpretation focuses on the parallels to the heart attack and then draws on strengths and wellness.)

Person-centered. Clients are ultimately self-actualizing, and our goal is to help them find the story that builds on their strengths and helps them find deeper meanings and purpose. Reflection of meaning helps clients find alternative ways of viewing the situation while interpretation/reframing are not used. Linking can occur through effective summarization.

Counselor: Charlis, that dream seems to mean something important to you. I hear the terrible fright, and I notice the rage of the sea. And you've had the dream many nights, and now you wonder what it means. (Reflection of meaning and reflection of feelings; interpretations would be very rare in this theory.)

Brief solution-focused counseling. Brief methods seek to help clients find quick ways to reach their central goals. The interview itself is conceived first as a goal-setting process and then methods are found to reach goals through time-efficient methods. Interpretation/reframing will be rare except for links of key ideas.

Counselor: You're facing new challenges since the heart attack and have some important decisions to make. Goal setting is important; which path do you want to take to get well? (mild interpretation with a move to goal setting)

Cognitive-behavioral theory. The emphasis is on sequences of behavior and thinking and what happens to the client, internally and externally, as a result. Often interpretation/reframing is useful in understanding what is going on in the client's mind and/or linking the client to how the environment affects cognition and behavior.

Counselor: The dream seems very close to what you face now. You have told me that you feel rage toward what happened to you, and now you are wondering which direction to take. Our next task is to work on some stress management strategies to help you find behaviors to cope with these challenges. Later, let's look at how this might relate to what's going on with your parents. (Here we see the counselor active in linking the dream with present issues.)

Psychodynamic theory. Individuals are dependent on unconscious forces. Interpretation/reframing are used to help link ideas and enable the client to understand how the unconscious past and long-term, deeply seated thoughts, feelings, and behaviors frame the here and now of daily experiences. Freudian, Adlerian, Gestalt, Jungian, and several other psychodynamic theories each tell different stories.

Counselor: You feel rage at your parents, and you can't tell them how you really feel. It frightens you. And now you find the people around you force you to keep quiet about your feelings, but the prospect of challenging them is terrifying. This links back to earlier stories you mentioned about not being able to depend on your parents. (emphasis on how the past affects the present)

Multicultural counseling and therapy (MCT). The person is situated in a cultural/environmental/contextual place, and we need to help clients interpret and reframe their issues, concerns, and problems in relation to their multicultural background. (See the RESPECTFUL model, Chapter 2). MCT is an integrative theory and uses all of the methods above, as appropriate, to facilitate clients' understanding of themselves and how cultural/environmental/contextual issues affect them personally.

Counselor: You felt frightened—I hear that. From what you've told me, sexual harassment was part of the stressors you faced before the heart attack. The cliff could be the hassles you had at the office, and returning to the job clearly is frightening at this point. I also hear a woman who has the courage to get out on those cliffs and face the challenges. We will have to work together to help you find some support here to cope with the challenges. (Feminist frame of reference and issues are interpreted in a multicultural context.)

All of the above provide the client with a new, alternative way to consider the situation. In short, interpretation renames or redefines "reality" from a new point of view. Sometimes just a new way of looking at an issue is enough to produce change. Which is the correct interpretation? Depending on the situation and context, any of these interpretations could be helpful or harmful. The first two responses deal with here-and-now reality whereas psychodynamic interpretation deals with the past. The feminist interpretation links the heart attack with sexual harassment on the job.

SUMMARY: HELPING CLIENTS RESTORY THEIR LIVES

Eliciting and reflecting meaning is a complex skill that requires you to enter the sense-making system of the client. Full exploration of life meaning requires a self-directed, verbal client willing to talk. The skill complex is most often associated with an abstract formal-operational interviewing style. However, all of us are engaged in the process of meaning-making and trying to make sense of a confusing world. With clients who are more concrete, you will still find that eliciting and reflecting meaning is useful. But these clients may not be able to see patterns in their thinking or be as self-directed and reflective as those who think at a more complex level. You may find that the more directive approach to meaning taken by the cognitive-behavioral therapists is more useful with clients who have difficulty reflecting on themselves.

With highly verbal or resistant clients, you may find that they like to spend all their time thinking, reflecting on meaning, and thus end up intellectualizing with little or no action to change their behaviors, thoughts, or feelings. Viktor Frankl was well aware of this possible problem and encouraged his clients to take action on their meanings. Meaning that does not move into the "real world" may at times become a problem in itself.

Meanings are organizing constructs that are at the core of our being. You will find that exercises with reflection of meaning, if completed in depth, will result in your having a more comprehensive understanding of your client than is possible with most other skills. Mastering the art of understanding meaning will take more time than other skills. The exercises in this chapter are designed to assist you along the path toward this goal.

When you interpret or reframe, first be sure that you have heard the client's story or concerns, and then draw from personal experience or a theoretical perspective to provide the client a new way of thinking and talking about issues. Focusing and multicultural counseling and therapy is the most certain way to bring multicultural issues into the interview. A woman, a gay or lesbian, or Person of Color may be depressed over what is considered a personal failure. By helping the client see the cultural/environmental/contextual nature of the issue, a new perspective will appear, providing a totally new and more workable meaning.

The effectiveness of an interpretation/reframe can be measured on the Client Change Scale. The new perspective is useful if the client moves in a positive direction. Each interviewing and counseling theory provides us with a new and different story about the interview. Drawing from theory for interpretation/reframing provides a more systematic frame for considering the client. However, logic and your personal experience and observations may be as effective as a theoretically oriented reframe.

Key points of the chapter are presented below.

■ KEY POINTS

Meaning	Meaning is not observable behavior, although it could be described as a special form of cognition that reaches the core of our being. Helping clients discern the meaning and purpose of their lives can serve as a motivator for change and provide a compass as to the direction of that change. Meaning organizes life experience and often serves as a metaphor from which clients generate thoughts, feelings, and behaviors. A person with a sense of meaning and a vision for the future can often work through and live with the most difficult issues and problems. Reflections of meaning are generally for more verbal clients and may be found more in counseling and therapy than in general interviewing.
The how of meaning	A well-timed reflection of meaning may help many clients facing extreme difficulty. It can help clarify cultural and individual differences, as the same words often have varying underlying meaning for each client. As meaning is often implicit, it is helpful to ask questions that lead clients to explore and clarify meaning. *Eliciting meaning:* "What does 'XYZ' mean to you?" Insert the key important words of the client that will lead to meanings and important thoughts underlying key words. "What sense do you make of it?" "What values underlie your actions?" "Why is that important to you?" "Why?" (by itself, used carefully) *Reflecting meaning:* Essentially, this looks like a reflection of feeling except that the words "meaning," "values," or "intentions" substitute for feeling words. For example, "You mean . . . ," "Could it mean that you . . . ," "Sounds like you value . . . ," or "One of the underlying reasons/intentions of your actions was. . . ." Then use the client's own words to describe his or her meaning system. You may add a paraphrase of the context and close with a check-out.

(continued)

KEY POINTS (continued)

	For example, you could reflect an immediate meaning this way: "Anish, you value service to others, and you've enjoyed working in the hospital as a volunteer." However, if there is conflict of values and meanings, the following could be added to confront the discrepancy between individual and family values. "On the other hand, your family values medical practice as it offers more money, but you want to work in research on cancer as you see that as the best way to help more people in the long run, and the financial rewards are not as important to you. Have I summarized the value conflict clearly?"
Interpretation/reframing	The counselor helps clients obtain new perspectives, new frames of reference, and sometimes new meanings, all of which can facilitate clients' changing their view and way of thinking about their issues. This skill comes primarily from the counselor's observations and occasionally from the client. *Theoretical interpretations:* These come from specific counseling theory such as psychodynamic and interpersonal, family therapy, or even Frankl's logotherapy. Clients tell the story or speak about their problems and issues. The counselor then makes sense of what they are saying from their theoretical perspective. "That dream suggests that you have an unconscious wish to run away from your husband." "Sounds like an issue of what we call boundaries—your husband/wife is not respecting your space." "I hear you saying that you don't know where you are going; it sounds like you lack meaning in your life." *Reframes:* These tend to come from here-and-now experience in the interview, or they might be larger reframes of major client stories. The reframes are based on your experience in providing the client with another interpretation of what has happened or how the story is viewed. Effective reframes can change the meaning of key narratives in clients' lives. The positive reframe is particularly important. "Charlis, what stands out to me at this moment is how able you are, and we can use your 'smarts' and ability to understand situations to find new, more comfortable directions." Positive reframes in the here and now are often the most useful.
Interpretation/reframing at the deepest level	Meaning affects interpretation. Viktor Frankl constantly reframed his experience in the German concentration camp, integrating here-and-now positive reframing with meaning. In the middle of the terror, he was able to enjoy the beauty of a sunset; he remembered his times with his wife; he was able to enjoy and focus on tasting and eating a small bit of bread. The major reframe of such traumatic experience, of course, is "I survived" or "You survived." Despite the traumatic experience (war, rape, accident), you are still here with the possibility of changing a part of the world.

COMPETENCY PRACTICE EXERCISES AND PORTFOLIO OF COMPETENCE

The concepts of this chapter build on previous work. If you have solid attending and client observation skills, can use questions effectively, and can demonstrate effective use of the encourager, paraphrase, and reflection of feeling, you are well prepared for the exercises that follow.

Individual Practice

Exercise 1: Identification of Skills

Read the following client statement. Identify the following counselor responses as paraphrases (P), reflections of feeling (RF), reflections of meaning (RM), or interpretations/reframes.

> I feel very sad and lonely. I thought Jose was the one for me. He's gone now. After our breakup I saw a lot of people but no one special. Jose seemed to care for me and make it easy for me. Before that I had fun, particularly with Carlos. But it seemed at the end to be just sex. It appears Jose was it; we seemed so close.

_____ "You're really hurting and feeling sad right now."

_____ "Since the breakup you've seen a lot of people, but Jose provided the most of what you wanted."

_____ "Sounds like you are searching for someone to act as the father you never had, and Jose was part of that."

_____ "Another way to look at it is that you unconsciously don't really want to get close; and when you get really close, the relationship ends."

_____ "Looks like the sense of peace, caring, ease, and closeness meant an awful lot to you."

_____ "You felt really close to Jose and now are sad and lonely."

_____ "Peace, caring, and having someone special mean a lot to you. Jose represented that to you. Carlos seemed to mean mainly fun, and you found no real meaning with him. Is that close?"

List possible single-word encouragers for the same client statement. You will find that the use of single-word encouragers, perhaps more than any other skill, leads your client to talk more deeply about the unique meanings underlying behavior and thought. A good general rule is to search carefully for key words, repeat them, and then reflect meaning.

_____   _____

_____   _____

_____   _____

Exercise 2: Identifying Client Issues of Meaning

Affective words in the preceding client statement include "sad" and "lonely." Some other words and brief phrases in the client statement contain elements that suggest more may be found under the surface. The following are some key words that you may have listed under possible encouragers: "the one for me," "care for me," "easy for me," "I had fun," "just sex," and "we seemed close." The feeling words represent the client's emotions about the current situation; the other words represent the meanings she uses to represent the world. Specifically, the client has given us a map of how she constructs the world of her relationships with men.

To identify underlying meanings for yourself, talk with a client, or someone posing as a client, observing his or her key words—especially those that tend to be repeated in different situations. Use those key words as the basis of encouragers, paraphrasing, and questioning to elicit meaning. Needless to say, this should be done with considerable sensitivity to the client and her or his needs. Record the results of your experience with this important exercise. You will want to record patterns of meaning-making that seem to be basic and that may motivate many more surface behaviors, thoughts, and feelings.

Exercise 3: Questioning to Elicit Meanings

Assume a client comes to you and talks about an important issue in her or his life (for instance, divorce, death, retirement, a pregnant daughter). Write five questions that might be useful in bringing out the meaning of the event.

_____

_____

_____

_____

_____

Exercise 4: Practice of Skills in Other Settings

During conversations with friends or in your own interviews, practice eliciting meaning through a combination of questioning and single-word encouragers, and then reflect the meaning back. You will often find that single-word encouragers lead people to talk about meaningful issues. Record your observations of the value of this practice. What one thing stands out from your experience?

_____

Exercise 5: Discernment: Examining One's Purpose and Mission

Using the suggestions of Box 11-2, work through each of the four sets of questions. You may do this by yourself, using a meditative approach and journaling. Or you may want to do this with a classmate or close acquaintance. Allow yourself time to think carefully about each area. Add questions and topics that occur to you—make this exercise fully personal.

What do you learn from this exercise about your own life and wishes?

_____

Exercise 6: Individual Practice in Interpretation/Reframing

Interpretations provide alternative frames of reference or perspectives for events in a client's life. In the following examples, provide an attending response (question, reflection of feeling, or the like) and then write an interpretation. Include a check-out in your interpretation.

Example A

"I was passed over for promotion for the third time. Our company is under fire for sex discrimination, and each time a woman gets the job over me. I know it's not my fault at all, but somehow I feel inadequate."

Listening response _____

_____

_____

Interpretation/reframe from a psychodynamic frame of reference (i.e., an interpretation that relates present behavior to something from the past) _____

_____

_____

Interpretation/reframe from a gender frame of reference _____

_____

_____

Interpretation/reframe from your own frame of reference in ways that are appropriate for varying clients _____

_____

_____

Example B

"I'm thinking of trying some pot. Yeah, I'm only 13, but I've been around a lot. My parents really object to it. I can't see why they do. My friends are all into it and seem to be doing fine."

Listening response _____

_____

_____

Reframe from a conservative frame of reference (one that opposes the use of drugs) _____

_____

_____

Reframe from an occasional user's frame of reference _____

_____

_____

Interpretation from your own frame of reference on this issue _____

_____

_____

The preceding examples of interpretations and reframes are representations of meaning and value issues that you will encounter in interviewing and counseling. What are the value issues involved in these examples and what is your personal position on these issues? Finally, how do you reconcile the importance of a client's responsibility for her or his own behavior with your position? What would you actually do in these situations?

Group Practice

Three group exercises are suggested here. The first focuses on the skill of eliciting and reflecting meaning, the second on the discernment process as it might be used in logotherapy, and the third on interpretation/reframing.

Exercise 7: Systematic Group Practice in Eliciting and Reflecting Meaning
Step 1: Divide into practice groups.
Step 2: Select a group leader.
Step 3: Assign roles for the first practice session.

▲ Client.
▲ Interviewer.
▲ Observer 1, who observes the client's descriptive words and key repeated words, using Feedback Form, Box 11-4.
▲ Observer 2, who notes the interviewer's behavior, using Feedback Form, Box 11-4.

Step 4: Plan. For practice with this skill, it will be most helpful if the interview starts with the client's completing one of the following model sentences. The interview will then follow along, exploring the attitudes, values, and meanings to the client underlying the sentence.

"My thoughts about spirituality are . . ."
"My thoughts about moving from this area to another are . . ."
"The most important event of my life was . . ."
"I would like to leave to my family . . ."
"The center of my life is . . ."
"My thoughts about divorce/abortion/gay marriage are . . ."

A few alternative topics are "my closest friend," "someone who made me feel very angry (or happy)," and "a place where I feel very comfortable and happy." Again, a decision conflict or a conflict with another person may be a good topic.

Establish the goals for the practice session. The task of the interviewer in this case is to elicit meaning from the model sentence and help the client find underlying meanings and values. The interviewer should search for key words in the client response and use those key words in questioning, encouraging, and reflecting. A useful sequence of microskills for eliciting meaning from the model sentence is (1) an open question, such as "Could you tell me more about that?" "What does that mean to you?" or "How do you make sense of that?"; (2) encouragers and paraphrases focusing on key words to help the client continue; (3) reflections of feeling to ensure that you are in touch with the client's emotions; (4) questions that relate specifically to meaning (see Box 11-2); and (5) reflecting the meaning of the event back to the client, using the framework outlined in this chapter. It is quite acceptable to have key questions and this sequence in your lap to refer to during the practice session.

Examine the basic and active mastery competencies in the self-assessment section and plan your interview to achieve specific goals.

Observers should study the feedback form especially carefully.

BOX 11-4 FEEDBACK FORM: REFLECTING MEANING

_____ (Date)

_____   _____
(Name of Interviewer)   (Name of Person Completing Form)

Instructions: Observer 1 completes the first part of this form, giving special attention to recording descriptive words the client associates with meaning and to key repeated words. In the second part, Observer 2 notes the interviewer's use of the reflection-of-meaning skill, giving special attention to questions that appeared to elicit meaning issues.

Part One: Client Observation
Key words/phrases:

_____

_____

_____

What are the main meaning issues of the interview?

Part Two: Interviewer Observation
List questions and reflections of meaning used by the interviewer, continuing on a separate sheet as needed.

1. _____

2. _____

3. _____

4. _____

5. _____

6. _____

Comment on the effectiveness of the reflection-of-meaning skill.

Step 5: Conduct a 5-minute practice session using the skill.

Step 6: Review the practice session and provide feedback for 10 minutes. The microsupervision process may include a group discussion of the place of values in the interview. The feedback forms are useful here. It is often tempting to just talk, but you might forget to give the interviewer helpful and needed specific feedback. Take time to complete the forms before talking about the session. As always, give special attention to the mastery level achieved by the interviewer. The client can complete the Client Feedback Form of Chapter 1.

Step 7: Rotate roles. Remember to share time equally.

Some general reminders. This skill can be used from a variety of theoretical perspectives. It may be useful to see if an explicit or implicit theory is observable in the interviewer's behavior.

Exercise 8: Discernment Practice

Take another person through the discernment procedure, working carefully with each step. Recall de Waal's statement on how to listen to the other person:

> Listen. Listen, with intention, with love, with the "ear of the heart." Listen not only cerebrally with the intellect, but with the whole of feelings, our emotions, imaginations, and ourselves. (de Waal, 1997, Preface)

Share the list of questions and ideas with your client. Ask your volunteer to suggest additional questions and issues that may be missing in this list. Have the client define which questions he or she may wish to discuss. Use all your listening skills as you help the client find personal direction and meaning. Take your time!

What do you and the client learn?

_____

_____

Exercise 9: Interpretation/Reframing Practice
Follow Steps 1–3 as outlined in Exercise 7 above. Use the interpretation/reframe form (Box 11.5).

Step 4: Plan. To practice with this skill, ask the client to think about and describe something that is frustrating at the moment. A few alternative topics are "having to move to a different residence hall," "having roommates with religious beliefs different from yours," "trying to adopt healthier eating habits," "taking a challenging course," and "finding that you tend to procrastinate."

Establish the goals for the practice session. The first task of the interviewer is to listen to the client story and learn how he or she thinks about or interprets the frustrating issue. The second task of the interviewer is to provide an alternative meaning or interpretation of the narrative; draw from personal experience or a theoretical perspective. Also, you may link critical ideas together (see linking examples in this chapter). Examine the basic and active mastery competencies in the self-assessment section and plan your interview to achieve specific goals.

BOX 11-5   FEEDBACK FORM: INTERPRETATION/REFRAME

_____ (Date)

_____   _____
(Name of Interviewer)   (Name of Person Completing Form)

Instructions: Observer 1 will complete the Client Change Scale in Chapter 9. Observer 2 will complete the items below.

1. Did the interviewer use the basic listening sequence to draw out and clarify the client's story or concern? How effectively?

2. Provide nonjudgmental, factual, and specific feedback for the interviewer on the use of the interpretation/reframe.

3. Did the interviewer check out the client's reaction to the intervention? Did the client move on the Client Change Scale?

The value of an interpretation or reframe depends on the client's reaction to it. Use the Client Change Scale (CCS) in Chapter 9 to assess the client's reaction to your reframe. Observers should study the feedback form before the practice exercise.

Step 5: Conduct a 5-minute practice session using the skill.

Step 6: Review the practice session and provide feedback for 10 minutes. Include a group discussion of the role of reframing in the interview. Don't forget to use feedback forms here and to give special attention to the mastery level achieved by the interviewer. The client can complete the Client Feedback Form of Chapter 1.

Step 7: Rotate roles. Remember to share time equally.

Some general reminders. When we use interpretation/reframing we are working primarily from the interviewer's frame of reference. Your goal is to help the client to restory or look at the frustrating problem or concern from a new perspective. To accomplish this goal you need to listen before you provide your interpretation or reframe. Respect clients' frame of reference before interpreting or reframing their words and frustrating situations in new ways. Provide clients with a new perspective or way of thinking about issues.

Portfolio of Competence

As you work through this list of competencies, think about how you would include the ideas related for reflection of meaning in your own Portfolio of Competence.

Use the following as a checklist to evaluate your present level of mastery. As you review the items, ask yourself, "Can I do this?" Check those dimensions that you currently feel able to do. Those that remain unchecked can serve as future goals. Do not expect to attain intentional competence on every dimension as you work through this book. You will find, however, that you will improve your competencies with repetition and practice.

Level 1: Identification and classification. You will be able to differentiate reflection of meaning and interpretation/reframing from the related skills of paraphrasing and reflection of feeling. You will be able to identify questioning sequences that facilitate client talk about meaning. You will be able to provide new ways for clients to think about their issues through interpretation/reframing.

❑   Identify and classify the skills.
❑   Identify and write questions that elicit meaning from clients.
❑   Note and record key client words indicative of meaning.

Level 2: Basic competence. You will be able to demonstrate the skills of eliciting and reflecting meaning and interpretation/reframing in the interview. You will be able to demonstrate an elementary skill in dereflection.

❑   Elicit and reflect meaning in a role-play interview.
❑   Examine yourself and discern more fully your life direction.
❑   Use dereflection and attitude change in a role-play interview.
❑   Use interpretation/reframing in the interview.

Level 3: Intentional competence. You will be able to use questioning skill sequences and encouragers to bring out meaning issues and then reflect meaning accurately. You will be able to use the client's main words and constructs to define meaning rather than reframing in your own words (interpretation). You will not interpret but rather will facilitate the client's interpretation of experience.

With interpretation/reframing, you will be able to provide clients with new and fresh perspectives on their issues.

❑ Use questions and encouragers to bring out meaning issues.
❑ When you reflect meaning, use the client's main words and constructs rather than your own.
❑ Reflect meaning in such a fashion that the client starts exploring meaning and value issues in more depth.
❑ In the interview, switch the focus as necessary in the conversation from meaning to feeling (via reflection of feeling or questions oriented toward feeling) or to content (via paraphrase or questions oriented toward content).
❑ Help others discern their purpose and mission in life.
❑ When a person is hyperreflecting on the negative meaning of an event or person, find something positive in that person or event and enable the client to dereflect by focusing on the positive.
❑ Provide clients with appropriate new ways to think about their issues, helping them generate new perspectives on their behavior, thoughts, and feelings.
❑ Provide a new perspective via interpretation/reframing, using your own knowledge from the interview, helping your clients use these ideas to enlarge their thinking on their issues.
❑ Use various theoretical perspectives to organize your reframing.

Level 4: Psychoeducational teaching competence

❑ Teach clients how to examine their own meaning systems.
❑ Facilitate others' understanding and use of discernment questioning strategies.
❑ Teach reflection of meaning to others.
❑ Teach clients how to interpret their own experience from new frames of reference and to think about their experiences from multiple perspectives.
❑ Teach interpretation/reframing to others.

DETERMINING YOUR OWN STYLE AND THEORY: CRITICAL SELF-REFLECTION ON REFLECTING MEANING AND INTERPRETATION/REFRAMING

Meaning has been presented as a central issue in interviewing, counseling, and psychotherapy. Interpretation has been presented as an alternative method for achieving much the same objective but with more interviewer involvement. What single idea stood out for you among all those presented in this chapter, in class, or through informal learning? What stands out for you is likely to be important as a guide toward your next steps. What are your thoughts on multicultural issues and the use of this skill? What other points in this chapter struck you as important? How might you use ideas in this chapter to begin the process of establishing your own style and theory? Are you able to find new meanings and reinterpret/reframe your own life experience? And, in particular, what have you learned about discernment and its relation to your own life?

What are your thoughts?

_____

_____

OUR THOUGHTS ABOUT CHARLIS

Eliciting and reflection of meaning are both skills and strategies. As skills, they are fairly straightforward. To elicit meaning, we'd want to ask Charlis some variation of the basic question, "What does the heart attack mean to you, your past, and your future life?" As appropriate to the situation, questions such as the following can address the general issue of meaning in more detail:

"What has given you most satisfaction in your job?"
"What's been missing for you in your present life?"
"What do you find of value in your life?"
"What sense do you make of this heart attack and the future?"
"What things in the future will be most meaningful to you?"
"What is the purpose of your working so hard?"
"You've said that you have been wondering what God is saying to you with this trial. Could you share some of your thoughts here?"
"What would you like to leave the world as a gift?"

Questions such as these do not usually lead to concrete behavioral descriptions. They may often bring out emotions, and they certainly bring out certain types of thoughts and cognitions. Typically, these thoughts are deeper in that they search for meanings and understandings. When clients explore meaning issues, the interview, almost by necessity, becomes less precise. Perhaps this is because we are struggling with defining the almost indefinable.

As part of our work with Charlis, we'd ask if she wants to examine the meaning of her life in more detail through the process of discernment. This is a more systematic approach to meaning and purpose defined in some detail in this chapter. If she wishes, we'd share the specific questions of discernment presented here and ask her which ones she'd like to explore. In addition, we'd ask her to think of questions and issues that are particularly important to her, and we would give these special attention as we work to help her discern the meaning of her life, her work, her goals, and her mission.

Reflection of meaning as a skill looks very much like a reflection of feeling or paraphrase, but the key words "meaning," "sense," "deeper understanding," "purpose," "vision" or some related concept will be present explicitly or implicitly. "Charlis, I sense that the heart attack has led you to question some basic understandings in your life. Is that close? If so, tell me more."

It can be seen that we regard eliciting and reflecting meaning as an opening for the client to explore issues where there often is not a final answer but rather a deeper awareness of the possibilities of life. At the same time, effective exploration of meaning becomes a major strategy in which you bring out client stories, past, present, and future. You will use all the listening, focusing, and confrontation skills to facilitate this self-examination. Yet the focus remains on the client's finding meaning and purpose in his or her life.

CHAPTER 12

Influencing Skills: Five Strategies for Change

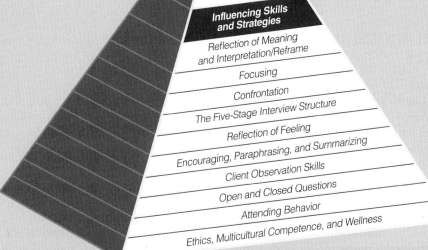

Influencing Skills and Strategies

Reflection of Meaning and Interpretation/Reframe

Focusing

Confrontation

The Five-Stage Interview Structure

Reflection of Feeling

Encouraging, Paraphrasing, and Summarizing

Client Observation Skills

Open and Closed Questions

Attending Behavior

Ethics, Multicultural Competence, and Wellness

Blessed is the influence of one true, loving human soul on another.

—George Eliot

How can influencing skills help you and your clients?

Chapter Goals

This chapter will provide you with the basics of five influencing skills—self-disclosure, feedback, logical consequences, information/psychoeducation, and directives. Each of these skills have specific strategies that may be useful in helping clients create new stories and generalize and act on what is discovered in the session.

Competency Objectives

Awareness, knowledge, and skills in the influencing skills will enable you to

▲ Help the client to look at the possible positive and negative results of alternative actions. (logical consequences)

▲ Share your own story, thoughts, or experiences briefly with clients. This may build a sense of equality in the session and encourage client trust and openness. (self-disclosure)

▲ Provide accurate data so that clients can learn how their behaviors, thoughts, and actions are seen by others and/or the interviewer. (feedback)

▲ Help clients look at the possible results of alternative actions. (logical consequences)

▲ Present new information and ideas to clients in a timely and appropriate fashion—for example, career information, teaching about sexuality, and results of test scores. (information/psychoeducation)

▲ Provide clients with specifics for action. Help them restory and take concrete action in their issues. Homework, meditation, and role-playing new ways of behaving are examples. (directives)

The client, Alisia, comes in with the complaint that she can't express herself. She feels helpless and believes that everybody "runs all over" her. She feels powerless and wants to talk about her difficulties in getting people's attention. She says that she has tried to communicate with her partner, her employer, and many others, but they don't seem to care. She is frustrated and angry. The case of Alisia will be discussed throughout this chapter to show how the influencing skills facilitate positive change.

INTRODUCTION: THE RELATIONSHIP OF LISTENING AND INFLUENCING SKILLS

Influencing is part of all interviewing and counseling. Being heard by another person greatly influences the way all of us think about ourselves and organize our lives. Confrontation, focusing, reflection of meaning, and interpretation/reframing have been identified already as skills of interpersonal influence. This chapter adds five more skills that place a different type of responsibility on the counselor, as all of them in one way or another directly seek to change the way the client thinks or acts.

Interpersonal Influence: Listening Skills and Influencing Strategies

Even a person-centered approach using only listening skills still influences what happens in the session. Through selective attention and the topics you choose to emphasize (or ignore) consciously or unconsciously, you influence what the client says. *You cannot* not *influence what happens in the interview.* Confrontation, focusing, reflection of meaning, and interpretation/reframing are strategies of interpersonal influence by which you may impact the client more directly.

Ethical practice demands respect for the client and awareness of the power relationship inherent in the interview. Interviewers and counselors, by their position, have perceived power. As you move to the direct action associated with the influencing skills, do not forget the foundation of listening and empathic understanding. Carefully developed listening skills are sometimes lost when one masters the influencing skills and strategies. Step back and remember that counseling and interviewing are for the client, not for you. Use influencing skills only with full client participation in the session.

Disclosure

Disclosure of what is going to happen in the session is an important part of maintaining a relationship and working alliance throughout the interview. By way of comparison, think of the effective dentist, nurse, or physician and how each one tells you ahead of time what to expect and whether the procedure will hurt. Our clients deserve the same respect. Disclosure tends to build comfort and trust even when the next step of the interview may not be comfortable. For example, if you have focused on listening and then decide to use influencing strategies, spend a moment acquainting the client with the change in style that is coming and its potential benefits. The general rule is to avoid surprises, although occasionally it is the very surprise that helps a client discover important new ideas.

Using Influencing Skills: Listen, Then Act

The "1-2-3" strategic model for using influencing skills presented in Box 12-1 is vital for maintaining client participation in the session. Keep listening skills as your most prominent style, even though you may be using a very directive intervention.

BOX 12-1 THE "1-2-3" PATTERN OF LISTENING, INFLUENCING, AND OBSERVING CLIENT REACTION

1. Listen

Use attending, observation, and listening skills to discover the client's view of the world. How does the client see, hear, feel, and represent the world through "I" statements and key descriptors for content (paraphrasing), feelings (reflection of feeling), and meaning (encouragers and reflection of meaning)?

2. Assess and Influence

An influencing skill is best used *after* you hear and understand the client's story. Timing is central—when is the client ready to hear or learn a new way to think

about what has been said? An influencing skill such as feedback, logical consequences, or a directive can be an abrupt change from the listening style unless the client is ready.

3. Check Out and Observe Client Response

Use a check-out as you offer an influencing skill ("How does that seem to you?"), then listen and observe carefully. If client verbal or nonverbal behavior seems incongruent or conflicted, return to the use of listening skills. Influencing skills can remove you from being totally "with" clients.

The effectiveness of your intervention can be readily assessed using the Client Change Scale (CCS, Chapter 9). It takes some practice, but eventually, you will assess client reactions to your leads in the here and now of the session. Being intentional demands that you flex and move with the client.

The Interpersonal Influence Continuum

The degree of interpersonal influence occurring in the interview varies from theory to theory. The word "influence" can be upsetting to a humanistic or person-centered counselor. By contrast, the many proponents of cognitive-behavioral theory aim to actively influence as much client change as possible. All theories agree that client involvement in the change process remains central.

It is possible to place all the microskills on a rough continuum of interpersonal influence. Figure 12-1 rates the attending skills and several of the influencing skills in terms of their impact in the interview. The task of the counselor is to open and close client discussion of a topic when appropriate. When a client is overly talkative or becomes upset, ask a closed question and then change the focus by asking an open question on another topic. When an interview is moving slowly, an interpretation or directive may add content and flow.

Think of the interpersonal influence continuum from time to time during your interviews. If you feel that you are coming on too strong and the client is resisting, it may be wise to move to lower levels of influence and focus more on listening. Similarly, if the client is bogged down, the careful use of an influencing skill may help organize things and move the interview along more smoothly.

Important for beginning and advanced interviewers is the moderate triad of skills: open and closed questions and focusing. If you can ask questions effectively and focus on varying topics, you have the ability to open and close almost any issue that your client presents. If a client has difficulty talking, an open question coupled with a slight change of focus helps open the interview. If a topic seems inappropriate or too emotional, a change of focus coupled with some closed questions will usually slow the pace. (Be sure to return to that emotional issue later.) You can achieve the same effects with other skills, but the moderate triad are "swing" skills in terms of their influence and do not seem to disrupt the interview flow when an interview needs to change direction or a topic needs to be discussed more fully.

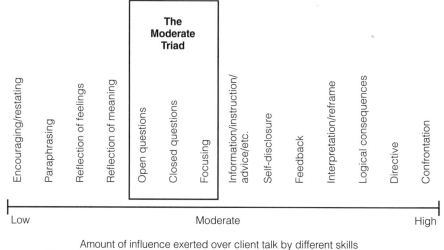

Amount of influence exerted over client talk by different skills
(All skills rest on a foundation of attending behavior and client observation.)

FIGURE 12-1   The interpersonal influence continuum.

EXAMPLE INTERVIEW: THE CASE OF ALISIA—HOW LISTENING SKILLS CAN INFLUENCE CLIENTS

The client, Alisia, comes in with the complaint that she can't express herself and people "run all over" her. The first interview began with a short rapport phase, and permission was obtained to record the session. The following transcript is an edited version of Alisia's first two interviews; listening skills were used almost exclusively. This is a *modified person-centered approach* in that questions are used more frequently than is typical of that theory. Effective listening skills and selective attending empower Alisia and enable her to see herself more as a self-in-relation, a person-in-community. Each of the influencing strategies presented in this and the next chapter will be applied to her issues.

Interviewer and Client Conversation	Process Comments
1. *Counselor:* Alisia, could you tell me what you'd like to talk about today?	The counselor personalizes the interview by using the client's name and the word *you* twice in the open question.
2. *Alisia:* I simply can't express myself. I've tried many times, and I can't get people to listen to me. Whether it is the boss, my partner, or the man at the garage, they all seem to run over me.	Alisia immediately identified her central issue. She appears to start the interview with acceptance and recognition of her problems (Level 3 on the Client Change Scale). Some clients need several interviews before the problem is defined this clearly. But she likely needs further understanding of this issues before change can be expected.
3. *Counselor:* Run over you?	Encourager focused on last few words.

(continued)

4. *Alisia:* Yeah, I keep finding that I'm so accommodating, that I'm always trying to get along with others. I was taught that I should please others. People like me for going along with them, but I never get what I want. I'm discouraged and disgusted with myself.	Alisia's body language is agitated, and her vocal tone moves to a higher pitch, which can indicate insecurity, fright, or general unsureness.
5. *Counselor:* Sounds as if you are really frustrated and disgusted about your inability to express yourself.	Reflection of feeling. Do you think the counselor should have changed the feeling word "discouraged" to "frustrated"?
6. *Alisia:* Right, it just goes on and on. . . . I never seem to change.	Resignation in her vocal tone, almost a sound of defeat.
7. *Counselor:* Could you give me a specific example of the last time you had these feelings of discouragement? What happened? What did you say? What did the other person do?	Open question searching for concreteness. If you look for one specific example, you will often obtain a much clearer understanding of client style and the depth of the problem.
8. *Alisia:* Well, I was at the garage. I had called in for an early morning appointment; I had to go to a meeting at 10:00. They said come in at 8:00, and so I was there on time. When I checked at 9:30, they hadn't even started yet. The service manager just smiled and said, "Sorry, lady, we couldn't get to it." He's one of those guys who really like to demean women. But I just looked down and didn't say anything—even though I really wanted to scream. I made another appointment, but my car still has that screwy, strange sound.	As Alisia shares the concrete example, she starts speaking in an angrier tone of voice and clenches her fist. Note that her lack of assertiveness shows here in a situation where she is not fully comfortable—the garage. But she is also able to point out that the service manager was likely being unfair to other women as well. She shows awareness of cultural/contextual issues and starts to get in touch with her underlying anger. This is further evidence of her functioning at acceptance and recognition, but no change would be recorded on the CCS.
9. *Counselor:* And after all this the car still isn't right. As I see you now, Alisia, you also seem to be getting angry. What's happening with you as you talk to me about this?	Paraphrase, reflection of feeling, open question oriented to the here and now.
10. *Alisia:* Angry. I'm just full of it. Men! I hurt too, deep inside. And I'm confused. (tears, but eyes flashing with determination) Everywhere I turn, it's there.	Listening skills also influence clients. Through sharing her story, Alisia becomes more aware of how she feels. The counselor is both surprised and personally touched by the depth of feeling and the meaning of all this to Alisia.
11. *Counselor:* I hear your anger and frustration. You're really angry and upset. You're tired of taking things as they are. The situation at the garage is just one instance of a pattern—something that repeats in various forms again and again. Have I heard you correctly?	Summarization with an emphasis on repeating patterns. Not all clients can see that they exhibit similar behaviors in different situations. A more concrete and less verbal client likely would be unable to realize that the situations are parallel.
12. *Alisia:* Yes, it's a pattern. The same day as the problem at the garage, my boss started leering at me again. I'm sick and tired of it. I used to think it was my fault, but now I'm wondering if men are the problem. The garage hassles me, the boss hassles me, and my partner does the same thing when he doesn't listen to me.	Notice how Alisia builds on the garage awareness to look at herself in other situations. Through effective use of listening skills, Alisia is starting to consider changes. This really is an expansion of acceptance/awareness (Level 3 on the Client Change Scale).

The counselor then spent time on Alisia's strengths and wellness assets. As part of the wellness search, the counselor encouraged Alisia to discuss several women heroes on whom Alisia would like to model herself. By the end of the interview, Alisia was still at Level 3 on the CCS, but understood her issues *and her strengths and resources* more fully. It may be necessary to build depth into each level of the Client Change Scale (horizontal development) before it is feasible to move to a higher level (vertical development). The second interview continues much the same as the first session, but near the end of the second interview, we hear the following:

Interviewer and Client Conversation	Process Comments
1. *Counselor:* So, Alisia, we've been talking for nearly an hour now. How do you put together all we've talked about? Have we missed something today?	The counselor could have summarized the session for Alisia, but uses the two questions as a way to involve the client in evaluation and planning for the future. The questioning process makes this version of person-centered counseling more active and influencing in style.
2. *Alisia:* I realize that much of what's been happening to me is a result of societal sexism. I learned in my family to do what "a good girl" should do and try to let the negative go and just be pleasant. But I'm not a girl; I'm a woman. I'm going to file harassment charges against my boss. And if we're going to stay together, I think I need to go with my partner for couples counseling. There! I feel better about myself right now.	She starts to see the need for behavioral change. Clients are not always so clear. This and the counselor statement above have been shortened from several client–counselor exchanges.
3. *Counselor:* You said a mouthful there, Alisia. That is a lot of things to do. Let's pick one or two of these possibilities and contract for what you might do next week as a start.	By suggesting contracting for change in behavior, the counselor has decided to move toward an active influencing approach and has presented a directive to the client. But even this directive includes Alisia in the planning of what is to happen.

Alisia is generating a new solution (Level 4 of the CCS), but still has some distance to go if her behavior is to change. Cognitively and emotionally, she is starting to touch on Level 4, but behavioral change will also be necessary to cement her newer thoughts and feelings. Drawing out her story carefully through the person-centered approach of these first two sessions provides a foundation for more active influencing later in this chapter.

What are some things that occur to you when working with a client such as Alisia, who has not allowed herself to express her thoughts and feelings more fully? Assuming a longer term relationship with her, what might be your goals and plan to help her?

What are your thoughts?

_____

_____

You may want to compare your thoughts with ours expressed at the end of this chapter.

INSTRUCTIONAL READING 1 AND EXERCISES: SELF-DISCLOSURE

If you use self-disclosure skills as described below, you can *predict* how clients will respond.

Self-Disclosure	Predicted Result
As the interviewer, share your own related personal life experience, here-and-now observations or feelings toward the client, or opinions about the future. Self-disclosure often starts with an "I" statement. Here-and-now feelings toward the client can be powerful and should be used carefully.	The client will be encouraged to self-disclose in more depth and may develop a more egalitarian relationship with the interviewer. The client may feel more comfortable in the relationship and find a new solution relating to the counselor's self-disclosure.

Should you share your own personal observations, experiences, and ideas with the client? Self-disclosure by the counselor or interviewer has been a highly controversial topic. Many theorists argue against counselors' sharing themselves openly, preferring a more distant, objective persona. However, humanistically oriented and feminist counselors have demonstrated the value of appropriate self-disclosure. Multicultural theory considers self-disclosure early in the interview as key to trust building in the long run. This seems particularly so if your background is substantially different from that of your client. For example, a young person counseling an older person needs to discuss this issue early in the session. Moreover, research reveals that clients of counselors who self-disclose report lower levels of symptom distress and like the counselor more (Barrett & Berman, 2001).

This brings up the important question of whether you have experienced the client's issues in some way. Many alcoholics are dubious about the ability of nonalcoholics to understand what is occurring for them. A client with a serious fertility problem often feels that no one can really understand her without experiencing her issues. Imagine that you are a heterosexual Christian Asian counselor working with a conservative Christian Latina struggling with lesbian issues. Your background is very different, and truly empathizing with the client may be a challenge. Open discussion, some self-disclosure on your part, and exploration of differences may be essential. Self-disclosure of who you really are can be helpful in those situations in which you have not "been there."

It appears that self-disclosure can encourage client talk, create additional trust between counselor and client, and establish a more equal relationship in the interview. Nonetheless, not everyone agrees that this is a wise skill to include among counselors' techniques. Some express valid concerns about counselors' monopolizing the interview or abusing the client's rights by encouraging openness too early; they also point out that counseling and interviewing can operate successfully without any interviewer self-disclosure at all.

Four dimensions of self-disclosure include the following:

1. *Listen.* Follow the "1-2-3" pattern—attend to the client's story, assess the appropriateness of your self-disclosure and share it briefly, and return focus to the client, while noting how he or she receives the self-disclosure.
2. *Use "I" statements:* Interviewer self-disclosure almost always involves "I" statements or self-reference using the pronouns "I," "me," and "my"—*or the self-reference may be implied.*

3. *Share and describe your thoughts, feelings, or behaviors briefly.* "I feel close to you." "I can imagine how much pain you feel." "My heart sings to hear you talk of that wonderful experience—it was a real change!" "My experience of divorce was hurtful." "I think your friends are taking advantage of you." "I also grew up in an alcoholic family and understand some of the confusion you feel."

4. *Use appropriate immediacy and tense.* The most powerful self-disclosures are usually made in the *here and now,* the present tense ("Right now I feel . . ."). "I am hurting for you at this moment—I care." However, variations in tense are used to strengthen or soften the power of a self-disclosure.

Making self-disclosures relevant to the client is a complex task involving the following issues, among others.

Genuineness in self-disclosure. To demonstrate genuineness, the counselor must truly and honestly have the feelings, thoughts, or experiences that are shared. Second, self-disclosure must be genuine and appropriate in relation to the client. Simply tell the client your own story. For example, if you are working with a client who grew up in an alcoholic family, and you yourself have had experience in your own family with alcohol, a brief sharing of your own story can be helpful. The danger of storytelling, of course, is that you can end up spending too much time on your own issues and neglect the client. Here the "1-2-3" influencing pattern, outlined above, is particularly important.

Immediacy, tense, and the here and now. The following examples show how the use of *here and now* bring immediacy to a session. These may be compared to the *there and then* of the past and future tense. But recall that all three approaches will be useful in the session—we need to know some past, and we also need to anticipate the future.

Alisia:	I am feeling really angry about the way I'm treated by men in power positions.
Counselor:	(present tense) You're coming across as really angry right now. I like that you finally are in touch with your feelings.
Counselor:	(past tense) I've had the same difficulty expressing feelings in the past. I recall when I would just sit there and take it.
Counselor:	(future tense) This awareness of emotion can help us all be more in touch in the future. I know it will continue to aid me.

Be careful when clients say, "What would you do if you were in my place?" Clients will sometimes ask you directly for opinions and advice on what you think they should do. "What do you think I should major in?" "If you were me, would you leave this relationship?" "Should I indeed have an abortion?" Effective self-disclosure and advice can potentially be helpful, *but* it is not the first thing you need to do. Your task is to help the client make her or his own decisions. The right solution for you may not be the right solution for the client, and involving yourself too early can foster dependency and lead the client in the wrong way. Note the following exchange.

Alisia:	How do you think I ought to present the idea of couples counseling to Chris?
Counselor:	I'm not in your position, and I haven't heard too much about Chris yet. First, could we explore your relationship in more detail? (past tense)
Counselor:	(If you feel forced to share your thoughts when you prefer not to, keep your comments brief and ask the client for her or his reflections.) I sense your hurt right now, Alisia. (here and now present tense) My own thought would be to share with Chris how you are hurting.

He may not be fully aware of how he's affecting you. But I'm not you. Chris may not hear that. What might be the outcome if you told him? (future tense)

Counselor: (after drawing out more information on the relationship) From what I've heard, it sounds wise to bring it up when Chris is in a good mood and able to listen to you. I think it is important to bring it up directly, and I admire the way you are thinking ahead. How does that sound to you? (future tense, moving to the here and now as Alicia explores the issue further)

Immediacy and timeliness. If a client is talking smoothly about something, counseling self-disclosure is not necessary. However, if the client seems to want to talk about a topic but is having trouble, a slight leading self-disclosure by the counselor may be helpful. Too deep and involved a self-disclosure may frighten or distance the client.

Individual Practice in Self-Disclosure

The structure of a self-disclosure consists of "I" statements made up of three dimensions: (1) the personal pronoun *I* in some form; (2) a verb such as *feel, think, have experienced;* and (3) a sentence objectively describing what you think or what happened. Box 12-2 provides an interesting reflection about when it is appropriate to self-disclose.

Exercise 1: Writing Self-Disclosure Statements

For each situation below, write one effective and one ineffective self-disclosure. Also, share one of your own stories briefly and include a check-out with each comment.

First, how would you self-disclose to Alisia after hearing her story?

_____

"My family is totally dysfunctional. You've heard my story. What do you think?"

_____

"I find myself afraid and insecure in large groups. It just doesn't feel right. What should I do?"

_____

INSTRUCTIONAL READING 2 AND EXERCISES: FEEDBACK

If you use feedback as structured below, you can *predict* how clients will respond.

Feedback	Predicted Result
Present clients with clear information on how the interviewer believes they are thinking, feeling, or behaving and how significant others may view them or their performance.	Clients will improve or change their thoughts, feelings, and behaviors based on the interviewer's feedback.

To see ourselves as others see us,
To hear how others hear us,
And to be touched as we touch others . . .
These are the goals of effective feedback.

BOX 12-2   WHEN IS SELF-DISCLOSURE APPROPRIATE?

Weijun Zhang

My good friend Carol, a European American, has had lots of experience counseling minority clients. She once told me that one of the first questions she asks her minority clients is "Do you have any questions to ask me?" which often results in a lot of self-disclosure on her part. She would answer questions not only about her attitudes toward racism, sexism, religion, and so forth, but also about her physical health and family problems. During the initial interview, as much as half of the time available could be spent on her self-disclosure.

"But is so much self-disclosure appropriate?" asked a fellow student in class after I mentioned Carol's experience.

"Absolutely," I replied. We know that many minority clients come to counseling with suspicion. They tend to regard the counselor as a secret agent of society and doubt whether the counselor can really help them. Some even fear that the information they disclose might be used against them. Some questions they often have in mind about counselors are "Where are you coming from?" "What makes you different from those racists I have encountered?" and "Do you really understand what it means to be a minority person in this society?" If you think about how widespread racism is, you might consider these questions legitimate and healthy. And unless these questions are properly answered, which requires a considerable amount of counselor self-disclosure, it is hard to expect most minority clients to trust and open up willingly.

Some cultural values held by minority clients necessitate self-disclosure from the counselor, too.

Asians, for example, have a long tradition of not telling personal and family matters to "strangers" or "outsiders," in order to avoid "losing face." Thus, relative to European Americans, we tend to reveal much less of ourselves in public, especially our inner experience. A mainstream counselor may well regard openness in disclosing as a criterion for judging a person's mental health, treating those who do not display this quality as "guarded," "passive," or "paranoid." Nonetheless, traditional Asians believe that the more self-disclosure you make to a stranger, the less mature and wise you are. I have learned that many Hispanics and Native Americans feel the same way, too.

Because counseling cannot proceed without some revelation of intimate details of a client's life, what can we do about these clients who are not accustomed to self-disclosure? I have found that the most effective way is not to preach or to ask, but to model. Self-disclosure begets self-disclosure. We can't expect clients to self-disclose if we don't sometimes do it ourselves.

According to the guidance found in most counseling textbooks here, excessive self-disclosure by the counselor is considered unprofessional; but if we are truly aware of the different orientation of minority clientele, it seems that some unorthodox approaches are needed.

The classmate who first questioned the practice asked with a smile, "Why are you so eloquent on this topic?" I said, "Perhaps it is because I have learned this not just from textbooks, but mainly from my own experience as both a counselor and a minority person."

Knowing how others see them is a powerful and impactful dimension in human change, and it is most helpful if the client solicits feedback. Feedback is an important influencing strategy if you have developed good rapport and enough experience with the client to know that he or she trusts you. These feedback guidelines are critical in counseling and interviewing:

1. *The client receiving feedback should be in charge.* Listen first, use the "1-2-3" pattern, and determine whether the client is ready for feedback. Feedback is more successful if the client solicits it.
2. *Feedback should focus on strengths and/or an issue the client can do something about.* It is more effective to give feedback on positive qualities and build on strengths. Corrective

feedback focuses on areas in which the client can improve thinking, feeling, and behaving. Corrective feedback needs to be about something the client can change or to help the client recognize and accept that something can't be changed.

3. *Feedback should be concrete and specific.* "You had two recent arguments with Chris that upset both of you. In each case, I hear you giving in almost immediately. You seem to have a pattern of giving up, even before you have a chance to give your own thoughts. How do you react to that?"

4. *Feedback should be relatively nonjudgmental and interactive.* Stick to the facts and specifics. Facts are friendly; judgments may or may not be. Demonstrate your nonjudgmental attitude through your vocal qualities and body language. "I do see you trying very hard. You have a real desire to accept the way Chris is and learn to live with what you can't change. How does that sound?" Compare the latter with "You give in too easily; I wish you'd try harder," or the all-too-common "That was a *good* job."

5. *Here-and-now, present-tense feedback can give real immediacy to the interview.* "Right now at this moment, I see you actually feeling new power—the way you sit, the look in your eyes. They all convey strength." Corrective feedback in the here and now can also be helpful, but must be approached with care. "Alisia, I really am touched by how hard you are trying. (here and now) I'd like to make a suggestion—would it be possible for you just to stand a little straighter and look him in the eye?" (there and then, future tense)

6. *Feedback should be lean and precise.* Don't overwhelm the client; keep corrective feedback brief. Most of us can hear only so much and can change only one thing at a time. Select one or two things for feedback and save the rest for later.

7. *Check out how your feedback was received.* Involve clients in feedback through the checkout. Their response indicates whether you were heard and how useful your feedback was. "How do you react to that?" "Does that sound close?" "What does that feedback mean to you?"

Positive feedback has been described as "the breakfast of Champions." Your positive, concrete feedback helps clients restory their problems and concerns. Wherever possible, find things about your client that are right. Even when you have to provide challenging feedback, try to include positive assets of the client. Help clients discover their wellness strengths, positive assets, and resources.

Corrective feedback is a delicate balance between negative feedback and positive suggestions for the future. When clients need to seriously examine themselves, corrective feedback may need to focus on things that clients are doing wrong or behavior that may hurt them in the future. Management settings, correctional institutions, schools, and universities often require the interviewer to provide corrective feedback in the form of reprimands and certain types of punishment. When you must give negative corrective feedback, keep your vocal tone and body language nonjudgmental and stick to the facts, even though the issues may be painful. *Praise* and *supportive statements* ("You can do it, and I'll be there to help") convey your positive thoughts about the client, even when you have to give troubling feedback.

Negative feedback is necessary when the client has not been willing to hear corrective feedback. For example, in cases of abuse, planned behavior that hurts self or others, and criminal behavior, negative feedback with the logical consequences is necessary and can be beneficial (see this skill on page 338). It is our responsibility to act in these situations. But listening to the client's point of view, even if he or she is a perpetrator, remains important.

Counselor: Alisia, I admire your ability to hang in with Chris and accept things as they are, but you really are giving away too much control. You need stronger boundaries. Times are changing, and it is okay for you to be assertive and own your own space. You have strength and ability. We can work on you becoming your own person. How does that feedback sound to you?

Feedback and the client who avoids certain topics. You may have a client who suddenly switches topic or gives you only a brief, vague response. Many clients have sensitive issues or topics that they don't want to explore. If the interview is just for one to three sessions, it is usually best to accept that behavior. But sometimes the issue really needs to be faced; meet the client and use confrontation skills as part of the feedback. Some examples:

"We've talked around the issue of dealing with Chris, but you never say whether you really want to stay together. On one hand, I hear you wanting to resolve issues, but on the other hand, you avoid expressing what you feel about the relationship and what you want."

"On one hand, Alisia, you really do seem to want to become more assertive, but then when we start to talk seriously about how you might actually change, you avoid the issue and turn away."

"Just now, I saw it again. We were starting to deal with real issues, and you changed the topic. What's going on?"

Here are some feedback examples with our client, Alisia:

Vague, judgmental, negative feedback	Alisia, I really don't think the way you are dealing with people who hassle you is effective. You come across as a weak person.
Concrete, nonjudgmental, positive feedback	Alisia, you have potential, I sense that you tried very hard to stand up for your rights in the garage. Now, you seem more sure of yourself than you used to be. May I suggest some specific things that might be helpful the next time you face that situation?
Corrective feedback	Your effort was in the right direction. You can do even more if we set up an assertiveness training session for you. As we do this, we'll likely come up with some useful new ideas to help you cope and use your strengths.

Box 12-3 summarizes evidence in support of the skills presented in this chapter.

BOX 12-3 RESEARCH EVIDENCE THAT YOU CAN USE

Self-Disclosure, Feedback

Interpretation has been extensively reviewed and researched by Hill and O'Brien (1999). Among their findings are that interpretations are well received by clients, even though they are used sparingly by the interviewer. They also state that the use of interpretations at a moderate depth tends to be favored. This would suggest that the concept of reframing, particularly as it relates to reflection of meaning, is useful. The language of reframing and reflection of meaning suggests "taking another perspective" and is less of an imposition from the interviewer.

In a carefully designed study on the therapeutic outcome of self-disclosure, Barrett and Berman (2001) found that clients whose therapists self-disclosed "not only reported lower levels of symptom distress but also liked their therapist more" (p. 596). The authors talked about psychodynamically oriented therapists who argue that the interviewer should be a "blank slate" and a neutral observer as contrasted with humanistic therapists who have long argued for appropriate self-disclosure. We should, however, keep in mind that self-disclosure (like interpretation/reframing) needs to be used only occasionally and with sensitivity to where the client is "at the moment."

Burkard et al. (2006) reported that European American psychotherapists' use of self-disclosure in cross-cultural counseling seemed to enhance the counseling relationship. They indicated also that they knew the role of racism/oppression in clients' lives and acknowledged their own racist/oppressive attitudes. Therapists reported that these self-disclosures frequently improved counseling because they helped clients to feel understood.

Feedback has been most investigated by group therapists. Moran, Stockton, Cline, and Teed (1998) have conducted a major review of the literature, and their findings are in agreement with the definitions and functions provided in this chapter. They also note that early feedback facilitates the group process but that

some group members have real difficulty in hearing feedback. For individual counseling, feedback should follow the guidelines in this chapter, and if clients resist feedback, continue listening and exploring the issue. A more solid relationship may come later on, and then feedback will be more appropriate. Timeliness is the issue!

Clients who are at risk for leaving the helping relationship may be more likely to stay if they receive effective feedback. Harmon et al. (2007) studied 1,347 at-risk clients who were deteriorating in the process of therapy. They found that effective feedback with appropriate immediacy improved client outcome.

Influencing Skills and Neuroscience

Influencing skills are almost always focused on the positive and the possibility of change. They provide many alternatives to empower clients and can be drawn from virtually all theoretical perspectives.

Our key task, from a neuroscience framework, is to energize positive emotions, which are located primarily in the frontal cortex, particularly the left executive, decision-making brain. For example, mindfulness meditation strengthens positive emotions (Kabat-Zinn, 2005). In times of severe stress or panic, the amygdala (primary location of sad, mad, and fear), combined with the intuitive right brain, can take over, and the individual becomes ruled by negative emotions. Damasio (2003, p. 12) points out that we need to build positive emotions and strength capacities to cope with the negative. As the old song goes: *Accentuate the positive; eliminate the negative!* Building on wellness and strengths will enable clients to cope with major challenges. Some even speculate that practitioners will be able to tailor specific treatments to modify brain circuits through a counseling wellness orientation and positive skills and strategies (Beitman & Good, 2006).

Individual Practice With Feedback

Exercise 2: Your Experience With Feedback

We have all experienced feedback on our performance. Some of this feedback has been helpful, and some has been painful. Allow yourself to recall a positive and a negative experience with feedback. This may have been feedback from friends, family, teachers, or a work supervisor. What do you notice personally about effective and ineffective feedback?

_____

Exercise 3: Writing Feedback Statements

Imagine that Alisia has just said, "I'm totally lost right now. I thought I had made progress on this issue, but this last week, I totally blew it again. I'm so discouraged with myself."

Write a positive feedback statement.

_____

What might a negative feedback statement be?

_____

Now try the more challenging issue of presenting corrective feedback in a positive way to Alisia.

_____

Review the above feedback statements considering the criteria of effective feedback. How many of these criteria did you meet in each situation?

Here are three more client statements. Imagine that you have heard a longer story and then provide positive and negative feedback comments and evaluate them as above:

"My partner is simply impossible. He demands so much of me. I just feel myself disappearing in the relationship. I need space."

"What do you think of me? I've done as much as I can in working through issues around my sexuality, yet others keep looking at me and talking behind my back."

"I have real difficulty with exams, no matter how hard I study. In the last test I worked really hard, read all the assignments, and thought I knew the material cold. But I still ended up with a weak C."

INSTRUCTIONAL READING 3 AND EXERCISES: LOGICAL CONSEQUENCES

If you use logical consequences as suggested below, you can *predict* how clients will respond.

Logical Consequences	Predicted Result
Explore specific alternatives and the logical positive and negative concrete consequence of each possibility with the client. "If you do this . . . , then . . ."	Clients will change thoughts, feelings, and behaviors through better anticipation of the consequences of their actions. When you explore the positives and negatives of each possibility, clients will be more involved in the process of decision making.

This strategy of logical consequences is particularly important in making decisions and is used in many theoretical approaches to the interview. It is most often a gentle strategy used to help people sort through issues when a decision needs to be made. It may be useful to rank alternatives when a complex decision is faced. In interviewing, assist clients to foresee consequences as they sort through alternatives for action: "If you do . . . , then . . . will possibly result."

The strategy of logical consequences is most often used in Adlerian counseling (Dreikurs & Gray, 1968; Sweeney, 1998) and decisional counseling (Chapter 13). The interviewer helps individuals explore alternatives, consider consequences of alternatives, and facilitate decision making among the possibilities. For example, an individual may come to the interview aware that changing jobs offers more pay but less aware of the effects of a move to a new city. Through systematic questioning and discussion, the interviewer can help the client clarify the factors involved in the decision. Potential *negative consequences* could include leaving a smoothly functioning and friendly workgroup, disrupting long-term friendships, moving to a new school, and other factors that may cause problems. *Positive consequences* might be a pay raise and the opportunity for further advancement, a better school system, and money for a new home.

In another use of logical consequences, the interviewer or counselor may need to help clients become aware of the potential negative consequences of their actions. Some examples include the client who is thinking of dropping out of school, the pregnant client who has not stopped smoking, or the client who wants to "tell off" a co-worker or friend. It is equally important to help clients anticipate the positive consequences—the results and rewards—of specific behaviors. The pregnant woman's baby is likely to be healthier if she stops smoking; the client who graduates from school will probably find a better job; and the person struggling with a difficult co-worker or friend may avoid more unpleasantness by simply keeping her or his mouth shut for the moment. Clients can make better decisions when they can envision the likely consequences of any given action.

Counselor: What is likely to happen if you continue smoking while you are pregnant?

Client: I know that it isn't good, but I can't stop, and I really don't want to.

Counselor: Again, what are the possible negative consequences of continuing to smoke?

Client: (pause) I've been told that the baby could be harmed.

Counselor: Right; is that something you want? What is the benefit of stopping smoking for the baby?

Client: No, I don't want to do harm. I'd be so guilty. But how can I stop smoking?

Counselor: Let's explore that. There are several ways, none of them easy. But let us consider . . .

In situations when the client has been required to come to the session, it is important to note that more power rests with the interviewer. The court may ask the interviewer to recommend actions that the legal system could take. The gentle logical consequence skill becomes more powerful. Warnings are a form of logical consequences and may center on *anticipation of punishment;* if used effectively, warnings may reduce dangerous risk taking and produce desired behavior. The counselor or correctional staff often need to help clients see clearly what might be ahead if they continue the present behavior.

Virtually all human behavior has costs and benefits. By involving the client in examining the pluses and minuses of alternatives, the counselor gives the client the power to make a better decision or at least to share his or her thinking more openly. Consider the following suggestions for using the strategy of logical consequences:

1. Through listening skills, make sure you understand the situation and the way your client understands it. After drawing out the situation, either you or the client can summarize what is happening.
2. Use questions and brainstorming to help the client generate alternatives for resolving issues. Where necessary, provide additional alternatives for consideration.
3. Work with the client to outline both the positive and negative consequences of any potential decision or action. In important cases, ask the client to generate a possible future story of what might happen if a particular choice is made. For example, "Imagine two years from now. What will your life be like if you choose the alternative we just discussed?"
4. As appropriate to the situation, provide clients with a summary of positive and negative consequences in a *nonjudgmental* manner.
5. Encourage client decision making as much as possible.

The following exchange with Alisia demonstrates a use of logical consequences. In this case, the decision is between keeping things as they are or introducing a new behavior.

Interviewer and Client Conversation	Process Comments
1. *Counselor:* Alisia, we know that you would like to be more assertive and speak up more for yourself. What are the likely positive consequences if you can do this?	The counselor paraphrases and then asks Alisia to identify positive consequences of change.
2. *Alisia:* Well, I've learned that if I don't speak up in the garage, nothing is going to happen. I suppose that I have nothing to lose by trying to be stronger. I guess the positive result would simply be something different.	Alisia responds well and notes that she has "nothing to lose" by trying something different.
3. *Counselor:* "Something different," sounds like you'd feel better about yourself and maybe even get the car fixed.	Encourager and paraphrase. The counselor suggests another potential positive consequence of change.
4. *Alisia:* That would be nice, but it is a little scary.	Change is not easy.
5. *Counselor:* I hear that. What are the negative consequences of continuing your past behavior?	Open question with a strong influencing dimension. Balancing the decision with negative consequences of not changing.
6. *Alisia:* Not good.	
7. *Counselor:* So the consequences of trying a change may make something happen for the good, and you have nothing to lose. On the other hand, the consequences of staying as you are, as you say, are "not good."	Summary of the positive and negative consequences of taking or nor taking an action. Decisional counseling, Adlerian counseling, and many other forms of cognitive counseling all use the logical consequences strategy.
8. *Alisia:* Right; well let's try something new. I've decided I'm ready for a change.	

Alisia also said that she wanted to take a stronger role with her partner. As she explored the negative consequences of speaking more forcefully, she realized the first negative consequence was that her partner might leave her. The financial challenge and the likely need to find a new and much less expensive apartment emerged as additional negative consequences. Alisia also feared being alone, as she has had other bad experiences with loss. These are common experiences around separation and unfortunately often result in an abused woman returning to her abusing spouse. On the more positive side, Alisia realized how good it would feel to speak up for herself, and she wasn't at all sure that her partner actually would leave. She would feel better about herself if she could learn to take decisive positions. As she balanced the positives and negatives, she decided it was time for her to speak up and see what happened. She and the counselor agreed to explore this issue in more detail in subsequent interviews.

The decisional balance sheet. With clients who have important decisions with several possibilities, it helps to write down the alternatives and the pluses and minuses of each. For example, in choosing a college or a job change, the several possibilities are listed and what the client likes and dislikes about each one can be seen visually. If there is one especially important issue, mark it with two "+" or "−" signs (Mann, 2001).

Combine focusing with logical consequences to help the client see issues in a broader context. What are the logical consequences for others if Alisia changes her style—the fellow in the garage, her partner, and others within the cultural/environmental/contextual sphere? The counselor could use varying focus dimensions to broaden Alisia's understanding. Here are some examples of focusing to help individuals see themselves as beings-in-relation, persons-in-community.

9. *Counselor:* Alisia, what are the implications for the service manager if you speak up more forcefully?	Focus on others—the service manager, in particular.
10. *Alisia:* Hmmm. Well, I imagine he would do one of three things. First, he might ignore me and continue, but I wouldn't allow that, as I want him to deal with me. Second, I bet he'll do a better job, and perhaps he will respect other women as well. The third possibility is that he will talk back to me rudely. But, if so, I'm going to talk to the manager of the garage. I'm fed up.	Here Alisia is looking at the logical consequences for the service manager if she becomes more assertive.
11. *Counselor:* And what are the consequences for your partner if you say that you want counseling because you want more equality in the relationship?	Focus on others—the partner. Open question.
12. *Alisia:* I think Chris will be put off; he is not very verbal and fears counseling. But I also know that he would like us to get along better.	Alisia can make a better decision if she anticipates what her decisions mean for others.
13. *Counselor:* So, Chris might accept it. It does sound as if you want to be a stronger woman. Your grandmother was a powerful role model, and you did say that talking back to your service manager might be a strike for women in general.	Focus on Chris, then on family, and finally, the cultural/environmental/contextual issue.

(continued)

14. *Alisia:* One thing that I'm learning here is that every woman has a responsibility to speak up. I need to be part of that.	We are seeing cognitive and emotional change to generating new solutions. But Alisia has to implement her thoughts and feelings in behavior to reach this level fully.
15. *Counselor:* It makes me feel good to hear you say that you see your responsibility for others. That may help you "hang in" when the going gets tough as you change your style.	Focus on the interviewer with a self-disclosure followed by pointing out the logical consequences for women in general as Alisia changes her style.
16. *Alisia:* I'm ready to start. What next?	

Individual Practice in Logical Consequences

Exercise 4: Writing Logical Consequences Statements

Using the five steps of the logical consequences skill, briefly indicate to the client, Alisia, what the logical consequences will be for her if she continues with her lack of assertiveness.

Summarize Alisia's problem in your own words using "if . . . , then . . ." terms.

_____

Ask Alisia specific questions about the positive and negative consequences of continuing her behavior.

_____

Provide Alisia with your own feedback on the probable consequences of continuing her behavior. Use "if . . . , then . . ." language.

_____

Summarize the differences between the feedback just given and Alisia's view when she says she doesn't want to change (this implies the use of confrontation).

_____

Encourage Alisia to make her own decision.

_____

Exercise 5: Logical Consequences Using Listening Skills

By using questioning skills you can encourage clients to think through the possible consequences of their actions. ("What result might you anticipate if you did that?" "What results are you obtaining right now while you continue to engage in that behavior?") However, questioning and paraphrasing the situation may not always be enough to make clients fully aware of the logical consequences of their actions. For each client and situation, write logical consequences statements that help the client understand the situation more fully.

A student who is contemplating taking drugs for the first time:

_____

A young woman contemplating an abortion:

_____

A student considering taking out a loan for college:

_____

An executive in danger of being fired because of poor interpersonal relationships:

_____

A client who is consistently late in meeting you and who is often uncooperative:

_____

INSTRUCTIONAL READING 4 AND EXERCISES: INFORMATION AND PSYCHOEDUCATION

If you use information and psychoeducation as shown below, you can *predict* how clients will respond.

Information and Psychoeducation	Predicted Result
Share specific information with the client (e.g., career information, choice of major, where to go for community assistance and services). Offer advice or opinions on how to resolve issues and provide useful suggestions for personal change. Teach clients specifics that may be useful—helping them develop a wellness plan, teaching them how to use microskills in interpersonal relationships, educating them on multicultural issues and discrimination.	If information and ideas are given sparingly and effectively, the client will use them to act in new, more positive ways. Psychoeducation that is provided in a timely way and involves the client in the process can be a powerful motivator for change.

Giving the client information, offering psychoeducation and your opinions, or making suggestions can be an important part of interviewing and counseling. However, be aware that advice giving is fraught with danger; unless clients actively seek the advice, they will rarely hear or heed even the best of suggestions. For example, try offering teens suggestions on how they should dress, drive, and/or handle alcohol. It is immensely difficult to convince a person to stop smoking. Children resist suggestions. Adults are told to lose or gain weight, get more exercise, and eat more fruits and vegetables, and yet we have real difficulty in listening to or following advice that may be critical to improving our physical well-being.

When listening to counselors who provide information and psychoeducation, the client needs to be in charge and actually want to hear and learn something new, and in many situations the information is welcome. Career and college counseling must provide students with career and college admissions information, and here the teen may actually listen. Students facing critical life decisions frequently want to know your opinions and advice. A family member caring for an older parent often desperately wants advice on how to handle this extremely challenging part of life, particularly around death and dying and hospice referral.

Psychoeducation involves a more formal and systematic set of strategies that can be influential in helping clients move to new places in their lives. Social skills training—basically teaching the microskills of this book to clients—has become a major part of most counselor and therapist options for treatment planning. In addition, microskills are a central part of most peer counseling training programs. Psychoeducation strategies are often taught in groups (meditation, relaxation training, assertiveness training, dating skills, multicultural awareness, etc.). But these same strategies are equally, if not more, effective if taught on the spot to clients as part of the session.

For example, consider the several dimensions of wellness discussed in Chapter 2. If you want to encourage clients to exercise, have better nutrition, expand their friendship group, or manage stress more effectively you need to wait for a timely and appropriate opportunity to share the values and health gains of a wellness program. You will also encounter clients who have obvious problems with their own sexism, racism, or other prejudice. When you wait and watch for a moment that is conducive to introducing information about these topics, you can teach and help individuals become more aware of their relationships with others.

All of the following situations may involve some balance of information giving and psychoeducation—clients going through a divorce, a person grieving the death of a family member, those who need information on how to obtain Social Security and health benefits, clients dealing with a family member who has Alzheimer's disease, or those struggling to find housing. While it is obviously vital to listen first and understand the client's issues and story, there are many times that entering in and providing key data and new perspectives can change the course of a client's life.

More challenging is the student who is not doing well in school or the office worker who consistently shows up late for work. Both know that change would be wise and what your advice is likely to be. It is all the more important to hear their story and point of view before attempting any advice. The use of logical consequences is likely to be more helpful here than advice.

In these situations, the strategy is clear and the issue is getting information across and motivating clients. It is especially important to listen first and learn the story *and discover the client's goal* before you offer information or psychoeducation. Change your teaching approach when you see clients roll their eyes, slump back in the chair, or look at the ceiling. However, most clients will look to you for important information, and as you build trust, they will become eager to hear your advice. Provide information, psychoeducation, and advice sparingly and only when the client is likely to need and accept it.

Be very cautious in giving advice, providing information, and engaging in psychoeducation. This skill area is literally addicting for some interviewers, and they will habitually take over the session, using most of the talk-time. We are sure that you can think of many instances when you were given advice or even psychoeducation that was inappropriate and controlling.

Here are some examples of sharing information with Alisia:

Career information: I'd like you to explore some job alternatives. I'm going to show you our career library, and we can examine other possibilities for the future. You are now in computer science, but not using all your talents. This chart shows a large increase in computer career opportunities coming in the next decade.

Sharing your thoughts: You asked for my advice about Chris. It's not my place to tell you what to do. But I do think it is time for you to sit down with him and have a serious talk. I've got some ideas for you that might help you express your thoughts and feelings in a way that could help Chris actually hear you. (This is closely related to self-disclosure but goes on in much more detail.)

Psychoeducation: One route toward handling difficult situations is to engage in attending behavior—listening carefully to the person whom you find difficult. We know that you tend to give in too easily, but if you listen and observe people carefully, you may find new ways to understand them and act more effectively. I'll teach you some of the basics of effective listening.

Psychoeducation: One of the most important items for health and wellness is a good exercise plan. One president of the American Medical Association commented that it is unethical for any physician not to recommend exercise. Thus, Alisia, it is critical that we develop an exercise plan to help you work with stress and maintain a healthy balance to your life. Let me tell you why, and then we can discuss it. Interested? Is this okay?

Individual Practice in Information and Psychoeducation

Exercise 6: Your Experience With Information and Advice

What is your own experience with information or advice given to you before you were ready? How do you personally respond when someone gives you advice or tells you what to do? Summarize below both negative and positive thoughts you have about this area.

_____

Exercise 7: Your Experience With Psychoeducation

You have likely been in a variety of workshops. Learning microskills themselves is a type of psychoeducation. Individual counseling may have focused on teaching you some skills. What has been your experience with psychoeducation, and what part might it play in your practice?

_____

Exercise 8: Writing Statements

Again, imagine that Alisia has just said, "I'm totally lost right now. I thought I had made progress on this issue, but this last week, I totally blew it again. I'm so discouraged with myself."

Write how you might use information or psychoeducation to involve Alisia fully in the process so that you do not foster dependency.

_____

Here are two client statements. Imagine that you have heard a longer story and then attempt to provide meaningful information or a psychoeducational plan. Again, what are the potential dangers of involving yourself too much?

"I don't know how to speak up in class. I just sit there and part of the grade is dependent on class participation."
"I feel so stressed. What can I do about it?"

_____

_____

INSTRUCTIONAL READING 5 AND EXERCISES: DIRECTIVES

If you use directive skills as structured below, you can *predict* how clients will respond.

Directives	Predicted Result
Direct clients to follow specific actions. Directives are important in broader strategies such as assertiveness or social skills training or specific exercises such as imagery, thought stopping, journaling, or relaxation training. They are often important when assigning homework for the client.	Clients will make positive progress when they listen to and follow the directives and engage in new, more positive thinking, feeling, or behaving.

While directives are useful in developing a new story or thinking in new ways, they are especially effective in helping a client move to behavioral action. Directives are particularly useful in the *restory* and *action* part of the interview. A positive new story may be sufficient for some clients, but many will profit from directive strategies outlining specific behaviors and actions they should take. Directive strategies tend to be drawn from various counseling theories that direct the client to follow a specific sequence of events designed to produce a likely result. One directive strategy discussed here—homework—has been shown to be especially important in producing results and follow-up from the interview. Homework will be discussed at the end of this section.

Effective directives require an expansion of the "1-2-3" pattern.

1. *Involve your client as co-participant in the directive strategy.* Rather than simply tell the client what to do, be sure that you have heard the story, issues, and problems sufficiently. Inform the client what you are going to do and the likely result. Some practitioners like to use surprises (e.g., Gestalt theory), and this strategy can be useful in some situations. But as a general rule, we urge *working with,* rather than *working on,* your client.
2. *Use appropriate visuals, vocal tone, verbal following, and body language.* Your attending behaviors need to flex in response to the needs of the client. Usually, a more forward and active behavioral style is needed when challenging an acting-out teen or an outgoing client. You may need a stronger persona with even clearer verbal and nonverbal behavior. With a more quiet and tentative client, appropriate attending may require being more still and tentative as you share new ways of thinking about issues. Directives given softly can be very effective.
3. *Be clear and concrete in your verbal expression and time the directive to meet client needs.* Directives need to be authoritative and clear but also stated in such a way that they are in tune with the needs of the client. Compare the following:

Vague:	Go out and arrange for a test.
Concrete:	After you leave today, contact the testing office to take the Strong-Campbell Interest Test (also known as the Strong Vocational Interest Inventory; another interest inventory is the Self-Directed Search or SDS). Complete it today, and the office will have the results for us to discuss during our meeting next week.
Vague:	Relax.
Concrete:	Sit quietly . . . feel the back of the chair on your shoulders . . . tighten your right hand . . . hold it tight . . . now let it relax slowly . . .

These examples illustrate the importance of indicating clearly to your client what you want to happen. Know what you are going to say and say it clearly and explicitly.

4. *Check out whether your directive was heard and understood.* Just because you think you are clear doesn't mean the client understands what you said. Explicitly or implicitly check to make sure your directive is understood. This is particularly important when a more complex directive has been given. For example, "Could you repeat back to me what I just asked you to do?" or "I suggested three things for you to do for homework this coming week. Would you summarize them to me to make sure I've been clear?" The Client Change Scale can be used to determine whether the client actually changed thoughts, feelings, or behaviors as a result of your directive.

As you start using directives, remember they can come across as "telling clients what to do." It is very important not to get too enthusiastic with these strategies. Very few of us like to be *told* what to do. Always remember to empower your clients so as to make them co-partners in the here and now of the interview and as you together select the skills and strategies that might help produce growth and the development of the *New*.

The counselor prepares Alisia ahead of time before giving a directive:

Counselor:	Alisia, I've heard your story about how frustrated you feel with that service manager. You have said that you'd like to try something new. But, as we start, would you like to try an imagery exercise to understand the situation a bit more fully? This is how it works. I'll ask you to relax, sit back, and allow yourself to recall the situation. It often helps to visualize the specifics of the situation, what was said, and so on—almost like a movie. Before we start, how does that sound? Let's talk a moment before trying this exercise. What issues come to your mind that we need to talk about?

The directive strategies mentioned below are tried and true and used in several different approaches to counseling and therapy. Test these strategies first by trying them on yourself. Later, work with a friend or classmate, and then have that person test the same directive strategy with you. If you practice the details of directives, you will have a better idea of their potential and how to pace and time the strategy. All of the following are well known and effective, but they must be in full attunement with client needs and wishes.

Homework

"Practice this exercise next week and report on it in the next interview."

"Alisia, next week I want you to try some of the skills we've worked on in the role-play. Focus on standing up straighter; be particularly attentive to your eye contact patterns—you've learned you tend to look down when there's the potential of conflict. Just this week, focus on these two changes."

Research increasingly shows that learning during the interview is easily lost if it is not immediately transferred to daily life.

The fifth stage of the interview is concerned with action—generalizing thoughts, feelings, and behaviors to the "real world" outside of the session. Working with the client to *do something different or new* during the coming week can be invaluable.

Homework assignments can include playing basketball with friends, starting a program of walking or running, or keeping a diary of foods eaten. If assertiveness is the issue, the homework may be to record the negative or faulty thought each time it occurs during the day.

A couple with difficulties may be asked to observe and count the number of arguments they have. They record the before, during, and after aspects of the argument to discuss with the counselor. No change is expected because the client is simply observing and recording what is going on.

With some clients, just observing themselves leads them to change their behavior! The client can also record what happened just before a thought and what happened afterward. Such records can be valuable in changing faulty thinking patterns. There are endless ways to involve clients in homework following the interview.

The Relaxation Response

> *The relaxation response is a physical state of deep rest that changes the physical and emotional responses to stress . . . and is the opposite of the fight-or-flight response.*
>
> —Herbert Benson

The goal in working with the relaxation response is to equip clients with an immediately accessible response as they encounter tense situations. When we are surprised or confronted with serious challenges, our heart rates go up, we breathe faster, our bodies tense, adrenaline and cortisol flow in, and we become "ready" to deal with whatever we face. However, this immediate tense reaction can become habitual and damaging to the body and brain. Look around and you will see many people whose bodies and lifestyles indicate constant tension. The relaxation response will enable clients to deal more effectively and healthfully with life challenges.

The relaxation response defined. In brief, clients learn to attend to body sensations and note the buildup of tension. Then before the tension takes over, they draw on the relaxation response. They may draw a deep breath, focus for a moment on a positive image, and then allow the muscles to relax. At the deeper level, clients learn to be attentive to the here and now almost constantly, have a relaxed, easy style, and flow through the day rather than fighting minute by minute. Few of us reach that wonderful, deeper state as our natural way of being, but with practice, we can come close to attaining a life that is more relaxed and *real*.

Relaxation training as the basis for the response. In 1934 Edmund Jacobson pioneered the scientific study of relaxation with his book *You Must Relax*. Jacobson used a sophisticated approach, oriented to treatment of both regular tension and severe psychological distress. He considered relaxation training a necessary part of any therapeutic situation. But it was Herbert Benson who popularized the use of relaxation techniques in medicine (Benson & Klipper, 2000). Benson's mind-body research has focused on the role of relaxation in reducing stress responses. Over time, the relaxation response and systematic relaxation can make an important mental health difference.

The relaxation response is a quickly learned strategy that clients can use immediately. The basic process for relaxation training is presented here. For the training to be effective, the client must be ready, and you will need a minimum of 15 minutes to engage in these fundamentals. Quick learning must be accompanied by daily, serious practice in reaching the here and now. Following are the basics of relaxation training:

1. Ask the client to sit quietly with closed eyes.
2. "Notice your breathing and focus on it. Discover the life-giving breath coming in and out." With practice, this one act is often sufficient to start the relaxation response. A deep breath loosens the body and prepares the client to deal realistically with the situation.
3. "Now, start with your feet and toes. Notice any tension in them; let the tension go, and notice how your muscles feel when they are relaxed."
4. "Next, move up the body, muscle area by muscle area, and relax the tension the same way, until your body is fully relaxed."
5. "Now that we have gone through the basics of relaxation, open your eyes slowly and pay attention to what is around you. When you get up, do so slowly and relaxed."
6. "How was this experience for you?" "How do you feel?"

With clients who are more tense, this relaxation awareness process may be ineffective. Ask them to tighten each muscle group separately, and then let it go. In this way, they can begin to learn the difference between tension and relaxation. When the tension/relaxation approach is mastered, then turn to the relaxation awareness training.

You can assign relaxation as homework, and then you need to follow up to see if the task was actually done. Daily homework is necessary for real success. The ultimate goal is for clients to be able to produce a here-and-now relaxation response when faced with stressful situations. Clients learn that they can release here-and-now tension by focusing on their breathing or on a part of their body even while working through the immediate stress. This makes a real difference in how clients respond when facing difficulty.

Mindfulness Meditation and Mindfulness

Jon Kabat-Zinn (2005) has also researched and promoted relaxation as a technique to help people cope with stress, pain, and anxiety. Kabat-Zinn calls his technique the "body scan," which he uses with mindfulness meditation. The body scan is basically the same as Benson's relaxation methods described earlier.

One of Allen and Mary's helpful life experiences occurred over several weeks when they participated in Kabat-Zinn's systematic program and learned mindfulness meditation. This technique is usually preferred by most practitioners over the body scan and systematic relaxation. But we do not believe that interviewers, counselors, and therapists should teach mindfulness unless they have sufficient training and have practiced it themselves. Mindfulness may require a lifestyle change for many.

Mindfulness meditation is derived primarily from Buddhist thought and practice. There is no "goal" except perhaps to live as much as possible in the immediate here and now. Similar in some ways to relaxation, practitioners usually lie comfortably on the floor or sit in a suitable chair, then close their eyes.

The focus becomes the *Now* and paying special attention to breathing, noting the breath come in and out. You may want to breathe in with one nostril and out with the other as this tends to help one focus on the *Now*. Thoughts and feelings will likely start running through your mind. Do not fight them; let them come, but as they enter your *Now* awareness, let them drift off. After practice, usually several weeks, you may find a near perfect "stillness" and awareness of the present moment. There is clear evidence that this state alone allows new neural connections to develop in positive areas of the brain. If you keep this up, you will eventually notice the here and now more fully throughout the day. You'll notice the beauty of the world in new ways. Your partner or lover will appear very differently to you because you are in the moment.

We recommend that you refer clients to honest experts who truly know mindfulness. There are charlatans out there. A safe, but secondary, alternative is to refer them to the Web site www.mindfulnesstapes.com where they can purchase materials and learn on their own. Kabat-Zinn (1990) has worked closely with the Dalai Lama and has participated in basic research showing that the brain does change with the positive, open approach of mindfulness.

Positive Imagery

Closely allied with psychoeducation, imagery is a popular technique to help clients relax and discover positive resources (Utay & Miller, 2006). All of us have experiences that are important for us, maybe a lakeside or mountain scene, or a snowy setting, or a quiet, special place. The image can become a positive resource to use when you feel challenged or tense. For example, Alisia likely feels real tension in her body when she encounters conflict. If she learns to notice her internal body tension, then she can immediately and briefly focus on a relaxing scene, take a deep breath, and deal with the challenging situation more effectively. When giving a guided imagery directive, time your presentation to your observations of the client.

"Imagine yourself in a pleasant and relaxation situation where you feel totally yourself. Relax, then close your eyes and enjoy that feeling. (long pause) What are you seeing, hearing, feeling?" (This itself may produce the relaxation response. Visualizing positive situations when under stress is calming.)

"What is your image of your ideal day/job/life partner?"

"Alisia, you seem a bit vague about the time you gave in to your parents. Close your eyes, visualize your parents. (pause) . . . Now what are they saying to you? What are you saying to them? How do you feel in this process?" (may bring out useful information, but is negatively oriented, which is sometimes necessary)

(better than the above) "Alicia, please tell me about a time when you were able to express yourself and hold your opinion." (The strength story is drawn out by the interviewer.) "Now imagine you are back in that situation. Close your eyes and tell me about it. What are you seeing, hearing, feeling?"

Imagery directives are often the most powerful directives and must be used with care. Images are particularly effective in helping clients experience the sensorimotor style. Many children and young people like the freedom and creativity allowed in this type of directive.

Imagery exercises need to be followed by a debriefing in which the interviewer discusses follow-up action with the client: "Now, Alisia, we've seen that this is comfortable for you. This is a real stress reducer. We've found that this relaxation response can be used in many settings when you find yourself upset. It can be used in the here and now of the situation over your car—or you can use this as a way to relax later, when you feel tense at any time."

Exploration of negative images is highly inappropriate and often unethical, unless the interviewer is fully qualified, and the time and situation are appropriate for the client. False memories can easily occur and do harm to the client.

Physical Exercise and Nutrition

A sound body is fundamental to mental health. Moving the body increases blood flow, and an exercise routine has been found to help reduce stress and depression. Proper eating habits and a regime of stretching and meditation make a difference in the life of your clients. Teaching clients how to nourish their bodies is becoming a standard part of counseling. We love and work more effectively if we are comfortable in our bodies. It has even been suggested that it is unethical not to include the recommendation for exercise in all treatment programs (Ratey, 2008).

Counselor:	Alisia, you seem stressed much of the time. What's happening with exercise in your life?
Alisia:	I simply don't have time, and when I think of it, I realize that I have some errands to run or someone calls me on the cell.
Counselor:	Evidence is clear that tension can be relieved by exercise. I'd like you to consider the possibility. What types of exercise have you enjoyed in the past?
Alisia:	Well, I used to run, and I did feel more "up" when I got out. But since I've moved to the city, I just don't seem to find time anymore.
Counselor:	As part of dealing with Chris and your various stressors, I think it is very important that we start some sort of exercise routine. You'll feel better and will be able to deal more effectively with those challenges if you take care of yourself.
Alisia:	Well . . . I should consider it. I know I felt better when I did run. (pause) But how?
Counselor:	Let's work on it. Tell me more about your schedule. (The session continues.)

It is obvious that we can't tell Alisia what to do. For example, "You should start exercising and running daily" simply won't work and will build client resentment. Helping clients change their behavior involves a more subtle use of directives in which the client is a full co-participant in the process. Regardless of what directive you want to provide on any topic, the client has to "buy in" and be central in the choice of action.

One of the best and most important wellness and stress prevention strategies is a well-designed exercise plan. How are you going to integrate these ideas into your own interviewing practice?

Thought-Stopping

This strategy is a brief but effective intervention. If you take the time to learn and practice thought-stopping on yourself, you gain a valuable tool to increase your self-esteem and effectiveness. Thought-stopping is useful for all kinds of client problems: perfectionism, excessive culture-based guilt or shame, shyness, and mild depression. This is one of our favorite strategies, and we have found it very helpful to us and our clients over the years. It stops our negative thinking about ourselves or someone who is troubling us.

Almost everyone engages in internalized negative self-talk. These stressful thoughts are said to yourself, perhaps several times a day. For example:

"Why did I do that?"
"I'm always too shy."
"I always foul up."
"I should have done better."
"Life is so discouraging for me."
"Nobody will listen to me."

These and other negative thoughts produce guilty feelings, procrastination when you "over think" the situation, fear that things will only get worse, anger that others "never get it right," repetitively thinking about past failures, or always needing the approval of others. Negative thoughts are basic to depression, and thought-stopping can be useful to the *mildly* depressed or sad. The following steps are key for learning and using thought-stopping:

Step 1. Learn the basic process. Relax, close your eyes and imagine a situation when you make the negative self-statement. Take time and let the situation evolve. When the thought comes, observe what happens and how you feel. Then tell yourself silently, "STOP." If you are alone, say it out loud and in a firm tone of voice.

Step 2. Transfer thought-stopping to your daily life. Place a rubber band around your wrist, and every time during the day that you find yourself thinking negatively, snap the rubber band and say to yourself, "STOP!"

Step 3. Add positive imaging. Once you have developed some understanding of how often you use negative thinking, after you say, "STOP," substitute a more positive statement. You may use positive imagery or think about an example when you had a positive experience, or use a brief, broader statement emphasizing general strengths. For example:

"I can do lots of things right."
"I am lovable and capable."
"I sometimes mess up—no one's perfect."
"I did the best I could."

Step 4. Make positive images and thoughts a habit. Remove the rubber band and use positive images and positive self-talk as a basis for building permanent self-esteem. (And consider using positive internal self-talk about other people who disturb or "bug" you. It can help.)

Role-Play Enactment

"Now return to that situation and let's play it out."
"Let's role-play it again, only change the one behavior we agreed to."

Alisia is encouraged to role-play the situation with Chris or the service manager. She plays herself as she usually behaves, while the counselor plays the role of the Other. This becomes the "baseline behavior" that counselor and client seek to change. After discussion, specifics for change are selected by the client, and another role-play follows. The role-play continues until the client has mastered new behaviors, and then the counselor plans with the client on action to be taken after the interview.

This is precisely the strategy that Allen used with psychiatric inpatients. He found that videotaping the interaction was especially helpful. Even a manic inpatient veteran could sit quietly, view himself on the screen, and select behaviors for change. This strategy by itself was often enough to enable the patient to return home.

Enactment via the Gestalt empty chair.

"Talk to your parent as if he or she were sitting in that chair. Now go to that chair and answer as your parent would."

This technique is similar to a role-play, but the client plays both people. The role-play concretizes the issue for both you and the client. The actual acting and movement help the client to get in touch with deeper sensorimotor emotions. And after the role-play, it is possible to use abstract formal operational analysis to consider and think about what occurred.

Enactment via Gestalt nonverbal strategies.
"I note that one of your hands is in a fist; the other is open. Have the two hands talk to one another."

Enactment via language change. Change *should* to *want to*. Change *can't* to *won't*.
"Alisia, you say you can't do anything with the shop head. Change 'can't' to 'won't,' because you are in charge of your own behavior. . . . Good . . . now say that again."

Words such as *can't* imply that a situation is out of control; by changing *can't* to *won't*, Alisia is being forced to be in charge of her own behavior. This environmental structuring approach may be particularly useful with self-directed, formal-operational clients who may need help in changing their thinking and behavior.

Free Association

"Take that feeling/image/issue and focus on it for a moment. Then, close your eyes, free associate, and let whatever comes to your mind flow in to it."

Originating in the psychoanalytic movement, this strategy enables clients to reflect back from the here and now to times in the past when they might have had similar thoughts and feelings to what you are observing now. This often provides a critical link that helps both you and the client understand the historical basis of the present issue.

A more straightforward approach is simply to say, "Stop for a moment and allow yourself to go inside. (pause) What occurs for you at this moment?" or "Stay with that feeling . . . magnify it. (pause) Now, what just flashed into your mind?"

Free association is best used with self-directed and reflective formal-operational clients. But note that it starts with here-and-now sensorimotor experience. Many clients will be able to identify feelings in specific places in their body that represent the flashback. We might expect Alisia to free associate back to a childhood experience in which efforts to express herself with parents or teachers were rejected roughly. Many of us repeat old parent-child interactions from the past in the present.

SUMMARY: INFLUENCING SKILLS

All the microskills and strategies emphasized in this chapter are aimed at client change—self-reflection, discovery of how others see the client, understanding the consequences and plus and minuses of every decision, providing the client with cogent and useful advice and information, and finally, directing the client through a variety of useful change strategies called directives.

It is tempting at this point to work to change clients while failing to listen to their stories. The relationship and the original client story and strengths remain important. Are you working toward the clients' goals as you seek to help them develop new narratives and stories, learn new behaviors, and discover more about themselves?

In short, use the skills of this chapter carefully. We urge that you consult with your clients and let them know that you are changing the focus of the session for a short time. Then return to careful listening and retain that relationship.

The Client Change Scale (CCS) used formally or informally will always help you assess the effectiveness of your interventions in the session. You can also use the CCS to assess the effectiveness and value of week-to-week sessions—and even a longer term course of counseling and therapy. We do suggest that this type of evaluation be done *with* the client. You will find that bringing your client into your evaluation plan is yet another way to help facilitation of change beyond the here and now of the interview.

Key points of Chapter 12 are summarized below.

■ KEY POINTS

"1-2-3" pattern	In any interaction with a client, first attend to and determine the client's frame of reference, then assess her or his reaction before using your influencing skills. Finally, check out the client reaction to your use of the skill.
Interpersonal influence continuum	The influencing and attending skill may be classified from low to high degrees of influence. Encouragers and paraphrasing are considered relatively low in influence, whereas confrontation and directives are considered more influential.
Moderate triad	The "swing skills" of the interpersonal influence continuum are focusing and open and closed questions. They provide a framework for determining the topic of conversation while keeping a balance between influencing and attending skills.
Self-disclosure	Indicating your thoughts and feelings to a client constitutes self-disclosure, which necessitates the following: 1. Use personal pronouns ("I" statements). 2. Use a verb for content or feeling ("I feel . . ." "I think . . ."). 3. Use an object coupled with adverb and adjective descriptors ("I feel happy about your being able to assert yourself . . ."). 4. Express your feelings appropriately. Self-disclosure tends to be most effective if it is genuine, timely, and phrased in the present tense. Keep your self-disclosure brief. At times, consider sharing short stories from your own life.
Feedback	Feed back accurate data on how you or others view the client. Remember the following: 1. The client should be in charge. 2. Focus on strengths. 3. Be concrete and specific. 4. Be nonjudgmental. 5. As appropriate, provide here-and-now feedback. 6. Keep feedback lean and precise. 7. Check out how your feedback was received. These guidelines are useful for all influencing skills.
Logical consequences	This skill predicts the probable results of a client's action, in five steps: 1. Listen to make sure you understand the situation and how it is understood by the client. 2. Encourage the client to think about positive and negative consequences of a decision. 3. Provide your data on the positive and negative consequences of a decision in a nonjudgmental manner. 4. Summarize the positives and negatives. 5. Let the client decide what action to take.

(continued)

KEY POINTS (continued)

Information and psychoeducation	Many times clients need the counselor's knowledge and expertise around key life issues. The counselor knows the community and resources available. He or she also knows the likely pattern and key issues of a divorce, the death of a family member, or other life issue. Psychoeducation is a more systematic way to teach clients of new life possibilities; this may range from training in communication skills to developing a successful wellness plan.
Directives	Involve your client with the choice of directives, even to the point of telling her or him what is to happen and what to expect as a result. Appropriately assertive body language, vocal tone, and eye contact are important, as are clear, concrete verbal expressions and checking out the degree of client participation.
What else?	Most of the influencing skills require you to be concrete and specific, although the ability to work abstractly is particularly important in interpretation/reframing. Remember to involve your client as a co-participant as you utilize influencing skills.

COMPETENCY PRACTICE EXERCISES AND PORTFOLIO OF COMPETENCE

Individual Practice

This chapter includes more possibilities for practice than you have time, unless you are using this book in a two-semester sequence. We strongly suggest that you take one (or two) of the most interesting and potentially useful exercises suggested here and try it (or them) out.

1. Try the directive strategy on yourself. Most can be done alone if you *take time* to really do it and study the process. What happens? What occurs for you? What did you learn? Would you like to continue and practice that method further?

   _____

   _____

2. Find a classmate or friend, get the person's permission, and the two of you try the strategy on each other. What happens? What occurs for you? What did you learn? Would you like to continue and practice that method further?

   _____

   _____

Group Practice

Exercise 9: Influencing Skills

Small-group work with the influencing skills requires practice with each skill. The general model of small-group work is suggested, but only one skill should be used at a time. Remember to include the Client Feedback Form from Chapter 1 as part of the practice session. This is a particularly important place to practice group supervision, sharing, and feedback.

Step 1: Divide into practice groups.

Step 2: Select a leader for the group.

Step 3: Assign roles for each practice session.

▲ Client.
▲ Interviewer, who will begin by drawing out the client story or issue using listening skills and then attempt one of the influencing skills. Feedback and self-disclosure may be used with other influencing skills.
▲ Observer 1, who will observe the client and complete the CCS Rating Form (see Chapter 9), deciding how much of an impact the interviewer's influencing skills have made.
▲ Observer 2, who will complete the Feedback Form (Box 12-4) that follows.

Step 4: Plan. In using influencing skills, the acid test of mastery is whether the client actually does what is expected (for example, does the client follow the directive given?) or responds to the feedback, self-disclosure, and so on in a positive way. For each skill, different topics are likely to be most useful. State goals you want to accomplish in each instance. Some ideas follow:

▲ *Self-disclosure.* A member of the group may present any personal issue. The task of the counselor is to share something personal that relates to the client's concern. Again, the check-out is important to obtain client feedback.
▲ *Feedback.* A most useful approach is for the interviewer to give direct feedback to another member of the group about his or her performance in the training sessions. Alternatively, the client may talk about an issue for which the counselor will provide feedback, sharing his or her perceptions of the situation as objectively as possible. A check-out should be included.
▲ *Logical consequences.* A member of the group may present a decision he or she is about to make. The counselor can explore the negative and positive consequences of that decision.
▲ *Information and psychoeducation.* The interviewer may give information or more systematic psychoeducation about a particular issue to the individual or group (the value of a wellness plan, dealing with a death in the family, or other possibility). The group gives feedback on the ability of the information giver to be clear, specific, interesting, and helpful. We suggest that you consider teaching microskills as communication skills to your individual or group.
▲ *Directives and specific strategies.* Select one of the strategies presented in this chapter and work through the specific steps. Involve your client in the process, and the two of you can select together the strategy that you would like to try. As part of the practice session, be sure to tell the client what to expect and the likely results.

Step 5: Conduct a 5- to 15-minute practice session using the skill. You will find it difficult to use particular influencing skills frequently, as they must be interspersed with attending skills to keep the interview going. However, attempt to use the targeted skill at least twice during the practice session.

Step 6: Review the practice session and provide feedback for 10 to 12 minutes. Remember to stop the tape to provide adequate feedback for the counselor.

Step 7: Rotate roles.

BOX 12-4   FEEDBACK FORM: INFLUENCING SKILLS

_____ (Date)

_____   _____
(Name of Interviewer)   (Name of Person Completing Form)

Instructions: The two raters will complete the form and then discuss their observations with the
practicing interviewer and the volunteer client.

1. Did the interviewer use the basic listening sequence to draw out and clarify the client's story or concern? How effectively?

2. Provide nonjudgmental, factual, and specific feedback for the interviewer on the use of the specific influencing skill or directive strategy.

3. As you view the totality of the session, where was the client at the beginning on the Client Change Scale? Where was he or she at the conclusion? What aspects of the skill or strategy impressed you as most useful and effective?

Portfolio of Competence

This chapter is about interpersonal influence, and it covers considerable material. You cannot be expected to master all these skills until you have a fair amount of practice and experience. At this point, however, it will be helpful if you think about the major ideas presented in this chapter and where you stand currently. Also, where would you like to go in terms of next steps?

Use the following as a checklist to evaluate your present level of mastery. As you review the items below, ask yourself, "Can I do this?" Check those dimensions that you currently feel able to do. Those that remain unchecked can serve as future goals. Do not expect to attain intentional competence on every dimension as you work through this book. You will find, however, that you will improve your competencies with repetition and practice.

Skill/Strategy	Level 1: Can you identify/classify the skill and write example statements?	Level 2: Can you demonstrate the skill in a role-played interview?	Level 3: Can you demonstrate your ability to use this skill in interviews with specific impact on your clients?	Level 4: Can you teach this skill to others?
Self-disclosure				
Feedback				
Logical consequences				
Information/psychoeducation				
Directives				

DETERMINING YOUR OWN STYLE AND THEORY: CRITICAL SELF-REFLECTION ON INFLUENCING SKILLS

You have encountered five skills and have had the opportunity for at least a brief introduction to each. With which of these skills do you feel most comfortable? Which might you seek to use? Which might you avoid? Perhaps most important, how do you feel about the idea of consciously influencing the direction of the interview?

What single idea stood out for you among all those presented in this chapter, in class, or through informal learning? What stands out for you is likely to be important as a guide toward your next steps. What are your thoughts on multicultural issues and the use of this skill? What other points in this chapter struck you as important? How might you use ideas in this chapter to begin the process of establishing your own style and theory?

What are your thoughts?

_____

_____

Given the complexity of this chapter and the many possible goals you might set for yourself, list three specific goals you would like to attain in the use of influencing skills within the next month.

What are your thoughts?

_____

_____

OUR THOUGHTS ABOUT ALISIA

Our orientation to the question of general direction and goals for counseling remains highly influenced by Carl Rogers. What does the client want to have happen? We really want to see the goals of counseling determined by the client. We would want to work with Alisia to ensure that she is determining the focus of counseling, so the listening skills of the first part of this book remain central to us.

But goal setting does not always have to be completed within a person-centered Rogerian tradition. While goals can come out of effective Rogerian work, our tendency would be to make them more explicit. We would likely first draw on decisional counseling theory and practice (see Chapters 8 and 13) and seek to outline with her some specific issues for exploration. And, somewhat contrary to Rogers, we would not hesitate to bring in issues that Alisia has not thought of, as you will see below. Along with goal setting, we would also keep some key questions in mind that we would raise with her. But determining whether to follow these up would be her decision.

Here are some possible directions we would consider for Alisia:

▲ Help her expand her consciousness as a woman in a sometimes oppressive world. Feminist counseling theory might be useful. The emphasis on positive women models would be characteristic of our work.
▲ Encourage her to balance negative thinking with positive assets and wellness strengths. This is one of our core values and a central part of the microskills framework and theory.
▲ Teach her stress management skills. These are outlined in detail in Chapter 14, where cognitive-behavioral counseling is presented.
▲ As many of these issues are related to gender, Allen would seek supervision if he worked with her and would refer her to a women's support group if the need might arise.

How do our thoughts compare with yours? Remember, counseling is for the client, and it is vital that we not impose our goals, thoughts, and feelings on her. We would focus on helping her make her own decisions—and here, once again, we note the central importance of listening!

SECTION IV

Skill Integration

How do you incorporate the microskills into your own conceptual framework and theory? What are your strengths? What is your own style of interviewing, counseling, or psychotherapy? This section provides a framework to help you integrate the many skills and concepts of this book.

Look for the following in this final section.

Chapter 13. Skill Integration: Putting It All Together   This chapter focuses on the complete interview and asks you to start analyzing your own interviewing behavior at a new level.

▲ *Decisional counseling.* The five stages of the interview can function as a framework for resolving issues and making decisions. Decisions underlie most issues in all theories of helping. Decisional counseling ideas will be useful in many settings and in other theoretical approaches.

▲ *Complete transcript of a decisional interview.* Here we focus on the presentation and analysis of an interview, including case conceptualization and pre-interview planning, interview summary, and planning for long-term treatment.

▲ *Your own interview transcript.* Most important is that you look at and analyze your own interviewing behavior and its impact on the client. This is the action and skill phase of this chapter and the entire book. *What is your style? What are you doing? In your own view, how effective are you?*

Chapter 14. Microskills and Counseling Theory: Sequencing Skills and Interview Stages
This chapter presents four theoretical approaches. If you have competence in the microskills, you are ready to engage in beginning work in the important theories of counseling and psychotherapy.

▲ *Person-centered counseling.* If you have completed a full interview using only the basic listening sequence, you have made a useful beginning in understanding and practicing this theoretical orientation. Chapter 14 adds the skills of reflection of meaning, focusing, and feedback to broaden this framework and bring it more in tune with Carl Rogers's original thinking.

▲ *Brief counseling.* This solution model is designed as a brief approach to human change. In a time of accountability, you may find this model useful in moving clients to action. Questioning is the central microskill.

▲   *Cognitive behavioral therapy*. CBT is currently the most researched therapy. CBT is rich with many alternative strategies to produce change. Stress management is emphasized here.

▲   *Motivational interviewing*. This is a relatively new approach that is rapidly gaining popularity. Developed as a therapy for work with persons suffering from alcoholism, it is now recognized as a highly useful strategy for all clients. You will find that your knowledge of microskills is particularly compatible with this method of helping.

Chapter 15. Determining Personal Style and Future Theoretical/Practical Integration   *Intentional Interviewing and Counseling* concludes with a focus on you and your plans for the future.

▲   *Level of competence*. How do you assess your abilities in using microskills and strategies?

▲   *Preferred interviewing style*. Which theories of counseling appeal to you most?

▲   *Sensitivity to multicultural issues*. How will you integrate an awareness of multicultural issues into your interviewing practice?

▲   *Natural style*. Putting it all together: Where are you going?

CHAPTER 13

Skill Integration: Putting It All Together

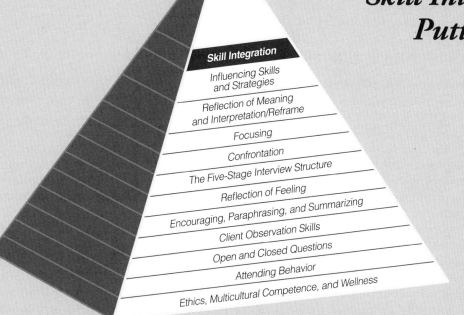

Not to decide is to decide.

—Harvey Cox

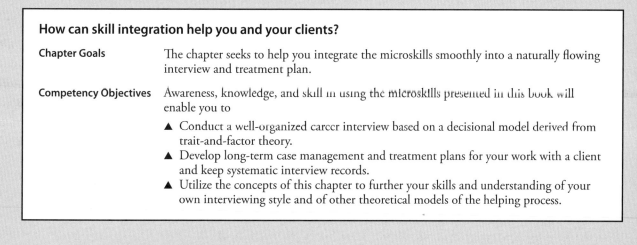

How can skill integration help you and your clients?

Chapter Goals

The chapter seeks to help you integrate the microskills smoothly into a naturally flowing interview and treatment plan.

Competency Objectives

Awareness, knowledge, and skill in using the microskills presented in this book will enable you to

▲ Conduct a well-organized career interview based on a decisional model derived from trait-and-factor theory.

▲ Develop long-term case management and treatment plans for your work with a client and keep systematic interview records.

▲ Utilize the concepts of this chapter to further your skills and understanding of your own interviewing style and of other theoretical models of the helping process.

Mary, the client, is a 36-year-old divorced woman with two children. She has worked as a physical education teacher for a number of years. Currently she is bored and stymied with her job as a PE teacher and wants help searching for a new career. She is thinking about doing something related to business but says that she is somewhat depressed by her situation and needs direction.

INTRODUCTION: ANALYZING THE INTERVIEW

This chapter presents a detailed analysis of a single interview with Mary and shows how you can use a similar process to examine your own work. It is possible to take the many skills and concepts of this book and use them systematically in an entire interview. The basic goal here is to facilitate an examination of your own style. As you read this chapter, think ahead to how you will conduct your own interviews, develop your own plan for interviews, keep notes, and integrate your own ideas into a long-term case management and treatment plan. The single practice exercise for this chapter suggests that you audio- or videorecord an interview and use it as a basis for conducting a comprehensive analysis of your own style and its impact on clients.

Decisional counseling is the theory/practice system that is demonstrated in detail in this chapter. This system will not appear in many of your theories of counseling/therapy texts. Nonetheless, it is likely the most widely practiced model of all, simply because so much of interviewing, counseling, and psychotherapy are about decisions and action. Decisional counseling is sometimes called trait-and-factor or problem-solving counseling. You will find that the increasingly popular motivational interviewing (Chapter 14) is basically a sophisticated decisional model.

Decisional counseling and a demonstration interview are presented in several segments in this chapter. We suggest that you record your own client interview and use the steps here as a framework for a final paper and presentation on your competence in interviewing. As a demonstration, a complete decisional interview is presented by Allen and Mary Bradford Ivey, with Mary as a client who has career issues. Personal concerns appear in the session as it progresses. The sections of this chapter include the following:

Instructional Reading 1: Decisional counseling. This system, sometimes termed "problem-solving counseling," will help you understand the decisional process that underlies the multiple theories of counseling.

Interview planning and case conceptualization. This section covers thinking through an intake file and then anticipating what might happen in the session.

Transcript of a complete interview. Read the interview and Allen's classification of skills and comments on the session transcript.

Instructional Reading 2: Interview transcript analysis and planning. This follows the transcript and focuses on a more detailed analysis of the interview. You will find ideas to further your examination of your own interview, referral and consultation, and case management/treatment planning.

INSTRUCTIONAL READING 1: DECISIONAL COUNSELING

Decisional counseling may be described as a practical model that recognizes decision making as undergirding most—perhaps all—systems of counseling. The basic point is simple: Whether clients need to choose among careers, decide to have a child, decide to get married, or search for new ideas on how to live more effectively, they are always making decisions. Decisional counseling facilitates the process of decision making underlying life patterns. Allen Ivey created this integration of decisional theory through microskills practice.

Many see Benjamin Franklin as the originator of the systematic decision-making model. He suggested three stages of problem solving: (a) identify the problem clearly, (b) generate alternative answers, and (c) decide what action to take. Another term for decisional counseling is problem-solving counseling. The essential issue is the same regardless of the terms we use: How can we help clients work through issues and come up with new answers?

Decisional counseling theorists believe that all theories of counseling and therapy—cognitive-behavioral, person-centered, and even psychoanalytic—deal with problem solving and decisions. The cognitive behavioral therapist will often teach clients how to make new decisions using a problem-solving approach. The person-centered counselor seeks to enable clients to make decisions for themselves through self-examination and self-reflection. The psychoanalytic therapist is searching for the underlying factors leading to decisions. All systems search for the creative *New* but use different language systems. Thus decisional counseling is a basic framework useful in many settings.

The Trait-and-Factor Legacy

The goal in decisional counseling is to facilitate decision making and to consider the many *traits and factors* underlying any single decision. Trait-and-factor theory has a long history in the counseling field, dating back to Frank Parsons's development of the Boston Vocational Bureau in 1908. Parsons pointed out that in making a vocational decision the client needs to (a) consider personal traits, abilities, skills, and interests; (b) examine the environmental factors (opportunities, job availability, location, and so on); and (c) develop "true reasoning on the relations of these two groups of facts" (Parsons, 1909/1967, p. 5). Since that time, proponents of trait-and-factor theory have searched for the many dimensions that underlie "true reasoning" and decision making.

Gradually trait-and-factor theory came to be seen as limited, and new decisional and problem-solving models have arisen (Brammer & MacDonald, 2002; Chang, D'Zurilla, & Sanna, 2004; D'Zurilla, 1996; Egan, 2007; Ivey, Simek-Morgan, Ivey, & D'Andrea, 2006; Janis & Mann, 1977). All of those models could be described as modern reformulations of Benjamin Franklin's original model and trait-and-factor theory, but they are based on more recent thinking and research. Decisional theory here is based on systematic microskills.

However, there is a common problem in the decisional models, particularly those close to the original trait-and-factor model in that insufficient attention is paid to the *emotional* aspects of the decision. Very few of us would be satisfied if our decisions reflected only left brain activity and cognitive processes. Decisions also require emotional energy and agreement with our long-term memory if they are to be meaningful and lasting. Think about bringing more feeling and emotion to the session when you work with clients and decisions.

The Balance Sheet and Future Diary: Bringing Emotions Into Decisional Practice

Emotions play a critical role in decision making, for it is how we feel about the alternative answers and possible solutions that leads us to a decision. Decision making involves "(1) the facts of the problem presented, (2) the option chosen to solve it, (3) the factual outcome to

TABLE 13-1   The five interview stages and decisional counseling

The Five Stages of Intentional Interviewing and Counseling	Decisional Counseling
1. *Relationship*—Initiating the session—Rapport and structuring ("Hello; this is what might happen in this session.")	These factors are obviously an important part of any problem-solving attempt.
2. *Story and strengths*—Gathering data—Drawing out stories, concerns, problems, or issues ("What's your concern?" "What are your strengths or wellness resources?")	Benjamin Franklin talked of defining the problem, whereas trait-and-factor theorists speak of the need to consider personal traits, abilities, skills, and interests. The environmental emphasis of trait-and-factor theory should be applauded. Decisional counseling adds more emphasis on strengths and draws out the story more fully.
3. *Goals*—Mutual goal setting—Establishing outcomes ("What do you want to happen?")	Defining goals specifically will make a difference in decisional counseling. Those with a problem or trait-and-factor orientation may spend insufficient time on goal setting.
4. *Restory*—Working—Exploring alternatives, confronting client incongruities and conflict, restorying ("What are we going to do about it?")	Both problem-solving approaches and decisional counseling often use brainstorming to find alternatives and employ "true reasoning" to discover the relationship of the person to the environment. Decisional counseling adds more emphasis on emotions and restorying.
5. *Action*—Terminating—Generalizing and acting on new stories ("Will you do it?")	Generalizing the decision to real life is not always stressed in trait-and-factor counseling but it is central in decisional counseling.

the solution, and, importantly, (4) the outcome of the solution in terms of emotion and feeling" (Damasio, 2003, p. 143). The last point reminds us that decisions are related to emotional outcomes—pleasure with a decision that works and pain when it doesn't. Damasio, a prominent neuroscientist, mapped decisions through brain scans. He found that emotional areas are active in making decisions. Scientifically, decisions are far more than just a rational process!

An Australian, Leon Mann, developed the Balance Sheet and Future Diary to help people "take decisions" (Mann, 2001; Mann, Beswick, Allouache, & Ivey, 1989). Motivational interviewing has recently adapted these strategies and finds them helpful in facilitating decision making, particularly among those with substance abuse issues (please visit Chapter 14, page 433, for a sample balance sheet). The strategies are simple and straightforward but can be a powerful addition to decisional counseling. In developing a balance sheet with a client, you list all the possible solutions and rate each possibility with a "+" or "−." Or you may use several pluses or minuses if you have several possibilities. This provides a visual map of the decisional territory and brings in awareness of emotions.

The balance sheet is coupled with the future diary in which clients anticipate how they will feel *emotionally* about each decision. The future diary can be used as a homework assignment for the client. However, we believe that it is preferable to work through the likely futures of decisions with the client in the here and now of the interview. You may wish to use guided imagery to help clients anticipate the future emotions associated with each decision.

Decision making is generally considered a rational process. But decisions need to be emotionally satisfying for the client. There are those who point out that it is irrational to consider people as rational beings. Whether or not this is true, it is clear that an intellectual decision is not enough—emotions must be considered as well.

Decisions, Problem Solving, and the Five Stages of the Interview

Decisional counseling concepts are basic to the five-stage model of the interview with which we have been working throughout this book. At this point, it may be helpful to see how these five elements fit with the decisional counseling models as shown in Table 13-1.

Decisional issues and problem-solving theory are clearly parallel. The simplest problem-solving model is defining the problem, generating or brainstorming alternatives, and then deciding among the alternatives. We have also seen that neuropsychology research follows a similar model, with special emphasis on emotion.

INTERVIEW PLANNING AND CASE CONCEPTUALIZATION

You are about to read the transcript of an initial interview session. However, before the interview was held, the interviewer, Allen, developed an interview plan, presented in Box 13-1. This plan was developed after a study of Mary's client intake file that consisted of a pre-interview questionnaire and case conceptualization. Mary had stated in her intake form, "I'd like to do something new with my career. I'm ready for something new—but what?" As you will note, the interview plan is oriented toward helping the client develop a career plan, but it facilitates the discussion of personal issues as well. The plan is a structure to remind us about what we need to consider as we start, but it will likely be changed as the interview progresses and new issues are brought up.

Mary role-plays a client who is 36 years old, divorced, with two children. She is working as a physical education teacher. She stated in her information file, completed during intake, that "I find myself bored and stymied in my present job as a PE teacher. I think it is time to look at something new. Possibly I should think about business. Sometimes I find myself a bit depressed by it all."

Watch to make sure the plan does not become rigid. You must not impose a plan or a case conceptualization on the client. Use cultural intentionality and flex with what actually does happen in the session.

Although the plan is tentative, it does illustrate that planning a session can provide you with a useful checklist to help ensure that you cover key issues in the session. It is all too easy to omit the critical action phase as both client and counselor are winding down. Allen's objectives were roughly realized in the session. However, if Mary had had different needs, the interview plan could have been scrapped, and a whole different approach would have been used, including changing from decisional theory to another, more appropriate framework.

BOX 13-1   FIRST INTERVIEW PLAN AND OBJECTIVES

Before the first interview, study the client file and try to anticipate what issues you think will be important in the session and how you might handle them. (This plan is Allen's assessment of his forthcoming interview with Mary.)

Relationship—Initiating the session—Rapport/structuring

Are any special issues anticipated with regard to rapport development? What structure do you have for this interview? Do you plan to use a specific theory?

Mary appears to be a verbal and active person. I note she likes swimming and physical activity. I like to run. . . . That may be a common bond. I think I'll be open about structure but keep the five stages in mind. It seems she may want to look into career choice, and she may be unhappy with her present job. I should keep in mind her divorce, as this may also be a personal issue. I'll use the basic listening sequence to bring things out and probably work with a decision-making model. She's probably an abstract formal-operational client. Therefore, in the first stages of the session, I'll need to listen to her stories and use mainly questions and reflection skills.

Story and strengths—Gathering data—Drawing out stories, concerns, problems, or issues

What are the anticipated problems? Strengths? How do you plan to define the issues with the client? Will you emphasize behavior, thoughts, feelings, meanings?

Mary seems to be full of strengths. I'll use her wellness strengths and work on finding out what she *can* do. I'll use the basic listening sequence to bring out issues from her point of view. I'll be most interested in her thoughts about her job and her plans for the future; however, I should be flexible and watch for other concerns. Mary may well bring up several issues. I'll summarize them toward the end of this phase, and we may have to list them and set priorities if there are many issues. Mainly, however, I expect an interview on career choice.

Goals—Mutual goal setting

What is the ideal outcome? How will you elicit the idealized self or world?

I'll ask her what her fantasies and ideas are for an ideal resolution and then follow up with the basic listening sequence. I'll end by confronting and summarizing the real and the ideal. It's worked for me in the past and probably will work again. As for outcome, I'd like to see Mary defining her own direction from a range of alternatives. Even if she stays in her present job, I hope we can find that it is the best alternative for her.

Restory—Working—Exploring alternatives, confronting client incongruities and conflict, restorying

What types of alternatives should be generated? What theories would you probably use here? What specific incongruities have you noted or do you anticipate in the client?

I hope to begin this stage by summarizing her positive strengths and wellness assets. It's too early to say what the best alternatives are. However, I'd like to see several new possibilities considered. Counseling and business are indicated in her case conceptualization and pre-interview form as two possibilities. They seem good. Are there other possibilities? The main incongruity will probably be between where she is and where she wants to go. I'll be interested in her personal life as well. How are things going since the divorce? What is it like to be a woman in a changing world? I expect to ask her questions and develop some concrete alternatives even in the first session. . . . I hope she will act on some of them following our first session. I think career testing may be useful.

Action—Terminating—Generalizing and acting on new stories

What specific plans, if any, do you have for suggesting transfer of training to the client? What will enable you personally to feel that the interview was worthwhile?

Drawing from the above, I'll feel satisfied if we have generated some new possibilities and can do some exploration of career alternatives after the first session. We can plan from there. I'd like it if we could generate at least one thing she can do for homework before our second session.

DEMONSTRATION INTERVIEW AND ANALYSIS: ALLEN AND MARY'S DECISIONAL SESSION

The initial interview illustrates the close relationship between career counseling and personal counseling (see Table 13-2). Note how the relationship between job change and personal issues develops during the session. Additionally, gender issues emerge as important multicultural factors that need to be considered. This appears when Mary starts developing a new story in Stage 4 of the session.

As in all interviews, you will find some interviewer responses that are less effective than others. We all make errors; it is our ability to learn from them and change that makes us become more effective interviewers and counselors. Stay alert to Allen's responses that you think could be phrased more effectively. What would you do differently? And what responses seem helpful to Mary?

Now you will read the interview as it actually happened, structured in the *relationship—story and strengths—goals—restory—action* framework. The verbatim transcript is supplemented by a skill-and-focus analysis of the session. The Process Comments column analyzes the effectiveness of skills used throughout the interview, with special attention given to the effect of confrontations ("C") on the client's developmental change. This framework could be useful to you as you present and analyze your own interviewing behavior.

TABLE 13-2 The Allen and Mary decisional interview

Skill Classifications				
Listening and Influencing	*Focus*	*C**	*Counselor and Client Conversation*	*Process Comments*
STAGE 1: Relationship: Initiate the session—Develop rapport and structuring ("Hello; this is what might happen in this person.")				
Open question	Client		1. *Allen:* Hi, Mary. How are you today?	
	Client, interviewer		2. *Mary:* Ah . . . just fine . . . How are you?	As Mary walked in, Allen saw her hesitate and sensed some awkwardness on her part. Note that she opens with two speech hesitations.
Information, paraphrase	Interviewer, client		3. *Allen:* Good, just fine. Nice to see you. . . . Hey, I noted in your file that you've done a lot of swimming.	
	Client, main theme		4. *Mary:* Oh, yeah, (smiling) . . . I like swimming; I enjoy swimming a lot.	Consequently, Allen decides to take a little time to develop rapport and put Mary at ease in the interview. Note that he focused on a positive aspect of Mary's past. It is often useful to build on the client's strengths even this early in the session.
Information, closed question	Main theme, interviewer, client		5. *Allen:* With this hot weather, I've been getting out. Have you been able to?	The distinction between providing information and a self-disclosure is illustrated at *Allen 5* and *Mary 6*. Allen only comments that he's been getting out, whereas Mary gives information and personal feelings as well.
	Client		6. *Mary:* Yes, I enjoy the exercise. It's good relaxation.	

* This column will record the presence of a confrontation.

(continued)

TABLE 13-2 (continued)

Skill Classifications				
Listening and Influencing	Focus	C	Counselor and Client Conversation	Process Comments
Paraphrase, reflection of feeling	Client		7. *Allen:* I also saw you won quite a few awards along the way. (Mary: Um-hmm.) . . . You must feel awfully good about that.	Mary's nonverbal behavior is now more relaxed. Client and counselor now have more body language symmetry.
	Client		8. *Mary:* I do. I do feel very good about that. It's been lots of fun.	
Information, closed question	Main theme		9. *Allen:* Before we begin, I'd like to ask if I can tape-record this talk. I'll need your written permission, too. Do you mind?	Obtaining permission to tape-record interviews is essential. If the request is presented in a comfortable, easy way, most clients are glad to give permission. At times it may be useful to give the tapes to clients to take home and listen to again.
	Client		10. *Mary:* No, that's okay with me. (signs form permitting use of tape for *Intentional Interviewing and Counseling*)	
Information giving	Client		11. *Allen:* As we start, Mary, there are some important things to discuss. We'll have about an hour today, and then we can jointly plan for future sessions. Today, I'd like to get to know you, and I'll try to focus mainly on listening to your concerns. At the same time, from your file, I know that some of your issues relate to women's issues. Obviously, I'm a man, and I think it is important to bring this up so that you will be more likely to feel free to let me know if I seem to be "off target" or misunderstand something. Feel free to ask me any questions you'd like around this or other matters.	Allen provides some additional structure for the session so that Mary knows what she might expect. He introduces gender differences and provides an opportunity for Mary to react and ask questions.
	Client, interviewer		12. *Mary:* I feel comfortable with you already. But a couple questions. One is that I'm interesting in the counseling field as a possibility— and the other is around the issue of living with divorce and being a single parent. What can you say about those?	Mary gives the okay, but then asks two questions. She leans forward when she asks them. This question is a surprise to the interviewer, and he should note that divorce and relationship issues may show themselves to be important later in the session.

(continued)

TABLE 13-2 (continued)

Skill Classifications				
Listening and Influencing	*Focus*	*C*	*Counselor and Client Conversation*	*Process Comments*
Self-disclosure, open question	Interviewer, client		13. *Allen:* Well, first, I'm divorced and have one child living with me while the other is in college. Of course, I'd be glad to talk about the counseling career and share some of my thoughts. What thoughts occur to you around divorce and counseling?	Keep self-disclosures brief and return the focus to the client. But be comfortable and open in that process.
	Client, main theme		14. *Mary:* That helps. Going through my divorce was the worst thing of my life. My children are so important to me. Allen, perhaps your experience with divorce will help you understand where I am coming from. Let's get started and look at what my career should be.	Mary smiles, sits back, and appears to have the information she was wondering about. She actually says, "Let's get started."

STAGE 2: Story and Strengths: Gather data—Draw out stories, concerns, problems, or issues ("What's your concern? What are your strengths or resources?")

Open question	Client		15. *Allen:* Could you tell me, Mary, where you would like to start?	In this series of leads you'll find that Allen uses the basic listening sequence of open question, encourager, paraphrase, reflection of feeling, and summary, in order. Many interviewers in different settings will use the sequence or a variation to define the client's problem.
	Client, problem/ concern, others		16. *Mary:* Well . . . ah . . . I guess there's a lot that I'd like to talk about. As I said, I went through . . . ah . . . a difficult divorce, and it was hard on the kids and myself and . . . ah . . . we've done pretty well. We've pulled together. The kids are doing better in school, and I'm doing better. I've . . . ah . . . got a new friend. (breaks eye contact) But, you know, I've been teaching for 13 years and really feel kind of bored with it. It's the same old thing over and over every day; you know . . . parts of it are okay, but lots of it I'm bored with.	As many clients do, Mary starts the session with a "laundry list" of issues. Though the last thing in a laundry list is often what a client wants to talk about; the eye-contact break at mention of her "new friend" raises an issue that should be watched for in the interview. As the session moves along, it becomes apparent that more than the career issue needs to be considered. Mary discusses a "pattern" of boredom. This is indicative of an abstract formal-operational cognitive/ emotional style.

(continued)

TABLE 13-2 (continued)

Skill Classifications				
Listening and Influencing	*Focus*	*C*	*Counselor and Client Conversation*	*Process Comments*
Encourage	Client		17. *Allen:* You say you're *bored* with it?	The key word "bored" is emphasized.
	Client, problem/ concern		18. *Mary:* Well, I'm bored, I guess . . . teaching field hockey and . . . ah . . . basketball and softball, certain of those team sports. There are certain things I like about it, though. You know, I like the dance, and you know, I like swimming—I like that. Ah . . . but . . . you know . . . I get tired of the same thing all the time. I guess I'd like to do some different things with my life.	Note that Mary elaborates in more detail on the word "bored." Allen used verbal underlining and gave emphasis to that word, and Mary did as most clients would: she elaborated on the meaning of the key word to her. Many times short encouragers and restatements have the effect of encouraging client exploration of meaning and elaboration on a topic. "I'd like to do some different things" is a more positive "I" statement.
Paraphrase	Client		19. *Allen:* So, Mary, if I hear you correctly, sounds like change and variety are important instead of doing the same thing all the time.	Note that this paraphrase has some dimensions of an interpretation in that Mary did not use the words "change" and "variety." These words are the opposite of boredom and doing "the same things all the time." This paraphrase takes a small risk and is slightly additive to Mary's understanding. It is an example of the positive asset search, in that it would have been possible to hear only the negative "bored." Working on the positive suggests what *can* be done. Note her response.
	Client, family, problem/ concern		20. *Mary:* Yeah . . . I'd like to be able to do something different. But, you know, ah . . . teaching's a very secure field, and I have tenure. You know, I'm the sole support of my two daughters, but I think, I don't know what else I can do exactly. Do you see what I'm saying?	Mary, being heard, is able to move to a deeper discussion of her issues. Note that Mary continues to talk within the formal-operational orientation. She discusses patterns and generalizations. If she were primarily concrete, she would give many more linear details and tell specific stories about her issues. She continues for most of the

(continued)

TABLE 13-2 (continued)

Skill Classifications				
Listening and Influencing	Focus	C	Counselor and Client Conversation	Process Comments
				interview in this mode of expression. She has equated "something different" with a lack of security. As the interview progresses, you will note that she associates change with risk. It is these basic meaning constructs, already apparent in the interview, that lie under many of her issues.
Reflection of feeling, followed by check-out	Client, problem	C	21. *Allen:* Looks like the security of teaching makes you feel good, but it's the boredom you associate with that security that makes you feel uncomfortable. Is that correct?	This reflection of feeling contains elements of a confrontation as well, in that the good feelings of security are contrasted with the boredom associated with teaching.
	Client, problem/ concern		22. *Mary:* Yeah, you know, it's that security. I feel good being . . . you know . . . having a steady income, and I have a place to be, but it's boring at the same time. You know, ah . . . I wish I knew how to go about doing something else.	Note that Mary often responds with a "Yeah" to the reflections and paraphrases before going on. Here she is wrestling with the confrontation of *Allen 21*. She adds new data, as well, in the last sentence. On the CCS, this would be acceptance and recognition (Level 3).
Summary, check-out	Client, family, problem/ concern	C	23. *Allen:* So, Mary, let me see if I can summarize what I've heard. Ah . . . it's been tough since the divorce, but you've gotten things together. You mentioned the kids are doing pretty well. You talked about a new relationship. *I heard you mention that.* (*Mary:* Yeah.) But the issue right now that you'd like to talk about is . . . this feeling of boredom (*Mary:* Ummm . . .) on the job, and yet you like the security of it. But maybe you'd like to try something new. Is that the essence of it?	This summarization concludes the first attempt at problem definition in this brief interview. Allen uses Mary's own words for the main things and attempts to distill what has been said. The positive asset search has been used briefly ("You've gotten things together . . . kids . . . doing well"). See other leads that emphasize client strength. Mary sits forward and nods with approval throughout this summary. The confrontation of the old job with "maybe you'd like to try something new" concludes the summary. Note the check-out at the end of the summary to encourage Mary to react.

(continued)

TABLE 13-2 (continued)

Skill Classifications				
Listening and Influencing	*Focus*	*C*	*Counselor and Client Conversation*	*Process Comments*
	Client, problem		24. *Mary:* That's right. That's it.	Mary again responds at Level 3 on the CCS, acceptance and recognition.

STAGE 3: Goals ("What do you want to happen?") Set goals mutually.

Open question	Main theme		25. *Allen:* I think it might be helpful if you could specifically define what some things are that might represent a more ideal situation.	In Stage 3, find where the client wants to go in a more ideal situation. You'll note that the basic listening sequence is present in this stage, but it does not follow in order, as in the preceding stage.
	Client, problem, others		26. *Mary:* Ummm. I'm not sure. There are some things I like about my job. I certainly like interacting with the other professional people on the staff. I enjoy working with the kids. I enjoy talking with the kids. That's kind of fun. You know, it's the stuff I have to teach I'm bored with. I have done some teaching of human sexuality and drug education.	Mary associates interacting with people as a positive aspect of her job. When she says "enjoy working with kids," her tone changes, suggesting that she doesn't enjoy it that much. But the spontaneous tone returns when she mentions "talking with them" and talks about teaching subjects other than team sports.
Paraphrase, open question	Client, main theme		27. *Allen:* So, would it be correct to say that some of the teaching, where you have worked with kids on content of interest to you, has been fun? What else have you enjoyed about your job?	The search here is for positive assets and things that Mary enjoys. Note the "what else?"
	Client, family, problem		28. *Mary:* Well, I must say I enjoy having the same summer vacations the kids have. That's a plus in the teaching field. (pause)	
Encourage			29. *Allen:* Yeah . . .	Mary found only one plus in the job. Allen probes for more data via an encourager. This type of encourager can't be classified in terms of focus.
	Client, others		30. *Mary:* You see, I like being able to . . . Oh, I know, one time I was able to do teaching of our own teachers, and that was really . . .	Mary brings out new data that support her earlier comment that she liked to teach when the content was of interest to her.

(continued)

TABLE 13-2 (continued)

Skill Classifications				
Listening and Influencing	*Focus*	*C*	*Counselor and Client Conversation*	*Process Comments*
			I really felt good being able to share some of my ideas with some people on the staff. I felt that was kind of neat, being able to teach other adults.	The "I" statements here are more positive, and the adjective descriptors indicate more self-assurance.
Closed question	Client, others		31. *Allen:* Do you involve yourself very much in counseling the students you have?	A closed question with a change of topic to explore other areas.
	Client, others		32. *Mary:* Well, the kids . . . you know, teaching them is a nice, comfortable environment, and kids stop in before class and after class, and they talk about their boyfriends and the movies; I find I like that part . . . about their concerns.	Mary responds to the word "counseling" again with discussion of interactions with people. It seems important to Mary that she have contact with others.
Summary, closed question	Client, problem, others		33. *Allen:* So, as we've been reviewing your current job, it's the training, the drug education, some of the teaching you've done with kids on topics other than phys. ed. (*Mary:* That's right.) And getting out and doing training and other stuff with teachers . . . ah, sharing some of your expertise there. And the counseling relationships. (*Mary:* Ummm.) Out of those things, are there fields you've thought of transferring to?	This summary attempts to bring out the main strands of the positive aspects of Mary's job. In an ongoing interview, a closed question on a relevant topic can be as facilitating as an open question. Note, however, that the interviewer still directs the flow with the closed question.
	Client, problem/ concern		34. *Mary:* Well, a lot of people in physical education go into counseling. That seems like a natural second thing. Ah . . . of course, that would require some more going to school. Umm . . . I've also thought about doing some management training for a business. Sometimes I think about moving into business . . . entirely away from education. Or even working in a college as opposed to working here in the high school. I've thought about those things, too. But I'm just not sure which one seems best for me.	Mary talks with only moderate enthusiasm about counseling. In discussing training and business, she appears more involved. Mary appears to have assets and abilities, makes many positive "I" statements, is aware of key incongruities in her life, and seems to be internally directed. She is clearly an abstract formal-operational client. For career success, she needs to become more concrete and action-oriented.

(continued)

TABLE 13-2 (continued)

| Skill Classifications | | | | |
Listening and Influencing	Focus	C	Counselor and Client Conversation	Process Comments
Paraphrase, closed question	Client, problem/concern		35. *Allen:* So the counseling field, the training field. You've thought about staying in schools and perhaps in management as well. (*Mary:* Um-hm, um-hm.) Anything else that occurs to you?	This brief paraphrase distills Mary's ideas in her own words.
	Problem/concern		36. *Mary:* No, I think that seems about it.	
Summary, open question, eliciting meaning	Client, problem/concern	C	37. *Allen:* Before we go further, you've talked about teaching and the security it offers. But at the same time you talk about *boredom*. You talk with excitement about business and training. How do you put this together? What does it mean to you?	This summary includes confrontation and catches both content and feeling. The question at the end is directed toward issues of meaning. The word "boredom" was underlined with extra vocal emphasis.
	Client, others		38. *Mary:* Uhhh . . . ah . . . If I stay in the same place, it's just more of the same. I see older teachers, and I don't want to be like them. Oh, a few have fun; most seem just *tired* to me. I don't want to end up with that.	Mary elaborates on the meaning and underlying structure of *why* she might want to avoid the occasional boredom of her job. When she talks about "ending up like that," we see deeper meanings. On the CCS, the client may again be rated at acceptance and recognition (Level 3). Though considerable depth of understanding and clarity is being developed, no change has really occurred. You will find that developmental movement often is slow and arduous. Nonetheless, each confrontation moves to more complete understanding.
Encourager/restatement	Client		39. *Allen:* You don't want to end up with that.	The key words are repeated.
	Client		40. *Mary:* Yeah, I want to do something new, more exciting. Yet my life has been so confused in the past, and it is just settling down. I'm not sure I want to risk it.	Mary moves on to talk about what she wants, and a new element—risk—is introduced. Risk may be considered Mary's opposing construct to security.

(continued)

TABLE 13-2 (continued)

Skill Classifications				
Listening and Influencing	*Focus*	*C*	*Counselor and Client Conversation*	*Process Comments*
Reflection of feeling	Client		41. *Allen:* So, Mary, risk frightens you?	This reflection of feeling is tentative and said in a questioning tone. This provides an implied check-out and gives Mary room to accept it or suggest changes to clarify the feeling.
	Client, problem/ concern		42. *Mary:* Well, not really, but it does seem scary to give up all this security and stability just when I've started putting it together. It just feels strange. Yet I do want something new so that life doesn't seem so routine . . . and . . . ah . . . I think maybe I have more talent and ability than I used to think I did.	Mary responds as might be predicted with a deeper exploration of feelings of fear of change. At the same time, she draws on her personal strengths to cope with all this.
Reflection of meaning, check-out	Client, problem/ concern	C	43. *Allen:* So you've felt the meaning in this possible job change as an opportunity to use your *talent* and take risks in something new. This may be contrasted with the feelings of stability and certainty where you are now. But *now* means you may end up tired and burned-out like some co-workers you have observed. Am I reaching the sense of things? How does that sound?	This reflection of meaning also confronts underlying issues that impinge on Mary's decision. It contains elements of the positive asset search or positive regard as Allen verbally stresses the word "talent."
	Client, problem/ concern		44. *Mary:* Exactly! But I hadn't touched on it that way before. I do want stability and security, but not at the price of boredom and feeling down as I have lately. Maybe I do have what it takes to risk more.	Mary is reinterpreting her situation from a more positive frame of reference. Allen could have said the same thing via an interpretation, but reflection of meaning lets Mary come up with her own definition. This reinterpretation of Mary's meaning represents generation of a new solution (CCS Level 4). She has a new frame of reference with which to look at herself. But this newly integrated frame is *not* problem resolution; it is a *step* toward a new way of thinking and acting. Allen decides to move to Stage 4 of the interview. It would be possible to explore problem definition and detail the goals more precisely, but we can take up these matters in later interviews.

(continued)

TABLE 13-2 (continued)

Skill Classifications				
Listening and Influencing	*Focus*	*C*	*Counselor and Client Conversation*	*Process Comments*
STAGE 4: Restory: Explore and create alternatives, confront client incongruities and conflict, restory ("What are we going to do about it?")				
Feedback	Client		45. *Allen:* Mary, from listening to you, I get the sense that you do have considerable ability. Specifically, you can be together in a warm, involved way with those you work with. You can describe what is important to you. You come across to me as a thoughtful, able, sensitive person. (pause)	Allen combines feedback on positive assets with some self-disclosure here and uses this lead as a transition to explore alternative actions. The emphasis here is on the positive side of Mary's experience. Allen's vocal tone communicates warmth, and he leans toward Mary in a genuine manner.
			46. *Mary:* Ummm . . .	During the feedback, Mary at first shows signs of surprise. She sits up, then relaxes a bit, smiles and sits back in her chair as if to absorb what Allen is saying more completely. There are elements of praise in Allen's comment.
Directive, paraphrase, open question	Client, main theme		47. *Allen:* Other job ideas may develop as we talk . . . ah . . . I think it might be appropriate at this point to explore some alternatives you've talked about. (*Mary:* Um-hm.) The first thing you talked about was you liked teaching drug education and sexuality. What else have you taught kids in that general area?	Allen starts exploring alternatives a little more concretely and in depth. The systematic problem-solving model—define the problem, generate alternatives, and set priorities for solutions—is in his mind throughout this section. He begins with a mild directive. "What else?" keeps the discussion open.
	Client, problem/ concern		48. *Mary:* Let's see . . . the general areas I liked were human sexuality, drug education, family life, and those kinds of things. Ah . . . sometimes communication skills.	
Closed question	Problem/ concern		49. *Allen:* Have you attended workshops on any of these topics?	Closed questions oriented toward concreteness can be helpful in determining specific background important in decision making.
	Client, problem/ concern, others		50. *Mary:* I've attended a few. I've enjoyed them. . . . I really did. You know, I've gone to the university and taken workshops in values clarification and communication skills. I liked the people I met.	Note that virtually all counselor and client comments have focused on the client and the problem. It is important to consider the client in each of your responses; too heavy an emphasis on the problem may cause you to miss the unique person before you. At the

(continued)

TABLE 13-2 (continued)

Skill Classifications				
Listening and Influencing	*Focus*	*C*	*Counselor and Client Conversation*	*Process Comments*
				same time, a broader focus might expand the issue and provide more understanding. Social work, for example, might emphasize the family and social context.
Reflection of feeling, information, check-out	Client, problem/ concern		51. *Allen:* Sounds like you've really enjoyed these sessions. One of the important roles in counseling, education, and business is training—for example, psychological education through teaching others skills of living and communication. How does that type of work sound to you?	Allen briefly reflects her positive feelings, and then shares a short piece of occupational information. This is followed by a check-out returning the focus to Mary.
	Client, problem/ concern		52. *Mary:* I think I would enjoy that sort of thing. Um-hmmm . . . it sounds interesting.	
Paraphrase, open question	Client, problem/ concern		53. *Allen:* Sounds like you have also given a good deal of thought to . . . ah . . . extending that to training in general. How aware are you of the business field as a place to train?	Mary's background and interest in a second alternative are explored.
	Client, problem/ concern, environmental context		54. *Mary:* I don't know that much about it. You know, I worked one summer in my dad's office, so I do have an exposure to business. That's about it. They all have been saying that a lot of teachers are moving into the business field. Teaching is not too lucrative, and with all the things happening here in California and all the cutbacks, business is a better long-term possibility for teachers these days. It just seems like an intriguing possibility for me to investigate or look into. The latest business cutbacks are scary, too.	Mary talks in considerably greater depth and with more enthusiasm when she talks about business. The important descriptive words she has used with teaching include "boring," "security," and "interpersonal interactions," while "interest" and "excitement" were used for training and teaching psychologically oriented subjects as opposed to physical education. Now she mentions cutbacks. Business has been described with more enthusiasm and as more lucrative. We may anticipate that she will eventually associate the potential excitement of business with the negative construct of risk and the lack of summer vacations and time to be with her children.

(continued)

TABLE 13-2 (continued)

Skill Classifications				
Listening and Influencing	*Focus*	*C*	*Counselor and Client Conversation*	*Process Comments*
Paraphrase, reflection of feeling	Client, problem/ concern	C	55. *Allen:* Mm-hmm . . . so you've thought about it . . . looking into business, but you've not done too much about it yet. Neither teaching nor business is really promising now, and that's a little scary.	This paraphrase is somewhat subtractive. Mary did indicate that she had summer experience with her father. How much and how did she like it? Allen missed that. The paraphrase involves a confrontation between what Mary says and her lack of doing anything extensive in terms of a search. The reflection of feeling acknowledges emotion.
	Client		56. *Mary:* That's right. I've thought about it, but . . . ah . . . I've done very little about it. That's all . . .	Mary feels a little apologetic. She talks a bit more rapidly, breaks eye contact, and her body leans back a little. Mary's response is at Level 2 on the CCS. She is only partially able to work with the issues of the confrontation.
Interpretation	Problem/ concern		57. *Allen:* And, finally, you mentioned that you have considered the counseling field as an alternative. Ah . . . what about that?	Allen omitted further exploration of business. If Allen had focused on positive aspects of Mary's experience and learned more about her summer experience, the confrontation (of thinking without action) probably would have been received more easily. As this was a demonstration interview, Allen sought to move through the stages perhaps a little too fast. Also, the counseling field is an alternative, but it seems to come more from Allen than from Mary. An advantage of transcripts such as this is that one can see errors. Many of our errors arise from our own constructs and needs. This intended paraphrase is classified as an interpretation, as it comes more from Allen's frame of reference than from Mary's.
	Problem/ concern, others		58. *Mary:* Well, I've always been interested, like I said, in talking with people. People like to talk with me about all kinds of things. And *that* would be interesting . . . ah . . . I think, too.	Mary starts with some enthusiasm on this topic, but as she talks her speech rate slows, and she demonstrates less energy.

(continued)

TABLE 13-2 (continued)

Skill Classifications				
Listening and Influencing	Focus	C	Counselor and Client Conversation	Process Comments
Encourage			59. *Allen:* Um-hmmm.	Use of encourager.
	Problem/concern		60. *Mary:* You know, to explore that. (pause)	Said even more slowly.
Encourage			61. *Allen:* Um-hmmm. (pause)	Allen senses her change of enthusiasm, is a bit puzzled, and sits silently, encouraging her to *talk more*. When you have made an error, and the client doesn't respond as you expect, return to attending skills.
	Problem/concern		62. *Mary:* But . . . I'd have to take some *courses* . . . if I really wanted to get into it.	One reason for Mary's hesitation appears.
Interpretation/reframe	Client, problem/concern		63. *Allen:* So putting those three things together, it seems that you want people-oriented occupations. They are particularly interesting to you.	This is a mild interpretation, as it labels common elements in the three jobs. It could be classified also as a paraphrase. Not all skill distinctions are clear.
	Client		64. *Mary:* Definitely . . . and that's where I am most happy.	Mary has returned to a Level 3 on the CCS.
Feedback	Client, problem/concern	C	65. *Allen:* And, Mary, as I talk, I see you . . . ah . . . coming across with a lot of enthusiasm and interest as we talk about these alternatives. I do feel you are a little less enthusiastic about returning to school. (*Mary:* Right!) I might contrast your enthusiasm about the possibilities of business and training with your feelings about education. There you talk a little more slowly and almost seem bored as you talk about it. You seem lively when you talk about business possibilities.	Allen gives Mary specific and concrete feedback about how she comes across in the interview. There is a confrontation as he contrasts her behavior when discussing two topics. Confrontation—the presentation of discrepancies or incongruity—may appear with virtually all skills of the interview. It may be used to summarize past conversation and stimulate further discussion, leading toward a resolution of the incongruity.
	Client, problem/concern		66. *Mary:* Well, they sound kind of exciting to me, Allen. But I just don't know how to go about getting into those fields or what my next steps might be. They sound very exciting to me, and I think I may have some talents in those areas I haven't even discovered yet.	Mary talks rapidly, her face flushes slightly, and she gestures with enthusiasm. She meets the confrontation and seems to be willing to risk more. This, however, may still be considered a Level 3 on the CCS, although there may be movement ahead.

(continued)

TABLE 13-2 (continued)

Skill Classifications				
Listening and Influencing	*Focus*	*C*	*Counselor and Client Conversation*	*Process Comments*
Feedback, information, logical consequences	Client, problem/concern	C	67. *Allen:* Um-hmmm. Well, Mary, I can say one thing. Your enthusiasm and ability to be open will be helpful to you in your search. Ah . . . at the same time, business and schools represent different types of lifestyles. I think I should give you a warning that if you go into the business area you're going to lose those summer vacations.	This statement combines mild feedback with logical consequences. A warning about the consequences of client action or inaction is spelled out. Mary is also confronted with some consequences of choice.
	Client, problem/concern, others		68. *Mary:* Yeah, I know that . . . and you know, that special friend in my life—he's in education—I don't think he would like it if I was, you know, working all summer long. But business does pay a lot more, and it might have some interesting possibilities. (*Allen:* Um-hmm.) . . . It's a difficult situation.	Confrontations often result in clients presenting new important concepts and facts that have not been discussed previously. A new problem has emerged that may need definition and exploration. Mary is still responding at Level 3 on the CCS, but Allen is obtaining a more complete picture of the problem and of the client.
Encourage/restatement	Problem		69. *Allen:* A difficult situation?	Again, the encourager is used to find deeper meanings and more information.
	Client, problem, others		70. *Mary:* Um-hmm. I guess I'm saying that . . . I'm . . . ah . . . you know, my friend . . . I don't think he would approve or like the idea of me having two weeks' vacation. (*Allen:* Uh-huh.) He wants me to stay in some field where I have the same vacation time I have now, so we can spend that time together.	Mary has more speech hesitations and difficulties in completing a sentence here than she has anywhere in the interview. This suggests that her relationship is important to her, and her friend's attitude may be important in the final career decision. Much career counseling involves personal issues as well as career choice. Both require resolution for true client satisfaction.
Interpretation/reframe, open question	Client, others, cultural/environmental/contextual	C	71. *Allen:* I hear you saying that your friend has a lot to say about your future. How does that strike you as an independent woman who has been on your own successfully for quite a while?	Here we see the introduction of gender relations as a cultural/environmental/contextual issue. Allen's reframing of the situation offers Mary a chance to explore her relationship with her friend from a different contextual perspective.
	Others		72. *Mary:* It really is . . . well, Bo's a special person. . . .	Mary's eyes brighten.

(continued)

TABLE 13-2 (continued)

Skill Classifications				
Listening and Influencing	*Focus*	*C*	*Counselor and Client Conversation*	*Process Comments*
Interpretation	Client, others		73. *Allen:* And, I sense you have some reactions to his . . .	Allen interrupts, perhaps unnecessarily. It might have been wise to allow Mary to talk about her positive feelings toward Bo.
	Client, problem/ concern		74. *Mary:* Yeah, I'd like to be able to explore some of my own potential without having those restraints put on me right from the beginning.	Mary talks slowly and deliberately, with some sadness in her voice. Feelings are often expressed through intonation. Here we see the beginning of a critical gender issue. Women often feel constraints in career or personal choices, and men in this culture often place implicit or explicit restraints on critical decisions. Feminist counseling theorists argue that a male helper may be less effective with these types of problems. What are your thoughts on this issue?
Interpretation/reframe, check-out	Client, problem/concern, others, cultural/environmental/ contextual	C	75. *Allen:* Um-hmm . . . In a sense he's almost placing similar constraints on you that you feel in the job in physical education. There are certain things you have to do. Is that right?	This interpretation relates the construct of boredom and the implicit constraint of being held down with the constraints from Bo. The interpretation clearly comes from Allen's frame of reference. With interpretations or helping leads from your frame of reference, the check-out of client reactions is even more important. The drawing of parallels is abstract formal-operational in nature.
	Client, problem/concern, others		76. *Mary:* Yes, probably so. He's putting some limits on me . . . setting limits on the fields I can explore and the job possibilities I can possibly have. Setting some limits so that my schedule matches his schedule.	Mary answers quickly. It seems the interpretation was relatively accurate and helpful. One measure of the function and value of a skill is what the client does with it. Mary changes the word "constraints" to the more powerful word "limits." Mary remains at Level 3 on the CCS, as she is still expanding on aspects of the problem.
Open question, oriented to feeling	Client		77. *Allen:* In response to that you feel . . . ? (deliberate pause, waiting for Mary to supply the feeling)	Research shows that *some* use of questions facilitates emotional expression.

(continued)

TABLE 13-2 (continued)

Skill Classifications				
Listening and Influencing	Focus	C	Counselor and Client Conversation	Process Comments
	Client, problem/concern		78. *Mary:* Ah . . . I feel I'm not at a point where I want to *limit things.* I want to see what's open, and I would like to keep things open and see what all the alternatives are. I don't want to shut off any possibility that might be really exciting for me. (*Allen:* Um-hmm.) A total lifetime of careers.	Mary determinedly emphasizes that she does not want limits.
Reflection of feeling, paraphrase	Client, problem/concern	C	79. *Allen:* So you'd like to have a life of exciting opportunity, and you sense some limiting . . .	A brief, but important, confrontation of Bo versus career.
	Client, problem/concern, others, cultural/environmental/contextual		80. *Mary:* He reminds me of my relationship with my first husband. You know, I think the reason that all fell apart was my going back to work. You know, assuming a more nontraditional role as a woman and exploring my potential as a woman rather than staying home with the children . . . ah . . . you know, sort of a similar thing happened there.	Again, the confrontation brings out important new data about Mary's present and past. Is she repeating old relationship patterns in this new relationship? The counselor should consider issues of cultural sexism as an environmental aspect of Mary's planning. This does not appear in this interview, but a broader focus on issues in the next session seems imperative. Other focus issues of possible importance include Mary's parental models, others in her life, a women's support group, the present economic climate, the attitudes of the counselor, and "we"—the immediate relationship of Mary and Allen. Thus far he has assumed a typical Western "I" form of counseling where the emphasis is on the client. Due to the development of new, more integrated data, this could be development of a more inclusive construct (Level 5 response on the CCS).
Summary	Client, problem/concern, others, cultural/environmental/contextual	C	81. *Allen:* There really are several issues that . . . you're looking at. One of these is the whole business of a job. Another is your relationship with Bo and your desire to find your own space as an independent woman.	The interview time is waning, and Allen must bring about a smooth ending and plan for the next session. He catches the incongruity that Mary faces between work and relationship. Allen fails to pick up

(continued)

TABLE 13-2 (continued)

| Skill Classifications | | | | |
Listening and Influencing	Focus	C	Counselor and Client Conversation	Process Comments
				fully on the cultural/ environmental/contextual focus. Many of Mary's issues relate to women's issues in a sometimes sexist world.
			82. *Mary:* (slowly) Um-hmmm . . .	Mary looks down, relaxes, and seems to go into herself.
Reflection of feeling	Client		83. *Allen:* You look a little sad as I say that.	This reflection of feeling comes from nonverbal observations and picks up on her facial reactions.
	Problem/ concern		84. *Mary:* It would be nice if the two would mesh together, but it seems difficult to have both things fit together nicely.	Mary is describing her ideal resolution. Here the interview could recycle back to Stages 2 and 3, with more careful delineation of the problem between job and personal relationships and defining the ideal resolution more fully. *A problem exists only if there is a difference between what is actually happening and what you desire to have happen.* This sentence illustrates the importance of problem definition and goal setting. Mary's response to the confrontation is 4 on the CCS. We have an important new insight, but insight is not action. She also needs to act on this awareness.
Information, directive	Problem		85. *Allen:* Well, that's something we can explore a little bit further. It seems this is an important part of the puzzle. Let's work on that next week. Would that be okay? I see our time is about up now. But it might be useful if we can think of some actions we can take between now and the next time we get together.	Many clients bring up central issues just as the interview is about to end. Allen makes the decision, difficult though it is, to stop for now and plan for more discussion later. Note that Mary is talking about her relationship mainly from an abstract formal-operational orientation.

STAGE 5: Action: Conclude, generalize, and act on new stories ("Will you do it?")

Summary, open question	Client, problem/ concern		86. *Allen:* We have come up so far with three things that seem to be logical: business, counseling, and training. I think it would be useful, though, if you were	Allen continues his statement and moves to Stage 5. He summarizes the career alternatives generated thus far and raises the possibility of taking a test.

(continued)

TABLE 13-2 (continued)

Skill Classifications				
Listening and Influencing	*Focus*	*C*	*Counselor and Client Conversation*	*Process Comments*
			to take a set of career tests. (*Mary:* Uh-huh.) That will give us some additional things to check out to see if there are any additional alternatives for us to consider. How do you feel about taking tests?	Note that he provides a check-out to give Mary an opportunity to make her own decision about testing.
	Client, problem/concern		87. *Mary:* I think that's a good idea. I'm at the stage where I want to check all alternatives. I don't want *anything* to be limited. I want to think about a lot of alternatives at this stage. And I think it would be good to take some tests.	Mary approves of testing and views this as a chance to open alternatives. She verbally emphasizes the word "anything," which may be coupled with her desire to avoid limits to her potential. Some women would argue that a female counselor is needed at this stage. A male counselor may not be sufficiently aware of women's needs to grow. Allen could unconsciously respond to Mary in the same ways she views Bo as responding to her.
Information	Client, problem/concern, interviewer		88. *Allen:* Then another thing we can do . . . ah . . . is helpful. I have a friend at a local firm who originally used to be a coach. She's moved into personnel and training at Jones. (*Mary:* Ummm.) I can arrange an appointment for you to see her. Would you like to go down and look at the possibilities there?	Allen suggests a concrete and specific alternative for action. Although Mary is predominantly formal-operational, she has tended to talk about issues and avoid action. This avoidance of action is also indicative of Level 3 on the CCS. Until Mary takes some form of concrete action or resolves the issue in her mind, she will remain at Level 2 or 3 on the CCS. If some action is taken on the issue during the coming week, then she will have moved at least partially to Level 4 on the CCS.
	Client, problem/concern		89. *Mary:* Oh, I would like to do that. I'd get kind of a feel for what it's like being in a business world. I think talking with someone would be a good way to check it out.	Stated with enthusiasm. The proof of the helpfulness of the suggestion will be determined by whether she does indeed have an interview with the friend at Jones and finds it helpful in her thinking.

(continued)

TABLE 13-2 (continued)

Skill Classifications				
Listening and Influencing	*Focus*	*C*	*Counselor and Client Conversation*	*Process Comments*
Feedback, open question	Client, problem		90. *Allen:* You're a person with a lot of assets. I don't have to tell you all the things that might be helpful. What other ideas do you think you might want to try during the week?	Allen recognizes he may be taking charge too much and pulls back a little. Although he is pushing Mary for action, he is now using her ideas. Too much direction and advice can make a client resistant to your efforts.
	Client, problem		91. *Mary:* What about checking into the university and ah . . . advanced degree programs? I have a bachelor's degree, but . . . maybe I should check into school and look into what it means to do more course work.	Mary, on her own, decides to look into the university alternative. This is particularly important, as earlier indications were that she was not all that interested. Note that real generalization is usually concrete *action*.
Summary	Client, problem, cultural/ environmental/ contextual		92. *Allen:* Okay, that's something else you could look into as well. (*Mary:* Uh-huh.) So let's arrange for you then to follow up on that. I'd like to see you doing that. (*Mary:* Um-hmmmm.) And . . . ah . . . we can get together and talk again next week. You did express some concern about your relationship with your friend, Bo, ah . . . would you like to talk about that as well next week? And as I look back on this session, one theme we haven't discussed yet is how being a woman with family responsibilities relates to all this. Maybe this is something to be explored next week as well?	Allen is preparing to terminate the interview. Fortunately, he does consider the women's issue. Probably this should have been done sooner in the session. Is this an issue with which he can help, or would you recommend referral?
	Client, cultural/ environmental/ contextual		93. *Mary:* I think so, they sort of all . . . one decision influences another. You know. It all sort of needs to be discussed. And thanks for bringing up the women's issue and my children. That's important to me.	An important insight at the end. Mary realizes her career issue is more complex than she originally believed. If you were Allen's supervisor, would *you* recommend a primary emphasis on career counseling or on personal counseling in the next session? Or perhaps some combination of them both? What else would you advise him to do?
Self-disclosure	Client, interviewer		94. *Allen:* Okay. I look forward to seeing you next week, then.	
			95. *Mary:* Thank you.	

INSTRUCTIONAL READING 2: INTERVIEW TRANSCRIPT ANALYSIS AND PLANNING

Following up, analyzing, and planning for the future are critical for case management and treatment. Instructional Reading 2 is focused on the following:

Skills and their influence on the client. What skills did Allen use and what was their effect on Mary, the client? What will be the impact of your skills on how the client reacts?

Referral and consultation. When shall you refer and how can you obtain feedback on your interviews?

Case management, case conceptualization, and planning for future interviews.

Skills and Their Influence on the Client

Table 13-3 presents a skill summary of Allen's interview with Mary. Note the different use of skills in each stage of this interview. Allen used no influencing skills in Stages 2 or 3 (*story and strengths,* and *goals*) but a more balanced use of skills in the other stages. In Stage 4 Allen used both influencing skills and confrontation of incongruity and discrepancies extensively. Stage 4, *restory,* could be considered the "working" phase of the interview.

In terms of a total balance of skill usage, Allen used a ratio of approximately two attending skills for every influencing skill. His focus remained primarily on the client, although the majority of his focus dimensions were dual, combining focus on Mary with focus on the problem or issue. He had four comments focusing on cultural issues.

We should note that Allen focused primarily on the client in the earlier phases of the interview and only in the later portions focused on the problem. Thus, he does seem to have mastered to some extent the ability to focus on the person and balance that effectively with the problem focus. An ineffective interviewer might have focused early on the problem and missed the unique person completely.

In each phase of the interview Allen fulfilled the objectives of each stage. In the relationship stage he was able to use client observation skills to note that Mary was somewhat tense at the beginning and then to flex and select a rapport-building exchange that enabled the client to be more comfortable in the interview. In the *story and strengths* stage Mary identified a specific issue she wanted to resolve, which Allen summarized at Allen 23. In Stage 3 he used the basic listening sequence to bring out some concrete goals (business, counseling), and through reflection of meaning (Allen 43) he summarized some of the key aspects of Mary's thinking about the issue.

In Stage 4 a number of incongruities were confronted, and with each confrontation Mary appeared to move a little deeper into some personal insights concerning her present and future. Note in particular Allen 71, which led Mary into the important area of her personal life and relationship with Bo. The slightly inaccurate confrontation at Allen 75 fortunately included a check-out, and thus Mary was able to introduce her important construct, substituting the word "limits" for Allen's "constraints." In the final stage Mary appeared ready and willing to take action. However, the real proof of the value of the interview will have to wait until the next meeting, which will show whether the generalization plan was indeed acted on.

Examining specific skills, we might note that Allen's open questions tended to encourage Mary to talk, that his paraphrases and reflections of feeling were often followed by Mary's saying "Yes" or "Yeah." In the important reflection of meaning at Allen 43, Mary responded with "Exactly! I hadn't touched on it that way before. . . ." (Mary 44). In short, Allen does seem to be able to use the skills to produce specific results with the client.

TABLE 13-3   Skill summary of Allen and Mary interview over five stages

	Skill Classifications																				
	Listening/Attending							Focus							Influencing						C
	Open question	Closed question	Enc./restatement	Paraphrase	Reflection of feeling	Reflection of meaning	Summary	Client	Problem/concern	Significant others	Family	Mutuality "we"	Interviewer	C/E/C	Interpretation/reframe	Logical consequences	Self-disclosure	Feedback	Info./adv./etc.	Directive	Confrontation
STAGE 1. *RELATIONSHIP:* Initiating the session 6 attending skills 3 influencing skills	2	2		2	1			6	2				3				1		4		
STAGE 2. *STORY AND STRENGTHS:* Gathering data 4 attending skills 0 influencing skills 2 confrontation skills	1			1			1	5	2		1										2
STAGE 3. *GOAL:* Mutual goal setting 14 attending skills 0 influencing skills 2 confrontation skills	3	3	2	2	1	1	2	8	6	2											2
STAGE 4. *RESTORY:* Working 16 attending skills 15 influencing skills 7 confrontation skills	4	1	3	4	3		1	15	14	4				3	6	1		3	3	2	7
STAGE 5. *ACTION:* Terminating 5 attending skills 3 influencing skills	2						3	5	4				2	1			1	1	1		
Total: 45 attending skills 21 influencing skills 11 confrontation skills	12	6	5	9	5	1	7	39	28	6	1	0	5	4	6	1	2	4	8	2	11

In general, Mary responded primarily at Level 3 on the Client Change Scale to the 11 confrontations in this session. However, most of the confrontations seemed to enable Mary to explore her issues in more depth. A particularly successful one was at 79, where Allen comments on Mary's desire to have a life of "exciting opportunity," but notes that she senses Bo's putting limits on her. Mary moves to discuss her own personal wishes in more depth. On the Client Change Scale, she has moved here to a "5," or a new way of thinking about her issues.

The work that clients do after the interview is as important as or more important than what they do in the session with you. The real impact of the interview and the confrontations will show in the next session and in Mary's life after interviewing is completed. You may note that you can use the Client Change Scale as a way to assess the effectiveness of your interviewing and counseling over time.

You may want to think back for a moment about this session. What did you like? What advice would you give Allen? What did he do right?

In examining this interview for competence levels, we find that Allen is able to identify and classify the several skills and stages of the interview. He is able to identify some of the impact of his skills on the client.

Allen also demonstrates Level 2, basic competence, which calls for the ability to use the basic listening sequence to structure an interview in five simple stages, and to employ intentional interviewing skills in an actual interview.

Level 3, intentional competence, is more difficult to assess. Let us examine this level in more detail. In terms of focus dimensions, Allen centered primarily on Mary and her concerns. Focus analysis is useful as it points out that he did not focus on others and the family (Bo and Mary's children, for example), although some theorists might say that family and Bo were the most important area of all. The relationship with Bo evolved with greater clarity later in the interview. At 71, 75, and 81, Allen brought in the cultural/environmental/contextual focus, enabling a beginning discussion of gender issues that clearly needs further work. Some would question whether a man can help a woman with these issues. Do you think that Allen could continue this topic, or would you recommend referral to a woman counselor?

Referral and Consultation

The word *referral* appears in the interview process notes. No interviewer has all the answers, and in the case of Mary, Allen thought referral to a women's group might be helpful, as many of her issues are common to women looking for career change. In addition, his planning notes (see Box 13-2) indicate the need for further exploration of possible careers outside the interview. An important part of individual counseling is helping your clients find community resources that may facilitate their growth and development. The community genogram (Chapter 10) helps interviewers and clients think more broadly and consider appropriate referral sources.

A key referral issue is whether this is a case for which interviewer expertise and experience are sufficient to help the client. Allen has considerable experience in career and personal counseling, and thus far the issues are within his expertise. But this is where supervision and case conferences can be helpful. Just because Allen or any other counselor thinks that he or she is working effectively may not be enough. Opening up your work to others' opinions is an important part of professional practice. Clients, of course, should be aware that you as counselor or therapist are being supervised.

If the clients' issues are clearly beyond our knowledge and experience, an appropriate referral needs to be arranged carefully. Referral at intake is fairly clear as the intake interviewer seeks to match client issues with staff expertise. Referral to another helper during an ongoing interview series is more complex. We do not want to leave our clients "hanging" with no sense of direction or fearful that their problems are too difficult. The client should be fully involved throughout the referral process and the new counselor or therapist informed of client needs. In such cases, we recommend that you maintain contact with the client as the referral process evolves, sometimes even continuing for a session or two until the change to a different counselor is complete.

BOX 13-2 SECOND INTERVIEW PLAN AND OBJECTIVES

After reviewing the preceding session, identify issues you anticipate will be important in the next session and plan how you might handle them. (This plan is Allen's assessment of his forthcoming interview with Mary.)

Relationship—Initiating the session—Rapport and structuring

Are any special issues anticipated with regard to rapport development? What structure do you have for this interview? Do you plan to use a specific theory?

Mary and I seem to have reasonable rapport. Mary seems to like emphasis on her strengths, and I need to keep that in my mind and not become enmeshed in problem-centered thinking. As I look at the first session, I note I did not focus on Mary's context, nor did I attend to other things that might be going on in her life. It may be helpful to plan some time for general exploration *after* I follow up on the testing and her interviews with people during the week. Mary indicated an interest in talking about Bo. Two issues need to be considered at this session in addition to general exploration of her present state. I'll introduce the tests and follow that with discussion of Bo. For examining the issues around Bo, I think a person-centered, Rogerian method emphasizing listening skills may be helpful.

Story and strengths—Gathering data—Drawing out stories, concerns, problems, or issues

What are the anticipated problems? Strengths? How do you plan to define the issues with the client? Will you emphasize behavior, thoughts, feelings, meanings?

1. Check with Mary on her career plans and how she sees her career concerns defined now. Use basic listening sequence.
2. Later in the interview, go back to the issue with Bo. Open it up with a question, then follow through with reflective listening skills. If she starts with Bo, save career issues until later. Be alert to this relationship from a women's and feminist perspective.
3. Mary has many assets. She is bright, verbal, and successful in her job. She has good insight and is willing to take reasonable risks and explore new alternatives. These assets should be noted in our future interviews.
4. Explore women's issues with her.

Goals—Mutual goal setting

What is the ideal outcome? How will you elicit the idealized self or world?

This may not be too important in this interview. We already have Mary's vocational goals, but they may need to be reconsidered in light of the tests, further discussion of Bo, and so on. It is possible that late in this interview or in a following session we may need to define a new outcome in which careers and her relationships are both satisfied.

Restory—Working—Exploring alternatives, confronting client incongruities and conflict, restorying

What types of alternatives should be generated? What theories would you probably use here? What specific incongruities have you noted or do you anticipate in the client?

1. Check on results of tests and report them to Mary.
2. Explore her reactions and consider alternative occupations.
3. Use person-centered, Rogerian counseling to explore her issues with Bo.
4. Relate careers to the relationship with Bo. Give special attention to confronting the differences between her needs as a "person" and Bo's needs for her. Note and consider the issue of women in a changing world. Does Mary need referral to a woman or a women's group for additional guidance? Would assertiveness training be useful?

Action—Terminating—Generalizing and acting on new stories

What specific plans, if any, do you have for suggesting transfer of training to the client? What will enable you personally to feel that the interview was worthwhile?

At the moment it seems clear that further exploration of careers outside the interview is needed. We will have to explore the relationship with Bo and determine Mary's objectives more precisely.

(continued)

BOX 13-2 (continued)

Overall case conceptualiza-tion, treatment plan, and case management	*How do you put the interview(s) and your knowledge of the client together for a coherent future?* *Impression.* Mary is a healthy, intelligent client who has done well so far in her life, particularly considering her apparent success as a single parent. This bodes well for successful counseling. *Challenges, identifiable issues, and problems.* The issues presented focus on midcareer change and the reasons for this wish make sense. I will follow the general plan above to help her reach her goals, which still need further definition. Issues with her friend, Bo, came out late in the session, and she may want to consider personal counseling in later sessions. *Case management issues.* While Mary seems very much together, a women's support group might be helpful. If this turns to personal counseling, I will have to change my counseling style and theory. *Specific goals for the immediate future.* Help her define goals more precisely, learn more about personal issues, but keep the main focus on career change.

Finally, sometimes the relationship simply doesn't seem to be working. In such situations, seek consultation and supervision immediately, as most often these challenges can be resolved. When you sense the relationship isn't working, focus on client goals and seek to hear the client's story completely. And ask her or him for feedback and suggestions as to what might be helpful. If referral to another interviewer becomes necessary, avoid blaming either the client or yourself.

Case Management, Case Conception, and Planning for Future Interviews

Box 13-2 contains Allen's interview plan for the second session. This plan derives from information gained in the first interview and organizes the central issues of the case, allowing for new input from Mary as the session progresses. An overall case management plan is included. It is also a framework for a longer term plan.

In short-term interviewing and counseling, the interview plan also serves as the treatment and case management plan. As you move toward longer term counseling, use the same outline but plan in greater detail. Consultation and supervision is very helpful to even the most experienced professional.

Case management has long been emphasized in social work, but only recently has it become more central in counseling, perhaps because of insurance company pressures and the greater demand for accountability in the helping fields. Documentation of progress made and plans for movement toward identifiable and measurable goals are important. But with accountability, share your plans and progress notes with your client as appropriate.

SUMMARY: INTEGRATING SKILLS

This chapter is designed to serve as a model for you in transcribing and generating your own analysis of the interview. In addition, it provides you with an opportunity to examine, criticize, and note effective behaviors of Allen as he worked with Mary. No interview is perfect. What counts is your ability to be intentional. "It's not the mistakes you make, but what you do to correct them that counts." This message came from Allen's plumber, but it certainly speaks to us in interviewing and counseling as well.

We believe that all interviewing and counseling could be described as decisional processes. This is so whether you are deciding your college major, selecting your career, or determining whether to obtain a divorce. The person-centered approach is focused on deciding what type of person you are and what you want to be. Logotherapy and Victor Frankl remind us of the need for deciding our life direction. Cognitive-behavior theory and practice focus on solving and resolving many types of problems and issues. Brief therapy is very much a decisional approach, but it places much more attention on the clear definition of life goals.

Key points of Chapter 13 are summarized below.

■ KEY POINTS	
Decisional counseling	Decisional counseling, a modern reformulation of trait-and-factor theory, assumes that most, perhaps all, clients are involved in making decisions. By considering the many traits of the person and factors in the environment, it is possible to arrive at a more rational and emotionally satisfying decision.
Decisional structure and alternative theories	The restorying model of the interview can be considered a basic decisional model underlying other theories of counseling and therapy. Once you have mastered the skills and strategies of intentional interviewing and the five-stage model, you will find that you can more easily master other theories of helping.
Interview analysis	Using the constructs of this book, it is possible to examine your own interviewing style and that of others for microskill usage, focus, structure of the interview, and the resultant effect on a client's cognitive and emotional developmental style.
Interview plan and note taking	It is possible to use the restorying model structure to plan your interview before you actually meet with a client. This same five-stage structure can be used as an outline for note taking after the interview is completed.
Treatment plan, case conceptualization, and case management	A treatment plan is a long-term plan for conducting a course of interviews or counseling sessions. Case conceptualization is your integration of client, issues, challenges, and goals for the future. Case management often leads to alternative change strategies beyond the original interview and may involve efforts with the family, community, and other agencies.

COMPETENCY PRACTICE EXERCISE AND PORTFOLIO OF COMPETENCE

Students often find that the highlight of their experience with this book is videotaping or audiotaping their own session, transcribing it, and analyzing what occurred. You will find that you have learned a great deal about what occurs in the interview, and your ability to discuss what you see will be invaluable throughout your interviewing or counseling career.

Practice Exercise in Case Conceptualization and Interview Analysis

This is the most important practice exercise in this book. It involves conducting a complete interview and preparing a transcript in which you demonstrate your interviewing style, classify your behavior, and comment on your development over the term.

BOX 13-3   INTERVIEW PLAN AND OBJECTIVES

	After studying the client file before the first session or after reviewing the preceding session, complete this form indicating issues you anticipate being important in the session and how you plan to handle them.
Relationship: **Initiating the session—Rapport and structuring**	Are any special issues anticipated with regard to rapport development? What structure do you have for this interview? Do you plan to use a specific theory? Skill sequence?
Story and strengths: **Gathering data—Drawing out stories, concerns, problems, or issues**	What are the anticipated problems? Strengths? How do you plan to define the issues with the client? Will you emphasize behavior, thoughts, feelings, meanings? In what areas do you anticipate working on problems?
Goals: **Mutual goal setting**	What is the ideal outcome? How will you elicit the idealized self or world?
Restory: **Working—Exploring alternatives, confronting client incongruities and conflict, restorying**	What types of alternatives should be generated? What theories would you probably use? What specific incongruities have you noted or do you anticipate in the client? What skills are you likely to use? Skill sequences?
Action: **Terminating—Generalizing and acting on new stories**	What are the specific thoughts, feelings, and behaviors that you and the client would like to generalize to real life? Wherever possible, work with the client to set up the specific goals of generalization.
Overall case conceptualization, treatment plan, and case management	How do you put the interview(s) and your knowledge of the client together for a coherent future? Impression. Challenges, identifiable issues, and problems. Case management issues. Specific goals for the immediate future.

The following steps are suggested:

1. Plan to conduct an interview with a member of your group, a friend, or an actual client. This interview should last at least 20 minutes (although most prefer a longer time) and should follow the basic restorying model. It should be an interview you are satisfied to present to others. Before you conduct the interview, be sure you have your client's permission to record the session.
2. Before this interview is actually held, fill in an Interview Plan and Objectives form for the session (see Box 13-3, for example).
3. Audiorecord or videorecord the interview.
4. Develop a transcript of the session. Place the transcript in a format similar to the one used by Allen in this chapter. Leave space for comments on the form.
5. Classify your interviewing leads by skill and focus; classify the client's focus as well.
6. Identify the specific stages of the interview as you move through them. Note that you may not always follow the order sequentially: indicate clearly that you have returned to Stage 2 from Stage 4 if that occurs.

7. Make process comments on the transcript. Use your own impressions plus the descriptive ideas and conceptual frames of this book.
8. Develop interview notes on your session using the five-stage structure of the interview.
9. Develop a second Interview Plan and Objectives for the next session (Box 13-3).

Portfolio of Competence

After you have completed the interview, please go back to the session you completed as you started this book and note how your style has changed and evolved since then. What particular strengths do you note in your own work? Are you meeting your client's needs?

As you review your own work, pay special attention to your understanding and use of cultural/environmental/contextual issues. Examine your interview from the perspective of someone from a different cultural group and gender from your own. How would he or she consider and evaluate your work?

We are not presenting a checklist of specific competencies for this chapter, but we would like you to reflect on your work and your changes during the semester. Please comment or put in your journal reflections on the process of completing a full interview.

_____

_____

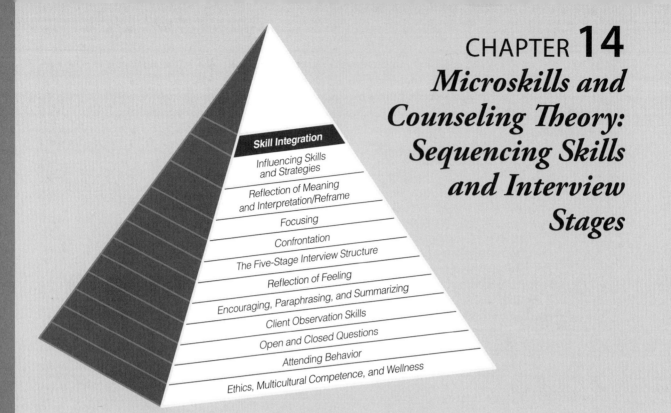

CHAPTER 14
Microskills and Counseling Theory: Sequencing Skills and Interview Stages

Our influence is determined by the quality of our being.

—Dale E. Turner

How can integrating microskills with counseling and therapy theory assist you and your clients?

Chapter Goals

This chapter presents four important approaches to counseling from the perspective of the five-stage interview—person-centered counseling, cognitive-behavioral therapy, brief solution-focused counseling, and motivational interviewing.

You will be able to apply key features of each theory in the interview. Microskills analysis and the *relationship—story and strengths—goal—restory—action* model will clarify how to use each theory.

Competency Objectives

Awareness, knowledge, and skills developed through this chapter will enable you to

▲ Conduct a beginning person-centered interview.
▲ Engage in some of the basics of cognitive-behavioral therapy with an emphasis on stress management.
▲ Practice the basics of brief solution-focused counseling.
▲ Practice key aspects of motivational interviewing.
▲ Realize that there are multiple paths to the resolution of client concerns and that the definition of change and the creative *New* will vary from theory to theory.

INTRODUCTION: MICROSKILLS, FIVE STAGES, AND THEORETICAL APPROACHES TO THE INTERVIEW

This chapter contains four different approaches to the interview, each one of them focusing, in a different way, on wellness and client positive strengths. Thus, this chapter is longer than others. We recommend that you read only one theoretical approach at a time, gain an understanding of it, and practice it in a session to at least a basic competence level before moving on to the next.

Another way to read this chapter is to select one or two theories here that appeal to you and learn these more fully. Read the theories of this chapter that are appropriate for your development as an interviewer and/or counselor. Finally, you may prefer to read the full chapter and focus on understanding. Later you can refer to these systems and develop intentional competence. Table 14-1 outlines the four approaches plus decisional counseling.

We face a time in the helping field when we are being asked for clear and measurable results from our work. We are also being pressured to be culturally aware and devise new ways to help clients make sense of their world. In addition, there is a great need to maintain the traditions of the past with a focus on human dignity. These are weighty demands on us and on the helping profession. Given the complexity of today's world, it is clear that being competent in only one or two approaches may be insufficient to reach the majority of your possible and ever-changing clientele.

INSTRUCTIONAL READING AND EXAMPLE INTERVIEW 1: PERSON-CENTERED COUNSELING

A major assumption of person-centered theory is that the client is competent and ultimately self-actualizing. The self-actualized person is able to constantly develop and achieve full personal potential. The task of the helper is to listen and help clients discover that inner, more ideal self. Primarily used are the listening skills and reflection of meaning, which seek to focus on internal strength and resilience.

A person-centered counselor is most often interested in focusing on the meaning and feelings of the client; the actual facts of the problem are considered less important. Decisions may be made and issues worked through, but it is how the client feels about *the self* that is most important. Therefore, the focus is much more on the person in the here and now and less on concerns and life issues. How this individual client relates to these issues and concerns is as important as or more important than resolving them. Questions are considered intrusive and generally should be avoided.

Person-centered theory is considered most appropriate for freely talking, abstract, self-directed/formal-operational clients who are best able to think through their own direction. These clients are able to reflect on their feelings. Person-centered work may also be helpful with clients who think more concretely. But your language needs to change and become more specific and focus more on specific, observable behavior. Give some attention to feelings, but remember that meaning issues may be more complex than most concrete clients can understand. Children are typically concrete in thought, but they can benefit from considerable paraphrasing, acknowledging feelings, and summarization, if you use their language system.

To illustrate, let us take the case study of Mary from Chapter 13. Let us assume that Mary comes to us with her laundry list of problems (Mary 16). You have seen how the decisional counselor, Allen, used the basic listening sequence to draw out client facts and

TABLE 14-1   Five major approaches to counseling and the five-stage interview structure: An outline for practice

Decisional Counselor Actions	Person-Centered Counselor Actions	Cognitive-Behavioral Counselor Actions	Brief Solution-Focused Counselor Actions	Motivational Interviewing Counselor Actions

Relationship: **Initiate the session—Develop rapport and structuring ("Hello; this is what might happen in this session").** All systems give special attention to developing the relationship and building a therapeutic alliance with a natural and personal style.

▲ Outlines purpose of session and what client can expect. ▲ May state what to expect in each stage of the interview.	▲ Tends not to discuss structure and moves immediately to direction established by client. ▲ May subtly point out importance of allowing client to direct the session.	▲ Structures session clearly (especially stress management). Emphasizes importance of client participation in the session and may state the importance of the client's defining specific goals for the session. ▲ Points out that specific thoughts and behaviors are the likely session focus.	▲ Clearly lets the client know what to expect—"What's your goal today?" "What has gotten better about the problem even before you got here?" ▲ Searches early for wellness strengths and positive assets.	▲ Structures session, stresses listening. "What would you like to talk about today?" ▲ Discovers motivation for change on 10-point scale. "How important is it for you to change?"

Story and strengths: **Gather data—Draw out stories, concerns, problems, or issues ("What's your concern? What are your strengths or resources?").** Attending, observation, and the basic listening sequence (BLS) are central in all five approaches.

▲ Uses BLS to draw out facts, feelings, and organization of client's problem or decisional issue. ▲ Draws out individual and multicultural strengths.	▲ Uses listening skills to draw out client concerns with a focus on the individual client and feelings. ▲ Maintains constant emphasis on positive regard and client strengths.	▲ Uses BLS to draw out concrete behaviors, thoughts, and feelings in specific situations. ▲ Focuses broadly on both individual and contextual issues. ▲ May provide information and instructions on various alternatives for cognitive-behavioral change.	▲ Draws out client story briefly, focusing on wellness and positive assets. Normalizes concerns and searches for contextual support systems. ▲ Seeks concrete examples of past successes.	▲ Uses BLS to widen awareness. "Change talk" focuses on the good and bad of behavior. "What do you like about (alcohol, drugs, etc.)?" "What's the down side?" ▲ Affirms client as a person. "You handled that well."

Goals: **"What do you want to happen?"** Each system helps the client find her or his own goals.

▲ Uses BLS to draw out client's ideal decisional and career goals. ▲ Makes client goals concrete and measurable.	▲ Reveals client goal through listening, but even Carl Rogers has been known to ask "What would you like to see happen?" ▲ Helps client define distinction between the real self and the ideal self.	▲ Continues search for concrete goals for behavior change and may seek to define goals more precisely here.	▲ Emphasizes clearly defining objectives that may solve the problem. ▲ Uses questions to facilitate the process—"When does the problem not happen? What are exceptions?"	▲ Positive and negative motivation for change explored on decisional balance sheet. Explores goals and values. ▲ "What would you like it to be like in the future?" "What are future consequences of change?"

(continued)

TABLE 14-1 (continued)

Decisional Counselor Actions	Person-Centered Counselor Actions	Cognitive-Behavioral Counselor Actions	Brief Solution-Focused Counselor Actions	Motivational Interviewing Counselor Actions
Restory: **Explore and create alternatives, confront client incongruities and conflict, restory ("What are we going to do about it?").** The distinctions between the five systems are reflected as each system confronts discrepancies and incongruity.				
▲ Considers the basic confrontation between the present decisional problem and the goal. ▲ Balances influencing and listening skills. Helps client see the impact of the decision via reframing and logical consequences. ▲ May use career testing, information giving, and other strategies to facilitate decisional process.	▲ Confronts the ideal self with the real self with hope of integration and self-actualization. ▲ Continues to use listening skills, although reflection of meaning may become central. May engage in brief self-disclosure. ▲ Maintains little focus on problem solving while helping client get a better sense of self.	▲ Considers basic confrontation between the present behavior and the goal behavior. ▲ Draws on multiple strategies for change and consults with client on each. ▲ May emphasize environmental factors related to behavioral change.	▲ Often combines Stages 4 and 5; emphasis is on finding specific ways to change what occurs in the real world. ▲ Uses wellness strengths as levers to change and generalization of new thoughts, feelings, and behaviors. ▲ Expects to involve client fully in the process of brainstorming and exploring alternatives.	▲ Elaborates and affirms change talk. "Give me an example of how it will be when you change." "What else can you do?" ▲ Responds to resistance—this is central as many resist change. Magnifies and explores discrepancies. Reframes resistance. ▲ Enhances confidence talk. Uses 10-point scale "confidence ruler." Reviews past successes.
Action: **Conclude, generalize, and act on new stories ("Will you do it?").** Many beginning *and* experienced human service professionals fail to plan for change beyond the interview. For change to occur, generalization plans need to be made.				
▲ Prescribes homework or action to follow up on the session. ▲ Uses techniques drawn from other theories to facilitate generalization.	▲ Historically has given little attention to generalization in the belief that significant changes in attitudes, thoughts, feelings, and meanings will eventually result in major changes.	▲ Gives the most attention of any theory to generalization. Expects client to leave with a clear behavioral change plan. ▲ Provides specific follow-up to ensure that change is maintained. ▲ May use relapse prevention strategies to ensure generalization to "real world."	▲ Uses strategies from other theories if the client has difficulty transferring learning from the session to daily life.	▲ Negotiates concrete change plan. "What specifically are you going to do?" "What is the first step?" ▲ Completes change plan worksheet. Examines client's support system. Obtains commitment.

feelings and then summarized them at Allen 23. The problem focus that Allen selected was on career and decisional issues.

If Mary were to see a person-centered counselor, very different things would happen. The person-centered counselor would have paused, looked at Mary supportively, and said "uh-huh" or sat quietly waiting to see what she would say next. This is in the belief that clients will talk about their most important issues if given enough time and support. Thus, the person-centered counselor may be expected to wait for Mary to initiate the conversational topic.

After a moment's pause, Mary speaks up:

Mary: Yes, sometimes I feel confused. I know I've done well, but where do I go next? Something seems to be missing. Here I am, 36, yet alone and feeling stalemated. What does it all mean?

Allen: Mary, you say something is missing; you feel alone and stalemated when you look at yourself from a deeper level. There's something missing for you . . . (pause) . . . there's something missing that's meaningful.

This is a reflection of meaning. The counselor is searching for Mary's underlying values and meanings in the belief that if she finds her true self, she will self-actualize and solve many issues spontaneously. Here you see the client-directed/formal-operational style that expects clients to be able to solve their own dilemmas.

Mary: Yes. . . . (pause) . . . (starts quietly crying) . . . I feel so alone. Nothing ever seems to work out. It's been so hard over these years. . . . Where am I? What should I do?

Allen: You've felt alone at the deepest level. You've had the strength and wisdom to work through many difficulties, but somehow, somewhere, something meaningful is missing for you. . . .

This is a complex statement, typical of those who adopt the person-centered style. Note the reflection of feeling at the beginning, followed by feedback that points out positive assets of the client; this is characteristic of the Rogerian concept of positive regard. The final portion of the statement orients itself again to meaning: the underlying, deeply felt issues that impel us to action, often without our awareness.

If you wish to extend your skills in person-centered theory, specifics of taking the theory into practice may be found in Ivey, D'Andrea, Ivey, and Simek-Morgan (2006). There you will find suggestions on how to include multicultural issues in this approach. You may also find it useful to read Carl Rogers's *On Becoming a Person* (1961) in which he provides the basics of his theory. Bozarth (1999) provides the most recent complete update of Rogerian theory.

Specifics for conducting a person-centered interview may be found in Table 14-1. From a skills perspective, the following guidelines are suggested:

1. Seek to eliminate or minimize questions.
2. Focus almost exclusively on the client. The words *you* and *your* and the client's name are central.
3. Search for and reflect underlying meaning and consider reflection of meaning along with paraphrasing, reflection of feeling, and summarization as the basic skills.
4. Constantly identify positives to help clients frame their experience in forward-moving ways. This is called "positive regard."
5. Use selected influencing skills of confrontation, feedback, and self-disclosure, but sparingly.
6. Note that most influencing skills are not part of this orientation.

Bringing Multicultural Issues Into the Person-Centered Approach

The focus on the individual client needs to remain, but you can use a double focus, helping Mary see her issues in a cultural context. For example, an early counselor response to Mary focuses very much on the pronoun "you"—"You've gotten things together. . . . You talked about a new relationship. . . . You'd like to talk about this feeling of boredom." This reflective comment leads Mary to talk about herself in a deeper fashion. However, you can hold true to client individuality and also focus on multicultural issues. Focusing on both person and situation can enrich client experiencing. (For detailed examples, see Ivey, D'Andrea, Ivey, and Simek-Morgan, 2006).

Gender is a multicultural issue that might enrich the person-centered discussion. Note how focusing on cultural/environmental/contextual issues as well as the client broadens the discussion and helps Mary see herself more completely.

Allen: Mary, you feel good about what you as a woman have done in a difficult situation. You have many strengths as a woman.

Mary: Exactly. I have many women friends who have supported me. I know I am not alone in this struggle. My mother's example and strength have also been important.

Listening and Selective Attention in Person-Centered Interviewing

The way you listen can and does influence the way clients respond. Thus, it is essential that you constantly examine your own behavior in the session. Choosing to listen exclusively to "I" statements in the person-centered mode affects the way clients talk about their issues. Focusing on culture, gender, and context also affects the way they respond. Return to Allen 17 (page 372) of the preceding chapter and examine how selective attention affected the progress of the session.

In dealing with the need to listen, you face two profound and important ethical and practical questions. How do you listen? How should you listen? The first question needs to be borne in mind constantly. The second question represents a value issue that perhaps is ultimately not answerable but that you nonetheless will be grappling with throughout your helping career. Your personal values influence how you listen. *"Our influence is determined by the quality of our being."*

A feedback form follows (see Box 14-1 on page 402). An observer and/or the client can provide feedback for you. You have the basics for demonstrating person-centered practice.

INSTRUCTIONAL READING AND EXAMPLE INTERVIEW 2: COGNITIVE-BEHAVIORAL THERAPY (CBT) AND STRESS MANAGEMENT

By Edna Brinkley and Allen Ivey*

Cognitive-behavioral therapy (CBT) works on the assumption that we can facilitate change in client behavior and that will be followed by changes in the client's thoughts and feelings. Also, we can change thinking patterns, and behavioral change can often be expected to result. Many feminist-oriented counselors use this method to help women become more direct and achieve their own goals. You will find that it is often needed to help career counseling clients communicate better at work or conduct better job interviews, and that many physicians may refer clients to a cognitive-behavioral specialist to learn the skills of life management through assertiveness training or stress management.

Throughout this text, we have mentioned that most and perhaps all of our clients suffer from stress, which in turn affects thoughts, feeling, behavior, and meanings. These same stressors lead to bodily issues, some of which are permanently damaging. For example, the constant stress placed on a child in a low-income environment negatively affects both body and mind—and can result in early death. College students also suffer from stress—exams, finances, parental and social pressures. The continuing microaggressions of racism, sexism, heterosexism, classism, and other forms of oppression are also major stressors.

While all theories seek to combat stress, cognitive-behavioral therapy is the one that most directly works on stress itself. Not all CBT is oriented to stress, but the example interview here

---

* Dr. Brinkley conducted this interview. The process notes and comments were written jointly by Dr. Brinkley and Dr. Ivey.

BOX 14-1   FEEDBACK FORM: PERSON-CENTERED INTERVIEW

_____ (Date)

_____   _____
(Name of Interviewer)   (Name of Person Completing Form)

Relationship: **Initiating the session—Rapport and structuring ("Hello; this is what might happen in this session").** How well did the interviewer establish rapport and how did he or she accomplish this objective? Was structuring the session kept at a minimum?

Story and strengths: **Gathering data—Drawing out stories, concerns, problems, or issues ("What's your concern? What are your strengths or resources?").** Was the interviewer able to conduct the session with a minimum of questions? Was at least one positive asset or wellness strength of the client identified? Were the client's emotions connected with the story adequately explored?

Goals: **Mutual goal setting—Establishing outcomes ("What do you want to happen?").** Were the client's ideal self/world and real self/world defined? How were the goals of the client expressed?

Restory: **Working   Exploring alternatives, confronting client incongruities and conflict, restorying ("What are we going to do about it?").** Was the interviewer able to assist the client in exploring himself or herself in more depth? Were meaning issues considered to help the client gain deeper understanding?

Action: **Terminating—Generalizing and acting on new stories ("Will you do it?").** This tends not to be emphasized unless the client brings the issue up.

General comments on interview and skill usage:

shows several techniques that can be used, regardless of your eventual theoretical orientation, to help clients deal with difficult issues.

Tonya, the client, is African American and a student in a predominantly White southern university. She is dealing with family health issues, some financial problems, and alienation from her White professors and students. The following is the third session where three different strategies for coping are presented. The issue of how to deal with racism in the classroom is presented, but was held for a more detailed discussion in a later session. The session has been thoroughly edited for clarification, but the problems discussed are indeed real.

Stage 1. *Relationship:* Initiating the Session—Rapport and Structuring

Interviewer and Client Conversation	Process Comments
Tonya: I'm really glad to be having a session today, Dr. B. I think I'm getting a better understanding of stress and how it's affecting my life, but I want to know what's going on inside me. I mean, the stuff you gave me last week to read said that stress is related to all sorts of diseases like diabetes, ulcers, and high blood pressure. My dad told me just last month that my mom has diabetes and hypertension, so I've got to be proactive and do something on my own. I figure, the more I know about how stress is affecting my body, the better off I'll be, right?	Obviously, there is already a good relationship/working alliance between Dr. Brinkley and Tonya. Tonya herself summarizes the last session and the key issues for today's session. Such rapport will not occur in all relationships, but this time, we can go immediately to the tasks of the day.
Dr. Brinkley: Yes, Tonya, you're so right. And actually the first thing on our agenda today is to continue talking about stress in your life. Let me start off by asking, when you think about stress, how do *you* define it? What comes to mind?	Paraphrase, and open questions. The definition question seems mild and almost not noticeable, but this looks to understanding concrete issues and behaviors underlying stress. "What comes to mind?" allows for free association and gives clients maximum room for discussion. We have a lot happening in Dr. B's comment.

Stage 2. *Story and Strengths:* Gathering Data—Drawing Out Stories, Concerns, Problems, or Issues

Interviewer and Client Conversation	Process Comments
Tonya: Well, . . . all I can think of is pressure. All the things I *should* do, and what I really *should* be doing a better job at than I am doing. Pressure from everywhere, school, my job, worrying about money. And this university of mainly White students—sometimes it's just plain difficult to be here and even go to class. I feel so visible and exposed. Even now, my body is actually starting to feel so tight and well, stressed. Sometimes I just want to run away, go somewhere, have a good cry. . . . I guess at the end of the day, all I can think of is I want to relax, and then I try, but my mind won't shut up. I keep on thinking about all the things I still have to do.	Classic signs of stress and how it impacts the brain and body with damaging cortisol.

(continued)

Interviewer and Client Conversation	Process Comments
Dr. Brinkley: Sounds like it's overwhelming for you right now?	Reflection of feeling and Dr. B's suggestion of a comprehensive word for all that's happening. "Overwhelmed," of course, is a complex and confusing emotional situation. The word "puts it together," but we also need to consider the specific feelings underneath it as the interview progresses. Considering those specific feelings will facilitate Tonya and Dr. B collaboratively coming up with personalized strategies for Tonya to try later outside of the session.
Tonya: Right, that's the word—overwhelmed.	
Dr. Brinkley: Let's stop for a moment and focus on some strengths and supports that we talked about last week. Those positives will be part of getting through this.	Directive with explanation of what's going to happen. This helps the client understand the "why" of the topic change. This is a prime example of how interviewers use a strength-based approach to help clients center themselves before really tackling the serious issues. However, if this were the first interview, Tonya would likely need to tell her story about challenges and issues before the interviewer could turn to concrete examples and the positive asset search.
Tonya: Thanks, because it feels like my body is starting to fall apart right now. Strengths . . . positives . . . you know that thought-stopping exercise from last week is working great. I do find saying "STOP" internally really helps when I start thinking too much. And last week you helped me realize I have a lot of support from my family. You reminded me that I am doing fine in school despite all the pressure. And it really helped when we talked about how I made it through coming out in high school when the other kids were giving me such a bad time. Perhaps I'm not as hopeless as I feel at times. . . . But still, I *should* be doing better.	Tonya is becoming aware of how her body relates to her mind and her thoughts. She also used the positive asset approach to recall that she is not alone in her difficulties but has supports and many strengths. The positive asset search and the wellness approach work!
Dr. Brinkley: You're right; you're not hopeless, especially with all those strengths, so let's always keep those in mind. (paraphrase, emphasis on strengths) I'd like to go back to something you said earlier about your mind not shutting up. Sounds like that's another stressor for you.	Example of using attending behavior to return the focus to the issues faced by Tonya. We now have a strengths base, and the client is more "centered." You will find that clients who learn that interviewers, counselors, and therapists respect them for their strengths are more open to examining and changing their weaknesses.
Tonya: Stressor? What's that?	

(continued)

Interviewer and Client Conversation	Process Comments
Dr. Brinkley: Stressors are simply internal or external conditions or situations. Internal because, for example, you said your mind won't shut up, and you end up feeling pressure. It's not anything going on outside of you; it's your mind that's precipitating the stress symptoms. So it's what you're thinking and saying to yourself. But for external, that's something outside of us. So stressors can be negative thoughts, events, people, situations, traffic, like that. All of these can impact your mind and body in negative ways.	Information/psychological education, including instructions about some elements of stress.
Tonya: Mind and body, huh. . . . That makes sense.	
Dr. Brinkley: And what's key here is our perception. Sometimes when our perception of something that needs our attention outweighs our perception of our ability to cope, we start to experience some different symptoms of stress. So stress is based on our perception, the meaning we give to different situations.	Information with elements of reframing—specifically the explanation about how our perceptions affect reality.
Tonya: My perception, huh? But how do I give meaning to a situation?	
Dr. Brinkley: Well . . . drawing on past experiences with a similar situation, beliefs and values from your family or cultural background, and maybe some fears about the future. What we're trying to do is to make the new situation make sense based on what we already know. And by the way, we are the only ones who can give meaning to those situations.	A complex interviewer comment with multiple dimensions. Note the focus on family and the cultural/environmental/contextual, linkage of these issues to current situations, but ultimately focusing on Tonya, the individual's need to make her own meanings.
Tonya: So I give meaning to it by trying to understand the situation and use my life as a guide, so I can then do something about it? So you mean, it's like, I *love* my calculus class; I always feel so psyched there.	
Dr. Brinkley: Exactly. The meaning you give to that class is excitement, you love it, and you look forward to it. Right?	Reflection of Tonya's meaning system, which helps her understand how she can take more direction in her own life.
Tonya: Yeah, I do. And after class, I'm still psyched, so I use that energy for my physics lab.	Don't we all wish that we had that excitement for school? Dr. Brinkley is lucky to have this client.
Dr. Brinkley: So for you, your calculus class energizes you, but for others, they'd rather have a root canal with no anesthesia than have to take it. So the meaning they give to this situation is possibly based on past experiences with terrible math professors. So their belief becomes, "This is going to be stressful, and I don't want to have anything to do with it."	Dr. Brinkley begins with an interesting confrontation illustrating where Tonya stands in contrast to others who are less enthusiastic about math. She explains various meaning systems developed in the past (there and then) and how they relate to the now. Then she offers the interpretation that fear of something being stressful leads many people to avoid tasks.

(continued)

Interviewer and Client Conversation	Process Comments
Dr. Brinkley: (continues) I mean, once your mind perceives something as a stressor, it takes only a few seconds for the stress response to kick in; it's automatic; we can't control it. And remember, your body doesn't know the difference between something real and something imagined, but it will react. So, with your sympathetic nervous system a lot of things happen very quickly just like on the handout here. (gives a handout on stress physiology) Take a look at the "Stressed" section; what do you see?	We first see an emphasis on CBT's automatic thoughts—perception of stress possibility leads to an automatic body response with resulting increase of stress and discomfort. This is followed by detailed information and instruction. Dr. B. is now working with Tonya using psychological education. "Teaching as treatment" is an important part of the successful interviewer's skills and strategies. Handouts are available on the accompanying CD-ROM.
Tonya: (reads some symptoms from handout) I know epinephrine and norepinephrine; they make my heart rate and blood pressure go up. And I've heard cortisol is called the stress hormone, but what does it do? I *should* learn how to control myself better, or my body's going to suffer.	More and more clients will be aware of these "technical" and brain-related terms. This type of conversation will become a standard part of interviewing and counseling in the future. See Appendix II for more detailed explanations of these and other key terms.
Dr. Brinkley: Cortisol takes fats and carbohydrates stored in the body and sends them along with blood from your stomach and kidneys to your brain, your muscles, and lungs, so that basically you are now ready for, well, to beat up on something or somebody or run away from them. But it's a good thing we have the cortisol, because even though it's the stress hormone, it's also an anti-inflammatory, because, well, we might get some scratches or cuts if we decide to beat up on someone.	Psychological education and information giving, pointing out the negatives and positives of cortisol. Such short discussions showing how the client's brain may react to various psychological interventions will likely be integral to counseling in the future.
Tonya: And all this is happening when I go see my professors?	Hooray—Tonya makes the important linkage herself with her own interpretation. She has changed the *meaning* of her stress reactions in class. This is a clear example of growth on the Client Change Scale. We have moved from Level 2 to Level 3, where we see the development of understanding. Level 4 will require change in behavior in the classroom and a much more efficient way of dealing with stressors in the future.
Dr. Brinkley: Yet there is no physical threat here. You want to go see a professor to get help with your classes, which is what a good student does. You've already had one professor say you don't belong in the department, plus you said last week, one of your professors, what did he say?	Paraphrase, open-ended question but still searching for concreteness and a clearer understanding of the situation.
Tonya: Oh, that American government class where we were discussing public policy? I remember now. The professor, he's White, and so is everybody else. Well, anyway, he made this statement and then said, "Let's	Students who find themselves in the minority in classrooms often talk about the difficulty of such situations and how unfair they think they are. We also see further cognitive/emotional

(continued)

Interviewer and Client Conversation	Process Comments
hear what the African American students have to say about this," and then he looked dead at me and stood there like he was expecting me to say something. What's he looking at me for anyway? I didn't say anything. I just stared back at him and looked around the room at my classmates, and they were all staring at me too. I think I see what you're getting at. In that class, in others too, there aren't any physical threats, but sometimes I feel like I'm being attacked in class, verbally, sometimes body language too. So it sounds like even when there's no physical threat, I still feel a threat to me, to who I am, how I think about myself, and so I feel I have to defend myself against all these negative images, these stereotypes. This is so tiring. (pause and aha moment) Then the stress reaction kicks in?	understanding, strengthening Tonya's Level 3 responses on the Client Change Scale.

Stage 3. *Goals:* What Do You Want to Happen?

Interviewer and Client Conversation	Process Comments
Dr. Brinkley: So, we've got a picture of a lot of stressors at your university. What can we do about it?	A very brief summary of the interview to date and the all-important goal-oriented open question.
Tonya: Well, you're the expert. What do we do? I'd really like to calm down, take care of my body as well as my mind, and have more fun.	Would that all our clients could have such clear goals. But you will find that when you have a good working relationship with clients, many of them will begin to understand the importance of clear objectives for each session. This is part of mutual goal setting and working together in relationship, rapport, and the therapeutic alliance.
Dr. Brinkley: That's our goal. Let's get at it!	

Stage 4. *Restory:* Exploring Alternatives, Confronting Client Incongruity and Conflict

Interviewer and Client Conversation	Process Comments
Dr. Brinkley: Today let's explore three strategies that will begin to make a difference in your life. We'll start with calming breathing, then we'll do some cognitive restructuring—that's examining your thinking, and finally, we'll end by looking at what exercise can do for you. Sound okay?	The interview plan is laid out by Dr. Brinkley with clear directives and with client agreement. Mutual goal setting fosters the relationship and helps counselor and client move toward attainment of results.

(continued)

Interviewer and Client Conversation	Process Comments
Tonya: Let's do it.	
Dr. Brinkley: Good. How are you feeling right now?	Open question, here and now.
Tonya: I'm a bit excited, but I also feel some tension in my stomach.	As you help clients learn the importance of bodily responses to life events, they will start to tell you about very important nonverbal reactions that we cannot see. Through Dr. B's help, Tonya has learned to tune in with her body and learn from it.
Dr. Brinkley: The strategy we're going to do will help you with that. One of the very best cognitive-behavioral strategies is awareness of breath, combined with visualization. Let's try some breathing exercises. First, take a deep breath and relax—close your eyes—what peaceful and calming image comes to your mind?	Dr. B gives a directive with one of the most popular CBT strategies. Interestingly, this comes from Eastern thought and very early work in relaxation training by Edmund Jacobson (1934) and popularized by Herbert Benson in *The Stress Response* (2000). For a more Eastern mindfulness approach with solid research evidence, consult Kabat-Zinn (2005).
Tonya: (pause) I see myself at the beach. It's a sunny day, and I'm all by myself, just feeling good. . . . (Tonya continues to discuss the image in some detail and the comfortable feelings she experiences when viewing it. She opens her eyes and smiles.)	
[Note: The full training in relaxation and imagery has been omitted here to save space.]	*Please review the specifics of relaxation and mindfulness in the Directives section of Chapter 12 and in the suggested references above.*
Dr. Brinkley: How are you feeling now at this moment?	Here-and-now question oriented to feeling.
Tonya: I do feel calm and relaxed.	
Dr. Brinkley: What words would you use to describe that feeling and image?	Open question, asking for a "name," plus the here-and-now experience facilitate the client "putting together" and taking the experience home and continuing the practice.
Tonya: "Peace, be still." That's what my grandmother always said when I was upset.	
Dr. Brinkley: Let's spend some more time with "Peace, be still." Focus on it for a while and allow it to become more fully you.	
Tonya: (after approximately three minutes) That really felt good. And you say this is something I can do when I feel stressed out. I can actually calm myself in the classroom when I feel uncomfortable.	This very brief exchange is easy to pass by. What has happened here is that Dr. Brinkley has started with *sensorimotor* body experience in the here and now, then moved to *concrete operational* naming. The body experience is

(continued)

Interviewer and Client Conversation	Process Comments
	behavioral while the naming is cognitive—thus cognitive-behavioral counseling. Following that, she imprinted the cognitive-behavioral experience through silence, thus making it more likely that Tonya will be able to draw on her grandmother's words in times of stress. And the exercise itself also becomes a multicultural experience. We can draw on our cultural and family backgrounds to help us through stressful experiences. What is your "Peace, be still"?
Dr. Brinkley: Yes you can. Now let's try something else. It's called cognitive restructuring. Tonya, remember last week when we discussed the need to look at your thinking process about events and people in your life?	Clear transfer to another strategy. With most clients, you will want to use only one strategy per session.
Tonya: Yes, I think you said we should look at how my words, my thoughts, my perception of situations can influence how I feel and then what I do.	Perceptions here are thoughts that come from our center of meaning; see Chapter 11 diagram, page 294.
Dr. Brinkley: Right. And we talked about continuing to self-monitor your triggers for anxiety and worry, and we identified that they were physical sensations and thoughts. Today I want to add something to that about thoughts and about our choice of words. We've talked about how our choice of words can influence how we feel, and that in turn can influence what we eventually do. Our choice of words can make us feel good or not so good. One way most of us feel good is when we think we have possibilities and control in our lives, rather than others controlling us.	Words give meaning to experience. Dr. Brinkley is reinforcing a new story about the way our bodies inform us about what is happening. Implicit is a focus on words and meaning. A person's body is the first alert system, moving even faster than words. But the words used and the thoughts associated with them also influence how the body reacts and feels.
Tonya: Yeah, I feel good when I have choices. I really hate feeling boxed in. But I feel I have no choice in that class.	
Dr. Brinkley: Since you say you hate feeling boxed in, what do you think about the possibility that sometimes the words we say to ourselves can decrease the choices we see for ourselves? For example, I noticed several times today that you used the word *should* to describe things. Did you notice that?	Paraphrasing, information giving, and focusing on the key word *should*. Watch your clients for "shouldy" language and "oughts." These are important words, because living a life dictated by the word *should* is very different from naming it as "my life," " I could," "I decided," or "by my own choice."
Tonya: Ahhh, well, no.	
Dr. Brinkley: That's okay for right now. But I do have another question for you that's related to our choice of words and using the word *should*. What do you think of the possibility that some of your thoughts may not be as truthful as your mind tells you?	Open-ended question.

(continued)

Interviewer and Client Conversation	Process Comments
Tonya: You mean what I think is real, may not really be real?	
Dr. Brinkley: Well, yes and no. No, because sometimes, what we think is happening, really is happening, like when you're anxious and troubled by the treatment you're getting in class. This is true. But when you think about this situation almost all the time, your thinking may not be as accurate. (information and instruction) Because we talk to ourselves a lot and if what we mostly say to ourselves is negative or in absolutes, like "always," "never," "I should," or "I have to," pretty soon those messages can become automatic, and we may begin to believe them and think that they're true.	Information/instruction with some interpretation/ reframing. Dr. B is returning once again to automatic thoughts and how the words we use determine our inner feelings and meanings.
Tonya: So you mean it's like I think I'm afraid in class, and thinking about it makes it so. I know my body gets tense and upset, but what about the words I say to myself?	
Dr. Brinkley: Well, for example, if we say words like *should, always, never,* those are called absolutes; they are either/or with no gray area. Yet is life really full of absolutes, no middle ground, no possibilities?	Information/psychological education.
Tonya: Hadn't thought about that before. I guess not.	
Dr. Brinkley: I mean, basically what you're doing is creating possibilities with your words and giving up the use of *have to* and *should.* One thing you may notice is that when you change the words, your emotions may change and things may seem a little easier. And this is related to your tension in class. Now usually our thinking, it's so automatic, we don't monitor our thoughts, and we end up feeling not so good. But if we begin to take a few extra seconds, notice our words and thoughts, and create openings or possibilities instead of absolutes and closures, we may begin to notice we feel differently. This is like when we talked earlier about using words that were meaningful to you to help you relax. You gave yourself an opening, a possibility.	Dr. B starts with more information/psychological education. This is followed by linkage, instruction about automatic thoughts, and how these thoughts relate to emotions and the body.
Tonya: So are you saying that I should, I *could,* change some of the words and thoughts, and that may change the way I feel?	Tonya paraphrases and provides a brief summary, indicating that she understands. When clients start repeating new material from the interview, we have a clear example of restorying.

(continued)

Interviewer and Client Conversation	Process Comments
Dr. Brinkley: It's something I want you to consider. Tonya, do you really want a "shouldy" life? Over the next week, see if you can catch yourself when you either say or think something in absolutes, particularly "shoulds." One thing that will help is an automatic thoughts journal. You'll be monitoring your thoughts and learning to observe your body. Watch for those feelings in your body, and note your thoughts at that time. What comes automatically with that tension? You'll write down what happened just before that feeling or thought, describe the thought, and then what happened afterward. I'll give you a handout that you can use as a framework. We can review it next week.	This is a common CBT procedure. The homework will help generalize new learnings to the real world. See Table 14-2 for an example of how Tonya might fill out an automatic thoughts chart. The automatic thoughts form can be found on the CD-ROM.

Tonya: Ummh. Okay. Oh, I forgot to tell you after that heart thing with my doctor, I went ahead and signed up for jazzercise classes at the community center near home. I even started jogging on their outside track. I am definitely feeling better. I don't feel as tense, and I can take stairs a lot better nowadays. They've also got personal trainers there, so I'm thinking about getting a few sessions with one of them to get a good program going.	In the first session, Tonya and Dr. Brinkley discussed the heart issue. Exercise is a key dimension of mental and physical health. This has become another standard treatment procedure for CBT practitioners.
Dr. Brinkley: Glad to hear you followed up on that. Sounds like you're starting to get the physical benefits of exercise and some psychological benefits too. The exercise is going to help with your mood. You've heard of endorphins right?	Positive feedback leading to more psychological education. Feedback and support; closed question as check-out.
Tonya: Oh yeah, I've heard of them. They're those natural opiates in my body, and they kick in when I exercise and make me feel really good. It's that runner's high I get when I'm out jogging. And I read that exercise helps fight cortisol, which is so destructive in my body when I get stressed.	This is a good example of what many clients will bring to the interview. They often have knowledge of health, psychological concepts, and neuroscience.
Dr. Brinkley: Exercise is also really great for decreasing depression and anxiety symptoms and chronic pain too. I'd put exercise in the top three stress relievers for all the benefits it gives us for stress relief and good health in general. (checks watch) Unfortunately, Tonya, we are nearing the end of our session.	Dr. B reinforces what has just been said through a very brief summary, to which she adds her own beliefs.

TABLE 14-2   Automatic thoughts monitoring record

What is the situation? Where are you? Who is there? What was said and done? (Be specific.)

We were discussing public policy in my American government class, and the professor said, "Let's hear what the African American students have to say about this." He then stood in front of the class with arms crossed and looked straight at me as if he were waiting for and expecting me to say something. The room was dead silent as he and most of my classmates stared at me. I stared back at them and didn't say anything.

What emotion(s) did you have? Strength level: 0 = no emotion, 10 = very strong emotion.	What physical feelings did you have? Where in your body did you feel any tension? Tension level: 0 = no tension, 10 = very tense.	What were your thoughts, your inner self-talk? (Verbatim if possible.)	What was your behavior following these thoughts?	What are some alternative thoughts that may more accurately describe the situation?	What are your emotions now? Strength level: 0 = no emotion, 10 = very strong emotion.	What physical feelings do you have now? Where in your body do you feel any tension? Did you notice any changes in body tension? Tension level: 0 = no tension, 10 = very tense.	What did you learn from this experience? What new behavior did you do?
Anger (10) Confusion (7) Hurt (7) Frustration (8)	Head hurt (8) Stomach in knots (6)	I hate this man! What a racist idiot! Who the hell does he think he is? I've got to drop this class; I don't care if it is required.	Talked it over later with a couple of the classmates who didn't stare at me. Talked it over with my girlfriend later in afternoon. Called my grandmother that night.	This guy is an idiot racist and just plain clueless. He probably has no idea how his statement affects anyone. I've got a really lousy professor.	Anger (5) Confusion (0) Hurt (0) Frustration (3)	My head still hurts (4) some because I still have some thoughts about the class in general and the professor. My stomach feels much better after talking it over and coming up with some alternative thoughts.	Realized talking to some classmates, girlfriend, and grandmother helped make me feel better, but didn't give me a new non-confrontational way to directly deal with this professor while I'm in his class. Decided I needed to talk to my academic advisor and the department chair to see what I can do.

Stage 5. *Action:* Generalizing and Action on New Stories

Interviewer and Client Conversation	Process Comments
Tonya: I'm starting to feel some of these benefits, both physical and psychological, so you don't have to convince me about exercise. And, you know, as I feel better, I think I can handle the situation in class with the professors and students—but I still need a lot of help there.	
Dr. Brinkley: I hear you. Could you sum up for me what you plan for the coming week?	It is best if the client can summarize the session. But be ready to supplement that summary as necessary.
Tonya: First, I'm going to practice calming myself when I feel tense—"Peace, be still" will remind me of better times and my grandmother. It'll help me relax. I think the change of words will be more difficult, but I'll give it a shot. I'm a very responsible person, and I guess I do have a lot of *shoulds* in my life. Perhaps the automatic thoughts charts will help in that. But, certainly, the exercise will continue. I've been amazed how much better I feel when I take time for it. I'm beginning to think it is basic for me if things are going to get better.	The word *perhaps* is important. Tonya has not fully bought into the automatic thoughts procedure. When talking about homework, observe the client's nonverbals as well. Be ready in the next session to check on homework and work on it to ensure that relapse is less likely to occur.
Dr. Brinkley: Great summary. I couldn't have said it better myself. Well, looking forward to seeing you next week.	
Tonya: Thanks, I'll be here.	

Dr. Brinkley did not bring up the issue of racism in the classroom until the third session. Because part of Dr. Brinkley's ethnic heritage is African American, it might be easier for her to discuss this issue with Tonya. However, remember that several factors must be considered when attempting to create and maintain an effective therapeutic relationship, such as Hays's (2007) multidimensional addressing factors, not just a possible shared ethnic heritage.

Additionally there are some general principles for dealing with racism or other forms of oppression that may show up in the session. First, don't shy away from talking about oppression. You will find some students who blame themselves or the general situation for the difficulties. "Professors have so many students and very few minorities. Maybe I should speak up for my own race, but it feels so uncomfortable." "If I weren't lesbian, they wouldn't bother me." "I know that it's hard for the administration. They'd like to help those of us who are disabled, but the legislature says no."

Name it! If the client is experiencing something that appears to be racism, heterosexism, any type of ableism, or some other form of oppression, name what is occurring and state that it is wrong! Interviewers, counselors, and therapists tend to be gentle people and like to avoid conflict. This is a place where you can help your clients. In the process of naming, you are changing cognitions in important ways.

BOX 14-2   FEEDBACK FORM: COGNITIVE-BEHAVIORAL AND/OR STRESS MANAGEMENT

_____ (Date)

_____   _____
(Name of Interviewer)   (Name of Person Completing Form)

Relationship: **Initiating the session—Rapport and structuring ("Hello; this is what might happen in this session").** How well did the interviewer establish rapport and how did he or she accomplish this objective? Was the session clearly structured?

Story and strengths: **Gathering data—Drawing out stories, concerns, problems, or issues ("What's your concern? What are your strengths or resources?").** Was at least one positive asset or wellness strength of the client identified? Were the client's emotions that are connected with the story adequately explored? If appropriate, were the client stressors clearly identified? Where appropriate, were the cultural/contextual issues explored?

Goals: **Mutual goal setting—Establishing outcomes ("What do you want to happen?").** Were highly specific goals for client cognitive-behavioral changed established? Was the client fully involved in goal setting?

Restory: **Working—Exploring alternatives, confronting client incongruities and conflict, restorying ("What are we going to do about it!").** What CBT strategies were used? Was specific cognitive-behavioral change emphasized? Could specific changes in the client be noted?

Action: **Terminating—Generalizing and acting on new stories ("Will you do it?").** Was a specific homework assignment agreed to by counselor and client?

General comments on interview and skill usage:

Name it as wrong! "This is something that happens to many students of color. It is wrong and it needs to be eliminated." "This is sexism, and this is something that is wrong and unfair." "It is not you who is at fault; it's the other person/the administrator/the school policy, etc."

What are we going to do about it? Dr. Brinkley's simple but important question speaks to mutually set goals. As helpers, we are there to *help* the client. Stress management and counseling are one route, but there is also the possibility of encouraging the client to take more direct action, putting the client in touch with support groups, and you yourself working on school, community, or campus issues of social justice. You likely cannot go out with the client to stop professors and students from engaging in oppressive acts, but you can support students and involve yourself in action as well.

Box 14-2 is a feedback form. At this point you can demonstrate beginning competence in stress-oriented CBT. Take your time. You have more ability than you realize.

INSTRUCTIONAL READING AND EXAMPLE INTERVIEW 3: BRIEF SOLUTION-FOCUSED COUNSELING

By Allen E. Ivey, Robert Manthei, Sandra Rigazio-DiGilio, and Mary Bradford Ivey

Brief solution-focused counselors believe that clients have their own answers and solutions readily available if we help them examine themselves and their goals. Once the goal is established, usually early in the session, the counselor focuses on ways to reach this goal. Rather than search out problems and examine them in detail, brief solution-focused counseling attends to the future and how to get there. It is possible to organize much of the pragmatic work of these brief solution-focused counselors within the microskills interviewing structure, but goal setting comes immediately along with establishing the relationship. Strengths are important, and stories about strengths will be given more attention than problems and concerns. In addition, you may find that several brief strategies and questions will be useful regardless of the position you will eventually take on whether to use this theoretical approach.

The famed psychiatrist Milton Erickson initiated the brief solution-focused approach, achieving seemingly miraculous results with his clients using sort-term methods. The best source for a foundation in brief work is *My Voice Will Go With You—The Teaching Stories of Milton H. Erickson* (1991). Erickson also brought the importance of stories to our field, and you will find this fascinating book an excellent and enjoyable read. A number of authorities have examined Erickson's work, added their own thoughts, and defined the basic principles of brief solution-focused counseling and therapy (e.g., Erickson & Keeney, 2006; Short, Erickson, & Erickson-Klein, 2005; Sklare, 2004; Winslade & Epston, 1997). Semmler and Williams (2000) have given multicultural issues and narrative approaches special attention.

As you begin to use brief methods, it is important to recall that building a relationship of rapport and trust remains essential. We may sometimes still need to hear the client's story in detail before moving to goals, as some clients really need to talk about concerns first. But keep this as short as possible. Drawing out stories of client strengths, resources, and positive experiences is important, because drawing out these strengths often serves as a key to solutions. You want to identify competencies existing *within the person* that will help the client find the solutions.

As you move to restorying, you may want to introduce metaphorical or real stories that relate to the client's life. Many counselors have memorized key stories that they frequently repeat to clients. Clients often use these to create new ways to view their issues and also find useful routes toward solutions.

How long is "brief"? Anticipate one to three interviews as typical. This approach may be a single interview, or it may extend to as many as 5 or 10 sessions. *The key word is "brief," emphasizing solutions rather than problems.* But the questions and methods remain useful in longer term counseling and therapy as well.

Whereas person-centered methods use very few or no questions, brief solution-focused counseling often makes questions the central skill. If you turn back to the use of questions in coaching and the GROW model (Chapter 4), you will see how coaching has been heavily influenced by the positive approach of brief solution-focused counseling. Both systems give attainment of goals central attention, but ethical coaching is *not* brief solution-focused counseling nor is it therapy of any kind.

We suggest that you consider sharing the list of specific solution-focused questions with a volunteer and, later, with real clients, particularly in the early stages of your practice. Try photocopying and using chapter questions and structure as a "worksheet" that you and the client share. As you gain experience and confidence with this method, you may wish to continue the sharing—counseling and interviewing can be more powerful and real in an egalitarian, co-constructed framework. We also recommend memorizing an array of questions to have available instantly when they might be helpful.

Stage 1. *Relationship:* Initiating and Structuring the Session

Basic questions. Begin solution-based thinking at the very beginning of the session. The positive asset search is central. Even as you listen to the client's story and/or reasons for coming to the interview, you can ask the following:

▲ What is your goal here today?
▲ Has anything changed since you decided to come to see me? Are things better in any way?
▲ What has gotten better about your concern/issue/problem? What made that happen?
▲ What's keeping it from getting worse?
▲ Are there any exceptions in this problem? When is the problem not so much of a problem?
▲ What do you do right? What have you been doing to keep this issue from really dragging you down?
▲ How can we keep that going?

Your own mind-set. Brief solution-focused counseling asks that you think differently about helping. Instead of focusing on defining "problems" or a long drawn-out exploration of "what happened and why," you need to focus on "solutions." This means establishing a positive expectation for both yourself and your client. Consequently, your task is to structure the session to achieve success in this important joint venture with your client.

Relationship. Traditional rapport and listening skills remain central, although you will want to use questioning skills as your intervention of choice. Some clients and those who may be culturally different from you may be suspicious of the frequent questions. You may wish to explain that you will be asking many questions and find out whether that is acceptable. Spend more time explaining what you are doing and more time on listening to stories to develop trust.

Structuring the session. Let the client know what is going to happen. Share what you are going to do in the session and why. For example:

> Many people can accomplish considerable progress in just a few sessions. What we are going to do here today is focus on solutions—the goals you want to achieve. Can you tell me what your goals are *for today?*

The words "for today" are important because they bring the client to the possible here and now rather than leading to a lengthy attempt to resolve everything at once. Some issues are too large to be resolved in a few sessions; your client may work on one primary issue now and leave the others for later. Returning to counseling is not failure; rather, it shows willingness to work on the many complexities of daily life. With children or adolescents, the wording may be better phrased as follows:

> Darryl, the teacher asked me to talk with you. Rather than talk about problems, I'd like to know how things might become better for you. Could you tell me one thing that you can do to feel better—happier about the rest of today? (or "Before we begin, I want to know something that makes you happy. Tell me about what you like to do.")

What else? Your client will not always respond in depth to your questions. De Shazer (1988, 1993) recommends the frequent use of "What else?" to prompt client thinking and the generation of more complete answers and solutions.

Children and adolescents (as well as adults) may initially respond negatively to your questions and even say "Nothing." Remember the importance of rapport and listening—with many clients, a sense of humor helps! With experience you will develop follow-up questions and help clients explore their issues in new ways.

Next—Stage 2 or Stage 3? If the problem was clearly defined during rapport and structuring and the goal is relatively clear, consider moving directly to mutually setting goals (Stage 3). *Do this especially if the individual is able to identify specific things that have gone better and/or times when the problem is "not a problem."* Examples of clearly defined concerns might be these:

▲ I'd like to stop arguing so much with my partner.
▲ My son doesn't sit still during meals.
▲ I'd like to be able to speak up at meetings more effectively.
▲ Our lovemaking has become too routine. I want my partner to warm up to me.
▲ I want more challenge in my work.

All these require some awareness of times when the problem is not a problem. For example, "When are you able to avoid arguments with your partner?" "When has lovemaking been real to you?" These *exceptions to the problem* may serve as levers for positive change.

On the other hand, if the concern is vaguely presented, "My relationship is falling apart"; if the client talks about multiple issues in a confusing array; and if the client has difficulty identifying goals, more time needs to be spent in gathering data (Stage 2). Brief methods work best on only one problem at a time: Other issues can be dealt with later.

The decision to omit gathering data (Stage 2) is not easy. Many clients need to tell you their story in detail before moving on to stating their goals. When in doubt, it is wise to be more conservative and spend time in exploration. But even here, you can occasionally ask goal-setting questions, building a foundation for more rapid change and client involvement.

Stage 2. *Story:* Gather Data and Search for Positive Assets

Basic questions. It is often a good idea to share a list of solution-based questions with your client and explore them together. As you become more familiar with the ideas, remember to continue to work *with* your client, not *on* her or him.

▲ Are there times when you do not have this problem? When does the problem not occur?
▲ What are the exceptions to the problem?
▲ What's different about the times when this problem does not occur?
▲ How do you get more positive results to happen?
▲ What are your strengths and resources?
▲ Suppose when you go to sleep tonight, a miracle happens, and the concerns that brought you in here today are resolved. But since you are asleep, you don't know the miracle has happened until you wake up tomorrow; what will be different tomorrow that will tell you a miracle has happened? (The miracle question; see de Shazer, 1988, p. 5.)
▲ Follow up the miracle question with "How will we know the issue has been resolved?" and "What are the first steps to keep the miracle going?"

These questions are common in brief solution-focused counseling, but to be fully effective, they require follow-up and exploration. Use the "What else?" question and positive asset search questions ("What is going right?" "What part of the problem is not a problem?"). If you get brief or sketchy client responses to the miracle or other questions, use your natural ability and listening skills to expand and draw out responses.

Being brief. The summary is particularly helpful in brief solution focused counseling to organize the session and serves as a foundation for clearer and more effective goal setting. Many clients will describe their problem using abstractions; be sure to clarify the problem/concern with specifics so that the abstractions are avoided or made concrete. But be sensitive to your client. Specifically, use the basic listening sequence to draw out details and be sure to summarize what has been said. Some people need to tell their stories just as much as or more than they need help changing their thinking, feeling, and behavior. You may tell the client that it may not be necessary to focus on all the details of the problem, and with the client's assistance, generate a brief narrative of the problem, concern, or issue.

Normalizing the narrative. It is normal for clients to have concerns, and it is normal to have difficult situations. Your task is to point out to clients that while we all have issues, our concerns are solvable. Your own nonverbal behavior and confidence are part of this process. Normalizing the narrative is *not* stating that very complex problems are always normal and expected parts of life; rather, normalizing the narrative means focusing on the idea that we all have concerns and it is indeed possible to do something about them. Care must be taken to avoid minimizing serious concerns. An eating disorder, an abusive family history, and racial or sexual harassment are difficult issues.

Cultural/environmental/contextual issues such as gender, race/ethnicity, and spirituality factors may be part of normalizing the narrative. The gay or lesbian client, for example, may begin the session by stating the problem as depression over constant harassment. As you hear this story, you note that the client is focusing on self as if the problem is internal. By focusing on cultural oppression, you help externalize and normalize the story.

The positive asset search and wellness. As you listen to the client's story search for strengths: positive assets, community assets, and cultural and/or spiritual strengths. An important part of normalizing client situations is enabling them to rediscover their wellness strengths and personal power. Sometimes client assets will provide an obvious solution the client may have missed earlier. The following questions (Chapter 4) are particularly useful to draw out strengths that can be used for solutions:

▲ Considering your ethnic/racial/spiritual history, can you identify some wellness strengths, visual images, and experiences that you have now or have had in the past? Can you recall a friend or family member who represents some type of hero in the way he or she dealt with adversity? What did he do? Can you develop an image of her?
▲ Tell me concretely about a special family member and what this person means to you. Family can include our extended family, our stepfamilies, and even those who have been special to us over time. For example, some people talk about a special teacher, a neighbor, an older person who was helpful.

Genuinely complimenting and giving feedback to the client on specific strengths and assets may be useful, but the client must accept wellness strengths as real, or the positive asset search may seem trite or disrespectful. Also some clients and some cultural groups consider direct compliments embarrassing. Indirect ways to compliment a client for her or his strengths include these:

▲ How did you know that?
▲ Where did you learn that?
▲ How did you figure that out?
▲ How did you develop that strength?

Scaling. On a scale of 1 to 10, with 1 meaning the concern is fully resolved and 10 meaning that the concern almost totally overwhelms you, where would you put yourself today? At this moment? In the session? Scaling provides an effective way for you and the client to communicate the current depth of the client concern.

Scaling serves as a temperature gauge so that you know how clients are feeling about their problems at any given moment. Scaling can evaluate whether you and the client are in synchrony, seeing things similarly. Use scaling periodically through the session and consider using it throughout all your interviews, regardless of theory. With children and younger adolescents, actually drawing a scale may be useful. The child can then point to where he or she is on the scale.

With experience and practice, you will want to expand your use of scaling. For example, you could have your client evaluate his or her present level of motivation for change ("How committed are you to solving the problem?"), the confidence of success ("How likely are you to succeed?"), or how he or she will deal with termination ("At what point do you feel that the problem is sufficiently resolved?").

Stage 3. *Goals:* Set Goals Mutually

This stage may not be needed if you and the client have set up clear and workable goals at the beginning of the session. However, about this time it is usually wise to revisit the original goals and make sure that the direction of the session(s) is clear.

Basic questions to ensure clear goals. We have heard your concern (summarize again, if necessary to keep interview on track and check accuracy). . . . Now, what specifically do you want to happen? Be as precise as possible.

Strengthen the following questions with variations on de Shazer's "What else?" "Can you add anything more?" "Any other thoughts?"

▲ What do you want to happen?
▲ How do you cope with the problem?
▲ What have you done so far that is helpful in achieving that goal?
▲ Let's focus on the exceptions. Tell me about the times when the concerns are absent or seem a little less burdensome.
▲ What is different about these times?
▲ How do you get that more positive result to happen?
▲ How does it make your day go differently?
▲ What did he or she do or say when it was better?
▲ How did you get her or him to stop?
▲ How is that different from the way you usually handle it?

These brief questions involve a change of pace and can add humor to the session.

▲ What do you do for fun?
▲ What would help you to feel that life is better? Name one thing that would help.
▲ Let's take a piece of the larger concern and work on that. Okay? We can't solve it all today, but we can make a piece of it a bit better.

Following goal definition, you can obtain very specific ideas about client wishes and desires when combining these with variations of "What else?"

▲ How will your life be different?
▲ Who will be the first to notice?
▲ What will he or she do or say?
▲ How will you respond?

Co-constructing concrete, achievable, clear goals with the client. Clients too often want to resolve all concerns simultaneously. Be sure you negotiate specific goals that can actually be reached. Help clients work toward resolving a smaller piece of the larger issue; a small change can lead over time to significant differences in a client's life. Some children and adolescents will have difficulty with goal setting; their life experience has been focused on what people in authority want from them. Patience and setting up concrete, achievable goals are important.

Stages 4 and 5. *Restory/Action:* Explore, Create, and Conclude

Brief solution-focused counseling combines these last two stages. When you have identified resources, found exceptions to the problem, and identified goals in the first three stages, you have already gone a long way to brainstorm and explore solutions. Constantly focus on the idea that something can be done. Your goal in this stage is to solidify and organize the solutions and move toward concrete action. Work on the clearly defined goals in specific manageable form. Every successful idea for solution needs to have a practical use outside the session.

Basic questions.
▲ What have you been doing right?
▲ What do you have to keep doing so that things continue to improve?
▲ What will tell you that things are going well?
▲ How can we take what we have learned today to daily life?

Thinking about change and "taking it home." We need to change negative talk to a new conversation about change and possibility. We need to transfer session learning to the real world. General guidelines from de Shazer (1985, 1993) include these:

▲ Note what the clients do that is good, useful, and effective. Find out what efforts they have been making and support their process of change. This is essentially the microskills positive asset search.
▲ Note exceptions to the problem. What is going on when the problem isn't happening? Be concrete and specific in this search.
▲ Promote the two above as they relate to clear, specific client goals.

In effect, de Shazer says *work on what we have already done.* If you did a good job with goals, exceptions, and other elements, the solution may already be in hand and may just need to be reemphasized.

Brief solution-focused counseling represents a contract and commitment to the clients. Do not leave them at this point. Stay with them until they accomplish *their goals.* Contract for specific follow-up in the next session or by phone. Assign a task that the client can use to ensure transfer from the interview. Concrete, achievable tasks, set up in small increments, move the client toward significant change. Homework is valuable to ensure learning.

As part of brief solution-focused counseling you could ask the child, adolescent, or adult client to tell the old story from a new frame of reference. Children can be asked to draw pictures of the old story and pictures of the new. The newly developed narrative becomes the cognitive and emotional framework for behavioral change. The skill of interpretation/reframing and focusing can help to describe the problem in new ways. For example, the old perspective may have focused on what other people are doing to make the client's life miserable. The new story focuses on what the client can do or has done to cope successfully with the situation. White and Epston (1992) suggest that counselors write down summaries of the client's new possibilities and share them in a letter sent to the client's home or at the next session. Finally, remind your client that he or she is welcome to come back at a future time for more work on this concern or any new issues that may arise.

Bringing multicultural issues into brief solution-focused counseling. A questioning style can be a problem if you have not built sufficient rapport and trust with your client. Establishing a natural and effective rapport is perhaps even more important in this approach than it is in others. Listen to the story until the client is ready; share your questions and interview plan with the client. Emphasis on positives will help make the solution approach culturally relevant. Box 14-3 illustrates the importance of using a multicultural approach to brief solution-focused counseling.

BOX 14-3   NATIONAL AND INTERNATIONAL PERSPECTIVES ON COUNSELING SKILLS

Can't You Be a Little Patient?
Weijun Zhang

I was once hired by an American multinational corporation to counsel its expatriate executives working in China. My first client, a middle-aged Caucasian male, had been in China for the past 3 years, functioning first as a manager of finance and then as the general manager of the joint venture. The reason he sought counseling involved his relationship with his local subordinates. "They are driving me nuts," as he initially put it.

When his eyes first met mine, I could tell right away that he had a big question mark in mind about my ability to counsel him. But after a few minutes of small talk, he seemed to be convinced that this Chinese guy was Westernized enough to be his counselor. Before I realized it, he burst into a string of complaints about his sour relationship with the local managers. Apparently he trusted me, for he revealed very specific facts of his situation and made no attempt to hide his hard feelings toward his Chinese colleagues.

Being a fan of the solution-focused approach, and feeling certain of knowing the mind-sets of White male businessmen, I wasted no time in starting to intervene when I heard him say, for the third time, "They are driving me nuts." I replied, "I see that your relationship with the local managers has really deteriorated. I suppose you know that you are not the only Western expatriate who is suffering from this problem. However, let me ask you a question here. Could you recall a period of time, no matter how short it is, when your relationship with your local partner was good, or normal, or not so bitter?"

He looked very surprised upon hearing this solution-oriented question, thought for a few seconds, murmured something like, "There should be, I suppose so . . . ," and then went on criticizing his Chinese co-workers with even more vigor.

I let him continue whining for about 3 minutes before I seized another opportunity to pose another solution question: "Since the operation of this joint venture has been going on for years without interruption, is it reasonable for me to assume that there have been times when you and your local managers have communicated?"

"Yes," he answered indifferently.

"Could you please give me one or two examples of this positive side of your relationship?" Reluctantly, he started to relate an incident in which he and his subordinates had had good cooperation. But when the story was barely half told, he shifted to focus on the negative and went on complaining again!

I am not a person who gives up easily. When I saw a chance to cut in on his grievance, I tried again to switch his focus to the positive side, in the hope that we could move to the goal-setting stage faster. My assumption was that since he was a busy executive in a bottom line–oriented company, he would surely favor short-term counseling and wanted to see concrete results quickly.

Much to my surprise, he burst into anger at my attempt and began to shout at me: "Why can't you let me finish my stories? Why can't you just listen? I thought I was lucky to have finally met one Chinese who can really understand what is going on here! Why can't you be a little patient?" Seeing I was taken aback by the outburst, he added, "By the way, do you know what PRC really stands for? The People's Republic of China, isn't it? Let me tell you what. It means you have to have PATIENCE and make RELATIONSHIPS in order to earn CASH!"

This left me embarrassed. This Caucasian client of mine, after being in China for 3 years, was now teaching his Chinese counselor, who had been in the States for about twice as much time, the importance of patience and relationships! Apparently, both of us had done a good job in adjusting to the local culture, though in opposite directions.

Brief solution-focused counseling can easily focus on the individual, with insufficient attention to broader contextual, multicultural, and social issues. However, balancing focus between the individual, the problem/issue, and the cultural/environmental/contextual may make brief work a most valuable addition to a multiculturally aware helping interview. Focusing on the cultural/environmental/contextual can be especially important. For example, you may be interviewing a woman who has experienced harassment in the workplace. If you focus solely on the problem and individual solutions, you may miss the most critical issue. The problem may be located not in the individual but in the system. If there is a family problem, it may be wise to focus on the family and not just on individual solutions.

The following example interview is a session conducted by Penny Ann John, a first-year graduate student at the University of Massachusetts, Amherst. We thank Penny for permitting us to share her work with you from Allen and Mary's first-year counseling skills class. As you will note, she worked with a verbal client volunteer with a fairly specific concern with some obvious solutions. As with all interviews, Penny's work is not perfect, but it is a fine example of how the positive asset search and the focus on exceptions and solutions can make a difference in the life of volunteer and real clients. The interview has been edited for clarity, but it remains the work of Penny.

Particularly note how Penny uses the basic listening sequence and search for positive assets and wellness as a vital part of her example. Balancing the questioning style of brief work with listening skills generally strengthens the interview. As you read this session, think how you might have handled the interview in accord with your own natural style of helping. It will not always be this easy and direct, but sometimes it is. For your first practice in brief solution-focused counseling, we suggest you find a classmate, friend, or family member. It will take some experience and practice to master these ideas with clients who have more complex issues or who may be resistant to the process.

Stage 1. *Relationship:* Initiate the Session—Rapport and Structuring

Penny shared a summary statement on brief solution-focused counseling with the client and talked about the process as she began the session. Carter is a close friend of hers, so she was able to jump right in after carefully explaining the process. Carter told Penny earlier that she wanted to explore the stress she was feeling as the end of the academic term was approaching.

Penny: Carter, we talked before about what we are going to do today, which is brief solution-focused counseling, and we are supposed to take an issue or concern for you and work through that and come up with some solutions. You said that you wanted to work on academic stress. Let us take a part of the larger issue—small parts of larger issues are often useful places to start. You've got a list of the questions just as I have here and, if you wish, add any questions I missed that you think are important.

Carter: Okay.

Penny: So, to start, suppose you tell me what your goal is for today. (Note immediate focus on goal setting.)

Carter: My goal for today is for us to brainstorm and come up with ideas to manage my stress because I am feeling really stressed out. (Of course, goal definition may not always be this quick and easy. Penny has a client who verbalizes well and who "buys in" to the brief model.)

Stage 2. *Story and Strengths:* Gather Data—Draw Out Stories, Concerns, Problems, or Issues

Penny: Okay, sure. What brings this topic to your mind today versus talking about this another time? (Focus on here and now and the reason for wanting to discuss it *now.*)

Carter: Well, I am a graduate student, and it is that time of the semester. Everything is coming to what seems like crunch time, and that is when I feel the most stress. There are so many things on my plate.

Penny: Right now it is stressful for you because it is coming toward the end of the semester . . . and you have a lot going on. Okay? With all of this going on at the moment, what might be positive about your situation right now? (reflection of feeling—"You feel X because Y"—followed by an open question oriented to strengths and solutions already existing in the client)

Carter: Well, you know I have got to say, I have talked to a lot of people lately, and they have a lot more to do . . . um hum . . . right. And I did not really realize that until I talked to them because I have been plugging right along and doing my papers, so I don't have everything to do all at once. And that made me feel a lot better.

Penny: Great. So you seem pretty organized. You seem like you are getting things done but still have some work to do, but you have been doing things right along. (positive feedback, paraphrase of problem and assets)

Carter: Yeah, I really have. I am not sure why I feel so stressed because I know I will have the time to do it, and I have been doing it so far, but I get, I still see the deadline at the end, and it is getting closer, so it feels a little stressful.

Penny: Have there been times in the past when you have felt this type of stress but have dealt with it in a positive way? (brief question seeking exceptions to the problem and past successes)

Carter: Sure, last semester or even when I was working. There have been times when it seemed like I had a lot to do. I had a very busy job. There are things I like to do when I have time to do them. I like to go dancing. I like to be active. I like to be social and that always . . . it is a real release for me. It is like freedom. You know. (*Penny:* Right.) And then you can forget about it for a while. (*Penny:* Uh-huh.) Then it is really good, and then I get rejuvenated, and I can come back and do what I need to do. Okay. It is just finding the time to do that.

Penny: Oh, that is really terrific. So, in the past when you have been stressed, you have gone out dancing, you have done social things, and you have done other things to keep your mind from it, and then you get more energized from it also. (positive feedback, paraphrase of strength in dealing with stress)

Carter: Yeah, it does really work.

Penny: And then you are able to focus. Oh, great. What is different about the times when you don't feel stressed out? (paraphrase, positive feedback in form of compliment, question searching for positive exceptions to the problem)

Carter: One of two things. Either I am using the technique to not be stressed out by doing all the social things I need and all of the good things and fun things I enjoy, or there is less to do. There is not a crunch time or a deadline time. I guess I feel the most organized in summer—I feel I have things under control, I feel less stressed.

Penny: Okay, so when you feel organized and have things under control, you feel less stressed. (brief summary or paraphrase/reflection of feeling)

Review the interview for focus, and note that virtually every one of Penny's comments focuses both on the client and on possible solutions to the problem.

Stage 3. *Goals*

Carter: Right, and when I finish a project and when I see that it is completed and do one thing at a time, I feel less stressed. When I try to do three things and none of them are completed but I have done all this work, it is still stressful. I guess when I finish and look at it and say oh, it's done. I did this, whatever task it may be.

Penny: You feel better when things are organized, and you complete your papers. When you do a part of each of your projects but don't finish any complete class project, it doesn't feel so good because you don't feel like you have completed anything. (paraphrase, reflection of feeling)

Carter: Right, and it might be more work, but it doesn't look that way because I can't check it off the list. You know?

Penny: Yeah. Say you woke up tomorrow and this stress was miraculously gone; what would it be like? What would it look like?

> This is a good time for the miracle question as we have an understanding of Carter's issues and her style. The miracle question often brings out new data, often unexpected, helping us find new solutions. We may find ourselves needing to totally redefine the problem or concern with data provided by the miracle question.

Carter: I would have everything done, and I would be going on vacation.

Penny: If you get everything done, then you could be on your vacation with your boyfriend, right? (paraphrase with check-out)

Carter: Yes, exactly.

> Sometimes the miracle question doesn't produce much in the way of useful data at first. In this instance it might have been more productive if Penny had been more specific and had asked, "What would you be doing differently?" Penny could also have followed up for more information on the ideal resolution, particularly as Carter said, "Yes, exactly." Penny was on track and could have asked for more concreteness. Another possibility: "Could you be more specific? What's the first thing you would notice that would be different if the stress were gone?" It takes time and practice to make the miracle question work. Also, this is a point when the client and counselor can look at questions together, seeking to elaborate mutually and make the miracle question more concrete, specific, and useful.

Penny: To recap, your goal has been to brainstorm and identify ways to deal with stress. We've identified some of the strengths in dealing with stress as your organization and your ability to do one thing at a time. And it helps, as you seem to be able to take time off and forget your studies for a while and enjoy yourself. Sounds like organization of your time these next few weeks will be important. (summary)

Carter: Yes, that's it. I guess my goal is to cool down a bit as I know I can do it. Then my boyfriend and I can be off on vacation for a week.

Stages 4 and 5. *Restory and Action:* Explore, Create, and Conclude

Penny: Let us talk about some of your strengths, your strengths in the way you can deal with this stress. (directive)

Carter: I don't know. I am pretty positive about things. I know that I will finish it, and I will get everything done. I think that is the strength. It is not a question of if I will do everything; it is just as I am in the moment, things get hectic. I am not a defeatist. I know I will get my work done, and I know I will graduate. I know I will, and I know what I need to do . . . um hum . . . and I know the things I like to do if I could carve out the time to do them and make sure that I take that time for myself. Then it will be better. So, I think that is a strength.

Penny: Yeah. Some of your strengths are that you are positive, that you know what you need to do, you know how to do it, and you know how to get there. I also heard you say you were organized before. What are some of your strengths in other parts of your life? (summary, open question that may lead to suggestions for dealing with stress)

Carter: I think those strengths also follow through in other areas of my life. That I am a positive person, that I like to try new things, and am adventurous. I really like life, and I think that has always helped me. Right. I think that is a strength and that I can do things.

Penny: You like to try new things, are positive, and enjoy life. It is interesting. . . . (paraphrase)

Carter: I just want to do everything very, very well, so sometimes that gets in the way. I want to do it perfectly or as perfectly as I possibly can. Sometimes I feel like I am not doing my best, and that bothers me. Even if it is stupid stuff. Even if I know I don't have to do the paper perfectly, I still try to. Right. Sometimes I just need to give myself a break.

Penny: You have some really great strengths, and with those strengths you were able to obtain your goal today, which was to brainstorm solutions to reduce your stress. Your positive attitude and willingness to do things and being organized and a risk taker will help you work through this stress. On a scale of 1 to 10 where do you see yourself in regard to your stress level at the moment? (Positive feedback with another compliment and summarization followed by scaling. Note that the client in the next statement responds to Penny's incomplete scaling question by defining the end points herself, a sign she understands the concept well.)

Carter: It is not too bad. Let's say 10 is the most and 1 was the least. I am probably a 5. I don't think it is that bad; talking about it makes it a lot easier. Like I said, it is more when it is in the moment, and I have had a crazy day. I was working all day, had my classes, I come home, and I have seven things to do, and there are three messages on the answering machine, and I think, I can't do everything. I probably could do most of them, and then I have to map it out and prioritize, but that is hard because it is so hard to say no. Especially when you want to do fun stuff. I guess it is a 5. Giving myself a break.

Penny: It seems like you are handling your stress pretty well. (positive feedback)

Carter: Thanks, yeah, it is not too bad. It is not as bad as it seems in those moments. You know?

Penny: Um-hum. So there are times when you have a higher stress level and times when you have a lower stress level. When are times when you are a 1? (Encourage, paraphrase, question, search for exceptions. Another useful question would be "So, what do you have to do to reduce your stress from a 5 to a 4?" This later question is an important part of the scaling process, particularly as it focuses on small change rather than total resolution.)

Carter: When I am not feeling any stress?

Penny: Yeah. (encourage)

Carter: When I can step away from responsibilities and play with my niece or go home and be with my family or out with my friends and dancing. I am a very active person, so I feel best when I am doing something outdoors or when I am moving or exercising. (Penny's question, focused on total removal of stress, works, but with many clients the smaller change from 5 to 4 would be more manageable.)

Penny: So, you feel stress-free when you are moving, or exercising, or playing, or are social. On a scale of 1 to 10, what would be your ideal stress level at this time? (Paraphrase. The client's stress patterns are examined by scaling.)

Carter: An ideal stress level would be a 3, because you can't always be playing. A certain amount of stress is good in order to be productive.

Penny: What do you want to happen precisely? (move toward generalization with an open question)

Carter: I have four weeks left of school. I think I need to map out the next four weeks in how I can balance my productive work time and my social outlets. I have six papers left to write and four weeks, so that averages to about one and a half papers per week. Now that I put it in that perspective, and I see it visually, it doesn't seem so bad after all.

Penny: No, it doesn't. You have time to do your work and have time for your social outlets. You seem to have a good handle on what you need to do and the amount of time you have to do it. With your strengths, it appears that you will get your work done. You are positive. You know you can get your work done. You are motivated and organized. I like the way you have put it all in perspective. (summary, positive feedback in the form of a compliment)

Carter: Yes, I feel relieved already.

Penny: Let's now generate a picture of what the next four weeks are going to look like for you. (directive, concretizing the plan)

Carter: Well, I have four classes and have my assistantship, which is 10 hours per week. I also have these six papers and plenty of time to exercise, and I can go out a few times over the next few weekends. My stress level has reduced tremendously already.

Penny: It sounds like you are doing a lot of things right already, and your future plan is very attainable. How about support networks? Do you have people around you for a support system? (positive feedback, open question focusing on support network)

Carter: Yes, I have my sister and my brother. I also have my housemates and classmates. They are the biggest support because they are going through the same thing with me. I feel like I have a good network around me.

Penny: You sure do. What positive assets in yourself can you also draw from? (encourage, return to focus on client and strengths)

Carter: My drive and motivation. My positive outlook and energy level.

Penny: It sounds like you have a lot to draw from. So what are you going to do differently tomorrow? (paraphrase, open question oriented toward generalization)

Carter: I will exercise first thing in the morning and get started on one of my papers. This week I have a lot of time off, so I can probably get a few papers done, and I can also go for a hike or something if the weather improves.

Penny: Well, this all sounds so clear to me. You know what you need to do. You have your timeline all mapped out and have a good balance of fun and work. Will you let me know in a week or so how it is all working out? (summary, open question)

Carter: Yes, I will call you next week.

Brief solution-focused counseling operates on the theory that clients have their own answers in their past success experiences and wellness strengths. Emphasis is on immediate goal setting and clarity. Questions are a major skill, although many practitioners consider listening carefully to the client's full story to be as critical. Skilled questioning is the major skill of brief solution-focused counseling. We recommend, at least in the early stages of your practice, that you have, and share with your client, a list of questions for each stage of the interview.

Think about referral or another approach to helping if the client does not respond to brief methods. Evidence of nonresponse include these: (a) The client presents with serious symptoms or problems; (b) the client is not able or willing to try the generated solutions;

BOX 14-4   FEEDBACK FORM: BRIEF SOLUTION-FOCUSED COUNSELING

_____ (Date)

_____   _____
(Name of Interviewer)   (Name of Person Completing Form)

Relationship: **Initiating the session—Rapport and structuring ("Hello; this is what might happen in this session").** How well did the interviewer establish rapport and how did he or she accomplish this objective? Were clear and achievable goals identified and set fairly soon after a relationship was established? If the goals were not immediately attainable, were smaller steps toward the larger goal discussed?

Story and strengths: **Gathering data—Drawing out stories, concerns, problems, or issues ("What's your concern? What are your strengths or resources?").** Was at least one positive asset or strength of the client identified? How completely did the interviewer draw out the story and/or issues?

Goals: **Mutual goal setting—Establishing outcomes ("What do you want to happen?").** Was the "miracle question" used? Was it effective? Were the original goals of the session reviewed and were the desired outcomes of the client really clear?

Restory/Action: **Working, terminating, and generalizing** How did the interviewer go about helping the client develop a concrete plan for action? Was homework agreed to by counselor and client? Did the interviewer help the client plan for generalization to daily life?

General comments on the interview and skill usage:

(c) the solutions are generated primarily by you rather than by the client; (d) the solutions are vaguely constructed and not carefully explored for positive and negative consequences; (e) the client may have difficulty making the solutions real in her or his life; (f) the client's context may not be receptive to certain solutions; (g) the client may want a long-term helping relationship.

A feedback form is shown in Box 14-4. An observer and/or the client can provide feedback for you.

INSTRUCTIONAL READING AND EXAMPLE INTERVIEW 4: MOTIVATIONAL INTERVIEWING

Counseling alcohol abusers, drug users, and clients with other forms of addictive behavior is recognized as extremely difficult and challenging, and out of these challenges came motivational interviewing (Miller & Rollnick, 2002). Particularly at issue is motivating the client to actually change behavior rather than just talk about it. Thus, capturing the interest, hopes, and ultimately the motivation to change is critical. In addition, research reveals that the system works and that motivational interviewing will be useful to clients who do not themselves have substance abuse issues.

Motivational interviewing (MI) could be described as a way to integrate the ideas of this book. MI is based on microskills such as those discussed in this book plus decisional theory and practice, in particular the work of Leon Mann (2001; Mann, Beswick, Allouache, & Ivey, 1989). In addition, it uses many of the methods and questions of solution-oriented counseling, is concerned with transfer of behavior from the interview to the real world, and uses the ideas of cognitive-behavioral relapse prevention. (See Box 14-5, Maintaining Change and Relapse Prevention Worksheet.) The "spirit" of motivational interviewing is based on collaboration with the client, evoking positive resources and motivation for change already in this client, and affirming client autonomy and self-direction. MI's four general principles for practice are to express empathy, develop discrepancy, roll with resistance, and support self-efficacy (Miller & Rollnick, 2002, pp. 33–42). The spirit of these principles should be basically familiar to readers. Self-efficacy is another term for intentionality, personal agency, and general wellness.

Let us now work through the five stages of the interview as they might be played out in an actual session of motivational interviewing. When you try MI for the first time, we suggest that you have these five stages with you as a guide. Feel free to share them with your client in the spirit of collaboration.

The client, Jerome, is 37, married with two children, and has serious issues with alcohol. Note how Farah, the counselor, gets down to business more rapidly than we usually suggest in this book. In that sense, motivational interviewing is similar to brief solution-focused counseling.

The interview is highly edited but with the commentaries on motivational interviewing plus your present knowledge, you can easily fill in the blanks of what is missing—the *relationship—story and strengths—goals—restory—action* model facilitates understanding of the workings of multiple theories.

Note how the counselor in motivational interviewing starts the session quickly and shortly begins seeking attainable goals.

BOX 14-5 MAINTAINING CHANGE AND RELAPSE PREVENTION WORKSHEET: SELF-MANAGEMENT STRATEGIES FOR SKILL RETENTION

I. CHOOSING AN APPROPRIATE BEHAVIOR, THOUGHT, FEELING, OR SKILL TO INCREASE OR CHANGE

Describe in detail what you intend to increase or change:

_____

_____

Why is it important for you to reach the above goal(s)?

_____

_____

What will you do specifically to make it happen?

_____

_____

II. RELAPSE PREVENTION STRATEGIES

A. Strategies to help you anticipate and monitor potential difficulties: Regulating stimuli

Strategy	*Assessing Your Situation*
1. Do you understand the relapse and change process, and that there will be challenging situations when it will be difficult to engage in new behaviors?	_____
2. What are the differences between learning the behavioral skill or thought and using it in a difficult situation?	_____
3. Support network? Who can help you maintain the skill?	_____
4. High-risk situations? What kinds of people, places, or things will make retention or change especially difficult?	_____

B. Strategies to increase rational thinking: Regulating thoughts and feelings

5. Are you aware that a slip, relapse, or mistake need only be temporary? "Relapse happens."	_____
6. What might be an unreasonable emotional response to a temporary slip or relapse?	_____
7. What can you do to think more effectively in tempting situations or after a relapse?	_____

C. Strategies to diagnose and practice related support skills: Regulating behaviors

8. What additional support skills do you need to retain the skill? Assertiveness? Relaxation? Communication microskills?	_____

_____

Permission to use this adaptation of the Relapse Prevention Worksheet was given by Robert Marx.

(continued)

BOX 14-5 (continued)

D. Strategies to provide appropriate outcomes for behaviors: Regulating consequences

9. Can you identify some probable outcomes of
   succeeding with your new behavior?   _____

   _____

10. How can you reward yourself for a job well done?   _____
    Generate specific rewards and satisfactions.   _____

III.   PREDICTING THE CIRCUMSTANCE OF THE FIRST POSSIBLE FAILURE (LAPSE)

Describe the details of how the first lapse might occur, including people, places, times, and emotional states.
This will be helpful to you in coping with the lapse when and if it comes.

_____

_____

_____

Stage 1. *Relationship:* Initiating the Session

Farah: Good morning, Jerome. Today we've got a little less than an hour. At the beginning the most important thing is that I listen to what you have to say. I'd like to know what your concerns are and what you'd like to see as a result of being with me. There will be some details I'll ask you about as well. Perhaps we could begin with your telling me what you'd like to talk about.

Needless to say, this introduction will vary with the client. MI does not stress multicultural issues, and we believe that the system would profit from more attention to individual and cultural variation. Thus we would suggest that you take whatever time is necessary to build sufficient rapport and trust before starting, although many clients would be "ready to go" with this brief introduction by Farah. However, if a true collaborative interview is to follow, it may be important in some cases to discuss differences of ethnicity/race or gender and share something of yourself.

Let us assume that Jerome comes to Farah with an issue around alcohol abuse. He shares his story in response to that beginning question. He tells Farah that his family has told him that he drinks too much and needs to get help. He says that he'd like to stop. MI theory recommends that we start working on change immediately.

The counselor, Farah, brings out a paper with a 10-point scale below and asks Jerome to indicate how important changing his behavior is to him. Jerome rated his interest as a 7—interested in change, but not fully committed.

How Committed Are You to Change and Reach Your Goals?

Not Interested or Motivated				Somewhat Interested and Motivated				Let's Start, I'm Ready! Highly Motivated	
1	2	3	4	5	6	7	8	9	10

The rating scale is an important "hook" in MI as it immediately introduces belief in the possibility of change, and the counselor also learns of the client's depth of motivation.

Another valuable by-product is that the interview will have a clear focus. Change has become the central goal at the very start. This rating scale can be useful in many sessions. Consider it a tool that will be useful to you and your clients in interviews with other theoretical orientations.

Stage 2. *Story and Strength:* Gathering Data—Drawing Out Stories, Concerns, Problems, and Issues

"Change talk" is a key phrase within MI. Through the use of open questions, reflective listening, and summarization the client is encouraged to provide detail and elaboration on the change issue. In particular, change talk focuses on what is positive, enjoyable, or useful in the behavior in which change is sought. This is followed by the "downside" and what goes wrong when the behavior occurs. This ultimately leads to a decisional balance as the pluses and minuses of change are discussed.

Farah: (elaborates the positives) Jerome, what do you particularly enjoy about drinking? What would you miss if you didn't drink?

Jerome: (elaborates the positives) Well, I get out of the house. Carlotta is always on me. It makes me feel good, and I forget about things for awhile. . . . (He continues.)

Farah: (elaborates the negatives) What is the negative side of drinking? What doesn't work so well for you?

Jerome: Actually, I worry sometimes about getting a little too angry when I drink. I get depressed the next day, say that I'm going to stop, but . . .

Throughout the session, the counselor maintains an open, nonjudgmental approach, seeking to empathically understand the world of the client. In addition, MI uses the word "affirmation," which is parallel to the positive asset search. The counselor is always looking for something right in the client that can be affirmed and recognized.

Farah: As I hear you, Jerome, I can see that you have thought a lot about what drinking does for and against you. I like the way you want to control yourself and your behavior more. I also heard that your children would like you to stop, and I sense your caring for them.

Further change questions focus on disadvantages of the status quo ("What is likely to happen if you don't stop drinking?"), advantages of change ("How would it be at home if your drinking stopped?"), optimism about change ("What are some supports that will help you maintain change?"), and intention to change ("Would you be willing to try stopping?")

Stage 3. *Goals*

The positive and negative motivation for change is explored on a decisional balance sheet (Mann, 2001; Mann et al., 1989). Their original decisional balance sheet (see page 365) has been adapted by MI and is a straightforward decisional procedure in which the client and counselor list together the pros and cons (costs and benefits) of changing behavior.

Jerome's balance sheet came out as follows:

Continue Drinking		Stop Drinking	
Pros	*Cons*	*Pros*	*Cons*
Get out of house.	Kids unhappy.	With kids more.	Miss drinking buddies.
Away from wife.	She may leave me.	Perhaps better with her.	Perhaps not.
Stops feelings of depression.	Get more depressed.	Maybe happier.	Must deal with my depression.
Problems disappear.	Boss suspects that I drink.	Lose weight.	Not crazy about sitting around house.
	Costly.	Healthier.	How will I deal with all these problems?
	Dad died of liver disease.	Live longer.	
	Not getting anywhere.	Job better, maybe.	

Two classic questions or variations help set up the final goal, "What would you like it to be like in the future?" and "What are the future consequences of change?" We should again mention that motivational interviewing has a wide array of specific questions and techniques to accomplish the aims of each part of the MI procedures.

Jerome: Well, I guess I really would like to stop drinking, and the positive consequences of stopping are all there on the balance sheet. Would that it were that easy! But I guess I do really want to try changing.

Stage 4. *Restory:* Working—Exploring Alternatives, Confronting Client Incongruities and Conflict, Restorying

A particular strength of MI is the way resistance is defined and made more central in the interview. First, resistance is the major impediment to change. For example, an alcoholic may speak of the desire to change but in truth is resistant and unmotivated to do the work that change requires—thus, the importance of increasing motivation to change. Resistance is discussed as follows (Miller & Rollnick, 2002, p. 98):

> Resistance arises from the interpersonal interaction between counselor and client. . . . research clearly demonstrates that a change in counseling style can directly affect client resistance, driving it upward or downward. . . .
>
> [Resistance] is observable client behavior that occurs within the context of treatment and represents an important sign of dissonance within the counseling process. In a way, it is a signal that the person is not keeping up with you; it is the client's way of saying, "Wait a minute; I'm not with you; I don't agree."

When this happens, it is time to listen and discover what is going on with the client. Rather than ignoring resistance, MI suggests that counselors need to be prepared to try several different responses until one is effective. For example, paraphrasing or reflecting the client's feelings is an acknowledgment that you are aware of how the client thinks and feels; this is often the best place to start. Shifting interview focus or reframing the discussion from another perspective may be helpful. MI even suggests agreeing with the client and pointing out that the client is in control. In short, if you are skilled in microskills, be intentional and shift your style to meet changing client needs.

The MI philosophy around resistance and listening is important for us all. And note the importance of intentionally shifting your skill or focus when you work with challenging clients. "If the first skill or strategy doesn't work, you've still got another."

One interesting way to work with resistance is magnifying discrepancies and incongruity even more than the client has stated thus far, or reframing them in a new perspective.

Jerome:	Carlotta worries too much.
Farah:	(magnifying discrepant behavior) She has no need for concern at all. You're in charge of your drinking, and she need not worry about you. (Rather than argue, the counselor extends Jerome's comment to another level, showing the limitations of this thought pattern.)
Farah:	(reframing) Do I hear you saying that you'd like her to worry less about you? You seem concerned how she thinks about you.

Confidence talk is a vital part of motivational interviewing. Behavior change takes confidence and belief in oneself. The idea here is to help the client focus on specifics of change and how he or she can deal with them effectively. The wellness and positive asset search are good places to enhance client self-efficacy and competence. This can include reviewing past successes, assessing the strength of community supports (perhaps via a community genogram) and focusing on personal strengths. Affirming the capability of the client to change is important, and that positive self-talk needs to enter Jerome fully before he can be expected to make major changes in his life.

After an emphasis on strengths and confidence, Jerome might be more optimistic around change.

Jerome:	I see what you're saying. I have done some good things at a high level in my life. My wife's family has been there to help over the years, and the community services you tell me about might be useful. And I keep visualizing the little ones. Just focusing on my children sometimes gives me more strength.

Motivational interviewing proposes the confidence ruler, another 10-point scale in which the client rates herself or himself on a 1 to 10 scale. The self-rating could be followed by asking the client to describe in concrete language what it is like to be at various levels on that scale.

Again, recall intentionality and the need to shift styles and tactics with the resistance client. Always have something else to try—and when all else fails—listen!

Stage 5. *Action:* Terminating—Generalizing and Acting on New Stories

Motivation interviewing uses a change plan worksheet. But you will find that the Maintaining Change and Relapse Prevention Worksheet of Box 14-5 covers the same territory and meets the same purpose. Taking the time to complete such a form is very worthwhile as a basis for follow-up and action from the interview.

As part of the process of planning for generalization, further positive reframing, confidence building, and listening will be essential. It is particularly important at this stage, of course, to watch for resistance.

The feedback form in Box 14-6 can be used to improve your skills as you practice motivational interviewing. An observer and/or the client can provide feedback for you.

BOX 14-6 FEEDBACK FORM: MOTIVATIONAL INTERVIEWING

_____ (Date)

_____   _____
(Name of Interviewer)   (Name of Person Completing Form)

Relationship: **Initiating the session—Rapport and structuring ("Hello; this is what might happen in this session").** How well did the interviewer structure the interview? How did he or she establish rapport? Was the client's motivation for change explored? Did the client complete the 10-point motivation for change score?

Story and strengths: **Gathering data—Drawing out stories, concerns, problems, or issues ("What's your concern? What are your strengths or resources?").** Was the basic listening sequence used to widen awareness with special attention to "change talk"? Were the positives and negatives of the area for potential change explored? Did the counselor offer sufficient client affirmations?

Goals: **Mutual goal setting—Establishing outcomes ("What do you want to happen?").** Were positives and negatives around change explored on the decisional balance sheet? Was there some future pacing with the client such as "What are the future consequences of change?"

Restory: **Working—Exploring alternatives, confronting client incongruities and conflict, restorying ("What are we going to do about it?").** Was the counselor able to maintain a client focus on change talk? How did the counselor handle resistance (be specific)? Was the client able to shift style and try another skill or tactic if resistance was met? Was listening continued? Was a "confidence ruler" brought in and were past strengths and successes of the client reviewed as support mechanisms?

Action: **Terminating—Generalizing and action on new stories. ("Will you do it?").** How did the interviewer go about helping the client develop a concrete plan for action? Was the Maintaining Change and Relapse Prevention Worksheet used? Did the interviewer really help the client plan for generalization to daily life?

General comments on interview and skill usage:

A Final Note on Motivational Interviewing

We are obviously not going to solve challenging issues such as alcohol abuse in one easy session. But the general structure of the interview above provides a basic model for how motivational interviewing can be effective. Consult Miller and Rollnick (2002) for more information and research on the effectiveness of this model.

If you complete a practice session using MI, you will see how it integrates many of the ideas proposed in this book. With the microskills, you have moved from basic attending and observation to beginning mastery of some tools and theories that can serve as a foundation for your entire professional career.

Motivational interviewing appears to be quite effective across many cultures. By its very nature, clarity, and focus on what can be done to produce change, it is an appropriate system to consider for working cross-culturally.

SUMMARY: INTEGRATING MICROSKILLS WITH COUNSELING

In addition to decisional counseling, this chapter has focused on four approaches to counseling—person-centered, cognitive-behavioral, brief solution-focused, and motivational interviewing. We believe that through using microskills you can engage in the basic strategies of these theories.

Your ultimate goal is to use this newly acquired knowledge with your clients. Now you can access a variety of interviewing styles to provide them with more alternatives and, thus, deliver more effective help. Furthermore, we encourage you to study the theories presented here and to engage in the intentional use of the tools they provide.

Following are the key points of Chapter 14.

■ KEY POINTS

Five approaches	Table 14-1 should be reviewed because it summarizes the structure of decisional counseling, person-centered helping, cognitive-behavioral therapy, brief solution-focused counseling, and motivational interviewing. Though all these approaches may be explained by their use of microskills and how the interview is structured, note that their emphases are quite different. Decisional counseling emphasizes careful listening to the story/problem/concern/challenge of the client before acting, whereas the solution approach emphasizes working on the problem as quickly as possible. Person-centered helping stresses listening to the client's feelings and story in detail, and thoughts and words are central. Cognitive-behavioral work, on the other hand, seeks very actively to encourage the client to change and adopt new behaviors, thoughts, and feelings, as you observed in the stress management session. Brief approaches focus on finding quick answers and using many questions whereas motivational interviewing appears to integrate most of the ideas of this book into a single package, potentially useful for particularly challenging clients.
Multicultural issues	Each style of interviewing requires different adaptations to be meaningful in multicultural situations. Particularly helpful in this regard is the concept of focus (Chapter 10). By focusing on the cultural/environmental/contextual dimensions, you can bring in these issues fairly easily to all helping approaches. However, you still must recognize that the aims of each approach may not be fully compatible with varying cultures. This same

(continued)

Key Points (continued)

	point, of course, should be made with the client regardless of cultural background. Some clients may prefer the Rogerian person-centered approach; others may want solutions and cognitive-behavioral action. Avoid stereotyping any client with prior expectations.
Cultural Intentionality	We are suggesting that the intentional interviewer and counselor will have more than one interviewing alternative available. At the same time, it is important that you select those approaches to helping that are most comfortable for you. Balancing your knowledge, skills, and interests as you counsel varying clients will be a lifetime process of learning for any helping professional.

COMPETENCY PRACTICE EXERCISES AND PORTFOLIO OF COMPETENCE

Practice Exercise

There is one central exercise for this chapter, a special type of practice interview. Seek out a classmate, friend, or colleague who is willing to work on a single issue for a half-hour to an hour. If that person has the interest, suggest that he or she read portions of this chapter to know what to expect. Alternatively, share the key points with him or her. This is an exercise in joint discovery. You may find that sharing the interview plan with your clients is useful in other forms of counseling and interviewing.

The best way to understand the four approaches in this chapter is to use each one separately in conducting an interview. As you meet with your volunteer client, share the interview plan and the key points of your plan. Consider working through your first interview with the book and notes on the table; both of you can use them for reference. You may wish to use the feedback form together as a way to summarize the specific steps.

Some topics that may be amenable to all four approaches include these:

Friendship, partner, or child difficulties
Relationship problems
School- or job-related issues
Family concerns
Getting started on career planning
Anger management
Expressing feelings more openly
Inadequate balance of work and leisure

In short, almost any topic can be useful, but the critical issue is finding a manageable part of the larger problem for a single practice session.

Portfolio of Competence

Developing and evaluating your skills and competence using each of the theories presented should be included in your Portfolio of Competence.

DETERMINING YOUR OWN STYLE AND THEORY: CRITICAL SELF-REFLECTION ON FOUR THEORETICAL ORIENTATIONS

How does the concept of theoretical orientation relate to your own developing style and theory? Which of the five approaches presented (including decisional counseling) most appeals to you? Do you agree with us that decisional counseling underlies most other approaches as a basic model?

We will not ask you to assess your competence in any of these approaches, as it is far too early, and you will want to work further with each one. Rather, please focus your attention on your early impressions and where you think you might go next in building competence in these or other theoretical orientations.

What single idea stood out for you among all those presented in this chapter, in class, or through informal learning? What stands out for you is likely to be important as a guide toward your next step. What are your thoughts on multicultural issues and the use of the microskills? What other points in this chapter struck you as important? How might you use ideas in this chapter to begin the process of establishing your own style and theory?

_____

_____

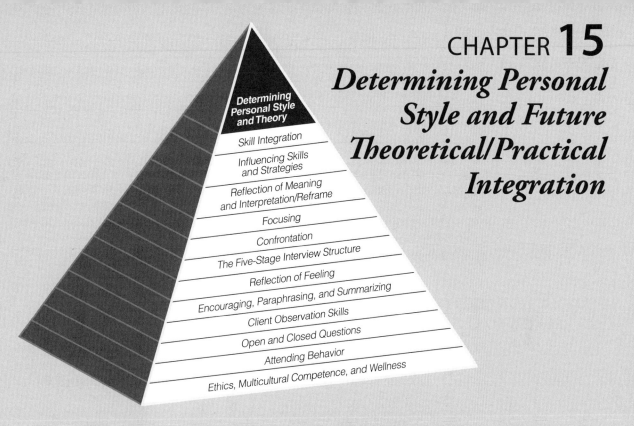

Determining
Personal Style
and Theory

Skill Integration

Influencing Skills
and Strategies

Reflection of Meaning
and Interpretation/Reframe

Focusing

Confrontation

The Five-Stage Interview Structure

Reflection of Feeling

Encouraging, Paraphrasing, and Summarizing

Client Observation Skills

Open and Closed Questions

Attending Behavior

Ethics, Multicultural Competence, and Wellness

CHAPTER 15

Determining Personal Style and Future Theoretical/Practical Integration

We make a living by what we get. We make a life by what we give.

—Winston Churchill

How can determining your own personal style help you and your clients? How do theoretical alternatives for practice relate to you and your clients?

Chapter Goals
The major purposes of this chapter are to help you review your work with microskills and consider your own orientation to interviewing, counseling, and psychotherapeutic practice. You can be most effective if you generate your own formulation of the helping process. What is your natural style and how does it relate to skills, interviewing structure, and alternative theoretical orientations? Also, how does your conception relate to the variety of clients that you will face?

Competency Objectives
Awareness, knowledge, and skills developed in integrating your skills will enable you to

▲ Be aware that interviewing and counseling are not simple but incredibly complex processes requiring movement, change, and constant growth.

▲ Realize that the skills and concepts of this book provide an effective framework to master this complexity of the interviewing process. You can meet the many challenges ahead from a solid base.

▲ Review the concepts of this book and evaluate your ability to master the basic competencies of microskills hierarchy and the interview.

▲ Begin the process of determining which theory or theories are most relevant to you.

▲ Consider how the rich array of theories and concepts available to you and your clients may be used to facilitate both your own and your clients' development.

> ▲ Make serious progress toward defining your own theoretical/practical construction of the nature of interviewing and counseling theory.
> ▲ Consider how your helping style and possible theoretical orientation will be relevant to the many differing types of clients you will encounter.

We are approaching the end of our journey. In this chapter we ask you to reflect on yourself, your values and personal meanings, and your skills. Now is a good time to assess your strengths and your competencies to work with individuals from various backgrounds; to determine how much have you learned about their different styles, values, and reasons for consultation; and to establish areas for further growth. This is also a good time to think about where you would like to go from here.

INTRODUCTION: IDENTIFYING AN AUTHENTIC STYLE THAT RELATES TO CLIENTS

Developing your own personal approach to interviewing, counseling, and therapy involves a multiplicity of factors. At this moment we are asking you to reflect on yourself, your values and personal meanings, and your skills. Where are your strengths? What areas need further development? Where would you like to go? What are your competencies for working with clients different from you in terms of personal style, values, type of concern presented, cultural difference, and other factors?

Some counselors and therapists have developed individual styles of helping that require their clients to join them in their orientation to the world. Such individuals believe they have found the one "true and correct" formula for counseling and therapy; clients who have difficulty with that formula are often termed "resistant" and "not ready" for counseling. These counselors do indeed have their own style, but their methods tend to be rigid and dogmatic. Such counselors and therapists can and do produce effective change, but they may be unable to serve many client populations. Missing from their orientation is an understanding of the complexity of humanity and the helping process.

Thus as you generate your own approach to the field, remember that clients may have widely different values and experiences from yours. They may wish to head in many directions different from your own. Consequently, it is urgent that you remain aware that the skills, strategies, and theories you favor may not be preferred by your client. If so, then you may wish to expand your skills and knowledge in areas where you are now less comfortable. It is their life, not yours.

And as you continually expand your competence, maintain your authenticity as a person. With study, patience, and experience, you will increase your abilities to work with those who are unlike you. The opportunity to learn from clients different from us is one of the special privileges of being an interviewer, a counselor, or a therapist. You will want to expand your understanding of cultural differences including race/ethnicity, gender, spiritual/religious orientation, disability, sexual orientation, age, and socioeconomic status.

This chapter returns to the beginning. We saw your natural style. Is it the same? How might it have changed? To that end, the microskills hierarchy will be reviewed, but this time as a framework for decision making on your part. Let us now remind you just how many skills and concepts you have learned in this book. A brief map showing how these skills and concepts play themselves out in different models of counseling and therapy should aid you in the search for your personal style.

The chapter concludes with a review of the five-stage structure of the interview, and the *relationship—story and strengths—goals—restory—action* model. This in turn leads to recommended readings for your future development in skills and theory.

INSTRUCTIONAL READING: DEFINING YOUR PERSONAL STYLE

The Microskills Hierarchy: A Summary

You have thus far been presented with 14 chapters of reading containing 34 major concepts and skill categories. Within those major divisions are more than 100 specific methods, theories, and strategies. Ideally, you will commit them all to memory and be able to draw on them immediately in practice to facilitate your clients' development and progress in the interview.

These skills, concepts, and understandings are but a beginning. We have not considered here details of personality development, testing, and the many other theories of counseling and therapy you will encounter in the future. Although these microskills and training concepts are used in multiple fields and by people around the world, only a brief introduction to the multitude of applications has been provided in this book.

How can you manage to retain and use all of this book's concepts and the five theories presented? Most likely, you cannot at this point. But recall the story of the Samurai (page 80). With time and experience, you will develop increased understanding and expertise. As you grow as an interviewer, counselor, or therapist, you will find the ideas expressed here becoming increasingly clear, as your mastery of skills and theories will likewise continue to increase. Use this book as a resource that can help you recall key concepts that you will use with your varied clients.

In addition, retaining and mastering the concepts of the microskills hierarchy may be facilitated by what is termed *chunking*. We do not learn information just in bits and pieces; we can learn it best by organizing it into patterns. The microskills hierarchy is a pattern that can be visualized and experienced. For example, at this moment you can probably immediately recall that attending behavior has certain major concepts "chunked" under it (the "three V's + B" of culturally appropriate visuals, vocal tone, verbal following, and body language). The basic listening sequence (BLS) will be easy to recall and essentially covers the first half of this book. You can probably also recall the purpose of open questions and perhaps which questions lead to which likely outcomes (for example, *how* questions lead to process and feelings).

With periodic review and experience, the many concepts will become increasingly familiar to you. If you complete a transcript examining and classifying your own interviewing style similar to the one presented in Chapter 13, the ideas of this book will become especially clear. As you must have noticed, the skills you have practiced instead of just read about are the ones you understand best and that have the most relevance for you. Reading is a useful introduction to counseling and interviewing, but the results of practice and experience will stick with you far into the future.

Table 15-1 summarizes the major concepts of the book; take some time to review your own competence levels in each of the 38 major areas and enter the results on the table. What is your present level of mastery and competence in each area?

TABLE 15-1   Self-assessment summary

Skill or Concept	Identification and Classification	Basic Competence	Intentional Competence	Psychoeduca-tional Teaching Competence	Evidence of Achieving Competence Level
1. Attending behavior					
2. Questioning					
3. Observation skills					
4. Encouraging					
5. Paraphrasing					
6. Summarizing					
7. Reflecting feelings					
8. Basic listening sequence					
9. Positive asset search—strengths, resources, wellness					
10. Empathy					
11. Five stages of the interview					
12. *Relationship—story and strengths—goals—restory—action*					
13. Confrontation					
14. Client Change Scale					
15. Focusing					
16. Reflection of meaning					
17. Noting concreteness and abstractions in self and clients					
18. Interpretation/reframing					
19. Logical consequences					
20. Self-disclosure					
21. Feedback					
22. Information/psychoeducation					
23. Directives					

(continued)

TABLE 15-1 (continued)

Skill or Concept	Identification and Classification	Basic Competence	Intentional Competence	Psychoeduca-tional Teaching Competence	Evidence of Achieving Competence Level
24. Analysis of the interview (Chapter 13)					
Theoretical/Practical Strategies 25. Wellness					
26. Ethics					
27. Multicultural awareness					
28. Family genogram					
29. Community genogram					
30. Decisional counseling					
31. Person-centered counseling					
32. Cognitive-behavioral counseling and stress management					
33. Brief solution-focused interviewing and counseling					
34. Motivational interviewing					
35. Coaching questions					
36. Neuroscience and how it may relate to the interview					
37. Teaching skills to clients					
38. Defining personal style and theory					

1. *Identify and classify the concept.* If it is present in an interview, can you label it? These concepts provide a vocabulary and communication tool with which to understand and analyze your interviewing and counseling behavior and that of others. The ability to identify concepts means most likely that you have chunked most of the major points of the skill together in your mind. You may not immediately recall all seven types of focus, but when you see an interview in progress you will probably recall which one is being used.

2. *Demonstrate basic competence.* At this level, you will be able to understand and practice the concept. Continued practice and experience form the foundation for later intentional mastery.

3. *Demonstrate intentional competence.* Skilled interviewers, counselors, and therapists can use the microskills and the concepts of this book to produce specific, concrete effects with their clients. Please visit the Ivey Taxonomy (Appendix I) and review specific aspects of intentional prediction. If you reflect feelings, do clients actually talk more about their emotions? If you provide an interpretation, does your client see her or his situation from a new perspective? If you work through some variation of the positive asset search, does your client actually view the situation more hopefully? If you conduct a well-formed, five-stage interview, where does the client stand in change using the Client Change Scale? Do behavior and thinking change?

   Intentional competence shows up not in your use of the skills and concepts but rather in what your client does. To demonstrate this level of skill, you need to be able to produce *results* due to your efforts in helping.

4. *Demonstrate psychoeducational teaching competence.* You are not expected to master teaching all the skills and concepts of this book at this point. However, you should have learned that you can teach attending behavior and the basic listening sequence to your clients in the interview. Some of you may have had the opportunity to take these skills and conduct teaching workshops with community volunteers, church groups, and peer counselor training programs.

For the longer term, we would suggest that you continue to think of the possibility of teaching skills to clients and their families. And if you become a professional, you most likely will find yourself teaching listening skills workshops at some point in your career. Box 15-1 provides a personal account about using the microskills throughout your career.

Alternative Theories of Interviewing and Counseling

Equipped with the foundation skills of listening, observing, influencing, and structuring an interview, you are well prepared to enter the complex world of theory and practice. Soon you will be encountering the 250 or more theories competing for your attention. This final discussion is oriented to assembling what you already know, bringing it together in one place, and looking toward the future.

If you can complete the five-stage interview structured around decisions, you have a good beginning understanding of the work of many professionals. Decisional counseling has been around for a long time. Frank Parsons in 1909 reminded us that much of our interviewing is about helping people make decisions. Some claim that decisional counseling (sometimes known as "problem-solving counseling") is the most widely practiced form of helping in current practice. The following is just a sample of places in which decisional or problem-solving counseling is practiced daily: employment counseling, placement counseling, AIDS counseling, alcohol and drug counseling, school and college counseling, and in the work of community volunteers and peer helpers.

Having completed a full interview using only listening skills in Chapter 8, you have had a beginning introduction to Carl Rogers's (1961) person-centered theory. There you likely discovered that many clients are self-directed, and with a good listener, they can do much to resolve their issues on their own. Mastery of this important strategy does not make you a person-centered counselor. You will want to study Rogers's work in detail.

BOX 15-1   NATIONAL AND INTERNATIONAL PERSPECTIVES ON COUNSELING SKILLS

Using Microskills Throughout My Professional Career
Mary Rue Brodhead, Executive, Canadian Food Inspection Agency

My first encounter with the microskills program was through my doctoral program in teacher education. I wanted to learn about counseling so that I could better reach students, particularly those at risk who often were the least likely to approach a counselor. What I learned very rapidly was that teachers often fail to listen to their students—in fact, one research study found that out of nearly 2,000 teacher comments, there was only one reflection of feeling and, of course, most comments focused on providing information. Often, even when teachers did listen, they weren't able to recognize the verbal and nonverbal cues that were reflecting the reality of their students.

After completing my degree, I entered teacher education and found that microskills lead to a more student-centered teaching. I also found that teachers who matched their students' cognitive/emotional style were the more effective. If a student is concrete, the teacher needs to provide specific examples and use concrete questions. If the student is more reflective, then formal operational strategies can be used. Bringing in emotional involvement via sensorimotor strategies enriched teaching.

I next lived on an island off Vancouver Island where my husband, Dal, and I worked with members of the Kwakiutl Nation. I trained teachers, but I was involved in counseling and established an alternative program, as part of the local School Board, for youth who had dropped out of high school. I actually taught them the same interviewing skills that you are learning in this book. Needless to say, the multicultural orientation here helped sensitize me to important differences among people.

One of our first activities was a trip in the Chief's fishing boat to gather Christmas trees for the old people, a cherished tradition in the community. This was the first time most of these young people had participated in the ritual. As they delivered the trees, the recipients, in an expression of gratitude, invited them in for something to eat and began to tell stories. These old people, thrilled to have an audience, would shower attention on these alienated youth who, in response, would listen with respect. And so an upward cycle of communication began. Eventually we raised money to buy tape recorders

and tried to capture these tales on tape (a commonplace activity now, but quite new in those days), leading to further strengthening of the students' listening and questioning skills. In this case, attending behaviors led to an increased sense of value and respect on both sides.

After three years in British Columbia, my husband and I returned to Ottawa where I worked in a federal government employment equity program. The task centered around providing culturally sensitive counseling and opportunities to members of the Canadian four "equity groups" (visible minorities, Aboriginal peoples, persons with disabilities, and women in nontraditional occupations) to develop the skills needed for career development and success in the Federal Public Service.

As part of my work, I used microskills to train government officials to listen and to really hear and understand the variety of perspectives, strengths, and styles of working found within members of our multicultural workforce. Here, I also learned the importance of language and eventually became reasonably fluent in French, an essential skill for success in bilingual, multicultural Canada. Our team developed a multicultural counseling course, used—all or in part—within 15 universities in Canada, that included many ideas presented here.

The agency where I now work is responsible for animal health, plant protection, and food safety. It is one of the "science departments," involving agriculture, fisheries, scientific research, and many other issues. Given the diversity of my workforce, my first task with my team has been to build a "culture of learning" where vast amounts of information and knowledge can be shared effectively and efficiently. Microskills are a key element in management training, and a good communication skills workshop can be vital in team building. And, of course, all our managers need to listen to and motivate those with whom they serve.

Looking back over my career, it is amazing to find that the basic microskills have been useful in my teaching, counseling, multicultural work, and governmental leadership positions. "Training as treatment" and "teaching competence" in microskills can help us all make a difference throughout our careers.

A few additional key suggestions are presented briefly in this chapter. Consider completing an interview using *no questions* at all, a very different task from that used in brief counseling approaches.

Chapter 14 provided you with the opportunity to engage in brief counseling and the newly popular motivational interviewing. There you may have found that your listening skills and the positive asset search enabled you to conduct a very different type of decision-making interview. Questions, of course, are the central skill of these two models.

Cognitive-behavioral therapy (CBT) and stress management present a very different approach to how clients make decisions about their lives. CBT is rich and full of many alternative strategies to produce change and create the *New* in the client. It is currently the most widely used and best-researched approach to client growth and development. At the same time, CBT tends to give insufficient attention to meaning and values issues in life, although that is currently changing. Note how Dr. Brinkley skillfully used meaning in Chapter 14.

Keep in mind the following key points as you think about the five theoretical/practical methods summarized above:

1. Different theoretical/practical systems use varying patterns of microskills. This could range from many questions in the CBT and brief approach to no questions at all in the person-centered orientation.
2. Theoretical/practical systems tend to focus on different areas of meaning and have different stories they would tell about the interviewing, counseling, and psychotherapeutic process.

   ▲ Decisional counseling focuses on making decisions about practical life issues and problems.
   ▲ Person-centered counseling emphasizes the client and self-actualization.
   ▲ Cognitive-behavioral counseling and therapy focuses on clients' thoughts and behaviors.
   ▲ Brief counseling uses client strengths, assets, and resources to resolve issues.
   ▲ Motivational interviewing appears to combine all of the above in an integrated fashion. It focuses on increasing client motivation for actual change.

We should add that you have learned a fair amount about Victor Frankl and the importance of reflection of meaning. When we consider that he was doing cognitive-behavioral practice *before* the term was invented and that he preceded Rogers in his emphasis on listening to the unique client and finding meaning, it is indeed sad that his theory and practice, which is still very much relevant and alive, seldom appears in our textbooks. However, see Ivey, D'Andrea, Ivey, and Simek-Morgan (2006) for a chapter presenting Frankl's ideas in more detail.

Table 15-2 illustrates once again in summary form how the microskills may be used in different approaches to the interview. The table shows that widely varying styles of helping may be understood and practiced via the microskills system.

Before you review Table 15-2, indicate below your preferred skills—the ones that you have most enjoyed and/or those that you favor and believe that you have mastered most completely. Rate each skill on a 3-point scale with 1 representing the skills you most prefer and would like to use most often; 2, those skills that you would commonly use; and 3, those skills you would prefer to use only occasionally. You can also fill in the circles as noted in the legend below the chart. Then compare your preferred skill pattern with Table 15-2 on the next page. Which theory (or theories) most closely matches your preferred skill pattern? Do the meaning issues that you select relate to the theories presented in that table? Does the result suggest areas that you may want to study further?

	Microskill Lead	Skill Preferences on 3-Point Scale			Fill in the Circles (see legend below)
BASIC LISTENING SKILLS	Open question	1	2	3	◯
	Closed question	1	2	3	◯
	Encourager	1	2	3	◯
	Paraphrase	1	2	3	◯
	Reflection of feeling	1	2	3	◯
	Summarization	1	2	3	◯
INFLUENCING SKILLS	Reflection of meaning	1	2	3	◯
	Interpretation/reframing	1	2	3	◯
	Self-disclosure	1	2	3	◯
	Feedback	1	2	3	◯
	Logical consequences	1	2	3	◯
	Information/psychoeducation	1	2	3	◯
	Directive	1	2	3	◯
	Confrontation (combined skill)	1	2	3	◯
SKILL FOCUS	Client	1	2	3	◯
	Main theme/problem	1	2	3	◯
	Other	1	2	3	◯
	Family	1	2	3	◯
	Mutuality	1	2	3	◯
	Counselor/interviewer	1	2	3	◯
	Cultural/environmental/contextual	1	2	3	◯
MEANING	Which issues of meaning would you most like to address?				
TALK-TIME	What amount of talk-time do you believe is appropriate?		High	Medium	Low
THEORY	Which theory (theories) do you prefer?				

Legend: ● Your favorite skill. ◐ Will use often. ◯ Will use occasionally.

Your Personal Style and Future Theoretical/Practical Integration

There are two major factors to consider as you move toward identifying your own personal style and integrating the many available theories—your own personal authenticity and the needs and style of the client. Unless a skill or theory harmonizes with who you are, it will tend to be false and less effective. It is also obvious that modifying your natural style and theoretical orientation will be necessary if you are to be helpful to many different clients.

In summary, remember that you are unique, and so are those whom you would serve. We all come from varying families, differing communities, and distinct views of gender, ethnic/racial, spiritual, and other multicultural issues.

TABLE 15-2 Microskills patterns of differing approaches to the interview

MICROSKILL LEAD	Decisional counseling	Person-centered	Logotherapy and meaning	Cognitive-behavioral assertiveness training	Brief solution-oriented	Motivational interviewing	Coaching (GROW Model)	Psychodynamic	Gestalt	Rational-emotive behavioral therapy	Feminist therapy	Business problem solving	Medical diagnostic interview	Eclectic/metatheoretical
BASIC LISTENING SKILLS														
Open question	●	○	●	◐	●	●	●●	◐	●	◐	◐	◐	◐	◐
Closed question	◐	○	●	●	◐	◐	◐	○	◐	◐	◐	◐	●	◐
Encourager	●	◐	●	◐	◐	●	●	◐	◐	◐	◐	◐	◐	◐
Paraphrase	●	●	●	◐	◐	●	●	◐	○	◐	◐	◐	◐	◐
Reflection of feeling	●	●	●	◐	◐	●	◐	◐	○	◐	◐	◐	◐	◐
Summarization	◐	◐	◐	◐	●	◐	●	◐	○	◐	◐	●	◐	◐
INFLUENCING SKILLS														
Reflection of meaning	◐	●	●	○	○	◐	◐	◐	◐	◐	●	○	○	◐
Interpretation/reframe	◐	○	◐	○	◐	●	○	●	●	●	◐	◐	◐	◐
Logical consequences	◐	○	◐	◐	◐	●	◐	○	○	●	◐	●	◐	◐
Self-disclosure	◐	◐	◐	○	○	◐	◐	○	○	○	●	○	○	◐
Feedback	◐	◐	◐	◐	◐	●	●	○	◐	◐	◐	◐	○	◐
Information/ psychoeducation	◐	○	○	●	○	◐	◐	○	○	●	◐	●	●	◐
Directive	◐	○	◐	●	○	◐	◐	○	●	●	◐	●	●	◐
CONFRONTATION (Combined Skill)	◐	◐	◐	◐	◐	●	◐	◐	●	●	●	◐	◐	◐
FOCUS														
Client	●	●	●	●	●	●	●	●	●	●	◐	◐	◐	◐
Main theme/problem	●	○	◐	●	●	◐	◐	◐	○	◐	◐	●	●	◐
Others	◐	○	◐	◐	◐	◐	◐	◐	◐	○	◐	○	○	◐
Family	◐	○	◐	◐	◐	◐	○	◐	○	◐	◐	○	○	◐
Mutuality	○	◐	◐	○	◐	○	○	○	○	○	◐	○	○	◐
Counselor/interviewer	○	◐	◐	○	○	○	○	○	○	◐	◐	○	○	◐
Cultural/ environmental context	◐	○	◐	◐	◐	○	◐	○	○	○	●	◐	○	◐
ISSUE OF MEANING (Topics, key words likely to be attended to and reinforced)	Problem solving	Relationship	Discernment	Behavior problem solving	Problem solving	Change	Strengths and goals	Unconscious motivation	Here-and-now behavior	Irrational ideas/logic	Problem as a "women's issue"	Problem solving	Diagnosis of illness	Varies
AMOUNT OF INTERVIEWER TALK-TIME	Medium	Low	Medium	High	Medium	Medium	Medium	Low	High	High	Medium	High	High	Varies

LEGEND

● Frequent use of skill   ◐ Common use of skill   ○ Occasional use of skill

BOX 15-2 YOUR NATURAL STYLE OF INTERVIEWING AND COUNSELING

The following issues are presented for you to consider as you continue to identify your natural style and future theoretical/practical integration of skills and theory.

Goals

What do you want to happen for your clients as a result of their working with you? What would you *desire* for them? How are these goals similar to or different from those of decisional, solution, person-centered, cognitive-behavioral, and motivational interviewing? What else?

Skills and Strategies

You have identified your competence levels in these areas. What do you see as your special strengths? What are some of your needs for further development in the future? What else?

Cultural Intentionality

With what cultural groups and special populations do you feel capable of working? What knowledge do you need to gain in the future? How aware are you of your own multicultural background? What else?

Theoretical/Practical Issues

What theoretical/practical story would you provide now that summarizes how you view the world of interviewing and counseling? Where next would you like to focus your efforts and interests? What else?

If you have presented and analyzed a transcript of an interview as recommended in Chapter 13, you have an excellent beginning for understanding yourself and how clients relate to you. If you are able to engage in the five theoretical/practical methods of this book, you have at least beginning competency in issues of decisions and solutions, meaning-making, and behavioral specifics.

At this point it may be useful to summarize your own story of interviewing and counseling. Box 15-2 asks you to review your goals, your special skills, and your plans for the future. Particularly important are your plans for the future. Where are you going next?

Key points of Chapter 15 are summarized below. They will help you remember main concepts.

■ KEY POINTS

Personal style and awareness of the client	Your arrival at the top of the microskills hierarchy offers you a time to reflect and examine your own style and its appropriateness for the highly diverse clients whom you will meet. This closing chapter suggests that you spend considerable time determining your own natural style. However, clients also have a natural style of their own; you will want to assess and respect their styles as well. You may decide to operate from a single theoretical orientation, or you may decide to offer your clients an array of skills, strategies, and theories. It is critical that you be aware of and respect both yourself and your clients as you develop a natural, integrated style of helping.
Competence with the microskills hierarchy	See Table 15-1 for a summary of the many skills and concepts in this book. Completing this table will help you develop an awareness of your skills and knowledge. Furthermore, the table can provide helpful insights into where you might want to go next in your development of skill competence.

(continued)

KEY POINTS (continued)

Alternative theories of helping	This book is designed to equip you to complete five types of interviews: decisional counseling, person-centered counseling, cognitive-behavioral counseling, brief solution-focused counseling, and motivational interviewing. If you complete the practice exercises, you will be able to conduct an interview using all five systems. Furthermore, with an understanding of skills and the decisional structure, you will be able to engage in other theories more directly and analyze how they function for client benefit. This chapter has also specifically shown how the basic listening sequence might be used differently with the same client, depending on the theoretical orientation of the counselor. The stories and issues each one emphasizes lead the client in varying directions. Multiple theories of interviewing and counseling can be viewed through the lens of the five-stage interview structure of the *relationship—story and strengths—goals—restory—action* model.

MICROSKILLS PRACTICE, SUPERVISION, AND LIFETIME GROWTH

The microskills framework was developed to clarify and ease the transition from the classroom to actual interviewing practice. Once we have learned the basic skills, we are well prepared for improving our counseling and psychotherapy work throughout our professional lives. The interview is a place where we all can improve. We hope you will find that the openness and specificity of the microskill practice sessions will continue as you move on beyond this text to your next steps.

When you move to field placements and internships in schools and community agencies, supervision of your sessions will be central to your learning. There you will share your work with professionals who can help and guide you toward further competence. Here, in your microskills practice sessions, you actually will be engaging in minisupervision sessions. Your classmates will provide you with feedback, on both your strengths and areas where you might develop further. Microskills practice with "microsupervision" will provide a preparatory experience in supervision as you develop a habit of sharing your interviewing work with others in an open atmosphere.

Supervision and openness are so important that they should be "introduced in the first class students take and then be intertwined throughout the curriculum by all faculty" (Miller & Dollarhide, in press). Microskills provides a vocabulary and system through which you can identify what you are doing and its effectiveness. Microskills supervision provides a comprehensive framework that can be used in many models of supervision ranging from person-centered through multicultural and psychodynamic (Daniels, Rigazio-DiGilio, & Ivey, 1997; Russell-Chapin & Ivey, 2004).

SUMMARY—AS WE END: THANKS, FAREWELL, AND GOOD LUCK!

We have come to the end of this phase of your interviewing and counseling journey. You have had the chance to learn the foundation skills and how they are structured in a variety of theoretical/practical interviews. Skills that once seemed awkward and unfamiliar are often now automatic and natural. As in the Samurai effect, you do not need to think of them constantly. The basic listening sequence, in particular, is likely part of your being at this point. Moreover, expect that the *relationship—story and strengths—goals—restory—action* framework will become part of your practice, regardless of your final theoretical orientation.

The next steps are yours. Many of you will be moving on to individual theories of counseling, exploring issues of family counseling and therapy, becoming involved in the community, and learning the many aspects of professional practice. Others may find this presentation sufficient for their purposes. We have designed this book as a clear summary of the basics; a naturally skilled person can use the information here for many effective and useful helping sessions.

We have selected a few key books that will help you follow up on ideas from our presentation. The books will take you in very different directions, but we have found them all helpful. Select one or two favorites to start and then expand from there. Enjoy delving more deeply into our exciting field.

We have enjoyed sharing this time with you. We have come a long way together, and we appreciate your patience. Many of the ideas for presentation in this book have come from students. We hope you will take a moment to provide us with your feedback and suggestions for the future. Again, e-mail us at info@emicrotraining.com and say hello. These pages will be constantly updated with new ideas and information. You have joined a never-ending time of growth and development.

The relationship is forever. . . . Find joy in helping.

—Benjamin Zander

Welcome to the field of interviewing and counseling! You are the key to the future.

Allen, Mary, and Carlos

SUGGESTED SUPPLEMENTARY READINGS

The literature of the field is extensive, and you will want to sample it on your own. We would like to share a few specific books that we find helpful as next steps to follow up ideas presented here. All of the materials build on the concepts of this book, but we have recommended some books that take different perspectives from our own.

Microskills

www.emicrotraining.com
   Visit this Web site for up-to-date information on microcounseling, microskills, and multicultural counseling and therapy. You likely will enjoy the many interviews with leaders of the field. There are links to professional associations, ethics codes, and many multicultural and professional sites.
Daniels, T., & Ivey, A. (2007). *Microcounseling* (3rd ed.). Springfield, IL: Thomas.
   The theoretical and research background of microskills is presented here in detail.
Evans, D., Hearn, M., Uhlemann, M., & Ivey, A. (2007). *Essential interviewing* (7th ed.). Belmont, CA: Brooks/Cole.
   Microskills in a programmed text format.
Ivey, A., Gluckstern, N., & Ivey, M. (1997). *Basic influencing skills* (3rd ed.). Hanover, MA: Microtraining Associates.
   More data on the influencing skills. Supporting videotapes are available.

Ivey, A., Gluckstern, N., & Ivey, M. (2005). *Basic attending skills* (4th ed.). Hanover, MA: Microtraining Associates.

Perhaps the most suitable book for beginners and those who would teach others microskills. Supporting videotapes are available (www.emicrotraining.com).

Zalaquett, C., Ivey, A., Gluckstern, N., & Ivey, M. (2007). *Las habilidades atencionales básicas: Pilares fundamentales de la comunicación efectiva.* Hanover, MA: Microtraining Associates. (Spanish edition of *Basic Attending Skills*)

Theories of Interviewing and Counseling With a Multicultural Orientation

Ivey, A., D'Andrea, M., Ivey, M., & Simek-Morgan, L. (2006). *Counseling and psychotherapy: A multicultural perspective* (6th ed.). Boston: Allyn and Bacon.

The major theories are reviewed, with special attention to multicultural issues. Includes many applied exercises to take theory into practice.

Thomas, R. (2000). *Multicultural counseling and human development theories: 25 theoretical perspectives.* Springfield, IL: Charles C Thomas.

Multiple orientations are presented in a comprehensive fashion.

Suggestions for Follow-Up on Specific Theories

Decisional Counseling

Brammer, L., & McDonald, G. (2004). *The helping relationship: Process and skills* (8th ed.). Boston: Allyn and Bacon.

An articulate discussion of skills and the helping process with a strong decisional flavor.

D'Zurilla, T., & Nezu, A. (2006). *Problem-solving therapy.* New York: Springer.

An entire counseling model derived from decision making.

Parsons, F. (1967). *Choosing a vocation.* New York: Agathon. (Originally published 1909)

It is well worth a trip to your library to read Parsons's work. You will find that much of his thinking is still up to date and relevant.

Brief Interviewing and Counseling

Connell, B. (2005). *Solution-focused therapy.* Beverly Hills, CA: Sage.

The basics of brief counseling in brief form.

Sklare, G. (2004). *Brief counseling that works* (2nd ed.). Beverly Hills, CA: Corwin.

School-focused and clear. Also available—videotape of a real interview with a child. (www.emicrotraining.com)

Person-Centered Counseling and Humanistic Theory

Frankl, V. (1959). *Man's search for meaning.* New York: Pocket Books. (Originally published 1946)

This is one of the most memorable books you will ever read. It describes Frankl's survival in German concentration camps through finding personal meaning. We recommend suggesting this book to your clients; it is an excellent counseling and therapeutic tool in itself, particularly for those in crisis.

Rogers, C. (1961). *On becoming a person.* Boston: Houghton Mifflin.

The classic book by the originator of person-centered counseling.

Cognitive-Behavioral Therapy

Alberti, R., & Emmons, M. (2008). *Your perfect right: A guide to assertiveness training* (9th ed.). San Luis Obispo, CA: Impact. (Originally published 1970)

The classic book by the originators.

Beck, J. S. (2005). *Cognitive therapy for challenging problems: What to do when the basics don't work.* New York: Guilford Press.

Davis, M., Eshelman, E., & McKay, M. (2008). *The relaxation and stress reduction workbook* (6th ed.). Oakland, CA: New Harbinger.

This is clear, direct, and easily translatable into microskills approaches to the interview.

Dobson, K. (Ed.). (2002). *Handbook of cognitive-behavioral therapies* (2nd ed.). New York: Guilford.

A comprehensive and specific presentation of key skills, strategies, and theories.

Wolpe, J., & Lazarus, A. (1966). *Behavior therapy techniques.* New York: Pergamon.

Old but in many ways still the best and worth searching for in the library.

Multicultural Counseling and Therapy

Fouad, N., & Arredondo, P. (2008). *Becoming culturally oriented: Practical advice for psychologists and educators.* Washington, DC: American Psychological Association.

The most comprehensive coverage of the necessary skills and competencies in the multicultural area.

Sue, D. W., Ivey, A., & Pedersen, P. (1999). *A theory of multicultural counseling and therapy.* Pacific Grove, CA: Brooks/Cole.

The first general theory of MCT with many implications for practice.

Sue, D. W., & Sue, D. (2008). *Counseling the culturally diverse* (5th ed.). New York: Wiley.

The classic of the field. This book helped launch a movement.

Integrative/Eclectic Orientations

Ivey, A. (2000/1986). *Developmental therapy.* Hanover, MA: Microtraining.

Two books that offer skill integration as well as a developmental theory. Useful follow-up from microskills training, particularly in the focus on sensorimotor, concrete, formal, and dialectic/systemic strategies.

Ivey, A., Ivey, M., Myers, J., & Sweeney, T. (2005). *Developmental counseling and therapy: Promoting wellness over the lifespan.* Boston: Lahaska/Houghton-Mifflin.

Lazarus, A. (2006). *The multimodal way.* New York: Springer.

The basic model for eclectic multimodal therapy. You will find the BASIC-ID model useful in conceptualizing broad treatment plans.

Psychodynamic Theory

Bowlby, J. (1988). *A secure base: Parent–child attachment and healthy human development.* New York: Basic Books.

Bowlby has become the most researched figure in the psychodynamic framework. His work on human attachment will continue to be the foundation for psychodynamic practice in the near future.

Miller, A. (1981). *The drama of the gifted child.* New York: Basic Books.

One route to understanding psychodynamic theory is looking at your own experience. This book reads well and can make a difference in the way you think about yourself as a counselor. Our students often thank us for referring them to Alice Miller.

The Ivey Taxonomy: Definitions and Predicted Results

Skill, Concept, or Strategy	Predicted Result When Using Skill, Concept, or Strategy
Ethics Observe and follow professional standards and practice ethically. Particularly important issues for beginning interviewers are *competence, informed consent, confidentiality, power,* and *social justice.*	Client trust and understanding of the interviewing process will increase. Clients will feel empowered in a more egalitarian session. When you work toward social justice, you contribute to problem prevention in addition to healing work in the interview.
Multicultural Competence Base interviewer behavior on an ethical approach with an awareness of the many issues of diversity. Include the multiple dimensions from the RESPECTFUL model (Chapter 2).	Anticipate that both you and your clients will appreciate, gain respect, and learn from increasing knowledge in ethics and multicultural competence. You, the interviewer, will have a solid foundation for a lifetime of personal and professional growth.
Wellness Help clients discover and rediscover their strengths through wellness assessment. Find strengths and positive assets in the clients and in their support system. Identify multiple dimensions of wellness.	Clients who are aware of their strengths and resources can face their difficulties and discuss problem resolution from a positive foundation.
Attending Behavior Support your client with individually and culturally appropriate visuals, vocal quality, verbal tracking, and body language.	Clients will talk more freely and respond openly, particularly around topics to which attention is given. Depending on the individual client and culture, anticipate fewer eye contact breaks, a smoother vocal tone, a more complete story (with fewer topic jumps), and a more comfortable body language.
Open and Closed Questions Begin open questions with often useful *who, what, when, where,* and *why.* Closed questions may start with *do, is,* or *are. Could, can,* or *would* questions are considered open but have the additional advantage of being somewhat closed, thus giving more power to the client, who can more easily say that he or she doesn't want to respond.	Clients will give more detail and talk more in response to open questions. Closed questions provide specific information but may close off client talk. Effective questions encourage more focused client conversations with more pertinent detail and less wandering. *Could, would,* and *can* questions are often the most open of all.

Skill, Concept, or Strategy	Predicted Result When Using Skill, Concept, or Strategy
Client Observation Skills Observe your own and the client's verbal and nonverbal behavior. Anticipate individual and multicultural differences in nonverbal and verbal behavior. Carefully and selectively feed back observations to the client as topics for discussion.	Observations provide specific data validating or invalidating what is happening in the session and provide guidance for use of various microskills and strategies. The smoothly flowing interview will often demonstrate movement symmetry or complementarity. Movement dissynchrony provides a clear clue that you are not "in tune" with the client.
Encouraging Encourage with short responses that help clients keep talking. They may be verbal (repeating key words and short statements) or nonverbal (head nods and smiling).	Clients will elaborate on the topic, particularly when encouragers and restatements are used in a questioning tone of voice.
Paraphrasing Shorten, clarify the essence of what has just been said, but be sure to use the client's main words when you paraphrase. Paraphrases are often fed back to the client in a questioning tone of voice.	Clients will feel heard. They will tend to give more detail without repeating the exact same story. If a paraphrase is inaccurate, clients will have an opportunity to correct the interviewer.
Summarizing Summarize client comments and integrate thoughts, emotions, and behaviors. Similar to paraphrase, but used over a longer time span.	Clients will feel heard and often learn how the many parts of important stories are integrated. The summary tends to facilitate a more centered and focused discussion. The summary also provides a more coherent transition from one topic to the next or as a way to begin and end a full session.
Empathic Response Experience the client's world as if you were the client. Understand his or her key issues and feed them back to clarify experience. This requires attending skills and using the important key words of the client but distilling and shortening the main ideas. Empathy is best assessed by the client's reaction to a statement.	Clients will feel understood and engage in more depth in exploring their issues.
Subtractive Empathy Interviewer responses give back to the client less than what the client says and perhaps even distort what has been said. In this case, the listening or influencing skills are used inappropriately.	Skill is used inappropriately and subtracts from client's experience. Clients will not feel understood.
Basic Empathy Interviewer responses are roughly interchangeable with those of the client. The interviewer is able to say back accurately what the client has said.	Clients will feel understood and engage in more depth in exploring their issues. Skilled intentional competence with the basic listening sequence demonstrates basic empathy.

(continued)

Skill, Concept, or Strategy	Predicted Result When Using Skill, Concept, or Strategy
Additive Empathy Interviewer adds meaning and feelings beyond those originally expressed by the client.	Clients will reach a better understanding of their own issues and engage in more depth in exploring these issues.
Reflection of Feeling Identify the key emotions of a client and feed them back to clarify affective experience. With some clients, the brief acknowledgment of feeling may be more appropriate. Often combined with paraphrasing and summarizing.	Clients will experience and understand their emotional state more fully and talk in more depth about feelings. They may correct the interviewer's reflection with a more accurate descriptor.
Basic Listening Sequence Select and practice all elements of the basic listening sequence—open and closed questions, encouraging, paraphrasing, reflection of feeling, and summarization. These are supplemented by attending behavior and client observation skills.	Clients will discuss their stories, problems, or concerns, including the key facts, thoughts, feelings, and behaviors. Clients will feel that their stories have been heard.

The Five Stages/Dimensions of the Well-Formed Interview

1. *Relationship:* Initiate the session. Rapport and structuring. "Hello, what would you like to talk about today?"	Clients will feel at ease with an understanding of the key ethical issues and the purpose of the interview. They may also know you more completely as a person and professional.
2. *Story and Strengths:* Gather data, draw out client stories, concerns, problems, or issues. "What's your concern?" "What are your strengths and resources?"	Clients will share thoughts, feelings, behaviors, and their stories in detail as well as strengths and resources.
3. *Goals:* Set goals mutually. "What do you want to happen?"	Clients will discuss and define goals, new ways of thinking, desired feeling states, and desired behavior changes. They may learn how to live more effectively with situations that cannot be changed (rape, death, an accident, an illness).
4. *Restory:* Explore and create, brainstorm and examine alternatives, confront client incongruities and conflict, restory. "What are we going to do about it?" "Can we generate new ways of thinking, feeling, and behaving?"	Clients may reexamine individual goals in new ways, solve problems from at least three generated alternatives, and start the move toward new stories and actions.
5. *Action:* Conclude. Plan for generalizing interview learning to "real life" and eventual termination of the interview or series of sessions. ("Will you do it?")	Clients will demonstrate change in behavior, thoughts, and feelings in daily life outside of the interview.

Skill, Concept, or Strategy	Predicted Result When Using Skill, Concept, or Strategy
Confrontation Supportively challenge the client: 1. Listen, observe, and note client conflict, mixed messages, and discrepancies in verbal and nonverbal behavior. 2. Point out internal and external discrepancies by feeding them back to the client, usually through the listening skills. 3. Evaluate how the client responds and whether it leads to client movement or change. If the client does not change, the interviewer flexes intentionally and tries another skill.	Clients will respond to the confrontation of discrepancies and conflict with new ideas, thoughts, feelings, and behaviors, and these will be measurable on the five-point Client Change Scale. If the client does not change, the interviewer flexes intentionally and tries another skill.
Focusing Use selective attention and focus the interview on the client, problem/concern, significant others (partner/spouse, family, friends), a mutual "we" focus, the interviewer, or the cultural/environmental/contextual (RESPECTFUL multicultural background, community, nation). You may also focus on what is going on in the here and now of the interview.	Clients will focus their conversation or story on the dimensions selected by the interviewer. As the interviewer brings in new focuses, the story is elaborated from multiple perspectives.
Reflection of Meaning Meanings are close to core experiencing. Encourage clients to explore their own meanings and values in more depth from their own perspective. Questions to elicit meaning are often a vital first step. A reflection of meaning looks very much like a paraphrase, but focuses beyond what the client says. Often the words "meaning," "values," "vision," and "goals" appear in the discussion.	Clients will discuss stories, issues, and concerns in more depth with a special emphasis on deeper meanings, values, and understandings. Clients may be enabled to discern their life goals and vision for the future.
Interpretation/Reframing Provide the client with a new perspective, frame of reference, or way of thinking about issues. Interpretations/reframes may come from your observations, they may be based on varying theoretical orientations to the helping field, or they may link critical ideas together.	Clients may find another perspective or meaning of a story, issue, or problem. Their new perspective could have been generated by a theory used by the interviewer, from linking ideas or information, or by simply looking at the situation afresh.
Self-Disclosure As the interviewer, share your own related personal life experience, here-and-now observations or feelings toward the client, or opinions about the future. Self-disclosure often starts with an "I" statement. Here-and-now feelings toward the client can be powerful and should be used carefully.	Clients may be encouraged to self-disclose in more depth and may develop a more egalitarian interviewing relationship with the interviewer. They may feel more comfortable in the relationship and find a new solution relating to the counselor's self-disclosure.

(continued)

Skill, Concept, or Strategy	Predicted Result When Using Skill, Concept, or Strategy
Feedback Present clients with clear information on how the interviewer believes they are thinking, feeling, or behaving and how significant others may view them or their performance.	Clients will improve or change their thoughts, feelings, and behaviors based on the interviewer's feedback.
Logical Consequences Explore specific alternatives and the logical positive and negative concrete consequence of each possibility with the client. "If you do this . . . , then . . ."	Clients will change thoughts, feelings, and behaviors through better anticipation of the consequences of their actions. When you explore the positives and negatives of each possibility, clients will be more involved in the process of decision making.
Information and Psychoeducation Share specific information with the client (e.g., career information, choice of major, where to go for community assistance and services). Offer advice or opinions on how to resolve issues and provide useful suggestions for personal change. Teach the client specifics that may be useful—helping them develop a wellness plan, teaching them how to use microskills in interpersonal relationships, educating them on multicultural issues and discrimination.	If information and ideas are given sparingly and effectively, clients will use them to act in new, more positive ways. Psychoeducation that is provided in a timely way and involves clients in the process can be a powerful motivator for change.
Directives Direct clients to follow specific actions. Directives are important in broader strategies such as assertiveness or social skills training or specific exercises such imagery, thought-stopping, journaling, or relaxation training. They are often important when assigning homework for the client.	Clients will make positive progress when they listen to and follow the directives and engage in new, more positive thinking, feeling, or behaving.
Skill Integration Integrate the microskills into a well formed interview and generalize the skills to situations beyond the training session or classroom.	Developing interviewers and counselors will integrate skills as part of their natural style. Each of us will vary in our choices, but increasingly we will know what we are doing, how to flex when what we are doing is ineffective, and what to expect in the interview as a result of our efforts.
Determining Personal Style and Theory As you work with clients, identify your natural style, add to it, and think through your approach to interviewing and counseling. Examine your own preferred skill usage and what you do in the session. Integrate learning from theory and practice in interviewing, counseling, and psychotherapy into your own skill set.	As a developing interviewer or counselor, you will identify and build on your natural style. You will commit to a lifelong process of constantly learning about theory and practice while evaluating and examining your behavior, thoughts, feelings, and deeply held meanings.

Counseling, Neuroscience, and Microskills

Experiences, thoughts, actions, and emotions actually change the structure of our brains. . . . Indeed, once we understand how the brain develops, we can train our brains for health, vibrancy, and longevity.

—J. Ratey

Interviewing, counseling, and psychotherapy can build new brain networks. You are entering our field at what likely will be its most exciting and productive time. In a few years, you will have knowledge and experience that makes these early words on neuroscience and its implications for counseling but a mere beginning. The bridge between biological and psychological processes is erasing the old distinction between mind and body, between mind and brain—*the mind is the brain.* We believe it is time to embrace a broader view that includes counseling and psychotherapy, neuroscience, molecular biology, and neuroimaging. We issue a call for action regarding the integration of these disciplines and the infusion of knowledge from such an integrated approach to practice, training, and research.

You likely have noticed frequent stories on television and in the popular media on brain research and its implications for the future. This research has reached a state of precision where it now has immediate meaning for you as an interviewer or counselor. Neuropsychology can be defined as "the study of relations between brain function and behavior" (Kolb & Whishaw, 2003, p. G16). Neuroscience should be considered the more basic and broader science. Brain research, including the cognitive sciences, has specific implications for your practice. Each microskill, used effectively, makes a difference. Add to this the wellness and positive asset search and the multiple strategies of varying theoretical approaches; all of these will give you an increasingly useful and relevant approach to interviewing, counseling, and psychotherapy.

Neuroscience and neuroimaging are lending strong support to our work in counseling and psychotherapy. As early as 1989, Kandel argued that because psychotherapy involves learning new ways of functioning, structural changes occurring in client brains would soon be detectable by neuroimaging machines that identify specifically what is going on inside the brain (Kandel, 2007). Making this prediction come true today are positron emission tomography (PET) scans and functional magnetic resonance imaging (fMRI). These highly sophisticated methods have found that cognitive and interpersonal therapy can change the structure of the brain (Brody et al., 2001; Goldapple et al., 2004; Martin et al., 2001). Clients, with the help of counseling (or medication at times) are capable of functionally "rewiring" the brain (Paquette et al., 2003).

NEUROSCIENCE RESEARCH AND INTERVIEWING, COUNSELING, AND PSYCHOTHERAPY: MODELING INTERDISCIPLINARY INTEGRATION

The whole brain is greater than the sum of its parts, and the brain is a constantly interacting system within itself and in relation to the cultural/environmental/contextual (CEC). Each component affects the total system of the holistic brain. Knowledge of the brain and awareness of the constant new knowledge being developed will lead you to more effective intentional interviewing and counseling. Of necessity, the following discussion breaks down the brain into specific parts, which are important for you to know, if you are to communicate with other professionals in the near future. Two important issues as you read this appendix:

1. This book presents a basic beginning to a very complex area of study. It is the first book to discuss the implications of a skills approach to neuroscience, interviewing, and counseling. This is only a beginning. A lot more is coming. Some key books and an audiotape for further study will follow this presentation.
2. *Again, the brain is holistic, and each part affects the others.* Of necessity, often discussion of a single part of the brain is presented. Always recall that this is an overview of complex interacting parts and that considerable further study is needed to complete the picture.

The key term for this new future is *neuroplasticity.* Simply put, the brain can change—it is not fixed, and it responds to external environmental events and actions initiated by the individual. The old idea that the brain does not change is simply wrong. Neuroplasticity means that even in old age, new neurons, new connections, and new neural networks are born and can continue development. "Neuroplasticity can result in the wholesale remodeling of neural networks. . . . A brain can rewire itself" (Schwartz & Begley, 2002, p. 16).

Particularly fascinating is *neurogenesis*, the development of completely new neurons, even in the aged. For example, there is evidence in adults that this occurs in the *hippocampus,* the main seat of memory (Siegel, 2007). This is where effective counseling can affect important change and new neural connections, but there are many areas of the brain where neurogenesis occurs. We develop new neural networks throughout the lifespan in response to new situations or experiences in the environment. Exercise is particularly important as a lifetime process to ensure brain and physical health (Ratey, 2008). Exercise increases blood flow and the release of positive neurotransmitters such as serotonin. Many of you reading this have experienced the serotonin "high" of running or other physical activity. This positive high through exercise needs to be part of your treatment regime. It is particularly helpful for depression due to serotonin release. If you are sad—walk or run! If you can't run, meditate and use relaxation training.

SOME BASIC BRAIN STRUCTURE

There are some key aspects of the brain that will enable you to understand and converse with physicians, neuropsychologists, and others that you will encounter in your career.

The frontal lobe (see Figure AII-1) is associated with executive functioning, abstract reasoning, and decision making. It is also the focus of much of motor behavior and the attentional processes. The *prefrontal cortex* is at the very front of this lobe. Clients with frontal lobe issues may show language problems, personality changes, apathy, or inability to plan.

The *parietal lobe* gives us our spatial sense, but it also serves as a critical link from the senses (see/hear/feel/taste/touch) to our motor abilities. The *temporal lobe* works with

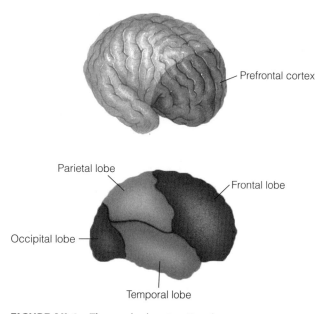

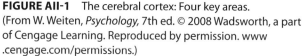

FIGURE AII-1   The cerebral cortex: Four key areas.
(From W. Weiten, *Psychology*, 7th ed. © 2008 Wadsworth, a part of Cengage Learning. Reproduced by permission. www .cengage.com/permissions.)

auditory processing and memory whereas the *occipital lobe* is for visual processing. You may observe clients with challenges in these areas.

At the base of the brain, the *cerebellum* is particularly important for motor coordination. It also has a role in several cognitive functions, including attention, language processing, and the sensory modalities. The *brain stem* connects to the spinal cord and is an important conduit for integrating the whole brain. It is critical for cardiovascular function and respiration, attention, and consciousness.

The Limbic System: The Social Brain

The limbic system is of prime importance to us as interviewers, counselors, and therapists as it helps us understand issues in emotion and memory (see Figure AII-2). The amygdala is recognized as the central emotional area, particularly concerning the negative emotions of fear, anger, sadness, and disgust. But it is also the energizer of emotive strength. Drawing information from other parts of the brain, the amygdala signals intensity.

Each emotional area appears to have its own set of connections in the brain. Studies on fear and gladness (positive memories, interactions, situations) have implications for clinical and counseling practice. At this point, data on mad (anger) and sad (depression) emotions have more limited practical implications for clinical practice. But all negative emotions seem to be affected by skilled counseling and therapy using a positive approach. Following are some early studies concerning location of emotions in the brain.

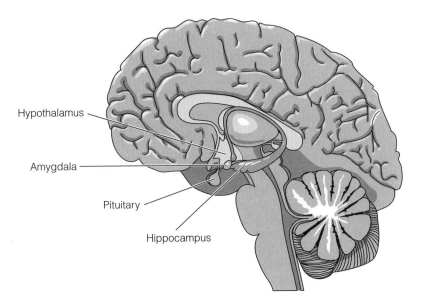

Hypothalamus

Amygdala

Pituitary

Hippocampus

FIGURE AII-2   The limbic system: The social brain.

▲ *Fear.* The amygdala is central in fear and other negative emotions. If you walk down a dark street and sense someone following you, expect the amygdala to activate. Awareness of fear was and is the most essential and basic emotion for our survival as individuals. The importance of the amygdala and its relationship to fear is one of the earlier findings of neuroscience (for example, see the study by Selden et al., 1991). Memory of fear is located in the hippocampus and can couple rapidly with the amygdala to bring out fright feelings. The right brain has connections to both the hippocampus and the amygdala and transfers information instantaneously to the decision-making left brain. But if time is short, the right brain takes over. Thus, notice how, when you are in danger, your heart races, your breathing speeds up, and your whole body takes over to deal with the situation.

Building on this foundation, recent neuroscience research has offered exciting findings. For example, it is now possible to identify specific neurons in the amygdala that affect anxiety, depression, and posttraumatic stress. These harmful networks remain in place unless treated effectively. Using classic behavioral methods derived from Pavlov's work, it is possible to change the power of these neurons "through presenting the feared object in the absence of danger." Medications will be able to do the same thing as counseling and therapy if targeted to specific "intercalated neurons" (Ekaterina et al., 2008).

In short, we have now some of the most concrete evidence that counseling and therapy focusing on strengths can enable a generation of new positive neural networks, right to the cellular level. This is accomplished through building strong, positive neural networks through many of the methods discussed in this book.

▲ *Sad.* In studies of brain scans of depressed patients, sadness produced altered activation in more than 70 different brain regions, particularly the amygdala, hippocampus, and prefrontal cortex (Freed & Mann, 2007).

▲ *Mad.* Depressed and angry patients were found to show lesser activity in the left ventromedial prefrontal cortex (where executive control is found) and the left amygdala as compared to depressed patients who did not show anger. One-third of depressed clients tend to show anger. Control subjects showed more activity in the prefrontal cortex and perhaps more ability to control anger (Dougherty & Fava, cited by Arehart-Treichell, 2004).

▲ *Glad.* Many areas of the brain are activated by positive emotions. The prefrontal cortex and the hippocampus are obviously important, but the nucleus accumbens sends out signals to the prefrontal cortex, making it possible to focus on the positives (Ratey, 2008b).

The *hippocampus* is our memory "organ" and works closely with the cerebral cortex. Energy from the amygdala tells the hippocampus which information should be remembered. When there is not enough interest or energy, no memory is produced. In contrast, a highly stressful event (war, rape, and so on) can overwhelm the whole system like a lightning bolt and result in destruction of neurons and distressed memory. New, negative neural networks take over. The research discussed above shows us the importance of wellness and positive assets as we seek to develop and strengthen positive memories in the hippocampus. Again, *effective interviewing and counseling can affect the brain in positive ways.*

Think of the *hypothalamus* as a "switching station" in which messages from inside and outside are transferred; it controls hormones that affect sex, hunger, sleep, aggression, and other biological factors. The *pituitary* is a "control" gland that relates to the hypothalamus. It also influences growth and blood pressure, sexual functioning, the thyroid, and metabolism.

The *medial prefrontal cortex* is another important part of the social brain, for our "mirror neurons" are primarily located here. For both self-observation and empathy/understanding of others, this section is vital. "It links the body, brain stem, limbic, cortical, and social processes into one functional whole" (Ratey, 2008, p. 278). You can get a better sense of how mirror neurons work if you notice what occurs in your body when you see an exciting ballgame or an involving movie. Many of us find ourselves tensing up and/or clenching our fists in close or exciting situations. We may even sway as the pass receiver grabs the ball and heads down the field. In good movies that touch you emotionally in some way, the same thing happens. Your heart rate goes up, and you may duck a swing from the villain as you sit on the edge of your seat.

There is a parallel to the above, which happens in a smaller way in the interviewing room—your body and the client's body react to each other in the here-and-now immediacy of the session. In addition, clients are often triggered to react in the interview as they have in the past as they tell and reenact their stories. Your observation skills will be helpful here as you get an upclose and personal view of what happened. But you, too, will be impacted by the client, and recent or old stories that you have lived may be awakened in you. Those feelings in your body can be a deeper clue as to what the client is experiencing—or worse, it may mean that you are completely off track in your own personal history. Melanie Klein, the famed psychoanalyst, termed these events *projective identification.* The activities of one individual project unconsciously to the other person's mirror neurons, and we literally do feel the other person's feelings.

The nucleus accumbens (not shown in the figure) is related to sexual functioning and the "high" from certain recreational drugs. It is particularly responsive to marijuana, alcohol, and related chemicals and thus is key in addiction. It provides a partial explanation to what we are working with in sex addition, as well, and speculatively, may relate to those who engage in stalking behavior. When you use motivational interviewing to help an addicted client, you are working against some very powerful parts of the brain. One of our great challenges is helping these clients examine and rewrite their stories and find new actions through healthy alternative highs to replace the strengths of addiction. When you find these clients developing new life satisfactions and interests (wellness), you are influencing them toward behavior that can result in new positive responses in the nucleus accumbens and other parts of their brain.

The distinction between left and right hemispheres is relatively well known, but the facts and issues beyond this generalization need examination. It is clear that the two sides work together, and their differences and similarities go beyond the common generalization of the linear (and somewhat boring) left brain and the intuitive and more interesting and fun intuitive right brain. In addition, the executive left brain is associated with positive emotions, while the right is more associated with the negative emotions—sad, mad, and fear.

Given the fact of real complexity, the left side controls operations that are more linear, cognitive, and logical, while the right side functions are considered intuitive, nonverbal, spatial, and more spontaneous and impulsive. (Keep in mind that left-handed people do have a different pattern.) The corpus callosum connects the two brain hemispheres and enables us to be human. Creativity and the production of the *New* often seem to appear to occur when the two hemispheres work in synchrony. However, it is the executive left brain that ultimately puts things together.

As mentioned, positive emotions are also more associated with the left hemisphere. The more primitive amygdala and right brain have greater control of the negative emotions and can work automatically beyond consciousness to wake us to objects that surprise us or we fear. Early humans needed to be prepared for constant danger; and thus, biologically, these negative emotions are deeper in the brain. The more positive emotions support human connections and working peacefully together.

NEURAL BASICS AND IMPLICATIONS FOR INTERVIEWING AND COUNSELING

The therapeutic relationship has the capacity to change the brain—and this can for be for good or ill, as a damaging session can have lifelong impact. The interactive and skill-oriented approach of the microskills is closely associated with human development in many positive ways. Add to this the impact of our theories and strategies of counseling and therapy, and we have a real chance to make a difference. We can help clients not only with current issues, but also provide long-lasting lessons in relationship and understandings. In this section, we will explore fundamental aspects of the neuron, the neural net, and neurotransmitters, each of which interviewing, counseling, and therapy can impact in significant ways. Again, keep in mind that this is an overview of a very complex interactive system, and many important details are not discussed.

We begin with the *neuron,* our main brain cell (see Figure AII-3). Neurons fire when we have any type of experience or stimulus, including the interview. The neuron and neural net is where learning from counseling and therapy ultimately takes place. You can have a large influence on the developing brain through neuroplasticity and your interviewing skills and strategies. For substantial learning and maintenance of information, *neural networks* are developed. "Neurons that fire together wire together." This means that enough neurons have gathered together in a network to produce significant change. In our language, we call that *learning* or *change.* And with effective therapy, new neuronal networks are developed. We also can call this *development* and seek to measure it on the Client Change Scale. As a person

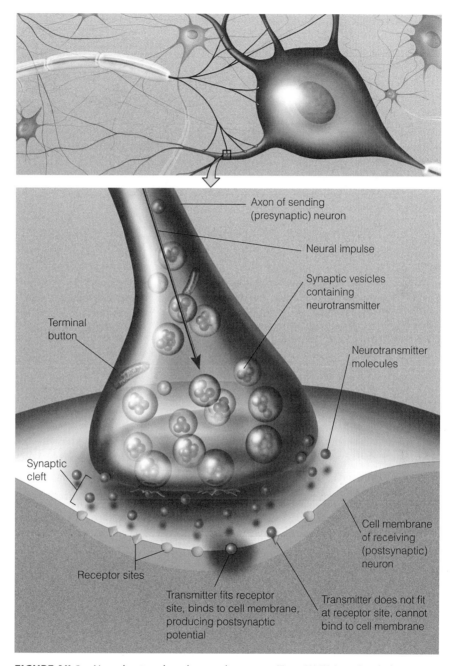

Axon of sending
(presynaptic) neuron

Neural impulse

Synaptic vesicles
containing
neurotransmitter

Neurotransmitter
molecules

Terminal
button

Synaptic
cleft

Cell membrane
of receiving
(postsynaptic)
neuron

Receptor sites

Transmitter fits receptor
site, binds to cell membrane,
producing postsynaptic
potential

Transmitter does not fit
at receptor site, cannot
bind to cell membrane

FIGURE AII-3   Neural network and example neuron. (From W. Weiten, *Psychology*, 7th ed. © 2008 Wadsworth, a part of Cengage Learning. Reproduced by permission. www.cengage.com/permissions.)

learns, we can now see such change in brain scans such as the PET and fMRI. Eventually, scans may become key diagnostic instruments and even show that your work has actually impacted specific areas of the brain.

TABLE AII-1   Neurotransmitters and possible treatment strategies

Neurotransmitter	Possible Impact Through the Interview
Glutamate—most important brain excitatory neurotransmitter, vital to neuroplasticity, movement, memory, and learning. Moderates neural firing. Monosodium glutamate (MSG) chemically close to glutamate and can cause problems for some.	Generally, we want to increase this central neurotransmitter. Stress management, wellness, and exercise are important. Preliminary evidence of glutamate abnormalities in schizophrenia. (Example Meds—glutamate uptake inhibitors)
Dopamine—attentional processes, pleasure, memory, reward system, fine motor movement. Addictive substances increase release. Low dopamine common in depression.	Stories and strengths and positive narratives should help dopamine production. All effective restorying should improve dopamine release as we move away from depression and ineffective behavior. (Meds—dopamine reuptake inhibitors [NDRIs] act as antidepressants)
Serotonin—vital to mood, sleep, anxiety control, and self-esteem. Implicated in depression, impulsiveness, and anger/aggression.	Think of the serotonin "high" of running. Get clients moving. It is hard to be depressed when one is exercising. Wellness, meditation, cognitive behavioral counseling, and finding clear visions and meaning for life should be helpful. Positive restorying and action following the interview is important. (Meds—SSRIs permit more intake.)
Norephinephrine (also known as adrenaline)—released immediately in stress, but also makes one sharper. Involved in heart rate and helps new memory transfer to long-term memory. Too much, damaging cortisol is released. Related to anxiety, depression, and bipolar diagnosis.	Again, get clients active and moving. But when needed, use stress management, decisional counseling, CBT, and so on to lessen stress. As always, telling one's story in a relationship of caring is ultimately calming. People can become addicted to an adrenaline high—you may have seen this in runners and even those overinvolved and excited at work. Finding meaning should help clients meet the challenges of life more effectively. (Meds—SSRIs and sometimes coupled with dopamine as an antidepressant)
GABA (gamma-aminobutyric acid)—inhibitory, prevents neurons from becoming too active, regulates neuron firing. Important in limbic system and amygdala. Stimulated by alcohol and barbiturates, which result in lowered sensitivity to stimuli.	Calming strategies of CBT stress counseling, meditation, the here-and-now emphasis are likely to be useful and increase the release of GABA. The basic listening sequence will help clients as you listen to their stories. (Meds—minor tranquilizers, antianxiety medications, lithium for low GABA)
Anandamide—impacts cannabinoid receptors, marijuana, affects nucleus accumbens, the brain's pleasure center. Involved in addictive behavior. Tetrahydrocannibinol (THC)—the active ingredient of marijuana—activates receptors.	Motivational interviewing likely the most effective strategy as it attacks addictive issues directly. The client has enjoyed the "high's" of drugs and needs alternative approaches to find positives and strengths in life. Referral to Alcoholics Anonymous or support groups focusing on other issues likely to be helpful (e.g., sexual, drugs, and other addictions). Marijuana may be helpful in Alzheimer's disease. (Meds—none are available, but some in trials)
Acetylcholine (ACh)—first neurotransmitter to be discovered. Affects memory, cognitive functioning, emotion, and aggression, central nervous system. Loss of ACh is a central indicator of Alzheimer's disease.	Exercise, meditation, social relationships, positive activities can slow Alzheimer's. Your work with families will be central to help make decisions and support the client appropriately. You will work with clients to help them deal this increasingly common challenge of life. (Meds—Aricept, cholestine inhibitors, but many new medications in advanced testing stages)
Enkephalins and endorphins—endogenous morphine-like peptides such as enkephalins and beta-endorphin are present within the central nervous system. Endorphins are released in response to pain or sustained exertion. They serve as internal analgesics and seem to have a role in appetite control.	Pain management is an important role in counseling and therapy. Modulate pain, reduce stress, and produce a sensation of calm. Meditation and mindfulness training and related counseling strategies have been found very useful for pain relief and are considered preferable to potentially addictive pain relievers, which usually have side effects. (Meds—an array of over-the-counter and prescription pain and headache relievers with codeine and morphine)

While this book has given attention to broad issues of neuroscience and the impact of counseling on the brain, we will take a risk and say that affecting neurotransmitters through *effective and quality* interviewing and counseling is where the "rubber ultimately hits the road." We do not yet have research that backs up this claim, but it is ultimately the neurotransmitters that influence the development of new neural networks. And as we have said, new neural networks through counseling is change—the creation of the *New*. Science and the art of counseling come together at this point.

Consider the presentation in Table AII-1 as a beginning presentation showing how your practice can influence neurotransmitters and produce change and create the *New*. Art becomes science, and science becomes art.

MICROSKILLS AND THEIR POTENTIAL IMPACT ON CHANGE

The microskills of attending, observation, and the basic listening sequence are vital for the communication of empathy. We start with the biological possibility of "feeling the feelings" of others because of mirror neurons. Through our childhood and later developmental experiences, we become more or less attuned to others. Neuronal structures of empathic understanding can pass away if not nourished. In turn, the teaching of empathy, particularly through the listening skills, may be especially helpful in human change. Moreover, if you are empathic with a client, you are helping that person become more understanding of others.

Restak (2003, p. 9) found that training volunteers in movement sequences produced sequential changes in activity patterns of the brain as the movements became more thoroughly learned and automatic. Systematic step-by-step learning, such as that emphasized in this book, is an efficient learning system also used in ballet, music, golf, and many other settings. If there is sufficient skill practice, changes in the brain may be expected, and increased ability in demonstrating these skills will appear in areas ranging from finger movements to dance—and from the golf swing to interviewing skills. Table AII-2 presents a summary of how various microskills relate to the learning process involved in counseling and psychotherapy.

TABLE AII-2   Key microskills concepts and neuropsychology

Microskills Concept	Some Issues Related to Neuroscience and Neuropsychology
Attending Behavior	Attention is measurable through brain imagining. When client and counselor attend to the story, the brain of both interviewer and client become involved. Factors in attention are arousal and focus. Arousal involves the reticular activating system, at the brain's core, which transmits stimuli to the cortex and activates neurons firing throughout many areas. Selective attention "is brought about by . . . a part of the thalamus, which operates rather like a spotlight, turning to shine on the stimulus" (Carter, 1999, p. 186). If you listen with energy and interest, and this is communicated effectively, expect your client to receive that affect as a positive resource in itself.
Questions	New histories and stories are written in the counseling session. The very asking of questions impacts old memories stored in the hippocampus. "The new history is influenced by current determinants of neural experience, and such factors are usually very different from those that affected the original experience a long time ago" Grawe (2007, p. 67).

(continued)

TABLE AII-2 (continued)

Microskills Concept	Some Issues Related to Neuroscience and Neuropsychology
Observation	Eberhardt (2005) summarizes useful data on nonverbal communication. ▲ Japanese have been found to be more holistic thinkers than Westerners. Expect different cognitive/emotional styles when you work with people who are culturally different from you—but never stereotype! ▲ As you learn to observe your client more effectively in the interview, your brain is likely developing new connections. Expect your multicultural learning to become one of those new connections. ▲ Blacks and Whites both exhibit greater brain activation when they view same-race faces and less when race is different. We tend to feel more comfortable when people are like us. This suggests that discussing racial and other cultural differences early in the session can be a helpful way to build trust. ▲ Expressions can transmit emotions to others. If you smile, the world does indeed smile with you (up to a point). Experiments in which tiny sensors were attached to the "smile" muscles of people looking at faces show the sight of another person smiling triggers automatic mimicry—albeit so slight that it may not be visible. The brain concludes that something good is happening out there and creates a feeling of pleasure.
Encouraging, paraphrasing, and summarization	Active listening is a key aspect of relationship—consider the importance of listening to wellness strengths as well as client challenges. Similarly, if you listen to problems only, expect the nerve cells to communicate that as well.
Reflection of feeling	Traditional categories of emotion (sad, mad, glad, fear) appear in brain imaging. The limbic system organizes bodily emotions and includes the amygdala, hypothalamus, thalamus, hippocampal formation, and cortex. Feelings of fear and anger are located in the amygdala, while it also transmits the *intensity of emotions*. Reflection of feeling is also basic to communicating empathy. The interviewer's mirror neurons "light up" when hearing the emotions and stories of the client.
Confrontation	What some call "creativity" may be located in the connections between the intuitive right brain and the linear left brain as well as the participation of the mainly unconscious limbic system. It is more clear that the executive left brain provides the final spark integrating creativity and the creation of the *New*. While not yet final, many believe that new learning (neuroplasticity) occurs when the two hemispheres synchronize their activity (Goodwin & Sherrard, 2008; Goodwin, Lee, Puig, & Sherrard, 2006; Goodwin, Puig, Lee, Goodwin, & Sherrard, in press). A confrontation that points out incongruities in a person's life is, by definition, an effort to open the person for a new way of thinking. Gentle and supportive confrontations often can reach underlying emotional structures as the empathic atmosphere provides the setting for creative new learning.
Focusing	Client selective attention is guided by existing patterns in the mind, and focusing is an intentional interviewing skill that can open up more possibilities for client thoughts, feelings, and actions. There are a number of regions of the prefrontal cortex that "are activated selectively during different aspects of attentional task preparation and execution" (Allport, 1998). Self-regulation and understanding of others (empathy) is also deeply affected by attentional systems. Focusing will also help clients learn new behaviors.
Reflection of meaning	Meaning is closely allied to emotion. Depression is marked by wide-ranging symptoms, but the cardinal feature of it is the draining of meaning from life. By contrast, those in a state of mania see life as a gloriously ordered, integrated whole. The area of the brain that is most noticeably affected in both depression and mania is an area on the lower part of the internal surface of the prefrontal cortex, the brain's emotional control center. It is exceptionally active in bouts of mania and inactive (along with other prefrontal areas) during depression. Ratey (2008, p. 41) has commented: "You have to find the right mission, you have to find something that's organic, that's growing that keeps you focused on and continues to provide meaning and growth and development for yourself. . . . Spirituality even lights up key centers in the brain. Meaning drives the lower centers and is connected to emotions and motivations areas. . . . If you can get *people* into a situation where they have the meaning direction provided by their mission or their job or their goal, they don't need medicine."

(continued)

TABLE AII-2 (continued)

Microskills Concept	Some Issues Related to Neuroscience and Neuropsychology
Interpretation/reframing and logical consequences, information/advice, and directives	Some clients enter therapy with negative emotions and amygdalas on overdrive, anxiously fearing what the therapy relationship will hold. In the language of neurobiology, the aim is to reduce this hyperactivity and bolster the activity of the nucleus accumbens, a brain area associated with pleasure (DeAngelis, 2005, p. 72). We need to activate cortical functions where positive thoughts and feelings are generated so that we can deal effectively with issues and problems. For example, cognitive therapy can encourage left brain activity to gain control over negative emotions (Carter, 1999, p. 41). The influencing skills and strategies, used effectively, provide clients with specific things they can do to build more positive thoughts, feelings, and behaviors. In this way, clients can deal more effectively with their issues.
	Under times of severe stress or panic, the amygdala can take over. Thus you will find many clients who fail to use more positive memories and personal skills to counteract negativity. Damasio (2003, p. 12) also points out that we need to build positive emotions to cope with the negative. Building on wellness and strengths will enable clients to cope with major challenges. Some even speculate that practitioners will be able to tailor specific treatments to modify brain circuits through counseling, medication, meditation, or other positive interventions (Beitman & Good, 2006).

SOCIAL JUSTICE AND STRESS MANAGEMENT

Stress is a concern in virtually all client issues and problems. It may not be the presenting issue, but be prepared to assess stress levels and provide education and treatment as needed, using some of the strategies of Chapters 13 and 14. Stress will show in body tension and nonverbal behavior. Cognitive/emotional stress is demonstrated in vocal hesitations, emotional difficulties, and the conflicts/discrepancies clients face in their lives.

A moderate amount of stress, *if not prolonged,* is required for development and for physical health. For example, it is well known that repeated stressing of a muscle through weight lifting and running breaks down muscle fibers, but after rest the rebuilding muscle gains extra strength. A similar pattern occurs for ballet and other physical exercise. If there is no stress, neither physical development nor learning will occur. Stress should also be seen as a motivator for focus and as a condition for change. The amygdala needs to be energized. For example, the skill of confrontation can result in stress for the client. But used as a supportive challenge, this stressor becomes the basis for cognitive and emotional change. In all of the above, note the word "moderate" and the need for rest between stressors.

Toxic and long-term stress is damaging. *"Cortisol* is the long-acting stress hormone that helps to mobilize fuel, cue attention and memory, and prepare the body and brain to battle challenges to equilibrium. Cortisol oversees the stockpiling of fuel, in the form of fat, for future stresses. Its action is critical for our survival. At high or unrelenting concentrations such as posttraumatic stress, cortisol has a toxic effect on neurons, eroding their connections between them and breaking down muscles and nerve cells to provide an immediate fuel source" (Ratey, 2008, p. 277).

"Poverty in early childhood poisons the brain. . . . neuroscientists have found that many children growing up in very poor families with low social status experience unhealthy levels of stress hormones, which impair their neural development. The effect is to impair language development and memory—and hence the ability to escape poverty—for the rest of the child's life" (Krugman, 2008, citing research by the American Association of Science, p. 15A). In an article entitled "Excessive Stress Disrupts the Architecture of the Developing Brain"

(National Scientific Council on the Developing Child, 2005), we obtain the following, among many other useful points:

1. In the uterus, the unborn child responds to stress in the mother, while alcohol, drugs, and other stimulants can be extremely damaging.
2. For the developing child neural circuits are especially plastic and amenable to growth and change, but again excessive stress results in lesser brain development, and in adulthood that child is more likely to have depression, an anxiety disorder, alcoholism, cardiovascular problems, and diabetes.
3. Positive experiences in pregnancy seem to facilitate child development.
4. Caregivers are critical to the development of the healthy child.
5. Children of poverty or who have been neglected tend to have elevated cortisol levels.

If you review the sections above, particularly those talking about the frontal cortex and limbic system, you can obtain some sense of what poverty and challenges such as racism and oppression do to the brain. Incidents of racism place the brain on hypervigilance, thus producing significant stress, with accompanying hyperfunctioning of the amygdala and interference with memory and other areas of the brain. We need to be aware that many environmental issues ranging from poverty to toxic environments to a dangerous community all work against neurogenesis and the development of full potential. And let us expand this to include trauma.

Let us recall that the infant, child, and adolescent brain can only pay attention to what is happening in the immediate environment. Again think of the varying positive and negative environments that your clients come from. One of the purposes of the community genogram is to help you and the client understand how we as individuals relate and related to individuals, family, groups, and institutions around us. The church that welcomes us helps produce positive development, while the bank that refuses your parents a loan, or peers that tease and harass you, harm development.

Clients need to be informed about how social systems affect personal growth and individual development Our work here is to help clients understand that the problem does not lie in them but in a social system or life experience that treated them unfairly and did not allow an opportunity for growth.

Finally, there is social action. What are you doing in your community and society to work against social forces that bring about poverty, war, and other types of oppression? Are you teaching your clients how they can work toward social justice themselves? A social justice approach includes helping clients find outlets to prevent oppression and work with schools, community action groups, and others for change.

LOOKING TO THE FUTURE

Neuroscience research provides an important biological foundation for understanding the impact of our work. *The very act of interviewing, counseling, and therapy produces changes in client memory (and your own). Always be aware that new ideas and learning are being constructed in the session.* We suggest that you continue to study and learn about brain structures and functions, as new findings may provide further support for our work and suggest specific guidelines for practice.

Brain research is not in opposition to the cognitive, emotional, behavioral, and meaning emphasis of interviewing and counseling. Rather, it will help us pinpoint types of interventions that are most helpful to the client. In fact, one of the clearest findings is that the brain needs environmental stimulation to grow and develop. You can offer a healthy atmosphere for client growth and development. We advocate the integration of counseling, psychotherapy, neuroscience, molecular biology, and neuroimaging, and the infusion of knowledge from such integrated fields of study to practice, training, and research.

RECOMMENDED FOR FURTHER STUDY

Listed below are materials that have informed and shaped our integration of neuroscience and counseling. The first five listed include three researchers (Sapolsky, Decety, and Jackson) and two theorist/clinicians (Grawe and Ratey). All of them present neuroscience in an accessible and interesting manner, although Klaus Grawe is more of a challenge. Several other key scholars and researchers are included later in this section, and there are many others well worth examining.

There is some debate regarding the use of Wikipedia as a source for research reports or scholarship. However, the Web sites in the listings below serve two important purposes: (1) they will acquaint you with the life work of the scholar, and (2) you will find key references for further reading. Knowing the life story and narratives of what many consider the *main influencers* will provide you with a historical and contextual sense of neuroscience. These brief Web presentations will also point you to future developments in the field.

Decety, J., & Jackson, P. (2004). The functional architecture of human empathy. *Behavioral and Cognitive Neuroscience Reviews,* Volume 3, Number 271-100. (Found in downloadable PDF form at (http://home.uchicago.edu/%7Edecety/publications/Decety_CDPS06 .pdf.) (See also http://en.wikipedia.org/wiki/Jean_Decety.) Decety and Jackson have what we consider the best article on empathy and neuroscience. They know counseling and therapy research and tie it clearly to workings of the brain. Read this article if you believe in empathy and the importance of relationship. Decety and Jackson were central in our discussion of empathy in this book. You will find many other PDF articles available on Decety's Web site.

Grawe, K. (2007). *Neuropsychotherapy: How the neurosciences inform effective psychotherapy.* Mahway, NJ: Erlbaum. Grawe, sadly, is deceased, but read his obituary at *http://www .psychotherapyresearch.org/displaycommon.cfm?an=1&subarticlenbr=50.* This is *the* book that best provides specifics of connections between neuroscience and therapy. *Neuropsychotherapy* had a great influence on us all as we wrote this book. Grawe's discussions of inconsistency and confrontation are breakthrough material. This is difficult and challenging reading, but it is profound in its implications for practice. Allen is on his third reading of the book and still finds things that he missed before. Reading it once is likely not enough, unless you have a hyperactive hippocampus that "gets it" the first time.

Ivey, A. (2009). *Neuroscience and counseling: Implications for microskills and practice* (Video). Framingham, MA: Microtraining Associates. (*www.emicrotraining.com*) Allen discusses his integration of neuroscience and counseling with a focus on microskills. You'll find expansions of the material that is in this text, plus an examination of neurotransmitters and other key issues.

Ratey, J. (2001). *A user's guide to the brain.* New York: Vintage. (http://en.wikipedia.org/wiki/ John_Ratey) Ratey presents the "four theaters of the brain" in an exceptionally clear format. It is the ideal follow-up or substitute for Sapolosky's lectures. You'll get the basics in understandable form.

Ratey, J. (2008). *Neuroscience and the brain: Implications for counseling and therapy* (Video). Framingham, MA: Microtraining Associates. Here you will find the "guts" of the *User's Guide* (listed above) in a video presentation. Ratey has been recognized as one of *America's Best Physicians* for seven years.

Sapolsky, R. (2005). *Biology and human behavior: The neurological origins of individuality*, 2nd Ed. (Audio). Chantilly, VA: The Teaching Company. (www.teach12.com, http://en.wikipedia.org/wiki/Robert_Sapolsky) This is where Allen began his study of neuroscience. Sapolsky is an outstanding presenter and provides the basics in a clear and understandable format. He is superb in the biological area and also supports an environmental/social justice approach to brain science. This series is updated frequently.

Following is a list of other key authors that you should investigate for theory, research, and practice. Please use Google or another search engine to find their most recent works and references. They will be updating and adding to their works often in the next few years. All are important. Where possible, we have selected Wikipedia as the source so that you can obtain an overview of their work.

Rita Carter—a medical writer who is expert at making brain issues clear. (www.ritacarter.co.uk)

Anthony Damasio—a foremost research scientist who writes beautifully. Especially strong on emotion and meaning. (http://en.wikipedia.org/wiki/António_Damásio)

Richard Davidson—offers key research and writing on meditation and multiple topics. (http://en.wikipedia.org/wiki/Richard_J._Davidson)

John Kabat-Zinn—the scholar who brought mindfulness meditation to center stage. (http://en.wikipedia.org/wiki/Jon_Kabat-Zinn)

Eric Kandel—a pioneer in neuroscience and practice who has written several books on meaning and health. (http://en.wikipedia.org/wiki/Eric_Kandel)

Michael Merzenich—a foremost researcher on brain plasticity and change. Through his company, Posit Science, he offers specific practical methods for improving the brain and developing new neural networks. (http://en.wikipedia.org/wiki/Michael_Merzenich)

Richard Restak—a prominent psychiatrist/neurologist who writes very clearly. (http://www.richardrestak.com/biography.htm)

Daniel Siegel—an excellent writer on interpersonal neurobiology and mindfulness. He works with Kabat-Zinn, and together they provide the key resources for using mindfulness in your practice. (http://en.wikipedia.org/wiki/Daniel_Siegel)

References

Adams, D. (2005). Cultural competency now law in New Jersey. Amednews.com, April 25. Retrieved December 25, 2008, from http://www.ama-assn.org/amednews/2005/04/25/prl20425.htm.

Aittasalo, M. (2008). Physical activity counselling in primary health care. *Scandinavian Journal of Medicine & Science in Sports, 18*(3), 261–262.

Alberti, R., & Emmons, M. (2008). *Your perfect right: A guide to assertiveness training* (9th ed.). San Luis Obispo, CA: Impact. (Original work published 1970)

Allport, A. (1998). Visual attention. In M. Posner (Ed.), *Foundations of cognitive science* (pp. 631–682). Cambridge, MA: MIT Press.

American Counseling Association. (2005). *Code of ethics and standards of practice.* Alexandria, VA: Author.

American Psychiatric Association. (2000). *Diagnostic and statistical manual of mental disorders*, Text Revision. Washington, DC: Author.

American Psychological Association. (2002). *Ethical principles of psychologists and code of conduct.* Washington, DC: Author.

American Psychological Association. (2003). Guidelines on multicultural education, training, research, practice, and organizational change for psychologists. *American Psychologist, 58,* 377–402.

Ammentorp, J., Sabroe, S., Kofoed, P., & Mainz, J. (2007). The effect of training in communication skills on medical doctors' and nurses' self-efficacy: A randomized controlled trial. *Patient Education and Counseling, 66,* 270–277.

Arehart-Treichel, J. (2001). Evidence is in: Psychotherapy changes the brain. *Psychiatric News, 36,* 33.

Arehart-Treichell, J. (2004, October 1). Brain activity offers clue to anger. *Psychiatric News, 39,* 19.

Arredondo, P., Ajamu, A., D'Andrea, M., Daniels, J., Ivey, A., Iwamasa, G., Leong, F., Kim, B., Martinez, A., Martinez, M., Orjuela, E., Parham, T., Santiago-Rivera, A., Vazquez, L., & Sue, D. W. (2000). *Culturally competent counseling and therapy: Five videotapes.* North Amherst, MA: Microtraining Associates.

Arredondo, P., Toporek, R., Brown, S., Jones, J., Locke, D. C., Sanchez, J., & Stadler, H. (1996). Operationalization of the multicultural counseling competencies. *Journal of Multicultural Counseling and Development, 24,* 42–78.

Asbell, B., & Wynn, K. (1991). *Touching.* New York: Random House.

Back, A. L., Arnold, R. M., Baile, W. F., Fryer-Edwards, K. A., Alexander, S. C., Barley, G. E., Gooley, T. A., & Tulsky, J. A. (2007). Efficacy of communication skills training for giving bad news and discussing transitions to palliative care. *Archives of Internal Medicine, 167,* 453–460.

Baker Miller, J. (1991). The development of a woman's sense of self. In J. Jordan, A. Kaplan, J. Baker Miller, I. Stiver, & J. Surrey (Eds.), *Women's growth in connection* (pp. 11–26). New York: Guilford.

Baker Miller, J., Stiver, I., & Hooks, T. (Eds.). (1998). *The healing connection: How women form relationships in therapy and in life.* Boston: Beacon.

Barrett, M., & Berman, J. (2001). Is psychotherapy more effective when therapists disclose information about themselves? *Journal of Consulting and Clinical Psychology, 69,* 597–603.

Barrett-Lennard, G. (1962). Dimensions of therapist response as causal factors in therapeutic change. *Psychological Monographs, 76,* 43. (Ms. No. 562)

Beck, A. (1976). *Cognitive therapy and the emotional disorders.* New York: International Universities Press.

Beck, A., & Beck, J. (1995). *Cognitive therapy: Basics and beyond.* New York: Guilford.

Beck, J. S. (1995). *Cognitive therapy.* New York: Guilford.

Beck, J. S. (2005). *Cognitive therapy for challenging problems: What to do when the basics don't work.* New York: Guilford.

Beek, Y., & Dubas, J. (2008). Age and gender differences in decoding basic and non-basic facial expressions in late childhood and early adolescence. *Journal of Nonverbal Behavior, 32,* 37–52.

Beitman, G., & Good, G. (2006). *Counseling and psychotherapy essentials.* New York: Norton.

Bensing, J. (1999a). *Doctor-patient communication and the quality of care.* Utrecht, Netherlands: Nivel.

Bensing, J. (1999b). The role of affective behavior. *Communication,* 1188–1199.

Benson, H. (2008). Home page of Benson and the Relaxation Response, www.relaxationresponse.org.

Benson, H., & Klipper, M. (1975). *The relaxation response.* New York: HarperTorch.

Bernhardt, S. (2001). *Emotional thought stopping.* Retrieved January 15, 2008, from www.have-a-heart.com/depression-II.html.

Blair, R. J. R. (2001). Neurocognitive models of aggression, the antisocial personality disorders, and psychopathy. *Journal of Neurology, Neurosurgery, and Psychiatry, 71,* 727–731.

Blair, R. J. R. (2003). Did Cain fail to represent the thoughts of Abel before he killed him? The relationship between theory of mind and aggression. In B. Rapacholi & V. Slaughter (Eds.), *Individual differences in theory of mind* (pp. 143–169). Hove, UK: Psychology Press.

Blair, R. J. R., & Cipolotti, L. (2000). Impaired social response reversal: A case of acquired sociopathy. *Brain, 123*, 1122–1141.

Blanchard K., & Johnson, S. (1981). *The one-minute manager.* San Diego, CA: Blanchard-Johnson.

Bozarth, J. (1999). *Person-centered therapy: A revolutionary paradigm.* Ross-on-Wye, UK: PCCS Books.

Brammer, L., & MacDonald, G. (2002). *The helping relationship* (8th ed.). Boston: Allyn and Bacon.

Brown, L. (1996). *Subversive dialogues: Theory in feminist therapy.* New York: Basic Books.

Budman, S., & Gurman, A. (2002). *Theory and practice of brief therapy.* New York: Guilford.

Burkard, A. W., Knox, S., Groen, M., Perez, M., & Hess, S. A. (2006). European American therapist self-disclosure in cross-cultural counseling. *Journal of Counseling Psychology, 53*, 15–25.

Bylund, C. L., Brown, R. F., Lubrano di Ciccone, B., Levin, T. T., Gueguen, J. A., Hill, C., & Kissane, D. W. (2008). Training faculty to facilitate communication skills training: Development and evaluation of a workshop. *Patient Education and Counseling, 70*, 430–436.

Camus, A. (1995). *The myth of Sisyphus.* New York: Vintage Books.

Canadian Counselling Association. (1999). *Code of ethics.* Ottawa, Ontario: Author.

Carkhuff, R. (2000). *The art of helping in the 21st century.* Amherst, MA: Human Resources Development Press.

Carless, D., & Douglas, K. (2008). Narrative, identity and mental health: How men with serious mental illness re-story their lives through sport and exercise. *Psychology of Sport and Exercise, 9*(5), 576–594.

Carstensen, L., Pasupathi, M., Mayr, U., & Nesselroade, J. (2000). Emotional experience in everyday life across the life span. *Journal of Personality and Social Psychology, 79*, 644–655.

Carter, R. (1999). *Mapping the mind.* Berkeley: University of California Press.

Chang, E. C., D'Zurilla, T. J., & Sanna, L. J. (2004). *Social problem solving: Theory, research, and training.* Washington, DC: American Psychological Association.

Cheatham, H., & Stewart, J. (1990). *Black families: Interdisciplinary perspectives.* New Brunswick, NJ: Transaction.

Contrada, R., Goyal, T., Cather, C., Rafalson, L., Idler, E., & Krause, T. (2004). Psychosocial factors in outcomes of heart surgery: The impact of religious involvement and depressive symptoms. *Health Psychology, 23*, 227–238.

Corey, G. (2005). *Theory and practice of counseling and psychotherapy* (7th ed.). Belmont, CA: Brooks/Cole–Thomson Learning.

Croce, A. 2003. Non-verbal communication causes cultural misconceptions. Retrieved December 25, 2008, from http://media.www.ramcigar.com/media/storage/paper366/news/2003/02/28/Campus/NonVerbal.Communication.Causes.Cultural.Misconceptions-382215-page2.shtml.

Crone, D., & Guy, H. (2008). "I know it is only exercise, but to me it is something that keeps me going": A qualitative approach to understanding mental health service users' experiences of sports therapy. *International Journal of Mental Health Nursing, 17*(3), 197–207.

Daley, A. (2008). Exercise and depression: A review of reviews. *Journal of Clinical Psychology in Medical Settings, 15*(2), 140–147.

Damasio, A. (2003). *Looking for Spinoza: Joy, sorrow, and the feeling brain.* New York: Harcourt.

D'Andrea, M., & Daniels, J. (2001). RESPECTFUL counseling: An integrative model for counselors. In D. Pope-Davis & H. Coleman (Eds.), *The interface of class, culture and gender in counseling* (pp. 417–466). Thousand Oaks, CA: Sage.

Daniels, T. (2007). A review of research on microcounseling: 1967—present. In A. Ivey & M. Ivey, *Intentional interviewing and counseling: Your interactive resource* (CD-ROM) (3rd ed.). Belmont, CA: Brooks/Cole.

Daniels, T. (2009). A review of research on microcounseling: 1967–present. In A. Ivey & M. Ivey, *Intentional interviewing and counseling: An interactive CD-ROM.* Belmont, CA: Brooks/Cole.

Daniels, T., & Ivey, A. (2007). *Microcounseling: Making skills training work in a multicultural world* (3rd ed.). Springfield, IL: Thomas.

Daniels, T., Rigazio-DiGilio, S., & Ivey, A. (1997). Microcounseling: A training and supervision paradigm. In E. Watkins (Ed.), *Handbook of psychotherapy supervision.* New York: Wiley.

Danner, D., Snowdon, D., & Friesen, W. (2001). Positive emotion in early life and longevity. *Journal of Personality and Social Psychology, 80*, 804–813.

Davidson, R. (2001). The neural circuitry of emotion and affective style: Prefrontal cortex and amygdala contributions. *Social Science Information, 40*.

Davidson, R. (2004). Well-being and affective style: Neural substrates and biobehavioral correlates. *The Philosophical Transactions of the Royal Society, 359*, 1395–1411.

Davidson, R., Pizzagalli, D., Nitschke, J., & Putnam, K. (2002). Depression: Perspectives from affective neuroscience. *Annual Review of Psychology, 53*, 545–574.

DeAngelis, T. (2005, November). Where psychotherapy meets neuroscience. *Monitor on Psychology,* http://www.apa.org/monitor/nov05/neuroscience.html.

Decety, J., & Jackson, P. (2004). The functional architecture of human empathy. *Behavioral and Cognitive Neuroscience Reviews, 3*, 71–100.

De Shazer, S. (1988). *Clues: Investigating solutions to brief therapy.* New York: Norton.

De Shazer, S. (1993). Creative misunderstanding: There is no escape from language. In S. Gilligan & R. Price (Eds.), *Therapeutic conversations.* New York: Norton.

Deutsch, B. (2002). The male privilege checklist. *Expository Magazine, 2*(2). Retrieved December 25, 2008, from http://www.expositorymagazine.net/maleprivilege_checklist.htm.

deWaal, E. (1997). *Living with contradiction: An introduction to Benedictine spirituality.* Harrisburg, PA: Morehouse.

Diaz, A. B., & Motta, R. (2008). The effects of an aerobic exercise program on posttraumatic stress disorder symptom severity in adolescents. *International Journal of Emergency Mental Health, 10*(1), 49–60.

Donk, L. (1972). Attending behavior in mental patients. *Dissertation Abstracts International, 33* (Ord. No. 72-22 569).

Donnelly, G. F. (2007). Happiness and the holidays: Reframing seasonal stress. *Holistic Nursing Practice, 21,* 283.

Draganski, B., Gaser, C., Busch, V., Schuierer, G., Bogdahn, U., & May, A. (2004). Neuroplasticity: Changes in grey matter induced by training. *Nature, 427,* 311–312.

Dreikurs, R., & Grey, L. (1968). *Logical consequences: A new approach to discipline.* New York: Dutton.

Duncan, B., Miller, S., & Sparks, J. (2004). *The heroic client.* San Francisco: Jossey-Bass.

D'Zurilla, T. (1996). *Problem-solving therapy.* New York: Springer.

Eberhardt, J. (2005). Imaging race. *American Psychologist, 60,* 181–190.

Egan, G. (2007). *The skilled helper* (8th ed.). Belmont, CA: Wadsworth.

Ekaterina, L., Popa, D., Apergis-Schoute, J., Fidacaro, G., & Paré, J. (2008). Amygdala intercalated neurons are required for expression of fear extinction. *Nature, 454,* 642–645.

Ekman, P. (1999). Basic emotions. In T. Dalgleish & M. Power (Eds.), *Handbook of cognition and emotion.* Sussex, UK: Wiley.

Ekman, P. (2003). *Emotions revealed.* New York: Holt Paperbacks.

Ekman, P. (2007). *Emotions revealed.* New York: Holt Paperbacks.

Ellis, A. (1999, March). *A continuation of the dialogue on counseling in the postmodern era.* Presentation to the American Counseling Association Convention, San Diego.

Epstein, R. M., & Hundert, E. M. (2002). Defining and assessing professional competence. *Journal of the American Medical Association, 287,* 226–235.

Erickson, B. A., & Keeney, B. (Eds.). (2006). *Milton H. Erickson, M.D.: An American healer.* Philadelphia: Ringing Rocks Press.

Etkin, A., Pittenger, C., Polan, H., & Kandel, E. (2005). Toward a neurobiology of psychotherapy: Basic science and clinical applications. *Journal of Neuropsychiatry and Clinical Neurosciences, 17,* 145–158.

Ewald, M., Derntl, B., Robinson, S., Fink, B., Gur, R. C., & Grammer, K. (2007). Amygdala activation at 3T in response to human and avatar facial expressions of emotions. *Journal of Neuroscience Methods, 161,* 126–133.

Farmer, R. F., & Chapman, A. L. (2008). *Behavioral interventions in cognitive behavioral therapy.* Washington, DC: American Psychological Association.

Farnham, S., Gill, J., McLean, R., & Ward, S. (1991). *Listening hearts.* Harrisburg, PA: Morehouse.

Fiedler, F. (1950). A comparison of therapeutic relationships in psychoanalytic, nondirective, and Adlerian therapy. *Journal of Consulting Psychology, 14,* 435–436.

Frankl, V. (1959). *Man's search for meaning.* New York: Simon and Schuster.

Frankl, V. (1978). *The unheard cry for meaning.* New York: Touchstone.

Fredrickson, B., Tugade, M., Waugh, C., & Larkin, G. (2003). A prospective study of resilience and emotion following the terrorist attacks on the United States on September 11, 2001. *Journal of Personality and Social Psychology, 84,* 365–376.

Freed, P. J., & Mann, J. J. (2007). Sadness and loss: Toward a neurobiopsychosocial model. *American Journal of Psychiatry, 164,* 28–34.

Freeman, E. M., & Couchonnal, G. (2006). Narrative and culturally based approaches in practice with families. *Families in Society, 87,* 198–208.

Fukuyama, M. (1990, March). *Multicultural and spiritual issues in counseling.* Workshop presentation for the American Counseling Association Convention, Cincinnati.

Gazzaniga, M. (2000). Cerebral specialization and interhemispheric communication: Does the corpus callosum enable the human condition? *Brain, 123,* 1293–1326.

Geary, B. B., & Zeig, J. K. (Eds.). (2002). *Handbook of Ericksonian psychotherapy.* Phoenix, AZ: Milton Erickson Foundation Press.

Gendlin, E., & Henricks, M. (Undated). *Rap manual.* (Mimeographed). Cited in E. Gendlin, *Focusing.* New York: Everest House.

Gergen, K., & Gergen, M. (2005). The power of positive emotions. *The Positive Aging Newsletter,* February, www.healthandage.com.

Gillen, K., Barton, K., Cane, V., Tomko, J., Fetherson, B., & Anderson, M. (2008). *Multicultural supervision issues.* Presentation at the Counselors Education and Supervision Conference, Columbus, Ohio.

Golby, A., Gabrelli, J., Chiao, J., & Eberhardt, J. (2001). Differential responses in the fusiform region to same-race and other-race faces. *Nature Neuropsychology, 4,* 845–850.

Goleman, D. (2000). *Working with emotional intelligence.* New York: Bantam.

Goleman, D. (2006). *Emotional intelligence: 10th anniversary edition: Why it can matter more than IQ.* New York: Bantam.

Goleman, D. (2007). *Social intelligence: The new science of human relationships.* New York: Bantam.

Goleman, D., Boyatzis, R., & McKee, A. (2004). *Primal leadership: Learning to lead with emotional intelligence.* Boston: Harvard Business School Press.

Goodwin, L. K., Lee, S. M., Puig, A. I., & Sherrard, P. A. D. (2006). Guided imagery and relaxation for women with early stage breast cancer. *Journal of Creativity in Mental Health, 1,* 53–66.

Goodwin, L. K., & Sherrard, P. (2005). *Guided imagery: A therapeutic intervention to facilitate breakthrough insight.* Unpublished paper, University of Florida, Gainesville.

Goodwin, L. K., & Sherrard, P. (2008). *Guided imagery: A therapeutic intervention to facilitate breakthrough insight.* Submitted for publication.

Grant, A. M. (2003). The impact of life coaching on goal attainment, metacognition and mental health. *Social Behavior and Personality, 31*(3), 253–264.

Grant, A. M. (2008, September). *Evidence-based coaching as applied positive psychology.* Presentation at Coaching: A New Horizon, Harvard Medical School and McLean Hospital, Boston.

Grant, A. M., & Cavanagh, M. (2007). Evidence-based coaching: Flourishing or languishing? *Australian Psychologist, 42*(4), 239–254.

Grawe, K. (2007). *Neuropsychotherapy: How the neurosciences inform psychotherapy.* London: Lawrence Erlbaum.

Green, L. S., Oades, L. G., & Grant, A. M. (2006). Cognitive-behavioural, solution-focused life coaching: Enhancing goal striving, well-being and hope. *Journal of Positive Psychology, 1*(3), 142–149.

Hackney, H., & Cormier, S. (2004). *The professional counselor* (5th ed.). New York: Pearson.

Hall, E. (1959). *The silent language.* New York: Doubleday.

Hall, J. A., & Schmid Mast, M. (2007). Sources of accuracy in the empathic accuracy paradigm. *Emotion, 7,* 438–446.

Harmon, S. C., Lambert, M. J., Smart, D. M., Hawkins, E., Nielsen, S. L., Slade, K., & Lutz, W. (2007). Enhancing outcome for potential treatment failures: Therapist-client feedback and clinical support tools. *Psychotherapy Research, 17,* 379–392.

Harvey, J., & Manusov, L. (Eds.). (2001). *Attribution, communication behavior and close relationships: Advances in personal relationships.* Cambridge, UK: Cambridge University Press.

Haskard, K., Williams, S., DiMatteo, M., Heritage, J., & Rosenthal, R. (2008). The provider's voice: Patient satisfaction and the content-filtered speech of nurses and physicians in primary medical care. *Journal of Nonverbal Behavior, 32,* 1–20.

Hays, P. (2007). *Addressing cultural complexities in practice: Assessment, diagnosis, and therapy.* Washington, DC: American Psychological Association.

Hersen, M., & Biaggio, M. (Eds.). (2000). *Effective brief therapies: A clinician's guide.* New York: Academic Press.

Hill, C. (2004). *Helping skills.* Washington, DC: American Psychological Association.

Hill, C., & O'Brien, K. (1999). *Helping skills.* Washington, DC: American Psychological Association.

Hill, C., & O'Brien, K. (2004). *Helping skills: Facilitating exploration, insight, and action.* Washington, DC: American Psychological Association.

Hillman, C. H., Erickson, K. I., & Kramer, A. F. (2008). Be smart, exercise your heart: Exercise effects on brain and cognition. *Nature Reviews Neuroscience 9,* 58–65.

Holland, J. M., Neimeyer, R. A., & Currier, J. M. (2007). The efficacy of personal construct therapy: A comprehensive review. *Journal of Clinical Psychology, 63,* 93–107.

Hubble, M., Duncan, B., & Miller, S. (1999). *The heart and soul of change.* Washington, DC: American Psychological Association.

Hunter, W. (1984). *Teaching schizophrenics communication skills: A comparative analysis of two microcounseling learning environments.* Unpublished doctoral dissertation, University of Massachusetts, Amherst.

Ishiyama, I. (2006). *Anti-discrimination response training (A.R.T.) program.* Framingham, MA: Microtraining Associates.

Ivey, A. (1973). Media therapy: Educational change planning for psychiatric patients. *Journal of Counseling Psychology, 20,* 338–343.

Ivey, A. (1991, October). *Media therapy reconsidered.* Paper presented at the Veterans Administration Conference, Orlando, FL.

Ivey, A. (1995). *The community genogram: A strategy to assess culture and community resources.* Paper presented at the American Counseling Association Convention, Denver.

Ivey, A. (2000). *Development therapy: Theory into practice.* North Amherst, MA: Microtraining Associates.

Ivey, A., D'Andrea, M., Ivey, M., & Simek-Morgan, L. (2001). *Theories of counseling and psychotherapy: A multicultural perspective* (5th ed.). Needham Heights, MA: Allyn and Bacon.

Ivey, A., D'Andrea, M., Ivey, M., & Simek-Morgan, L. (2006). *Theories of counseling and psychotherapy: A multicultural perspective* (6th ed.). Boston: Allyn and Bacon.

Ivey, A., Gluckstern, N., & Ivey, M. (2006). *Basic attending skills* [Manuals and videos]. Framingham, MA: Microtraining Associates. (First edition 1974)

Ivey, A., Ivey, M., Myers, J., & Sweeney, T. (2005). *Developmental counseling and therapy: Promoting wellness over the lifespan.* Boston: Lahaska/Houghton Mifflin.

Ivey, A., & Matthews, W. (1984). A meta-model for structuring the clinical interview. *Journal of Counseling and Development, 63,* 237–243.

Ivey, A., Normington, C., Miller, C., Morrill, W., & Haase, R. (1968). Microcounseling and attending behavior: An approach to pre-practicum counselor training [Monograph]. *Journal of Counseling Psychology, 15,* Part II, 1–12.

Ivey, A., Pedersen, P., & Ivey, M. (2001). *Intentional group counseling: A microskills approach.* Pacific Grove, CA: Brooks/Cole.

Ivey, A., Simek-Morgan, L., Ivey, M., & D'Andrea, M. (2006). *Theories of counseling and psychotherapy: A multicultural perspective* (6th ed.). Boston: Allyn and Bacon.

Jacobson, E. (1934a). *Progressive relaxation.* Chicago: University of Chicago Press.

Jacobson, E. (1934b). *You must relax.* New York: McGraw-Hill.

Janis, I., & Mann, L. (1977). *Decision making: A psychological analysis of conflict, choice, and commitment.* New York: Free Press.

Jordan, J., Walker, M., & Hartling, L. (Eds.). (2004). *The complexity of connection: Writings from the Stone Center's Jean Baker Miller Training Institute.* New York: Guilford.

Kabat-Zinn, J. (2000). *Full catastrophe living.* New York: Delta Paperbacks.

Kabat-Zinn, J. (2005a). *Coming to our senses: Healing ourselves and the world through mindfulness.* New York: Hyperion.

Kabat-Zinn, J. (2005b). *Wherever you go, there you are: Mindfulness meditation in everyday life.* New York: Hyperion.

Kain, C. (1992, Fall). We cannot turn our backs on AIDS. *American Counselor.*pa

Kandel, E. (2007). *In search of memory: The emergence of a new science of mind.* New York: Norton.

Kauffman, C. (2008, September). Introduction to coaching: The G.R.O.W. model. Presentation at Coaching: A new horizon, Harvard Medical School and McLean Hospital, Boston.

Kim, B., Hill, C., Gelso, C., Goates, M., Asay, P., & Harbin, J. (2003). Counselor self-disclosure: East Asian American client adherence to Asian cultural values, and counseling process. *Journal of Counseling Psychology, 50,* 324–332.

Kim, J.-U. (2008). The effect of a R/T group counseling program on the Internet addiction level and self-esteem of Internet addiction university students. *International Journal of Reality Therapy, 27*(2), 4–12.

Kolb, B., & Whishaw, I. (2003). *Fundamentals of human neuropsychology* (5th ed.). New York: Worth.

Kübler-Ross, E. (1969). *On death and dying.* New York: Macmillan.

Kuntze, J., van der Molen, H. T., & Born, M. (2007). Progress in mastery of counseling communication skills: Development and evaluation of a new instrument for the assessment of counseling communication skills. *European Psychologist, 12,* 301–313.

LaFrance, M., & Woodzicka, J. (1998). No laughing matter: Women's verbal and nonverbal reactions to sexist humor. In J. Swim & C. Stangor (Eds.), *Prejudice: The target's perspective.* San Diego, CA: Academic Press.

Lane, P., & McWhirter, J. (1992). A peer mediation model: Conflict resolution for elementary and middle school children. *Elementary School Guidance and Counseling, 27,* 15–23.

Lazarus, A., & Fay, A. (2000). *I can if I want to.* New York: FMC Books.

Levant, R. (2008). *Depression in men.* Presentation at the American Psychological Association, Boston.

Li, J., & Lambert, V. A. (2008). Job satisfaction among intensive care nurses from the People's Republic of China. *International Nursing Review, 55,* 34–39.

Libert, Y., Marckaert, I., Reynaert, C., Delvaux, N., Marchal, S., Etienne, A.-M., Boniver, J., Klastersky, J., Scalliet, P., Slachumuylder, J., & Razavi, D. (2007). Physicians are different when they learn communication skills: Influence of the locus of control. *Psycho-Oncology, 16,* 553–562.

Liu, W., Pickett, T., & Ivey, A. (2007). Middle-class privilege: Entitlement, social class bias, and implications for training and practice. *Journal of Multicultural Counseling and Development, 35,* 194–206.

Loftus, E. (1997, September). Creating false memories. *Scientific American,* 51–55.

Loftus, E. (2003). Our changeable memories: Legal and practical implications. *Nature Reviews: Neuroscience. 4,* 31–34.

Lucas, L. (2007/2008). The pain of attachment—"You have to put a little wedge in there": How vicarious trauma affects child/teacher attachment. *Childhood Education, 84,* 85–91.

Luszczynska, A., Gerstorf, D., Boehmer, S., Knoll, N., & Schwarzer, R. (2007). Patients' coping profiles and partners' support provision. *Psychology & Health, 22,* 749–764.

Mahoney, M., & Freeman (Eds.). (1985). *Cognition and psychotherapy.* New York: Springer.

Mallen, M. J., Vogel, D. L., & Rochlen, A. B. (2005). The practical aspects of online counseling; ethics, training, technology, and competency. *Counseling Psychologist, 33*(6), 776–818.

Mann, L. (2001). Naturalistic decision making. *Journal of Behavioural Decision Making, 14,* 375–377.

Mann, L., Beswick, G., Allouache, P., & Ivey, M. (1989). Decision workshops for the improvement of decision making skills. *Journal of Counseling and Development, 67,* 237–243.

Masuda, T., & Nisbett, R. (2001). Attending holistically versus analytically: Comparing the context sensitivity of Japanese and Americans. *Journal of Personality and Social Psychology, 81,* 922–934.

Mayo, C., & LaFrance, M. (1973). *Gaze direction in interracial dyadic communication.* Paper presented at the Eastern Psychological Association meeting, Washington, DC.

McGoldrick, M., & Gerson, R. (1985). *Genograms in family assessment.* New York: Norton.

McGoldrick, M., Giordano, J., & Garcia-Preto, N. (2005). *Ethnicity and family therapy* (3rd ed.). New York: Norton.

McIntosh, P. (1988). *White privilege and male privilege: A personal account of coming to see correspondences through work in women's studies.* Wellesley, MA: Wellesley College Center for Research on Women.

McMinn, M. (1996). *Psychology, theology, and spirituality in Christian counseling.* Wheaton, IL: Tyndale.

Meara, N., Pepinsky, H., Shannon, J., & Murray, W. (1981). Semantic communication and expectation for counseling across three theoretical orientations. *Journal of Counseling Psychology, 28,* 110–118.

Meara, N., Shannon, J., & Pepinsky, H. (1979). Comparisons of stylistic complexity of the language of counselor and client across three theoretical orientations. *Journal of Counseling Psychology, 26,* 181–189.

Meichenbaum, D. (1994). *A clinical handbook/practical therapy manual for assessing and treating adults with post-traumatic stress disorder (PTSD).* Waterloo, Ontario: Institute Press.

Microsoft (2009). *MSN Encarta Dictionary.*

Miller, K. I. (2007). Compassionate communication in the workplace: Exploring processes of noticing, connecting, and responding. *Journal of Applied Communication Research, 35,* 223–245.

Miller, S., Duncan, B., & Hubble, M. (2005). Outcome-informed clinical work. In J. Norcross & M. Goldfried (Eds.), *Handbook of psychotherapy integration* (pp. 84–104). Oxford, UK: Oxford University Press.

Miller, W., & Rollnick, S. (2002). *Motivational interviewing: Preparing people for change.* New York: Guilford.

Monk, G., Winslade, J., Crocket, K., & Epston, D. (1997). *Narrative theory in practice: The archaeology of hope.* San Francisco: Jossey-Bass.

Moos, R. (2001, August). *The contextual framework.* Presentation at the American Psychological Association, San Francisco.

Moran, K., Stockton, R., Cline, R., & Teed, C. (1998). Facilitating feedback exchange in groups: Leader interventions. *Journal for Specialists in Group Work, 23,* 257–268.

Myers, J. E., & Sweeney, T. J. (2004). The indivisible self: An evidenced-based model of wellness. *Journal of Individual Psychology, 60,* 234–245.

Myers, J. E., & Sweeney, T. J. (Eds.). (2005). *Counseling for wellness: Theory, research, and practice.* Alexandria, VA: American Counseling Association.

National Association of Social Workers, (1999), *Code of ethics.* Washington, DC: Author.

National Organization of Human Service Professionals. (2000). Ethical standards of human service professionals. *Human Service Education, 20,* 61–68.

Nwachuku, U., & Ivey, A. (1991). Culture-specific counseling: An alternative approach. *Journal of Counseling and Development, 70,* 106–151.

Nwachuku, U., & Ivey, A. (1992). Teaching culture-specific counseling use in microtraining technology. *International Journal for the Advancement of Counseling, 15,* 151–161.

Office of the Surgeon General. (1999). *Mental health, culture, race, and ethnicity.* Washington, DC: U.S. Department of Health and Human Services.

Ogbonnaya, O. (1994). Person as community: An African understanding of the person as intrapsychic community. *Journal of Black Psychology, 20,* 75–87.

Pack-Brown, S., & Williams, C. (2003). *Ethics in a multicultural context.* Thousand Oaks, CA: Sage.

Parsons, F. (1967). *Choosing a vocation.* New York: Agathon. (Originally published 1909)

Petersen, C., & Seligman, M. (2004). *Character, strengths, and virtues: A handbook and classification.* Oxford, UK: Oxford University Press.

Pfiffner, L., & McBurnett, K. (1997). Social skills training with parent generalization: Treatment effects for children with attention deficit disorder. *Journal of Consulting and Clinical Psychology, 65,* 749–757.

Pos, A., Greenberg, L., Goldman, R., & Korman, L. (2003). Emotional processing during experiential treatment of depression. *Journal of Clinical and Consulting Psychology, 73,* 1007–1016.

Posner, M. (Ed.). (2004). *Cognitive neuropsychology of attention.* New York: Guilford.

Power, S., & Lopez, R. (1985). Perceptual, motor, and verbal skills of monolingual and bilingual Hispanic children: A discrimination analysis. *Perceptual and Motor Skills, 60,* 1001–1109.

Probst, R. (1996). Cognitive-behavioral therapy and the religious person. In E. Shafranski (Ed.), *Religion and the clinical practice of psychology* (pp. 391–408). Washington, DC: American Psychological Association.

Ratey, J. (2008a). *Spark: The revolutionary new science of exercise and the brain.* New York: Little, Brown.

Ratey, J. (2008b). *Neuroscience and the brain.* (Transcript from video interview). Framingham, MA: Microtraining Associates.

Restak, R. (2003). *The new brain: How the modern age is rewiring your mind.* New York: Rodale.

Rigazio-DiGilio, S., Ivey, A., Grady, L., & Kunkler-Peck, K. (2005). *The community genogram.* New York: Teachers College Press.

Rigazio-DiGilio, S., Ivey, A., & Locke, D. (1997). Continuing the postmodern dialogue: Enhancing and contextualizing multiple voices. *Journal of Mental Health Counseling, 19,* 233–255.

Roessel, L. L. (2007). Protect your patients' rights with advance directives. *Nurse Practitioner, 32,* 38–43.

Rogers, C. (1957). The necessary and sufficient conditions of therapeutic personality change. *Journal of Consulting Psychology, 21,* 95–103.

Rogers, C. (1961). *On becoming a person.* Boston: Houghton Mifflin.

Roiter, W. (2008, September). *Executive coaching: Becoming an MVP.* Presentation at Coaching: A New Horizon, Harvard Medical School and McLean Hospital, Boston.

Rosen, S. (1991). *My voice will go with you—The teaching stories of Milton H. Erickson.* New York: Norton.

Roysircar, G., Arredondo, P., Fuertes, J., Ponterotto, J., & Toporek, R. (2003). *Multicultural competencies.* Washington, DC: Association for Multicultural Counseling and Development.

Salovey, P., & Mayer, J. (1990). Emotional intelligence. *Imagination, Cognition, and Personality, 9,* 185–211.

Sanderson, W. (2002). Are evidence-based psychological interventions practiced by clinicians in the field? Editorial column in *Medscape Mental Health 7*(1).

Schlosser, L. (2003). Christian privilege: Breaking a sacred taboo. *Journal of Multicultural Counseling and Development, 31,* 44–51.

Schwartz, J., & Begley, S. (2003). *The mind and the brain: Neuroplasticity and the power of mental force.* New York: Regan.

Selden, N., Everitt, B., Jarrard, L., Robbins, T. (1991). Complementary roles for the amygdala and hippocampus in aversive conditioning to explicit and contextual cues. *Neuroscience, 42,* 335–350.

Seligman, M. (1998). *Learned optimism: How to change your mind and your life.* New York: Pocket Books.

Seligman, M. (2004). *Authentic happiness.* Old Tappen, NJ: Free Press.

Semmler, P., & Williams, C. (2000). Narrative therapy: A storied context for multicultural counseling. *Journal of Multicultural Counseling and Development, 28,* 51–62.

Sharpley, C., & Guidara, D. (1993). Counselor verbal response mode usage and client-perceived rapport. *Counseling Psychology Quarterly, 6,* 131–142.

Sharpley, C., & Sagris, I. (1995). Does eye contact increase counselor-client rapport? *Counselling Psychology Quarterly, 8,* 145–155.

Sherrard, P. (1973). *Predicting group leader/member interaction: The efficacy of the Ivey Taxonomy.* Unpublished doctoral dissertation, University of Massachusetts, Amherst.

Short, D., Erickson, B. A., & Erickson-Klein, R. (2005). *Hope and resiliency: Understanding the psychotherapeutic strategies of Milton H. Erickson, MD.* Bethel, CT: Crown House.

Shostrom, E. (1966). *Three approaches to psychotherapy* [Film]. Santa Ana, CA: Psychological Films.

Siegel, D. (2007). *The mindful brain.* New York: Norton.

Singer, T., Seymour, B., O'Dougherty, J., Kaube, H., Dolan, R., & Frith, C. (2004). Empathy for pain involves the affective but not sensory components of pain. *Science, 303,* 1157–1161.

Sklare, G. (2004). *Brief counseling that works: A solution-focused approach for school counselors and administrators.* Beverly Hills, CA: Corwin.

Snyder, C., & Lopez, S. (2002). *Handbook of positive psychology.* Oxford, UK: Oxford University Press.

Spence, G. B., & Grant, A. M. (2003, July). *Individual and group life coaching—Findings from a randomised, controlled trial.* Paper presented at the First Australian Evidence-Based Coaching Conference, Sydney, Australia.

Sternberg, K., Lamb, M., Hershkowitz, I., Esplin, P., Redlich, A., & Sunshine, N. (1996). The relationship between investigative utterance types and the informativeness of child witnesses. *Journal of Applied Developmental Psychology, 17,* 439–451.

Stewart, R., Jackson, A., Neil, D., Jo, H., Hill, M., & Baden, A. (1998). *White counselor trainees: Is there multicultural counseling competence without formal training?* Poster session presented at the Great Lakes Regional Conference of Division 17 of the American Psychological Association, Bloomington, IN.

Sue, D. W., Carter, R. T., Casas, J., Fouad, N., Ivey, A., Jensen, M., LaFromboise, T., Manese, J., Ponterotto, J., & Vazquez-Nutall, E. (1998). *Multicultural counseling competencies.* Thousand Oaks, CA: Sage.

Sue, D. W., Ivey, A., & Pedersen, P. (1996). *A theory of multicultural counseling and therapy.* Pacific Grove, CA: Brooks/Cole.

Sue, D. W., & Sue, D. (2007). *Counseling the culturally diverse: Theory and practice* (5th ed.). New York: Wiley.

Sweeney, T. J. (1998). *Adlerian counseling: A practitioner's approach* (4th ed.). Muncie, IN: Accelerated Development.

Sweeney, T. J., & Myers, J. E. (2005). Optimizing human development: A new paradigm for helping. In A. Ivey, M. B. Ivey, J. E. Myers, & T. J. Sweeney (Eds.), *Developmental strategies for helpers* (2nd ed., pp. 39–68). Amherst, MA: Microtraining.

Tamase, K. (1991). Factors which influence the response to open and closed questions: Intimacy in dyad and listener's self-disclosure. *Japanese Journal of Counseling Science, 24,* 111–122.

Tamase, K., & Kato, M. (1990). Effect of questions about factual and affective aspects of life events on an introspective interview. *Bulletin of Institute for Educational Research* (Nara University of Education), *39,* 151–163.

Tamase, K., Otsuka, Y., & Otani, T. (1990). Reflection of feeling in microcounseling. *Bulletin of Institute for Educational Research* (Nara University of Education), *26,* 55–66.

Tamase, K., Torisu, K., & Ikawa, J. (1991). Effect of the questioning sequence on the response length in an experimental interview. *Bulletin of Nara University of Education, 40,* 199–211.

Tillich, P. (1964). The importance of new being for Christian theology. In J. Campbell (Ed.), *Man and transformation.* Princeton, NJ.

Torres-Rivera, E., Pyhan, L., Maddux, C., Wilbur, M., & Garrett, M. (2001). Process vs. content: Integrating personal awareness and counseling skills to meet the multicultural challenge of the twenty-first century. *Counselor Education and Supervision, 41,* 28–40.

Tyler, L. (1961). *The work of the counselor* (2nd ed.). East Norwalk, CT: Appleton and Lange.

University of Massachusetts Memorial Medical Center, Behavioral Medicine Clinic. (2004). *Treatment plan.* (Unpublished document). Griswold Mental Health Clinic, Palmer, MA: Author.

Utay, J., & Miller, M., (2006). Guided imagery as an effective therapeutic technique: A brief review of its history and efficacy research. *Journal of Instructional Psychology, 33,* 40–43.

Van der Molen, H. (1984). *Aan verlegenheid valt iets te doen: Een cursus in plaats van therapie* [How to deal with shyness: A course instead of therapy]. Deventer, Netherlands: Van Loghum Slaterus.

Van der Molen, H. (2006). Social skills training and shyness. In T. Daniels & A. Ivey (Eds.), *Microcounseling* (3rd ed.). Springfield, IL: Thomas.

Van der Molen, H., Hommes, M., Smit, G., & Lang, G. (1995). Two decades of cumulative microtraining in the Netherlands: An overview. *Educational Research and Evaluation: An International Journal on Theory and Practice, 1,* 347–387.

Van Velsor, P. (2004). Revisiting basic counseling skills with children. *Journal of Counseling and Development, 82,* 313–318.

Wade, S., Borawski, E., Taylor, H., Drotar, D., Yeates, K., & Stancin, T. (2001). The relationship of caregiver coping to family outcomes during the initial year following pediatric traumatic injury. *Journal of Consulting and Clinical Psychology, 69,* 406–415.

Walker, M., & Rosen, W. (Eds.). (2004). *How connections heal.* New York: Guilford.

Weston, D. (2007). *The political brain.* New York: Public Affairs.

White, M., & Epston, D. (1990). *Narrative means to therapeutic ends.* New York: Norton.

Whiting, J. B. (2007). Authors, artists, and social constructionism: A case study of narrative supervision. *American Journal of Family Therapy, 35,* 139–150.

Witkiewitz, K., & Marlatt, G. (2004). Relapse prevention for alcohol and drug problems. *American Psychologist, 59,* 224–235.

Young, K. S. (2007). Cognitive behavior therapy with Internet addicts: Treatment outcomes and implications. *CyberPsychology & Behavior, 10*(5), 671–679.

Zalaquett, C. P., Foley, P., Tillotson, K., Hof, D., & Dinsmore, J. (2008). Multicultural and social justice training for counselor education programs and colleges of education: Rewards and challenges. *Journal of Counseling and Development, 86,* 323–329.

Zalaquett, C., Ivey, A., Gluckstern, N., & Ivey, M. (2008). Las habilidades atencionales básicas: Pilares fundamentales de la comunicación efectiva. [Manuals and videos]. Framingham, MA: Microtraining Associates.

Zhan-Waxler, C., Radke-Yarrow, M., Wagner, E., & Chapman, J. (1992). Development of concern for others. *Developmental Psychology, 28,* 128–136.

Name Index

A
Adams, D., 43
Aittasalo, M., 54
Allouache P., 366, 429
Allport, A., 282, 468
Amir, O., 186
Anderson, M., 39
Arehart-Treichel, J., 26, 463
Arredondo, P., 43, 282
Asbell, B., 125

B
Back, A. L., 77
Barrett, M., 331, 337
Barrett-Lennard, G., 306
Barton, K., 39
Beek, Y., 131
Begley, S., 26, 29, 460
Beitman, G., 337, 469
Bensing, J., 77, 160, 186
Benson, H., 348, 408
Berman, J., 331, 337
Berra, Y., 122
Beswick, G., 366, 429
Blair, R. J. R., 208
Blanchard, K., 215
Born, M., 23
Boyatzis, R., 6
Bozarth, J., 400
Brain, L., 181
Brammer, L., 365
Brinkley, E., 401–411, 412, 413
Brodhead, M. R., 445
Buck, P. S., 199
Burkard, A. W., 337
Busby, H., 89
Bylund, C. L., 77

C
Camus, A., 302
Cane, V., 39
Carkhuff, R., 204
Carless, D., 54
Carstensen, L., 186
Carter, R., 54, 77, 160, 242, 307, 467, 469, 472
Chang, E. C., 365
Cheatham, H., 214

Churchill, W., 439
Cipoletti, L., 208
Cline, R., 337
Contrada, R., 307
Cox, H., 363
Cozolino, L., 186
Crespi, T., 208
Croce, A., 125
Crocket, K., 17
Crone, D., 54
Currier, J. M., 17

D
Dalai Lama, 11
Daley, A., 54
Damasio, A., 54, 175, 185, 337, 366, 469, 472
D'Andrea, M., 47, 204, 365, 400, 446
Daniels, J., 47
Daniels, T., 16, 23, 77, 96, 160, 186, 203, 450
Danner, D., 185
Davidson, R., 54, 185, 472
DeAngelis, T., 469
de Bono, E., 265
Decety, J., 208, 282, 471
de Shazer, S., 417, 418, 420, 421
Deutsch, B., 45
de Waal, E., 320
Diaz, A. B., 54
Donk, L., 77
Dougless, K., 54
Draganski, B., 132
Dreikurs, R., 339
Dubas, J., 131
Duncan, B., 19, 204, 306
D'Zurilla, T. J., 365

E
Eberhardt, J., 135, 468
Egan, G., 204, 207, 365
Ekaterina, L., 462
Ekman, P., 125, 131, 174
Eliot, G., 325
Ellis, A., 134
Epstein, R. M., 35, 106
Epston, D., 17, 415, 421

Erickson, B. A., 415
Erickson, K. I., 54
Erickson, M., 415
Erickson-Klein, R., 415
Etkin, A., 27

F
Farnham, S., 302
Fetherson, B., 39
Fiedler, F., 306
Frankl, V., 292, 302, 304–305, 306, 312, 314, 393, 446
Franklin, B., 365, 366
Fredrickson, B., 185
Freed, P. J., 462
Freeman, A., 305
Friesen, W., 185
Fuertes, J., 43
Fukuyama, M., 301

G
Garcia-Preto, N., 277
Gazzaniga, M., 242
Gendlin, E., 204
Gergen, K., 185
Gergen, M., 185
Gerson, R., 277
Gill, J., 302
Gillen, K., 39
Giordano, J., 277
Gluckstern, N., 4, 22, 23, 65n
Gluckstern-Packard, N., 101, 162
Golby, A., 132
Goldman, R., 186
Goleman, D., 6
Good, G., 337, 469
Goodwin, L. K., 242, 468
Grant, A. M., 112
Grawe, K., 130, 131–132, 240, 242, 467, 471
Green, L. S., 112
Greenberg, L., 186
Grey, L., 339
Guidara, D., 131
Guy, H., 54

H
Haase, R., 4, 22, 65
Hall, E., 75, 125, 131

Subject Index

confrontation (*continued*)
  Client Change Scale and, 243, 250–253, 258–259
  competencies related to, 262–263
  conflict identification and, 243–244, 245
  definition of, 241, 257
  evaluating effectiveness of, 243, 244, 249–252
  example interview illustrating, 252–256
  feedback form on, 261
  helping clients through, 240–241
  incongruity clarification and, 243–244, 246–248
  international conflicts and, 248–249
  key points about, 256–257
  mediation process and, 252, 253
  multicultural issues and, 248–249, 257
  neuroscience and, 242, 468
  practice exercises on, 258–262
  research evidence on, 242
  self-reflection on, 263–264
  steps in process of, 243–250
  summary of, 256–257
congruence, 209
consciousness, 25
consultation, 392
content, reflection of, 158–159
contextual issues, 50–51
control, wellness and, 53
conversational distance, 75
conversational styles, 135–137
coping self, 52–53
corpus callosum, 464
corrective feedback, 335, 336
cortisol, 469
"could" questions, 99, 101, 103
counseling
  alternative theories of, 444, 446, 450
  five major approaches to, 398–399
  interviewing, psychotherapy, and, 13–14, 28
  national and international perspectives on, 18
  theoretical interpretations in, 310–312
creative self, 53
cross-cultural situations. *See* multicultural issues
cultural background, 61
cultural identity, 52
cultural intentionality, 21, 29, 48, 240, 437, 449
cultural strength inventory, 107
culture
  definitions of, 21
  focus on, 280–281
  See also multicultural issues

D
deaf people, 66
death and dying theory, 249–250, 257
decisional balance sheet, 366, 432–433

decisional counseling, 364–367
  definition of, 393
  demonstration interview, 369–387
  emotions and feelings in, 365–367
  five-stage interview model and, 366, 367
  interpretation/reframing and, 311
  logical consequences and, 339, 366
  overview of actions in, 398–399
  prevalence of uses for, 444
  principal focus of, 446
  supplementary readings on, 452
  trait-and-factor theory and, 365
denial, 250, 251
dereflection strategy, 305–306
developmental counseling and therapy (DCT), 188–189
Developmental Counseling and Therapy: Promoting Wellness Over the Lifespan (Ivey, Ivey, Myers, & Sweeney), 135
Diagnostic and Statistical Manual of Mental Disorders-TR (APA), 18
dialectic/systemic emotions, 189, 191
direct challenge, 247
directives, 346–353, 355
  free association, 353
  homework, 347–348
  mindfulness meditation, 349–350
  nutrition, 351
  physical exercise, 351
  positive imagery, 350
  practice exercises on, 355–357
  relaxation response, 348–349
  role-play enactment, 352–353
  suggestions for using, 346–347
  thought-stopping, 351–352
disabled people, 66
discernment, 295, 302, 303–304
disclosure, 326. *See also* self-disclosure
discrepancies
  confronting, 240, 246–247
  goal setting, 140
  identifying, 243–244, 245, 257–258
  nonverbal behavior, 132–133, 138–139
  observing, 124–125, 132–133, 138–140, 141, 143
  verbal behavior, 139
discrimination
  accumulative stress and, 156
  See also prejudice; racism
diversity
  listening skills and, 161–162
  multiculturalism and, 41–43, 200
  See also multicultural issues
dopamine, 160, 466
dual relationships, 40

E
eating habits, 351
eclectic orientations, 453

eliciting meaning, 293–294, 295–296, 301, 313
emotional intelligence, 6–7
emotions
  decisional counseling and, 365–367
  expression of, 175, 181
  identification of, 190
  key words related to, 161–162, 173, 184, 198
  language of, 174, 191–192, 198
  layers of, 175
  meaning and, 307
  neuroscience research on, 460–464
  positive, 185, 187, 337, 464
  processing of, 186
  styles of, 188–189, 191, 193
  transmission of, 160
  wellness and, 53
  See also reflection of feeling
empathic listening, 160
empathy, 7, 203–209
  authenticity and, 209
  competency in, 234–235
  concreteness and, 207
  congruence and, 209
  definition of, 203–204, 226
  feedback form on, 230–231
  human experience and, 206
  immediacy and, 207
  nonjudgmental attitude and, 207–209
  positive regard and, 205
  practice exercises on, 228–229
  research evidence on, 208
  respect/warmth and, 207
  three types of, 204
Empowering Black Males (Lee), 105
encouragement
  child counseling and, 152
  listening process and, 151, 157–158, 164
  practice exercises on, 165–167
encouragers, 151, 152, 157–158, 159, 164
endorphins, 466
enkephalins, 466
environmental focus, 280–281
essential self, 52
ethical witnessing, 283
ethics, 34–40
  coaching and, 109–110
  codes of, 36
  competence and, 35–36, 56
  confidentiality and, 37–38, 56
  diversity and, 41–42
  informed consent and, 36–37, 56
  key points related to, 56
  power and, 39–40, 56
  practice exercises on, 57, 59
  sample practice contract, 38
  social justice and, 40, 56
  technology and, 39
Ethics Updates Web site, 36